Jean Cornier and Franz Pursche (Eds.)

Particle Technology and Textiles

Also of interest

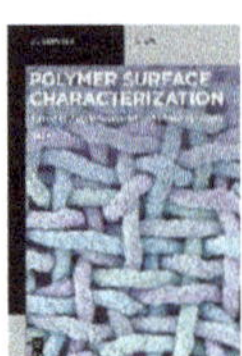

Polymer Surface Characterization
Edited by Luigia Sabbatini, Elvira De Giglio, 2022
ISBN 978-3-11-070104-3, e-ISBN (PDF) 978-3-11-070114-2

Green Chemistry.
Principles and Designing of Green Synthesis
Syed Kazim Moosvi, Waseem Gulzar Naqash,
Mohd. Hanief Najar, 2021
ISBN 978-3-11-075188-8, e-ISBN (PDF) 978-3-11-075189-5

Active Materials
Edited by Peter Fratzl, Michael Friedman, Karin Krauthausen,
Wolfgang Schäffner, 2021
ISBN 978-3-11-056181-4, e-ISBN (PDF) 978-3-11-056206-4

Smart Polymers.
Principles and Applications
José Miguel García, Félix Clemente García, José Antonio Reglero Ruiz,
Saúl Vallejos, Miriam Trigo-López, 2022
ISBN 978-1-5015-2240-6, e-ISBN (PDF) 978-1-5015-2246-8

Encyclopedia of Color, Dyes, Pigments
Edited by Gerhard Pfaff, 2022
Volume 1 Antraquinonoid Pigments – Color Fundamentals
ISBN 978-3-11-058588-9, e-ISBN (PDF) 978-3-11-058807-1
Volume 2 Color Measurement – Metal Effect Pigments
ISBN 978-3-11-058684-8, e-ISBN (PDF) 978-3-11-058710-4
Volume 3 Mixed Metal Oxide Pigments – Zinc Sulfide Pigments
ISBN 978-3-11-058686-2, e-ISBN (PDF) 978-3-11-058712-8

Particle Technology and Textiles

Review of Applications

Edited by
Jean Cornier and Franz Pursche

2nd Edition

DE GRUYTER

Editors
Pharm.D. Jean Cornier
vdlconsult
Seumestr. 7
81379 München
Germany
j.cornier@arcor.de

Dr.-Ing. Franz Pursche
RWTH Aachen University
Institut für Texiltechnik of RWTH Aachen University
Otto-Blumenthal-Str. 1
52074 Aachen
Germany
franz.pursche@ita.rwth-aachen.de

ISBN 978-3-11-067076-9
e-ISBN (PDF) 978-3-11-067077-6
e-ISBN (EPUB) 978-3-11-067086-8

Library of Congress Control Number: 2022946144

Bibliographic information published by the Deutsche Nationalbibliothek
The Deutsche Nationalbibliothek lists this publication in the Deutsche Nationalbibliografie; detailed bibliographic data are available on the Internet at http://dnb.dnb.de.

Cover image: AXIO-IMAGES/iStock/Getty Images Plus
Typesetting: Integra Software Services Pvt. Ltd.
Printing and binding: CPI books GmbH, Leck

www.degruyter.com

Foreword

For more than 20 years now, I have been working in the research field of fibers, textiles as well as their products as head of the Institut für Textiltechnik (ITA) of RWTH Aachen University. Particle-modified fibers and textiles are part of our daily research, whereby the knowledge and know-how about this technology are scattered. It was with great pleasure that I heard that a worldwide collaboration of colleagues of highest scientific quality, who are very well known to me, is working on a joint book publication on this topic.

The authors take up many aspects relevant to our daily life, starting with technical developments in the agricultural sector and ending with legal framework conditions in the biggest economies of this world. Particularly noteworthy is the high sustainability relevance of the contents of this book, in which particle technology might be a key factor in the future.

Currently, I am not aware of any other book publications that shed light on a similar aspect of textiles and nanoparticles. This book, and the way the content is presented, is unique in the market. Textiles are an integral part of almost every aspect of our daily lives. Nowadays, high-tech textiles are used in almost all industrial sectors. In the book, the essential part of particle technology in these key industries is highlighted in detail, and an interface between research and application is created. Especially the high application relevance of the chapter contents is a unique point of this book. Each chapter places a high degree of focus on end-use applications, further reinforcing industry and societal relevance.

Our technology-driven society has benefited from leaps in innovation in textile technology for decades, and our daily lives today would not be possible without particle-modified fibers. Many highly innovative companies develop new high-tech products every day that influence our daily lives. We are in a time of rapid technological change. Particles are a key driver of innovation in this process, and their potential has not yet been fully exploited. I am convinced that by studying this book, you will gain access to a broad knowledge base that will open new aspects for you in your daily work in research and industry as well as in your daily life and interaction with textiles.

Thomas Gries
Chair of Textile Engineering
Director of Institut für Textiltechnik (ITA) of RWTH Aachen University
Otto-Blumenthal-Straße 1, 52074 Aachen

https://doi.org/10.1515/9783110670776-202

Preface

Increasing customer demand for durable and functional textiles manufactured in a sustainable manner has created an opportunity for micro-/nanomaterials to be integrated into textile substrates. The textile industry is now at the forefront of applying particle technology into its products. As this revolution progresses, the possibilities offered by new technologies for modification of particles at the micro- and nanoscale to provide new properties and features are growing quickly. This gives great opportunities to the textile industry for new applications but also challenges in assuring consumers and users that products are safe and in being able to demonstrate reduced environmental impact and progress toward circular economy.

Particle Technology – Applications in Textiles aims at applying particle-enabled technologies to the development and production of innovative textile products. Particles have a large surface area-to-volume ratio and high surface energy. Therefore, their application during textile processing techniques, such as finishing, coating and dyeing, enhances the product performance manifold and imparts unachieved functionality and performance. Many properties, including shrinkage, strength, electrical conductivity and flammability, can be affected. Particle-treated textiles have already led to many inventive commercial applications that will be reviewed in this book.

A broad range of topics are addressed, from the description of micro-, nanosized materials, their unique properties, formulation, production, toxicity and safety, to their numerous technical applications. Research and development in this field requires a multidisciplinary approach, involving materials, chemical and textile engineers, biologists, medical and legal experts.

The goal of this unique book is to present an overall picture of the use of micro- and nanoparticles in textiles. It is designed to be a reference textbook on the application of particle science in the development of new structures for use in the development of innovative textile products. Focus is placed on the description, research and manufacturing of candidate new materials, as well as their translation and use into marketable products by industry. We review the most interesting and promising developments in this emerging but fast-developing field.

Part I gives *a general introduction into particles used in textiles and their functions and application areas in textiles.*

In **Part II**, *a systematic review of the various particles used in textiles* is provided with a description of currently used structures (like fillers for fibers, coatings for finishing, micro- and nano-fibers), their production and their applications. Emphasis is also laid on the new properties, improving functionality of the materials.

In **Part III**, *an overview of the numerous applications of particles in technical textiles* is presented. It also gives details about the use of particle-enabled textile products in 11 different industrial sectors: in sport articles, agriculture, construction,

https://doi.org/10.1515/9783110670776-203

environmental protection and energy efficiency, protective equipment, filtration, materials handling/conveying, cosmetics, medicine, smart textiles and sensor fibres.

Part IV addresses *new emerging production technologies* like nanofiber electrospinning and 3D-printing.

In **Part V**, *toxicology and safety aspects of micro- and nanoparticles* are reviewed. A special focus is also put on *micro-/nanoplastics* released by textile materials.

It is followed in **Part VI** by an exhaustive presentation *of regulatory issues* for translation to the market of the most promising particle-enabled materials.

The European REACH legislation is detailed, and several nanoregister initiatives set up by various European countries are presented. Focus is also put on the China's governance, legislation and legal framework. The occupational health aspects and the safety legislation relevant to the production and use of materials are also addressed.

In **Part VII**, we report about *testing on micro/nanotechnology for textiles* with emphasis about product quality and certifications.

Part VIII reviews important aspects related to *circular economy and sustainability in textile technology as well as recycling efforts and issues*. The challenges and opportunities for the use of particles are described.

Part IX lays emphasis *on industrial innovation and market opportunities for nano- and micro-enabled textiles*. Focus is put on textile industry in China, the world's largest and biggest textile products' manufacturer and exporter. Moreover, general issues brought by the industrial commercialization of micro-/nanotechnologies are also addressed. Perceived limitations of particle technology in the textile industry and future directions are identified.

Finally, **Part X**, the last chapter, gives *a short outlook about the challenges and opportunities of particle technologies in textiles*.

The chapters are written by leading researchers mainly working at textile R&D institutes, research centers, and national and international regulatory agencies. The authors come from a range of backgrounds, including academia, industry, and national and international laboratories, from Europe and China.

It is expected that this book will become a standard work for textile scientists, and the textile industry, but also a reference work for scientists, researchers and students, as well as for agencies, government and regulatory authorities.

This book aims to bring inspiration for scientists, new ideas and for textile developers, innovation in industry, and finally will result in the further development of guidelines for agencies and government authorities to establish safety rules in using this new promising technology. The book will also stimulate breakthroughs in textile sciences, leading to improved applications and products.

Jean Cornier
Munich, Germany
Franz Pursche
Aachen, Germany

Contents

Foreword — V

Preface — VII

About the editors — XIII

List of contributors — XV

I Definition of particles and their application areas

Merle Orth, Martin Dylon, Franz Pursche, Thomas Gries
1 Definition of particles and their application areas — 3

II Fillers in fibers and textiles

Aurélie Cayla, Eric Devaux, Fabien Salaün, Jeanette Ortega, Stefan Li, Franz Pursche
2 Particle fillers for fibers — 13

Boris Mahltig
3 Finishing of textile materials — 45

Aurélie Cayla, Eric Devaux, Fabien Salaün, Jeanette Ortega, Thomas Gries
4 Microfibers and nanofibers — 67

III Applications in technical textiles

Avinash P. Manian, Thomas Bechtold, Tung Pham
5 Sports textiles/apparel — 95

Mark Pätzel, Kyra Scheerer, Thomas Gries
6 Applications in agricultural textiles — 123

Kira Heins, Martin Scheurer
7 Particle applications in technical textiles in construction — 131

Jamal Sarsour, Thomas Stegmaier, Goetz Gresser
8 Ecotech textiles in environment protection and energy efficiency: current examples from research — 147

Rahel Krause, Carolin Schwager, Jan Jordan
9 Protective textiles — 177

Leonie Beek
10 Filtration — 197

Annett Schmieder, Christoph Müller
11 Technical textiles for machine elements — 213

Maroua Ben Abdelkader, Nedra Azizi, Fabien Salaün, Mustapha Majdoub, Yves Chevalier
12 Cosmetotextiles — 235

Michael Doser
13 Particles for medical textiles — 275

Robert Tadej Boich, Vadim Tenner
14 Application of nanotechnology in smart textiles — 287

Jeanette Ortega, Thomas Gries
15 Electronic nanotechnologies in textiles — 295

IV **Emerging production technologies**

Thomas Schneiders
16 Electrospinning of micro-/nanofibers — 305

Sofie Huysman
17 3D printing — 329

V Toxicology/safety/ecological toxicity and environmental impact

Dana Kühnel, Harald F. Krug, Andreas Mattern, Anita Jemec Kokalj
18 Human and environmental hazard of nanomaterials used in textiles — 343

Ellen Bendt, Maike Rabe, Sabrina Kolbe, Susanne Küppers, Stefan Brandt, Jens Meyer, Malin Obermann, Karin Ratovo,
19 Micro/nanoplastics — 373

VI Governance/legislation/EU legal framework-REACH

Frauke Averbeck, Angelina Gadermann
20 Nanomaterials and REACH — 387

Abdelqader Sumrein
21 Nanoregisters in Europe — 395

Luis Almeida, Delfina Ramos
22 Occupational health aspects (EU) — 407

Ning Cui
23 China's governance, legislation and legal framework — 421

VII Testing on micro/nanotechnology for textiles

Edith Classen
24 Test methods and labels for testing on micro/nanotechnology for textiles — 437

VIII Sustainable practices

Dominic Berndt, André Matthes, Holger Cebulla
25 Circular economy and recycling — 455

Amrei Becker, Thomas Gries
26 Sustainability in the textile industry — 473

IX Industrial innovation and market opportunities for nano- and micro-enabled textiles

Ning Cui
27 Industrial innovation and market opportunities for nano- and micro-enabled textiles — 483

X Future prospects for particle technology in textiles

Franz Pursche, Thomas Gries
28 Future prospects for particle technology in textiles — 499

Index — 503

About the editors

Jean Cornier is presently a consultant to several companies in the areas of life science, new technologies and business development. Located in Munich, Germany, he obtained his state diploma of "Doctor of Pharmacy" – Pharm.D. – from the University of Caen, France, and a M.Sc. degree in pharmaceutical medicine from the University of Duisburg-Essen, Germany. Since 1986, he has worked in the space industry as expert in Materials and Life Science research and projects, was participant in commercialization initiatives supported by the European and German space agencies as well as in several EU-funded projects in biotechnology and civil security research. He was at the origin of the first clothing physiology project to gain new insights into the interaction of body, clothing and climate under weightlessness performed on the International Space Station. He has wide knowledge about nanotechnologies and their applications, and is the co-editor of two reference books on Nanopharmacy and Nanocosmetics published in 2017 and 2019.

Franz Pursche is currently group leader at the Institut für Textiltechnik (ITA) of RWTH Aachen University in the field of monofilament technologies. He obtained his Master of Science at the RWTH Aachen University in 2013 in the field of business engineering with a specialization in process technology. Between 2014 and 2019, he worked as a scientist at the ITA in the field of man-made fibers and received his doctorate in engineering on the topic of carbon fiber production. In 2019, Mr. Pursche worked as a development engineer for melt-spun yarns at PHP Fibers GmbH. Since 2020, he works intensively on technical fibers and monofilaments and their functionalization with nanoparticles. In this research field he is responsible for the acquisition and processing of bilateral and public funded research projects with an annual volume of over 3 Million €.

Both co-editors would like to thank Dr. Robert Brüll, co-founder of the company Fibrecoat GmbH, Aachen, Germany for his decisive support in the creation of this book.

https://doi.org/10.1515/9783110670776-205

List of contributors

Jean Cornier
vdlconsult
Seumestr. 7
81379 München
Germany
j.cornier@arcor.de

Franz Pursche
Institut für Textiltechnik of RWTH Aachen University
Otto-Blumenthal-Str. 1
52074 Aachen
Germany
franz.pursche@ita.rwth-aachen.de

Merle Orth
Institut für Textiltechnik of RWTH Aachen University
Otto-Blumenthal-Straße 1
52074 Aachen
Germany
merle.orth@rwth-aachen.de

Martin Dylon
Institut für Textiltechnik of RWTH Aachen University
Otto-Blumenthal-Straße 1
52074 Aachen
Germany
martin.dylong@rwth-aachen.de

Thomas Gries
Institut für Textiltechnik of RWTH Aachen University
Otto-Blumenthal-Straße 1
52074 Aachen
Germany
thomas.gries@ita.rwth-aachen.de

Boris Mahltig
Hochschule Niederrhein
Faculty of Textile and Clothing Technology
Webschulstr. 31
41065 Mönchengladbach
Germany
boris.mahltig@hs-niederrhein.de

Aurélie Cayla
GEMTEX–Laboratoire de Génie et Matériaux Textiles
Univ. Lille ENSAIT
F-59000 Lille
France
aurelie.cayla@ensait.fr

Eric Devaux
GEMTEX–Laboratoire de Génie et Matériaux Textiles
Univ. Lille ENSAIT
F-59000 Lille
France
eric.devaux@ensait.fr

Fabien Salaün
ENSAIT
GEMTEX - Laboratoire de Génie et Matériaux Textiles
Université de Lille nord de France
F-59000 Lille
France
fabien.salaun@ensait.fr

Jeanette Ortega
Institut für Textiltechnik of RWTH Aachen University
Otto-Blumenthal-Straße 1
52074 Aachen
Germany
jeanette.ortega@ita.rwth-aachen.de

Stefan Li
Institut für Textiltechnik of RWTH Aachen University
Otto-Blumenthal-Straße 1
52074 Aachen
Germany
yuwei.li@rwth-aachen.de

https://doi.org/10.1515/9783110670776-206

Avinash P. Manian
Research Institute of Textile Chemistry and Textile Physics
University of Innsbruck
Hoechsterstrasse 73
6850 Dornbirn
Austria
avinash.manian@uibk.ac.at

Thomas Bechtold
Research Institute of Textile Chemistry and Textile Physics
University of Innsbruck
Hoechsterstrasse 73
6850 Dornbirn
Austria
thomas.bechtold@uibk.ac.at

Tung Pham
Research Institute of Textile Chemistry and Textile Physics
University of Innsbruck
Hoechsterstrasse 73
6850 Dornbirn
Austria
tung.pham@uibk.ac.at

Mark Pätzel
Institut für Textiltechnik of RWTH Aachen University
Otto-Blumenthal-Straße 1
52074 Aachen
Germany
mark.paetzel@ita.rwth-aachen.de

Kyra Scheerer
Institut für Textiltechnik of RWTH Aachen University
Otto-Blumenthal-Straße 1
52074 Aachen
Germany
kyra.scheerer@rwth-aachen.de

Kira Heins
Institut für Textiltechnik of RWTH Aachen University
Otto-Blumenthal-Str. 1
52074 Aachen
Germany
kira.heins@ita.rwth-aachen.de

Martin Scheurer
Institut für Textiltechnik of RWTH Aachen University
Otto-Blumenthal-Str. 1
52074 Aachen
Germany
martin.scheurer@ita.rwth-aachen.de

Jamal Sarsour
German Institutes of Textile and Fiber Research
Koerschtalstrasse 26
73770 Denkendorf
Germany
jamal.sarsour@ditf.de

Thomas Stegmaier
German Institutes of Textile and Fiber Research
Koerschtalstrasse 26
73770 Denkendorf
Germany
thomas.stegmaier@ditf.de

Goetz Gresser
German Institutes of Textile and Fiber Research
Koerschtalstrasse 26
73770 Denkendorf
Germany
goetz.gresser@ditf.de

Rahel Krause
Institut für Textiltechnik of RWTH Aachen University
Otto-Blumenthal-Strasse 1
52074 Aachen
Germany
rahel.krause@ita.rwth-aachen.de

Carolin Schwager
Institut für Textiltechnik of RWTH Aachen University
Otto-Blumenthal-Strasse 1
52074 Aachen
Germany
carolin.schwager@rwth-aachen.de

Jan Jordan
Institut für Textiltechnik of RWTH Aachen University
Otto-Blumenthal-Strasse 1
52074 Aachen
Germany
jan.jordan@ita.rwth-aachen.de

Leonie Beek
Institut für Textiltechnik of RWTH Aachen University
Otto-Blumenthal-Str. 1
52074 Aachen
Germany
leonie.beek@ita.rwth-aachen.de

Annett Schmieder
Technische Universität Chemnitz
Reichenhainer Str. 70 |R. D303
09126 Chemnitz
Germany
annett.schmieder@mb.tu-chemnitz.de

Christoph Müller
Technische Universität Chemnitz
Reichenhainer Str. 70 |R. D303
09126 Chemnitz
Germany
christoph.mueller@mb.tu-chemnitz.de

Maroua Ben Abdelkader
Laboratoire des Interfaces et Matériaux Avancés (LIMA)
Faculté des Sciences
Université de Monastir
bd de l'Environnement
5019 Monastir
Tunisia
azizinedra@yahoo.fr

Nedra Azizi
Laboratoire des Interfaces et Matériaux Avancés (LIMA)
Faculté des Sciences
Université de Monastir
bd de l'Environnement
5019 Monastir
Tunisia
azizinedrayahoo.fr

Mustapha Majdoub
Laboratoire des Interfaces et Matériaux Avancés (LIMA)
Faculté des Sciences
Université de Monastir
bd de l'Environnement
5019 Monastir
Tunisia
mustaphamajoub@gmail.com

Yves Chevalier
Université Claude Bernard Lyon 1
CNRS UMR 5007
Laboratoire d'Automatique de Génie des Procédés et de Génie Pharmaceutique
43 bd 11 Novembre
69622 Villeurbanne
France
yves.chevalier@univ-lyon1.fr

Michael Doser
Stellv. Leiter Technologiezentrum Biomedizintechnik und stellv. Vorstand
Deutsche Institute für Textil- und Faserforschung Denkendorf (DITF)
Körschtalstraße 26
73770 Denkendorf
Germany
michael.doser@ditf.de

Vadim Tenner
Institut für Textiltechnik of RWTH Aachen University
Otto-Blumenthal-Str. 1
52074 Aachen
Germany
vadim.tenner@ita.rwth-aachen.de

Robert Boich
Institut für Textiltechnik of RWTH Aachen University
Otto-Blumenthal-Str. 1
52074 Aachen
Germany
robert.boich@ita.rwth-aachen.de

Thomas Schneiders
Institut für Textiltechnik of RWTH Aachen University
Otto-Blumenthal-Str. 1
52074 Aachen
Germany
thomas.schneiders@ita.rwth-aachen.de

Sofie Huysman
Researcher "Plastic characterisation, Processing & Recycling"
Centexbel Technologiepark 70
9052 Zwijnaarde
Belgium
shu@centexbel.be

Ning Cui
Senior Eng.
China Textile Academy
No.3 Yanjingli Middle Street
Chaoyang District
Beijing
China
cuining1@cta.gt.cn

Dana Kühnel
Helmholtz Centre for Environmental Research – UFZ
Department Of Bioanalytical Ecotoxicology
Permoserstrasse 15
04318 Leipzig
Germany
dana.kuehnel@ufz.de

Harald F. Krug
NanoCASE GmbH
St. Gallerstr. 58
9032 Engelburg
Switzerland
hfk@nanocase.ch

Andreas Mattern
Helmholtz Centre for Environmental Research – UFZ
Dept. Bioanalytical Ecotoxicology
Permoserstrasse 15
04318 Leipzig
Germany
andreas.mattern@ufz.de

Anita Jemec Kokalj
Biotechnical Faculty
Department of Biology
University of Ljubljana
Večna pot 111
1000, Ljubljana
Slovenia
anita.jemec@bf.uni-lj.si

Ellen Bendt
Fachbereich Textil- und Bekleidungstechnik
Hochschule Niederrhein
Webschulstr. 31
41065 Mönchengladbach
Germany
ellen.bendt@hs-niederrhein.de

Frauke Averbeck
Bundesanstalt für Arbeitsschutz und Arbeitsmedizin – BAuA
Bundesstelle für Chemikalien
Chemikalienbewertung und Risikomanagement
Friedrich-Henkel-Weg 1 – 25
44149 Dortmund
Germany
averbeck.frauke@baua.bund.de

Angelina Gadermann
Bundesanstalt für Arbeitsschutz und Arbeitsmedizin – BAuA
Bundesstelle für Chemikalien
Chemikalienbewertung und Risikomanagement
Friedrich-Henkel-Weg 1 – 25
44149 Dortmund
Germany
gadermann.angelina@baua.bund.de

Abdelqader Sumrein
Directorate of Submissions and Interaction
European Chemicals Agency
P.O. Box 400
00121 Helsinki
Finland
abdelqader.sumrein@echa.europa.eu

Luis Almeida
Department of Textile Engineering and Centre for Textile Science and Technology
University of Minho
Guimarães
Portugal
lalmeida@det.uminho.pt

Delfina Ramos
Centre for Research and Development in Mechanical Engineering (CIDEM)
School of Engineering of Porto (ISEP)
Polytechnic of Porto
Portugal and Algoritmi Centre
School of Engineering
University of Minho
Guimaraes
Portugal
dgr@isep.ipp.pt

Edith Classen
Life Science and Care
Hohenstein Institut für Textilinnovation gGmbH
Schlosssteige 1
74357 Boennigheim
Germany
e.classen@hohenstein.com

André Matthes
Professur Textile Technologien
Fakultät für Maschinenbau
Technische Universität Chemnitz
Reichenhainer Straße 70 | R. D205
09126 Chemnitz
Germany
andre.matthes@mb.tu-chemnitz.de

Dominic Berndt
Professur Textile Technologien
Fakultät für Maschinenbau
Technische Universität Chemnitz
Reichenhainer Straße 70 | R. D205
09126 Chemnitz
Germany
dominic.berndt@mb.tu-chemnitz.de

Cebulla Holger
Professur Textile Technologien
Fakultät für Maschinenbau
Technische Universität Chemnitz
Reichenhainer Straße 70 | R. D205
09126 Chemnitz
Germany
holger.cebulla@mb.tu-chemnitz.de

Amrei Becker
Institut für Textiltechnik of RWTH Aachen University
Otto-Blumenthal-Straße 1
52074 Aachen
Germany
amrei.becker@ita.rwth-aachen.de

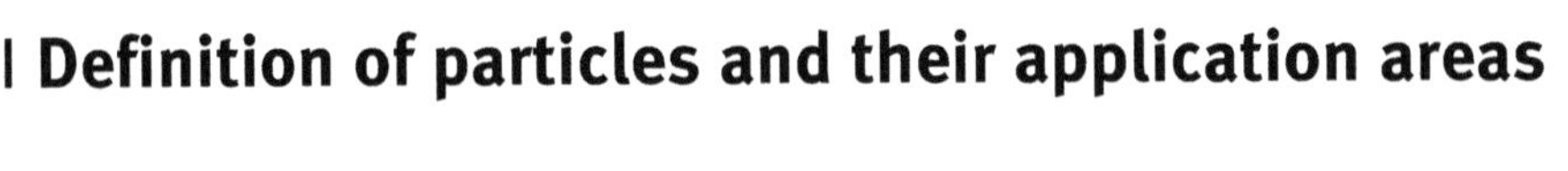

I Definition of particles and their application areas

Merle Orth, Martin Dylon, Franz Pursche*, Thomas Gries

1 Definition of particles and their application areas

Keywords: particles, nanomaterials, application, nanoparticles, definition

1.1 Introduction

The textile industry is a major factor in the development and industrialization of developing countries due to outsourcing and the effects of globalism [1]. The demand for new and functional textiles and the need of new materials and technology to produce these textiles is rising. The textile industry has always been a pioneer in new developments in production, economy and trade [2]. High-tech materials and new fabric constructions improve textiles and can equip textiles with unique, customized properties. Some desired features like an antimicrobial effect, UV-protection and flame retardancy are important for many textile applications. In addition, textiles are used in many different application areas from clothing to engineering. Therefore, new technologies imparting new characteristics to textiles have the potential to be used in various other applications too [3].

With the advent of new technologies, which are able to add special functions and new features to fabrics, comes a revolution to the textile industry. These new technologies lead to the possibility and improvements for natural and synthetic textile finishing, smart textiles and highly functional textiles. The use of micro-, nanomaterials plays a major role in these technological advancements due to their surface properties which have the potential to be more effective than traditional additives and materials in improving and adding new functions to a textile. For instance, characteristics like antimicrobial properties, UV-protection and a self-cleaning ability can be achieved by using conventional nanomaterials like metal oxide agents or carbon-based materials with a nanoscale structure [3].

Historically particle technology can be traced back to the ancient Egyptians [4]. The earliest use of nanoparticles are glazes for early Chinese porcelain. The Romans used nanosized gold clusters to create different colors [5]. Glass in rose was made with nanogold particles in the seventeenth century [6]. Particle technology was

***Corresponding author: Franz Pursche,** Institut für Textiltechnik of RWTH Aachen University, Otto-Blumenthal-Straße 1, 52074 Aachen, Germany, e-mail: franz.pursche@ita.rwth-aachen.de
Merle Orth, Martin Dylong, Thomas Gries, Institut für Textiltechnik of RWTH Aachen University, Otto-Blumenthal-Straße 1, 52074 Aachen, Germany

https://doi.org/10.1515/9783110670776-001

early used in every aspect of the daily life. Particles have been used from bread making in form of flours and grains to the discovery and usage of particles as colorants and all the way to the use as filter materials [7]. The beginning of modern nanotechnology was set by Richard Feynman in his lecture titled “There’s Plenty of Room at the Bottom” in December 1959 by showing that matter at nanometer dimensions can be used in order to significantly improve material properties. In the following years numerous advances were made in nanotechnology and its applications in the textile industry [8].

Today textiles and textile structures are used in many areas starting from aerospace applications, automotive textiles, food packaging, furnishing to protective textiles and medical applications. Nanotechnology has a huge potential in improving these textiles and opening new possibilities. Nanotechnology is already being used for novel applications for fashion and industry. Scent-embedded textiles, stay-clean textiles, textiles with displays and bulletproof lightweight textiles are just some examples of current usage of nanoparticles in the textile industry. The integration of electronic technology into the clothing industry has the potential to be a key breakthrough technology. These nanoparticles have even the potential to be used in other textile areas too [9].

By the definition of [10] and the EU-Commission a “particle” is a small piece with defined physical boundaries [11]. There are commonly five different forms of particles as shown in Table 1.1. All of these particles have a different form. This form is described in the so-called aspect ratio of the particle. The aspect ratio describes the ratio between particle length to particle thickness. The ratio is often measured microscopically. The knowledge of the particle form suffices in practice [12].

In order to describe these particles, the prefix “nano” is used. “Nanos” is a Greek word and means “dwarf.” Furthermore, it is used in SI units, for example, $1\ nm = 10^{-9}\ m$. Therefore the nanoscale is defined from 1 nm up to 100 nm in nanotechnology [10, 13].

Size and form are important parameters for nanoparticles especially for their function. Therefore, the five different forms in Table 1.1 are classified in three groups of forms with its own definition. There is (a) the nanoparticle which has all three outer dimensions in nanoscale like the sphere and the cube. Then there is (b) the nano-fiber with two outer dimensions in nanoscale like the cuboid and (c) the nanoplates which has only one outer dimension in nanoscale like platelets and fibers [10].

These particles usually occur in groups as isolated or separated units. Due to their surface energy, these particles interact with one another in groups. These groups of particles are divided in their respective kind of interactions. All of these groups of particles are based on primary particles. Primary particles are the components of these groups. The two groups are agglomerates and aggregates [10].

An agglomerate is an accumulation of particles which are weakly bound together. The resulting surface area is nearly as large as the sum of the single components it consists of. These agglomerates are bound together by weak forces, for example, Van-der-Waals interactions. Whereas an aggregate consists of firmly

Table 1.1: Form of particles [12].

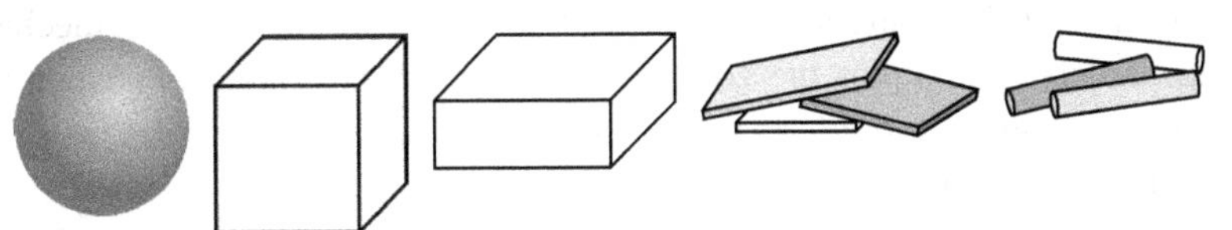

Form	Sphere	Cube	Cuboid	Platelets	Fibers
Aspect ratio	1	~1	1,4-4	5–100	>10
Examples	Glass spheres, silicate, spheres	$CaCO_2$ $CaSO_4$	SiO_2 $BaSO_4$	Mica, Talcum, Kaolin, Graphite, $Al(OH)_3$	Glass fibers, Asbestos, Wollastonite, Cellulosic fibers, Carbon fibers

bound or merged particles. The surface area of an aggregate is much smaller as the sums of the single particles it consists of. Aggregates are hold together by various strong interactions like strong covalent or ionic forces. Furthermore, they can be merged by physical alterations of the materials [10].

These various variants of particles in groups are shown in Figure 1.1.

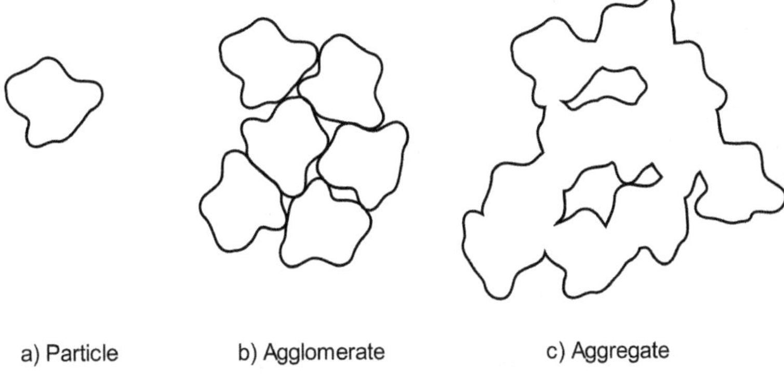

Figure 1.1: Particle variants [10].

The EU Commission released a definition which is summarizing the above-mentioned definitions of nanoparticles and their various forms:

> "Nanomaterial" means a natural, incidental or manufactured material containing particles, in an unbound state or as an aggregate or as an agglomerate and where, for 50% or more of the particles in the number size distribution, one or more external dimensions is in the size range 1 nm–100 nm [11].

These objects in nanoscale have properties which are enabling them to be important components in various materials and systems. Nanoparticles have these properties due to the fact that materials in bulk form are governed by classical mechanics and small particles and atoms are governed by quantum mechanics. Between these two domains the nanosize particles are located and show these particular material behavior. They even have the potential to improve the properties of a material significantly better than conventional pendants. These so-called nanoobjects exhibit not only pure extrapolations of their properties in their bigger form but also they even have new properties due to their nanoscale which are understood as newly emerged and transformative properties. For example ceramic particles can be deformable in nanoparticle size and copper becomes transparent in nanoform [10, 14–16].

There is great commercial potential for these materials in the textile industry, by using their ability to permanently functionalize fabrics and textiles, which in comparison with conventional methods, for example, the coating of fibers with particles of bigger size, do not wash or wear out. Nanoparticles have the ability to be durable due to their high surface area and a great "adhesion" to textiles and fibers. Nanotechnology therefore has a big impact on the functionality and performance of materials [13].

Typical properties of nanoparticles are as follows [13]:

1. A large specific surface area in terms of the same mass in comparison with micron-sized particles. For example, titanium oxide: micron-sized particle 10.1 m^2/g; nanoparticle 329,1 m^2/g [13, 17].
2. The surface layer has the majority of atoms or molecules. Therefore the cohesion between the bulk of the material and the nanoparticles on the surface is weaker.
3. High coverage of available space due to small interparticle distances.
4. Specific color development and scattering of radiation caused by the smaller dimensions of these particles than UV radiation wavelengths (10–400 nm) and those of visible-range electromagnetic waves (380–740 nm).
5. The critical length (mean free path, scattering length) allows for different material properties (conductivity, diffusion) in comparison with or higher than the nanoparticles' dimensions.
6. In comparison with bulk materials nanoparticles have a smaller energy band gap.
7. Decreasing particle size results in higher toxicity especially in the nanoscale range.

Furthermore they enhance or even add various characteristics in fibers and textiles, for example, water repellence, self-cleaning ability, hydrophilicity, antibacterial characteristic, UV protection/absorption, antistatic abilities, wrinkle resistance, electroconductivity, moisture-absorption, flame retardancy/fire proof, increased durability and the ability to control the release of active agents, medical products or fragrances [3]. These functions greatly improve the textiles. Various nanomaterials have other

Table 1.2: Particles and their added functionalities.

	Hydrophobic	Self-cleaning	Antibacterial	UV-protective	Antistatic	Wrinkle resistant	Electro-conductiv	Flame retardant	Increased durability	Controlled realease
Nanoclay				X				X		X
Carbon nanotubes	X	X			X	X	X	X	X	
Carbon black					X		X		X	
Graphite nanofibers	X				X		X			
Titanium dioxide	X	X	X	X	X	X		X		
Zinc oxide	X	X	X	X	X	X		X		
Magnesium oxide			X	X			X		X	
Aluminium oxide			X	X			X		X	
Silicon dioxide	X	X	X	X		X		X	X	X
Gold			X							
Silver	X	X	X			X				

(continued)

Table 1.2 (continued)

	Hydrophobic	Self-cleaning	Antibacterial	UV-protective	Antistatic	Wrinkle resistant	Electro-conductiv	Flame retardant	Increased durability	Controlled realease
Iron(III) oxide			X	X	X					
Nickel(II) oxide			X						X	
Copper(II) oxide			X	X						

characteristics, which improve certain areas of the fabric. In Table 1.2 the most common nanoparticles and their funcational properties are listed.

1.2 Conclusion and outlook

Nanoparticles are matter at nanoscale dimension. By use of nanoparticles, material properties can be drastically changed and a specific functionalization can be achieved. Nowadays, nanoparticle technology is used in a variety of application areas, starting from smart textiles for sports applications and ending in the aviation industry. However, the use of nanoparticles in textiles has not been completely explored. The following chapters of this book provide insight into the great potential of using nanoparticles in textile applications.

References

[1] Köksal, D.; Strähle, J.; Müller, M.; Freise, M.: Social sustainable supply chain management in the textile and apparel industry – A literature review [online]. Sustainability, 2017; 9(1), 100. Verfügbar unter: doi:10.3390/su9010100.

[2] Gries, T.; Veit, D.; Wulfhorst, B.: Textile Fertigungsverfahren [online]. 2018. Verfügbar unter: doi:10.3139/9783446458666.

[3] Ul-islam, S.; Butola, B. S.; Hg.: Nanomaterials in the Wet Processing of Textiles. Hoboken NJ: WileyScrivener Publishing, 2018. Advanced materials series. ISBN 9781119459842.

[4] Seville, J.; Wu, C.-Y.; Hg.: Particle Technology and Engineering. Elsevier, 2016. ISBN 9780080983370

[5] Sharon, M.: History of Nanotechnology. From Pre-historic to Modern Times. Hoboken, NJ, USA: John Wiley & Sons, Inc, 2019. Advances in Nanotechnology & Applications. ISBN 9781119460084.

[6] Johnston, J. H.; Lucas, K. A.: Nanogold synthesis in wool fibres: Novel colourants [online]. Gold Bulletin, 2011; 44(2), 85–89. Verfügbar unter: doi:10.1007/s13404-011-0012-y.

[7] Lee, S.; Henthorn, K. H.; Hg.: Particle Technology and Applications. CRC Press, 2017. ISBN 9781138077393

[8] Sawhney, A. P. S.; Condon, B.; Singh, K. V.; Pang, S. S.; LI, G.; Hui, D.: Modern applications of nanotechnology in textiles [online]. Textile Research Journal, 2008; 78(8), 731–739. ISSN 0040-5175. Verfügbar unter: doi:10.1177/0040517508091066.

[9] Joshi, M.; Bhattacharyya, A.: Nanotechnology – A new route to high-performance functional textiles [online]. Textile Progress, 2011; 43(3), 155–233. ISSN 0040-5167. Verfügbar unter: doi:10.1080/00405167.2011.570027.

[10] Deutsches Institut für Normung e.V.: DIN CEN ISO/TS 80004-2, DIN CEN ISO/TS 80004-2, 2017-09, Nanotechnologien_- Fachwörterverzeichnis_- Teil_2: Nanoobjekte (ISO/TS_80004-2:2015); Deutsche Fassung CEN_ISO/TS_80004-2:2017. Berlin: Beuth Verlag GmbH, 2017.

[11] Office, P.; Commission Recommendation of 18 October 2011 on the definition of nanomaterialText with EEA relevance, 2011.

[12] Maier, R. D.; Schiller, M.; Hg.: Handbuch Kunststoff Additive. München: Carl Hanser Verlag GmbH & Co. KG, 2016.
[13] Mishra, R.; Militky, J.; Baheti, V.; Huang, J.; Kale, B.; Venkataraman, M.; Bele, V.; Arumugam, V.; Zhu, G.; Wang, Y.: The production, characterization and applications of nanoparticles in the textile industry [online]. Textile Progress, 2014; 46(2), 133–226. ISSN 0040-5167. Verfügbar unter: doi:10.1080/00405167.2014.964474.
[14] Sherman, J.: Nanoparticulate Titanium Dioxide Coatings and processes for the production and use thereof. US6653356B2, United States, 2003.
[15] Patra, K.: Application of nanotechnology in textile engineering: An overview [online]. Journal of Engineering and Technology Research, 2013; 5(5), 104–111. Verfügbar unter: doi:10.5897/JETR2013.0309.
[16] Qian, L.; Hinestroza, J.: Application of nanotechnology for high performance textiles Journal of Textile and Apparel, Technology and Management 4(1), 2004.
[17] Allen, N. S.; Edge, M.; Sandoval, G.; Ortega, A.; Liauw, C. M.; Stratton, J.; Mcintyre, R. B.: Interrelationship of spectroscopic properties with the thermal and photochemical behaviour of titanium dioxide pigments in metallocene polyethylene and alkyd based paint films [online]. Polymer Degradation and Stability, 2002; 76(2), 305–319. Verfügbar unter: doi:10.1016/S0141-3910(02)00027-7.

II Fillers in fibers and textiles

Aurélie Cayla*, Eric Devaux, Fabien Salaün, Jeanette Ortega, Stefan Li, Franz Pursche

2 Particle fillers for fibers

Keywords: functional fiber, nanoparticles, additivation, nanofillers, fiber production, fiber processing, functional properties, classification

Incorporation of nanofillers in polymeric systems to fabricate fiber nanocomposites has been a popular approach in the last three decades for imparting functional attributes to textiles. In the first section of this chapter, basic characteristics of nanoparticles for nanocomposite fibers are summarized, reviewing recently published papers. The second section is devoted to the nanocomposite fiber's-spinning methods including melt-spinning, wet-spinning, dry-spinning and electrospinning. The choice of one method over another is often dictated by the physicochemical properties of the materials as well as the average diameter targeted. The last section on polymer micro- and nanocomposites focuses on the main applications including thermal comfort, flame retardant (FR), antimicrobial, electrical, and energy harvesting properties.

2.1 Introduction

Fibers and textile materials in the broadest sense of the word have been a part of human life since ancient times. The development of textiles has undergone several industrial or technological "revolutions" in recent centuries. The invention of machines has allowed since the eighteenth century to make it affordable for a more significant number of people, with an increase in production rates allowing the transition of artisanal textiles to industry. The discovery and use of artificial or synthetic fibers during the twentieth century corresponds to the second textile revolution. These versatile fibers with low production costs and good mechanical or chemical properties have quickly acquired an important market share. The end of the twentieth century marks the beginning of the third textile revolution. Indeed, if in the 1980s, the main functions of textiles were to protect or correspond to a social means of representation or belonging, the development of new materials and/or processes has

***Corresponding author: Aurélie Cayla,** Univ. Lille ENSAIT, GEMTEX – Laboratoire de Génie et Matériaux Textiles, F-59000 Lille, France, e-mail: aurelie.cayla@ensait.fr
Eric Devaux, Fabien Salaün, Univ. Lille ENSAIT, GEMTEX – Laboratoire de Génie et Matériaux Textiles, F-59000 Lille, France
Jeanette Ortega, Stefan Li, Franz Pursche, Institut für Textiltechnik of RWTH Aachen University, Otto-Blumenthal-Straße 1, 52074 Aachen, Germany

https://doi.org/10.1515/9783110670776-002

allowed making it more technical. The beginning of the twenty-first century has been marked by the emergence of new technologies that add significant functions and properties to textiles, whether at the scale of the fiber, the yarn or the fabric. This stage is marked by the emergence of intelligent, multifunctional textiles designed, considering ecological and sustainable development requirements, where nanotechnology plays a key and significant role in this evolution.

Nanotechnology, considered an emerging technology, is defined as "the application of scientific knowledge to manipulate and control matter at the nanoscale to make use of properties and phenomena related to size and structure, distinct from those associated with individual molecules or atoms or basic materials" [1]. The scale considered is between 1 and 100 nm, and properties that are not extrapolations of larger-scale material properties will typically, but not exclusively, appear at this scale. In addition, the lower limit according to this definition (about 1 nm) is included to avoid individual atoms or small groups of atoms being designated as nano-objects or nanostructure elements [2].

Nanotechnology is recognized as one of the most promising directions of technological development in the twenty-first century, as predicted by Richard Feynman as early as 1959 [3]. However, the concepts developed during his speech did not find an echo until 1974 when Norio Tanigucchi defined the notion of "nanotechnology" [4]. The term "nano" means one billionth or 10^{-9}; and a nanometer is one billionth of a meter. Numerous manufacturing systems or means allow the fabrication of nanostructures by controlling the chemical composition, size, morphology or shape, and crystalline (or amorphous) structure. The use of nanotechnology in the field of polymers and moreover in the field of textiles is an activity that arouses a growing research interest both at the fundamental and industrial levels. Polymeric nanocomposites are a class of polymers reinforced with small quantities, typically from 1 to 10% by mass, of particles possessing at least one dimension of nanometric size. The first generation of nanocomposites dates back to the 1970s, when inorganic molecules were dispersed in a polymer matrix by sol–gel process by the US Air Force. The second generation appeared in the 1980s, where the choice of inorganic reinforcements was oriented toward minerals and clays. The third generation, which appeared in the 1990s, corresponds to the synthesis of low-density nanostructures containing silicon or polyhedral oligomeric silsesquioxanes. Nanocomposites do not differ from traditional composites by a reduction in the size of the incorporated fillers. This new generation of materials presents superior properties for equivalent filler content because the interactions between the matrix and the fillers now occur at the nanoscale.

These further developments have led researchers to search for new fabrication systems of various nano-sized materials with improved properties (strengths, lightness, compatibility, chemical reactivities, etc.), that is, nanoclays, carbon nanotubes (CNTs), nanofibers and graphene [5]. Manufacturing methods are divided into two main families: bottom-up and top-down. The study of these nanomaterials, allowing a better understanding of their properties, opens new perspectives for years to come in order to improve the quality of life. Other important aspects are the potentially increased toxicity, a negative environmental impact and recycling/sustainability limitations. Since nanoparticles represent another material system, which is introduced into an existing system to increase functionality, recycling becomes more difficult as both systems have to be separated for clean recycling. This is not always possible, but on the other hand not always necessary. The advantages in using nanoparticles as fillers must outweigh these disadvantages, which could impair commercialization and acceptability.

Nanomaterials are gradually appearing on the market of commercialized products and are gradually starting to become commodities and are used in many advanced technologies.

Thus, polymeric nanocomposites have attracted significant scientific interest because of their specific properties (strength, modulus, thermal stability, thermal and electrical conductivity, gas barrier) compared to raw plastic materials. Their applications in the textile field are in fast growth. This chapter focuses on the development of polymer nanocomposites as an advanced materials for textile applications such as new fibers. It is divided into three main parts: (i) the types and classification of particle fillers for fibers, (ii) nanocomposite fibers processing and (iii) the improvement of fiber properties.

2.2 Type and classification of particle fillers for fibers

2.2.1 Size and morphologies

Particle fillers are of great interest for the improvement or the contribution of new functionalities to the fibers. Their incorporation in mass in the fibers leads to modifications of intrinsic properties resulting from the type of additives used. The latter can be of a very different chemical nature, but another impacting key point is their morphology and their size. The first point concerns the size. The scale of the particles introduced into the fibers is relatively constrained because the fibers have diameter variations between 10 and 50 µm. Therefore, a large part of the fillers used consists of nanoobjects which facilitate their implementation and avoid the irregularity of the fiber diameters due to the size of the additives. Nanoobjects are materials with one, two or three external dimensions in the nanoscale (from 1 to 100 nm) [5, 6].

Nanoobjects are classified according to three types according to the number of nanoscales (Figure 2.1): (i) 0D or nano dimensions along the three directions, corresponding to nanoparticles (nanospheres), (ii) 1D or nano dimensions in two directions for nanowires (nanofibers, nanotubes, etc.) and (iii) 2D for nano dimensions in one direction, such as nanoplates.

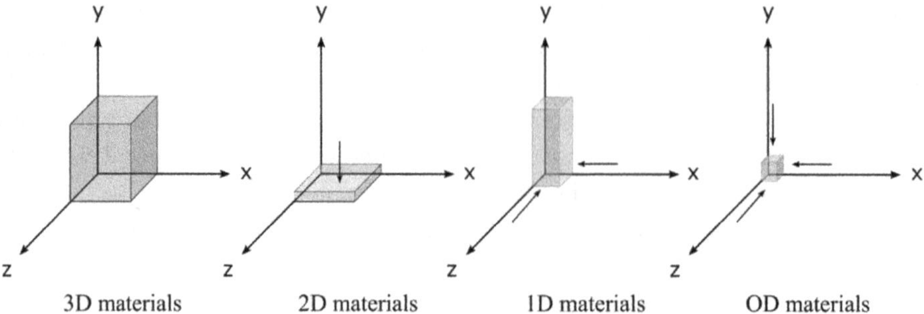

Figure 2.1: Illustration of the types of nanoobjects according to the number of nanometric scales (modified after [7]).

The shape has an important impact on the final property, especially for electrochemical energy devices [8] or for conductive polymer composites' (CPCs) electrical and thermal conductivity [8, 9].

Through nanofillers, the functionalized fibers in bulk are then considered nanocomposites themselves because of the dispersion of nanoparticles in their structural frameworks. A nanocomposite is a composite material whose dispersed phase is made up of particles of which at least one of the three dimensions is of the order of a nanometer or a few tens of nanometers at most (nanoobjects), wherein the small size offers some level of controllable performance that is expected to be better than in conventional composites [10]. In nanocomposite, the addition of new functionalities is ensured by the mass addition of nanofillers in the polymer. The advantage of nanoscale fillers consists of a high specific surface area and an increased interface between the polymer and the filler. The unique properties of nanocomposites come from the interfaces that control the interactions between the matrix and the filler. Nanoparticles improve the properties of the materials at a lower ratio than composites with larger microparticles, allowing a decrease of weight. At the nanoscale, properties are no longer governed by physical laws, gravitational forces and macromechanisms [11]. This is due to the large surface area to volume ratio, which has a dramatic effect on interfacial-dependent properties such as catalytic reactivity, electrical resistivity, adhesion, gas storage, and chemical reactivity [11]. In addition to the quality of the filler/matrix interface induced by the specific surface area, the quality of the nanocomposite depends on the adhesion between the two (impregnation and wettability) and the arrangement of the nanoobjects in the fiber (Figure 2.2). Dispersion and distribution are two significant elements directly affecting the intrinsic properties of the fiber.

(1) Bottom-up and (2) top-down are two main approaches to synthesisize nanofillers. Bottom-up, also called the constructive method, allows building a nanoparticle from atoms and their agglomerates/clusters. As illustrated in Figure 2.3, the primary methods are spinning, chemical vapor deposition (CVD), pyrolysis, spraying and

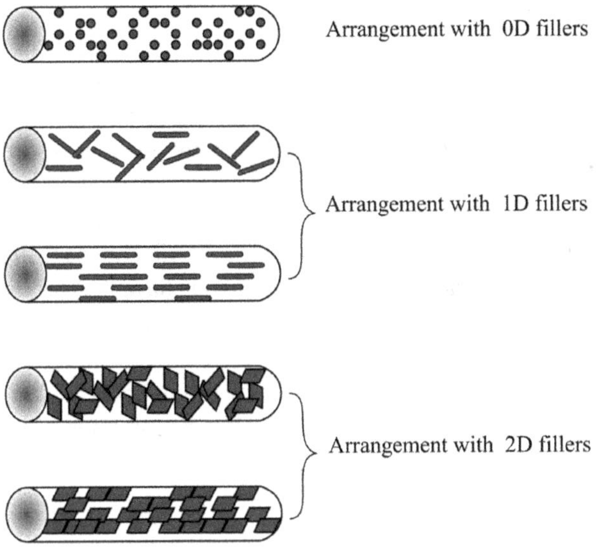

Arrangement with 0D fillers

Arrangement with 1D fillers

Arrangement with 2D fillers

Figure 2.2: Arrangement of fillers in filament according to the fillers size and shape (adapted from [10]).

biosynthesis. Top-down, also called the destructive method (like mechanical milling, nanolithography, laser ablation and sputtering), reduces a bulk material to nanometric scale particles [12].

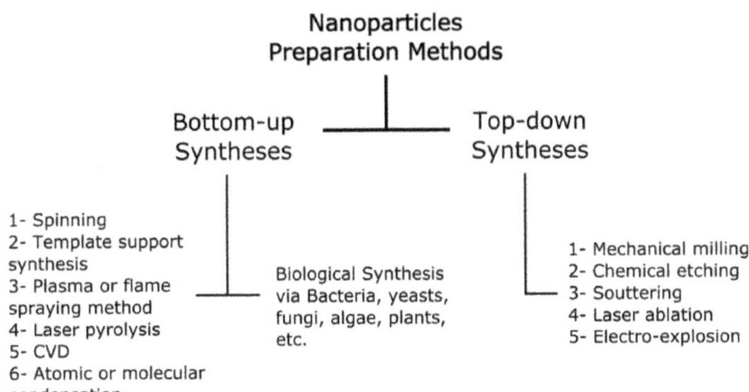

Figure 2.3: Main methods to produce fillers according to the bottom-up and top-down classifications (adapted from [5]).

2.2.2 Classification of fillers

Particles can be classified according to different criteria such as shape, the number of nanometric scales, their chemical nature and whether they are biobased or not. Based on the chemical nature of the particles, two significant families are to be distinguished (Figure 2.4), that is, (i) the fillers of inorganic origin such as carbonaceous, metallic, metallic oxides, mineral clays and ceramics and (ii) the organic fillers which one finds the natural polymers such as the cellulosic derivatives, the chitosan, the lignins or the proteins. Protein- and polysaccharide-based materials have been of increasing interest to researchers due to their bio-sourced nature and functional properties. Among these materials, the main ones used are chitin, chitosan, cellulosic derivatives, keratin or casein. The main difference between proteins and polysaccharides is in their chemical composition, where proteins are based on amino acids while polysaccharides are based on sugar chemistry [13].

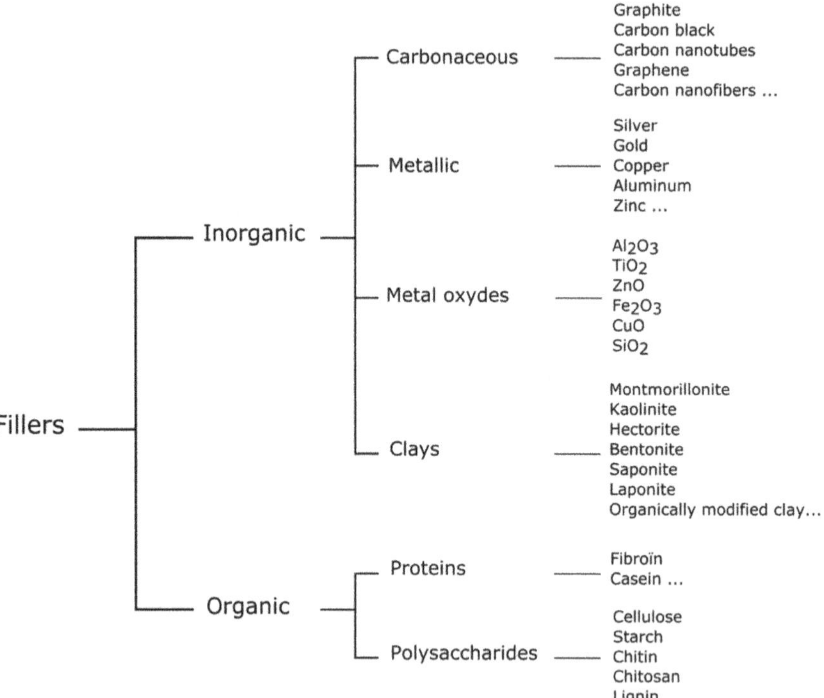

Figure 2.4: Nonexhaustive list of the main fillers used in textile applications.

2.2.2.1 Inorganic fillers – carbon-based fillers

Since antiquity, carbon has been mainly known in two forms: graphite and diamond. In 1985, Kroto et al. [14] discovered the third allotropic form of carbon called fullerene. Following this came the identification of fullerenes with tubular structures called CNTs by Iijima in 1991 [15], following his study on the conditions of synthesis of the form of the most known fullerenes: the C60 fullerenes. Currently, different carbonaceous particles are used in textile materials such as graphite, carbon black (CB), carbon fibers, CNT and graphene. These fillers are often used for their mechanical, electrical and thermal properties. However, mainly their stability and low cost (except for CNT and graphene) make them excellent candidates for nanocomposites in textile applications. The choice between them is often also related to the processing capability, and the size and quantity of fillers required to obtain the desired property are often limiting factors. For example, the addition of filler leads to a very significant increase in melt-viscosity that no longer allows transformation into a multifilament form [16]. Table 2.1 summarizes the main characteristics of carbonaceous fillers.

Table 2.1: Main characteristics of carbon nanofillers [17–20].

	Graphite	Carbon black	Carbon nanotubes	Carbon nanofibers	Graphene
Shape	**Plate** **–**	**Particle** **0D**	**Fibrillar** **1D**	**Fibrillar** **1D**	**Plate** **2D**
Surface area (m^2/g)	10–20	80–90	1,300	20–200	1,500
Electrical resistivity (Ω cm)	2.9×10^{-2}	$10–10^{-2}$	$1.2 \times 10^{-4}–5.1 \times 10^{-6}$	0.2–20	5×10^{-4}
Thermal conductivity (W/m K)	100–390	300	3,320	10–1,000	4,800–5,300
Young's modulus (TPa)	0.5	/	1.28	0.8	1

Among these carbonaceous fillers, the two most studied in textile applications are CNT and CB. These additives are mainly used to contribute to electrical conductivity and therefore for applications in smart textiles (passive or active). The CNTs have been attracting research for more than 15 years. The restrictions on their use due to their production costs are beginning to diminish. There are two prominent families of CNT: (i) single-walled CNTs, which have a diameter of the order of nanometers and their length is of the order of micrometers and (ii) the multiwall nanotubes are composed of different concentric layers of carbon stacked, ordered and spaced of 3.4 Å, whose diameter varies between 2 and 25 nm and their length between 20 and 80 μm [21]. Two leading families of synthesis exist to make these nanotubes: (i) high-temperature synthesis with electric arc ablation, laser ablation and solar oven synthesis and (ii) medium temperature synthesis or CVD. The objective of these syntheses is to produce a high amount of CNTs with the highest degree of purity and alignment, good uniformity and the lowest possible costs.

In comparison, CB is obtained by thermal decomposition, or incomplete combustion of hydrocarbons (liquid or gaseous) under controlled conditions. They have been classified according to the ASTM standard thanks to a generic code Nxyz, where x represents a range of diameters and yz a number referring to the aggregation capacity. A high value allows particles (with a spherical shape with a mean diameter in the range of 20 to 70 nm) to form agglomerates with a diameter of up to 500 µm [18]. Graphene is one allotrope of carbon in sheet form (monolayer of carbon atoms) with a thickness of 1 nm [22]. Graphene attracts more and more attention and according to recent studies transparent conductive fibers [23] can be achieved. All these carbonaceous fillers can be used pristine, but are also the subject of numerous studies on their modification (functionalization) in order to modify their interactions (dispersion, adhesion, etc.) and also to combine properties [23–26].

2.2.2.2 Inorganic fillers – metal-based fillers

The use of metallic materials in textiles brings very sought-after properties such as electrical conductivity or antibacterial properties. Among all the metals that can be found at the nanoparticle scale (between 10 and 100 nm): aluminum (Al), cadmium (Cd), cobalt (Co), copper (Cu), gold (Au), iron (Fe), lead (Pb), silver (Ag) and zinc (Zn) [12]. Destructive or constructive methods produce metal nanoparticles, and one of the main restrictions is environmental sensitivity (oxygen, moisture, UV and temperature).

Silver-based particles are mostly used for their antimicrobial and self-cleaning activities. In reality, silver nanoparticles are composed of a large percentage of silver oxide due to their large ratio of surface to bulk silver atoms [27]. A variety of preparation techniques have been reported for the synthesis of silver particles; notable examples include laser ablation, gamma irradiation, electron irradiation, chemical reduction, photochemical methods, microwave processing and synthetic biological methods [28].

Gold nanoparticles are widely used in medical applications (diagnostics, therapeutics). The spherical shape presents interesting optoelectronic properties, large surface-to-volume ratio, biocompatibility and low toxicity [29]. Through their possibility of multiple functionalizations, they show multiple advantages for various applications, the cost remaining a limiting point [30]. For example, in the textile area, Au particles can improve colorfastness during washing and abrasion [31].

2.2.2.3 Inorganic fillers – metal oxide-based fillers

Unlike metal fillers, metallic oxide particles have high reactivity and efficiency, and they are generally used more due to cost issues and lower toxicity. Titanium oxide is a standard filler used in a textile application for UV protective, self-cleaning and antibacterial properties [32], but the main application is for the white color of fibers due to its brightness and high refractive index [33]. TiO_2 exists naturally in three crystalline forms: anatase, rutile and brookite, and the synthesis of its nanoparticles can be done by various methods, including the sol–gel process, CVD and hydrothermal [34].

Another metal oxide is zinc oxide (ZnO), a semiconductor and biocompatible nanoparticle used as antimicrobial, photocatalytic moisture management, hydrophobicity, flame retardancy and UV-blocking properties [35]. To produce ZnO fillers, different techniques can be used such as thermal hydrolysis, hydrothermal processing, sol–gel method, vapor condensation method, spray pyrolysis and thermochemical techniques [36]. ZnO (as with powder) can be present in a wide diversity of shapes, such as nanocombs, nanorings, nanohelixes/nanosprings, nanobelts, nanowires and nanocages, with even more shapes being possible [37].

Less common fillers as aluminum oxide (Al_2O_3) nanoparticles are those used in textile materials to produce functional properties, such as antibacterial, flame-retardant, thermal conductivity and higher tensile strength properties [38, 39]. Copper oxide (CuO) for antibacterial activity is also studied [40].

2.2.2.4 Inorganic fillers – clays

Clay is a large family, where the most commonly used textile application is phyllosilicates (2:1 layered silicates) like mica, talc, montmorillonite (MMT), vermiculite, hectorite and saponite. They are from the smectite group composed of layers of two tetrahedrally coordinated silicon atom groups fused to an edge-shared octahedral sheet of either aluminum or magnesium hydroxide [41]. Silicates are minerals composed of tetrahedral units (SiO_4) with a silicon atom in the center and oxygen atoms at the four corners. They form an extremely large class of materials that can be divided into families. As it is usual to use crystallochemical criteria to establish this distribution, it is then convenient to consider the arrangement of the units (SiO_4) between them to establish a classification.

Silicates are clay minerals that occur as nanoscale particles: around 1 nm thick and their length varies from tens of nanometers to more than 1 µm, depending on the layered silicate. Phyllosilicates are lamellar by definition, due to the arrangement of the atoms in tetrahedral or octahedral planes, but can also be found in tubular or fibrillar form at the microscopic level. The surface area of each nanoclay particle is around 750 m^2/g with an aspect ratio around 50 [42]. When a lamellar clay is introduced into a thermoplastic matrix, different morphologies can be generated. They are three in number and are the result of different factors (polarity and rheology of the matrix, processing conditions).

- There is no insertion of polymer chains in the interlayer space. The clay is dispersed in the polymer matrix as primary particles or agglomerates. Therefore, the interpillar distance of the clay remains the same as its original value, and it is a microcomposite.
- The clay is dispersed in the polymer matrix in the form of tactoids, which are stacks of parallel clay platelets at ~10 Å separation containing from a few to as many as 1,000 sheets. Polymer chains are then inserted in the interleaf space, leading to an increase in the interleaf distance while maintaining the layered structure of the clay, meaning an intercalated nanocomposite.
- The clay sheets are wholly separated (delaminated) and homogeneously dispersed in the polymer matrix, and therefore it is called exfoliated nanocomposite.

Surface modifications are generally carried out beforehand to modify the affinity and thus the morphology of the clays with the polymers [43].

They are commonly used as constituents of FR systems in different FR modes and impermeability to liquids/gases in the textile industry, especially MMT [44–46]. MMT belongs to the phyllosilicate family (type 2:1), used mainly in textile. It has the formula $(Na, Ca)O,3(Al, Mg)_2 \cdot Si_4OIO(OH)_2 \cdot nH_2O$ and its structure is shown in Figure 2.5. The elementary structure of MMT results from the combination of two tetrahedral layers of silica and an octahedral layer of aluminumm or magnesium ions. The repetition of this unit in the direction of the plane forms a sheet comparable to a platelet whose lateral dimensions can reach several hundreds of nanometers and whose thickness is of the order of 1 nm. This geometry gives the clay a very high specific surface (800 m^2/g) and thus an extensive surface area for interaction with the polymer. Thus, MMT has been used so much in the realization of nanocomposites. In its natural state, MMT usually contains hydratable Na^+ or K^+ compensating cations that make it hydrophilic. It is necessary to convert its normally hydrophilic surface into an organophilic surface to disperse the clay in other polymeric matrices. Two main techniques are mentioned in the literature [47]: (i) cationic exchange (most straightforward and the most used, which consists of replacing the interlayer cations with organic molecules) (Figure 2.5) and (ii) organosilane grafting.

2.2.2.5 Organic fillers – protein-based fillers

Proteins are complex macromolecules with substantial molar weight variation. They are composed of amino acid chains, constituting the monomeric unit, united by peptide bonds. These constituent amino acids belong to a restricted group of 20 different amino acids. Chains with less than 100 amino acid units are called polypeptides.

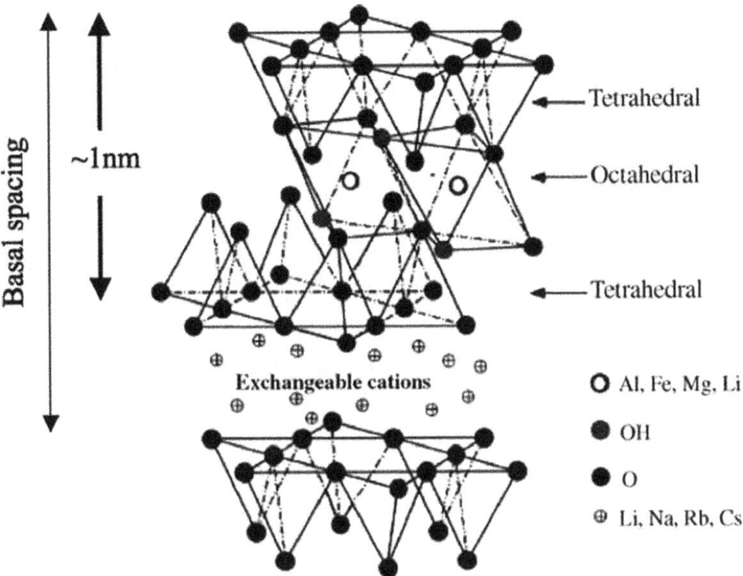

Figure 2.5: Structure of 2:1 phyllosilicates (reprint from [48]).

Beyond that, they are called proteins. The nature, number and sequence of these different amino acids lead to many combinations, and the number of other proteins on earth is estimated to be 1,014. In this respect, proteins differ from polysaccharides, where only a few monomers are involved in their structure. For example, cellulose and starch, the most common polysaccharide polymers, contain a single monomer, glucose. Proteins, which are of natural origin, have been attracting great interest for several years as biomaterials and especially as additives in thermoplastics. They can be of multiple origins: plant proteins (soy, gluten, and so on), animal proteins (collagen, gelatin, milk proteins (lactoserum and casein)). Proteins can have unique properties but generally cannot be recovered and are then considered waste, leading to economic loss and environmental pollution [49]. As an example, residual soy proteins are a by-product of the soy oil industry and are composed of a mixture of albumins and globulins, 90% of which are storage proteins with globular structure and can be used by electrospinning in poly(vinyl alcohol) for filtration [50]. These different proteins can be used as additives in addition to other nanofillers to promote dispersion. For example, soy proteins used with CNTs [51] or silk with TiO_2 [52]. But also, proteins can provide functionality to fibers, such as the antibacterial properties of casein which represents 80% of milk proteins [53].

2.2.2.6 Organic fillers – polysaccharide-based fillers

– Cellulose-based fillers

Polysaccharides are being further developed as nanofillers in materials from natural resources, especially nanocelluloses. Cellulose, as raw material, is abundant and biodegradable and can be extracted from wood, algae and bacteria. This assembly of polymeric glucose chains is composed of crystalline and amorphous domains, which depends in particular on the initial resource and the extraction treatment carried out [54]. Nanocellulose, also known as cellulose nanocrystals, nanocrystal of cellulose or cellulose nanowhiskers, is nanocellulose with high strength, usually extracted from cellulose fibrils in several ways. These particles are tiny and correspond to crystals of great purity, with a diameter between 5 and 10 nm and a length between 100 and 500 nm [55]. These extractions can be carried out via mechanical, chemical or even enzymatic methods, often combining the latter. The most common is acid hydrolysis. It is only the crystalline part of the cellulose chains which is used. The amorphous regions of cellulose act as weak points of the structure and are directly responsible for the transverse cleavage of microfibrils into short single crystals under acid hydrolysis. Indeed, the coexisting amorphous regions are destructured under certain conditions while the crystalline areas remain intact. The amorphous region has much faster hydrolysis kinetic than the crystalline one. These crystals occur naturally and are known to be stable and physiologically inert [56]. Manufacturing methods from textile waste allow reintroducing cellulose nanocrystals extracted from cotton fibers into the manufacturing loop [57–59].

– Lignin-based fillers

Lignin is one of the naturally occurring macromolecules in plants and trees. It constitutes 15–30% of the dry weight of wood plants together with cellulose and hemicellulose [60]. Its primary function in the plant cell is to provide mechanical strength and prevent any mechanical shock. Lignin is the second most abundant polymer from biomass after cellulose, and the main one based on aromatic subunits, its availability on Earth is more than 300 billion tons, increasing exponentially every year. Alternatively, industrial lignin is primarily a by-product of the wood pulping and paper making industries; its amount is estimated to be 70 million tons/year worldwide [61, 62]. However, less than 2% is recovered for utilization as a chemical product, and the rest is considered waste and primarily burnt for recovering energy. The lignocellulosic biomass is composed of three major components: cellulose (35–83%), hemicellulose (0–30%) and lignin (1–43%) based on dry weight [63]. Lignin has a complex structure created by a three-dimensional cross-linked network whose base unit is phenylpropane, instigating from three phenolic precursors (monolignols) known as sinapyl *p*-coumarin and coniferyl alcohols, respectively. The phenolic subunits that originate from these monolignols are *p*-hydroxyphenyl (H, from coumaryl alcohol), guaiacyl (G, from coniferyl alcohol) and syringyl (S, from sinapyl alcohol) moieties. Combining these three monomers leads to a highly branched polyphenolic polymer with a three-dimensional structure. Lignin composition and content depend on the plant species and its environment, its variation is more prominent in hardwoods than in softwoods.

Lignin is extracted from different lignocellulosic parts by physical and chemical and/or biochemical treatments. Technical lignin is classified based on different chemical nature. Standard pulping processes are based on the splitting of ether and ester linkages. As a consequence, technical lignin differs significantly from the plant lignin. Based on the chemical processes, they are classified into two main categories such as sulfur-bearing and sulfur-free processes. The successful introduction of lignin in new bio-based materials is highly dependent on its structure and purity. Extraction processes represent the critical point to use lignin in industrial applications. However, despite these limitations, the large amount of available lignin has driven numerous efforts and research to develop its use for industrial applications. The difference among organosolv (OL), ionic liquid (IL) and Klason (KL) processes was studied for extracting lignin from the same botanical source, for example, poplar wood. Each process yielded a different amount of lignins that are 5.5 wt% for milled lignin, 3.9 wt% for OL, 5.8 wt% for IL and 19.5 wt% for KL. Thus, the amount of recovered lignin is mainly dependent on the extraction process.

Only 2% of the lignin production from the pulp industry is used for value-added products. Lignin can lead to high-value products thanks to fragmentation (oxidation or pyrolysis). However, the production of high-value chemicals from lignin is complex to industrialize. Another promising domain of applications is to use lignin from its initial state in the polymer system as an additive, filler, reinforcement agent and so on. Moreover, chemical functionalization can be undertaken to tune its properties according to its structure. Crosslinking with other polymers is also possible via its hydroxyl groups to give novel materials. Lignin is known for enhancing the biodegradability of polymers where it has been incorporated.

2.2.2.7 Nano- and microcapsules

Over the last 30 years, microencapsulation as technology has led to much research in the textile field. One of the main features of this technology is the possibility to combine the properties of inorganic materials with those of organic materials, which is complex with other technologies. Microencapsulation is not limited to a single product or product component and can be defined as a process by which fine particles of a solid, liquid or gaseous compound are trapped by an "inert" membrane, which isolates and protects them from the external environment. This "inert" character is related to the reactivity of the membrane toward the encapsulated substance and the matrix in which the capsules will be dispersed. Thus, the transformation of a product in this form improves the compatibility of the active ingredients with the environment and improves their thermal stability [64].

The choice of a membrane or a process is made concerning the application, the final use envisaged and according to the mechanical or thermal constraints of the process of implementation chosen for the incorporation in the fibers. Thus, two main categories of membranes are to be distinguished: (i) semiporous membranes when a controlled release of the substance is desired and (ii) impermeable membranes when the final application requires maintaining the substance trapped to obtain the envisaged property [65].

Depending on the physicochemical and thermomechanical properties of the membranes and their size, the particles can be integrated into the fibers by melt-spinning or wet-spinning or by electrospinning. Whatever the process, the ratio between the average diameter of the particles and that of the fibers must be small enough to limit the loss of mechanical properties of the textile [66].

In general, aminoplast membranes obtained by in situ polymerization will be preferred for melt-spinning incorporation into polyolefin fibers [66, 67]. Indeed, the other types of membranes degrade during the process. For polymers requiring higher processing temperatures, such as polyamides or polyesters, silica membranes obtained by the sol–gel process can be used [68]. Membranes based on acrylic chemistry are introduced by wet-spinning [69].

In an in situ polymerization, the monomers and/or prepolymer are dispersed in a single phase (in the continuous or dispersed phase). In the case of melamine–formaldehyde (MF) microcapsules, when the emulsion step is realized under high agitation and in the presence of surfactants, agitation is reduced, and membrane formation is initiated by temperature rise or pH adjustment resulting in a change in the solubility of the oligomers in water and initiating membrane formation [70]. Polymers and oligomers diffuse at the interface during synthesis to form the capsule wall [71]. Monomers and prepolymers are soluble in the continuous phase; however, the polymer must not be soluble to allow its diffusion at the interface and the formation of particles. Core/shell capsules are obtained when the oligomers resulting from the polymerization are not soluble in the dispersed phase.

MF microcapsules have been successfully commercialized for the last two decades because of their low price, simple fabrication route, good thermal and mechanical properties, high fire resistance and acid and alkaline pH resistance. The residual formaldehyde is considered the main drawback of this process and can initiate predegradation during the melt-spinning process when the temperature is higher than 170 °C. The brittle rigid MF shell was used to entrap most of the active principle having a potential action in functional textiles.

Sol–gel is the abbreviation for "solution-gelling." It is a process by which essentially inorganic materials are synthesized. Sol (solution) is a stable dispersion of polymers or colloidal particles with diameters of (1–100 nm) in a solvent prepared at low pH. This solution is prepared through condensation reactions of silica precursors already hydrolyzed to polymerize and form sol nanoparticles. The stability of the suspension is ensured by the Van-der-Waals-type interactions occurring with small particles. During the reaction progresses, the smaller polymer chains are consumed, and solvents and monomers diffuse less in the particle pores. The material densifies and stabilizes when some parts depolymerize and polymerize again, resulting in well-compact gel and releasing some solvent that needs to be eliminated by drying. The formation of long macromolecular chains induces an increase in the solution viscosity leading to a transition to a gel characterized by an infinite three-dimensional network structure. Various physico-chemical parameters such as the solution composition (solvents, precursors), temperature, pH or drying, influence the sol–gel transition process (gelling). These parameters determine the reaction mechanisms and kinetics and thus the final properties of the material. Different studies used the sol–gel method to encapsulate active substance for fiber applications by inorganic shells. Therefore, inorganic materials have attracted significant attention, thanks to their excellent properties, such as high thermal conductivity, non-flammability and durability.

Microencapsulation by heterophase polymerization is divided into three main categories: (i) suspension polymerization, (ii) emulsion/miniemulsion polymerization and (iii) dispersion polymerization which is rarely used in textile field.

The suspension polymerization includes two phases: (i) the dispersed phase containing the active agent, water-insoluble monomers and initiator; (ii) the continuous phase involving solvent and reactant of shell materials. Active substance droplets are obtained under high shearing rate of the mixture. The stability of the emulsion is ensured by the presence of protective colloid, for example, poly(vinyl alcohol), polyvinyl pyrrolidone or cellulosic derivative. This method is mainly used to create polystyrene, poly(methyl methacrylate) and poly(vinyl chloride) shells [72].

The emulsion polymerization involves emulsifying monomers and creating a cross-linked system in chemical, thermal or enzymatic ways. The mean diameter of the resulting particles ranges between 50 nm to 1 μm and is related to the mass ratio of monomer/aqueous phase, surfactant and initiator concentration, and polymerization temperature [73]. The first step consists of the formation of monomer micelles in the presence of a surfactant. The polymerization reactions are initiated by the addition of a water-soluble initiator leading to free radical formation and diffusion through the swollen monomer micelles. Polymethyl methacrylate, polystyrene, polyvinyl acetate and poly(ethyl methacrylate) are the main shells synthesized through this way.

2.3 Nanocomposite fiber processing

The currently applied spinning technologies are all based on the extrusion of a polymer through a spinneret, allowing for the continuous production of single- or multi-filament materials. The polymer is either mechanically stretched as it is wound onto a spool (spinning), blown by a pressurized gas or subjected to centrifugal forces. The properties of the final product of each spinning technology depend on the process and its parameters, the material selected and the post-processing performed. In this context, spinning methods are classified into three groups according to the solidification mechanism of the extruded material: (i) solution spinning (evaporation or coagulation); (ii) emulsion spinning (phase separation) and (iii) melt-blowing/spinning (cooling) [74]. The elaboration of nanocomposite fibers uses these conventional technologies for forming of the man-made fibers, which are developed in this section.

2.3.1 Melt spinning route

2.3.1.1 Compounding step

The melt spinning process is the most commonly used process for making synthetic filaments. It allows the processing of thermoplastic polymers (also filled with nanofillers) at high production rates without using solvents. Compared with the solvent technique with high energy consumption, the melt-processing technique is more practical on an industrial scale. However, in the case of nanocomposites, a preliminary step of formulation of the blend is necessary, which is called compounding. This preliminary step allows the dispersion and distribution of the nanofillers in the polymer matrix via an extrusion process. Main parameters influence the dispersing effect as screw configuration and speed, length/diameter ratio, and temperatures [75]. Melt compounding is designed in a single or twin-screw extruder to optimize polymer homogenization and dispersing fillers. Depending on the thermoplastic/filler combination used, the parameters must be optimized particularly to the targeted properties [76]. The use of in situ polymerization can also allow the functionalization of the polymer via the incorporation of fillers with the monomer before the polymerization. Some authors compare the effect of the preparation approach and observe variations on the thermal, morphological and mechanical behavior of the final nanocomposites. For example, a polyamide 11 filled with clay obtained by in-situ polymerization presented higher values of Young's modulus, tensile strength and heat distortion temperature due to a better dispersion of the nanoclays compared to melt compounding [77].

2.3.1.2 Melt spinning process

Once the thermoplastic polymer is functionalized with the nanofillers, the nanocomposite can be shaped into a filament. The most widely used spinning process is melt spinning due to its economy and variety of processable polymers. Figure 2.6 shows the basic prinpicle of the process [78]. Some polymers like Polyethylene Terephthalate

(PET) need to be dried to remove any moisture detrimental to the process. The extrusion temperatures used are adjusted according to the viscosity of the polymer, the extrusion pressure and the extrusion speed. A spinning pump feds the molten polymer to the spinnerets. Then the filaments are cooled by ambient air, which is directed perpendicular to the production direction. The filaments then converge to a single point for the application of spin finish in an oiler. The application of spin finish during the spinning process is necessary for several reasons [78]:
- Lubrication to facilitate the sliding of the yarn on the metal parts,
- The antistatic role in suppressing the repulsion between filaments due to the electric charges developed during the various rubbings,
- The cohesion between the filaments during the spinning process and the following operations (unwinding, twisting, weaving, etc.).

Two spinning possibilities are available after the sizing (a blend of oil and surfactant) of the filaments:
- The high-speed spinning process where the filaments are directly collected at the exit of the spinnerets on very high-speed rollers (over 3,000 m/min)
- The lower speed spinning process is called the spin-draw process, where the filaments are first implemented at speeds between 300 and 3,000 m/min. Then they are drawn and, in some cases, directly twisted. The drawing is applied between two rotating and heated rollers: the feed roller and the drawing roller.

Four categories of filaments according to the orientation of the macromolecular chains can be listed [78]:
- LOY (low-oriented yarn) for production speeds below 1,500 m/min;
- POY (partially oriented yarn) for production speeds around 3,000 m/min;
- FDY (fully drawn yarn) for production speeds above 5,000 m/min.

Melt spinning allows the elaboration of mono or multifilaments ranging from 1 to more than a thousand microns. The stretching step by rollers provides the filament's desired physical, mechanical and thermal properties. In the case of nanocomposites, drawing also allows the orientation and alignment of the nanofillers in the production axis of the filaments [79].

2.3.2 Solution spinning

Solution spinning (or solvent spinning) consists of dissolving the polymer in a solvent which is then passed through the spinnerets to create the filaments. The removal of the solvent can take place using two possible routes: (i) wet-spinning and (ii) dry-

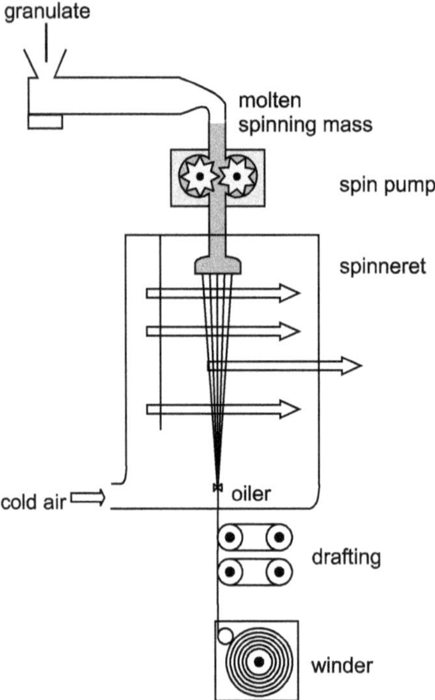

Figure 2.6: Principle of melt spinning [78].

spinning. Wet-spinning allows precipitating the filaments in a bath to solidify them. The dies are then immersed in a chemical bath, and the filaments are formed by coagulation. Dry-spinning does not use a bath. The filaments are cooled by air at the exit of the spinnerets, allowing the evaporation of the solvent.

In both cases, the filaments are then drawn and collected at a speed of a few tens of meters per minute. Preparation of the polymer/additive mixtures is also necessary before the spinning process to disperse the nanofiller [80, 81]. Natural polymers such as cellulose (artificial fibers) [82, 83], and also thermoplastics with thermal transitions not suitable for melt-processing [84, 85] are processed using this technique.

2.3.3 Wet-spinning

The oldest process is wet-spinning, where the filaments are directly precipitated in a coagulation bath (Figure 2.7) [78]. This so called precipitation or coagulation bath consists of one or more nonsolvent(s) of the polymer, which is miscible with the solvent. Phase inversion phenomena then take place with diffusive transfers from solvent to non-solvent(s) and vice versa to let the polymer solidify and the filament(s) form. This process is not universal and strongly depends on the nature of the polymer that is used. However, after coagulation, it is often customary to perform stretching to improve the mechanical properties of the filament(s), washing and heat-drying to remove the remaining solvent before winding. Finishing steps such as sizing or coating can also be added. Acrylic [86], rayon or aramids are notably spun with this technique.

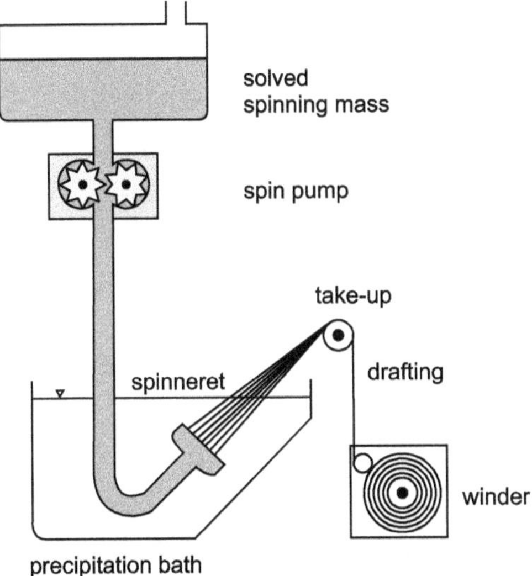

Figure 2.7: Principle of wet spinning [78].

2.3.4 Dry-spinning

Dry-spinning allows the solvent to be extracted by pulsed or radiant hot air (Figure 2.8). The critical variables in dry spinning are heat transfer, mass transfer and stress on the filament [87]. The same steps used for wet-spinning can be added before winding. For example, acetate, triacetate, polyvinyl chloride or spandex filaments are spun using this method. Dry spun fibers have a cross-section in "dogbone" shape due to the diffusion of this residual solvent.

2.3.5 Electrospinning

The production of electrospun nanofibers loaded with nanoparticles or active ingredients, metals, ceramics and so on has recently attracted a lot of research work from scientists and researchers around the world. This process allows the production of

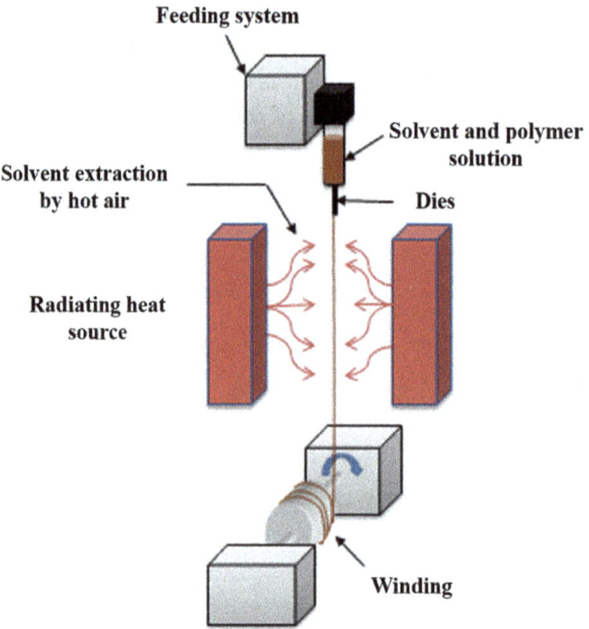

Figure 2.8: Schematic representation of the dry-spinning process.

nanofibers with an average diameter between 50 and 1,000 nm, using solutions of natural and/or synthetic polymers. Electrospun fibers exhibit a high surface-to-volume ratio, superior surface functionality and mechanical properties when the polymer fiber diameter is reduced to the submicron or nanometer scale. This type of material has many applications in different fields requiring high specific surface area, controlled porosity or new functionalities, such as air and water filtration, tissue engineering, sensors, drug delivery, wound dressing, technical textiles and energy applications.

The main techniques used to produce nanofibers are solution or melt electrospinning (Figure 2.9). These processes are based on the application of an electrical potential applied between the physical emission zone of the polymer solution or the molten polymer and an electrically conductive surface acting as a collector on which the fibers are deposited. The electric field generated allows the fluid to be stretched into the shape of a Taylor cone. When the electrical voltage exceeds a critical value, the surface tension is no longer sufficient to maintain the cone, and the polymer jet is initiated and drawn toward the collector. In melt electrospinning, a temperature above the melting point of the polymer is applied to ensure its flow through the nozzle, and the fiber is obtained after stretching by elongation during the time of flight.

The key factors in the process are the type of polymer and the processing line, which includes the extruder, the metering pump and the filter unit with die and

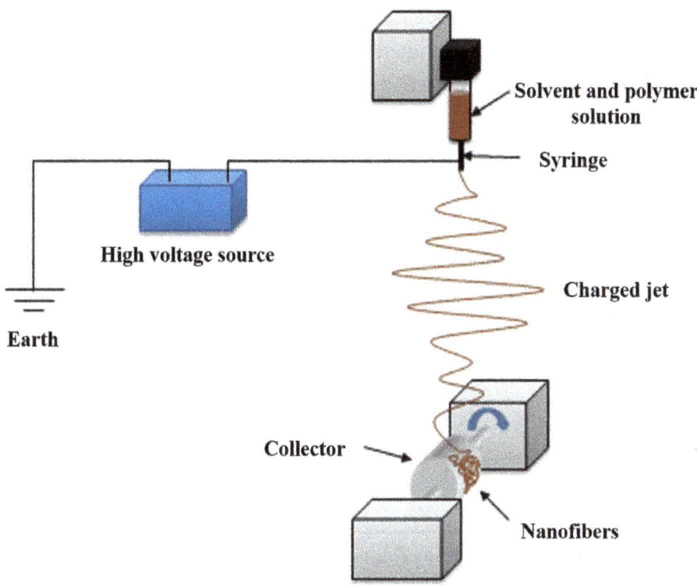

Figure 2.9: Schematic representation of the electrospinning process.

filter media. The die is a plate with a number of holes from which the polymer filaments exit before being cooled by gas or air to a temperature below the melting point of the polymer as it exits the die. The choice of the temperature must allow the realization of a flow by elongation on the melted polymer. The cooling process and the drawing rate control the crystallinity and mechanical properties of the resulting fibers. A second hot drawing may be used to improve the orientation of the macromolecular chains, thus improving the properties of the fibers. A cold drawing is requested to improve the mechanical properties of the thermoplastic fibers by promoting the orientation and the crystallinity of the polymers.

This process is mainly used, when it is difficult to find a proper solvent at room temperature for the polymer. Even if the process has a high efficiency, the mean diameter of the fibers obtained is usually higher than those obtained by a solution electrospinning process. Furthermore, such a process requires specific equipment and much higher voltages due to the low electrical conductivity and high viscosity of molten polymers.

The principle of solution electrospinning consists in the uniaxial stretching of a viscoelastic polymer solution in an electric field. The equipment consists of a syringe pump equipped with a needle, a high-voltage power supply and an electrically conductive collector. A syringe is filled with a polymer solution, the syringe pump is adjusted to a certain flow rate and an appropriate voltage is applied between the needle and the collector. The process parameters are adjusted to ensure the formation of a Taylor cone at the tip of the needle. The fine jet is extruded from the Taylor cone, when the electrostatic forces are greater than the surface tension and viscous forces. Evaporation of the solvent during the flight process allows the formation of solid nanofibers recovered from the collector surface.

2.3.6 Conclusion

All these methods of implementation are selected according to the type of polymers to be implemented but also the fineness to be reached as well as the production capacities. Table 2.2 presents a comparison of the different methods.

Table 2.2: Selection of fiber manufacturing processes and their characteristics (adapted from [88]).

	Fiber diameter	Advantages	Drawbacks
Melt electrospinning	<100 nm to 500 μm	Ability to print patterns; solvent-free; low cost; diameter is proportional to flow rate.	Low output; difficulty in designing the process; limited number of polymers; requires good thermal stability of polymers.
Solution electrospinning	<50 nm to 10 μm	Simple to realize; low cost; suitable for many polymers; sub-micron diameters	Low output; random fiber; solvent release
Melt spinning	1 μm to 3000 μm	High output; very consistent production; fibers use in many textile technologies; industrially successful	Requires drawing, low diameter obtained at high stretching; significant cost and difficulties to attain sub-micron diameter fibers
Solution spinning	1 μm to 200 μm	Use for thermally unstable polymers; various system configurations; industrially successful	Solvent removal system; requires removal, coagulation baths.
Melt blowing	<500 nm to 10 μm	High output; industrially successful	High cost to establish, difficult to control fiber morphology

The use of nanoscale additives in fibers can modify or add properties to textile materials. Nanofillers can be of multiple natures and open doors to new functionalities for a wide range of applications. The implementation of these nanocomposites in the form of fibers is realized via conventional textile industry processes but requires adaptations. The nanofillers in polymers can lead to numerous modifications in the process parameters. For example, incorporating cellulose nanowhiskers in PLA can enormously decrease the stretchability, mainly due to a restriction of the polymer chain mobility due to the nanoscale additives [89]. Special attention to the thermal stability of the nanofillers must be considered in melt spinning and chemical stability in solvent spinning.

2.4 Improvement of fiber properties

In recent times, filler-based fiber composites have attracted a great deal of scientific interest due to its unique properties over conventional fiber materials, such as superior strength, modulus, thermal stability, thermal and electrical conductivity, and gas barrier. They are thus finding real and fast-growing applications in wide-ranging sectors requiring a wide range of functionalities, such as antimicrobial, FR, gas barrier, shape memory, energy-scavenging, sensors, as well as medical applications, such as in tissue engineering and wound dressings, for creating a new range of smart and innovative textiles. Like other industrial fields, the development of these new kinds of materials has shown tremendous promise in developing high-performance multifunctional materials with prospects for application in specialty fibers and technical textiles.

2.4.1 Influence of additives on the mechanical properties

Contrary to the viscosity, which increases with most fillers, whether nano or micrometric, the material's mechanical properties are modified differently depending on the type of additive incorporated. Indeed, it is established that conventional fillers negatively affect the mechanical properties while the presence of nanofillers improves them in some cases.

The influence of nanofillers on the mechanical properties of polymer matrices has also been widely commented on in the literature since the 1990s. For example, the incorporation of MMT in PA-6 increases the modulus of the material, thanks to its good dispersion in the polymer matrix. Since then, the number of studies on "nanocomposite" materials, combining polymer matrices and nanofillers, has continued to grow. These materials combine the ease of processing, flexibility and lightness of thermoplastics with inorganic nanofillers' thermal and chemical stability and mechanical resistance.

Due to the rheological and mechanical limitations of melt spinning, additive contents rarely exceed 5% by mass. The main parameters conditioning the mechanical properties of polymer/additive blends are the particle size and shape of the fillers, their content and their interactions with the polymer matrix (interfacial tension, adhesion and dispersion state). The mechanical properties depend strongly on the final shape of the material (films, sheets or fibers) and consequently on the chosen processing method. The use of coupling agents is a technical solution to improve the properties of filled polymers. However, this method requires the preparation of the materials on a molecular scale, and the solutions developed are generally not adapted to all types of matrices. In most of the cases, the evolution of the mechanical properties is related to the cohesion between the polymer and the nanofillers, which create void zones near the interfaces during processing and limit the improvement of the mechanical properties under high stretching.

2.4.2 Thermal comfort

Heat retention in textile materials using the heat-storage method with far-infrared radiation can be obtained from ceramic-embedded yarn manufacturing technology. Some commercialized inorganic ceramics that are embedded in heat-storage textile goods, such as Solar-α by Unitika storage fiber embedded with ZrC, Thermotron® (Unitika), Megatron® (Toray) and Reothermo® (Asahi Kasei), propose some fabrics containing ceramic-imbedded yarns or heat-storage filaments using ZrC which were developed for bedding textiles by Kuraray, Mitsubishi-rayon and KB Seiren [90]. Ceramic materials, such as charcoal, ZnO, Al_2O_3, SiO_2 and TiO_2, were also used to enhance the far-infrared emission related to heat-storage and heat-release properties [91, 92].

2.4.3 Flame-retardant properties

The use of nanoparticles allows to improve the thermal stability and the flame-retardant properties of textile fibers. They participate in the formation of a protective barrier layer, the promotion of carbonization and the scavenging of free radicals. Nevertheless, to allow the improvement of fire-retardant properties, it is necessary to control the morphology and composition of the nanoparticle, the compatibility between the nanoparticle and the polymer, the dispersion of the nanoparticles in the polymer matrix during processing and the migration rate of the nanoparticles toward the surface during the degradation of the material. In addition, without the use of active FR, these polymer nanocomposites do not pass some regulatory fire tests, due to the fact that they do not self-extinguish, even though they have good thermal stability, lower heat release rate, higher char low oxygen index value. Thus, one of the approaches taken is to combine nanoparticles with conventional FRs to achieve regulatory fire safety performance. On the other hand, the use of nanoparticles allows to reduce the concentration of common FR compounds, to improve the physical and mechanical properties and to develop multifunctional textiles [93]. Nanoparticles, such as clays, CNTs or nanofiber, are used to delay the depolymerization of the polymer and thus reduce the amount of released heat, therefore improving the fire behavior of almost all conventional polymers [94].

2.4.4 Antimicrobial properties

Many commercialized products with antimicrobial activity are produced by incorporating organic or inorganic agents through melt spinning. The active products must have good thermal stability to allow the production of a yarn. Amicor's® acrylic fiber uses an organic agent like triclosan to provide antibacterial properties. Metal and metal oxide nanoparticles are effective against bacteria, viruses and microorganisms due to their nanosize which increases both the reaction rate and the contact area, although the antimicrobial mechanism is not yet fully understood. Among these, silver can also be incorporated into synthetic monofilaments, for example, Trevira Bioactive® and Meryl® Skinlife, made of polyester and polyamide, respectively, which have permanent antimicrobial properties [95].

2.4.5 Enhanced electrical properties

The polymers constituting the chemical fibers are electrically insulating, but addition of conductive fillers allows to realize materials with a more or less high electrical conductivity. The principle is based on the formulation of CPC. The insertion of conductive fillers in a polymer matrix can generate the passage from an insulating state to a semiconducting or conductive state of the material from a certain concentration of fillers introduced inside the latter. We speak of electrical percolation and electrical percolation threshold. Many models have been developed on this phenomenon, the main and most used is the static model introduced by Kirkpatrick and Zallen [96, 97]. Three main classes of materials can be considered for the realization of CPC: metallic, carbon and intrinsic conductive polymers as additives. In the literature it is mainly carbonaceous fillers (CNT, CB, graphene) that are added to the thermoplastic polymer in the melt process to achieve synthetic fibers with antistatic or semiconducting properties [98]. Even if the solvent process allows to reach higher levels of conductive fillers, the contribution of electrical conductivity in fibers is very useful for smart textiles, heating textiles, electromagnetic barriers and sensors and actuators.

2.4.6 Energy harvesting

One of the solutions is to generate electrical energy from the recovery of energy energy harvesting. This solution would potentially allow to generate energy continuously, thus eliminating the use of batteries and making the textile autonomous or on a smaller scale by prolonging the life of the battery. Different properties can be exploited, such as the photovoltaic effect, thermoelectricity and so on. Recently, the concept of piezoelectricity is strongly developed by the industry and the military for the generation of electrical energy. Piezoelectric materials are capable of transforming a small movement into a large movement and into a large electrical potential.

The application of piezoelectric textiles in energy harvesting is extensive. Piezoelectric textiles made from polymer fibers have several advantages over traditional ceramic piezoelectric materials, including flexibility and comfort, and can be more effectively used in wearable piezoelectric devices. In piezoelectric materials, micro and nanofibers improve the alignment of the dipoles, which results in a higher piezoelectric potential. A high piezoelectric coefficient is responsible for the extensive study of the piezoelectric properties of PVDF and copolymers. In medical devices, PLA with biodegradable and biocompatible properties has excellent potential. PAN nanofibers prepared by electrospinning have piezoelectric properties that are comparable to PVDF. In addition, adding nanoparticles can significantly enhance the piezoelectric properties of nanofibers. Smart textiles with micro and nanostructures are widely used in nanogenerators, sensors and energy storage. In order to recover energy from surrounding movements, piezogenerators, devices generating piezoelectric energy, must be equipped with a piezoelectric material as well as conductive materials, playing the role of electrode and conducting the electricity produced. In the literature, two materials in particular are widely referenced for their piezoelectric effect: lead zirconate titanate (PZT) and polyvinylidene fluoride (PVDF). Nanoparticles and polymers are used in electrospinning to produce nanocomposites with enhanced piezoelectric properties. Therefore, nanocomposite-based energy harvesting has received increasing attention in recent years. In most cases, PVDF is chosen as a polymer because of its low cost, flexibility and biocompatibility. The addition of certain additives (ZnO, GO, PZT, $BaTiO_3$ and inorganic salts (NaCl, KBr, LiCl, KCl), clay, CNT, nanowires or nanofillers such as palladium and silver) to the polymer solutions prior to the electrospinning process plays a distinct role in influencing the crystallization kinetics, crystallite size, crystal morphology, degree of crystallinity and β-phase content of the matrix by enhancing the α to β transformation.

2.5 Conclusion

This chapter reviews the fundamental aspects, production, processing and applications of nanoparticle or microparticle-based nanocomposite fibers and nanofibers. It has been possible to observe that many nanomaterials have been used in a variety of polymer matrices of multifunctional fibrous nanocomposites while maintaining good mechanical properties for the final application. The implementation of these materials is specific to each of the established blends and specifications. Thus, several production techniques are listed in the literature, such as melt spinning, dry-spinning, wet-spinning, dry jet spinning and electrospinning. The obtained properties of nanocomposite fibers depend on the type of nanofiller, its loading and dispersion state, the type of polymer, the spinning technique and the spinning method.

Several potential applications of nanocomposite fibers and nanofibers have already been demonstrated. However, the current main difficulties for commercializing these fibers is the limited supply of different nanofillers when large quantities are needed for continuous and bulk production of nanocomposite fibers and their costs. Toxicity and processing issues are other factors that hinder the commercialization of these materials. Nevertheless, given the enormous application potential and distinct advantages, extensive research and development work are currently being conducted to address the problems associated with nanocomposites. Development is currently underway to resolve the above mentioned issues and fully explore these up-and-coming materials in various industrial applications.

References

[1] ISO TS 27687: Nanotechnologies. Terminology and definitions for nano-objects. Nanoparticle, nanofibre and nanoplate, 2008.
[2] ISO TS 80004-1: Nanotechnologies – Vocabulary Part 1: Core Terms, 2010.
[3] Feynman, R. P.: Plenty of room at the bottom: Proceedings of the Annual Meeting of the Amercian Physical Society. Pasadena, 1959 Dec 29.
[4] Taniguchi, N.: On the Basic Concept of Nanotechnology. Tokyo, 1974.
[5] Khan, I.; Saeed, K.; Khan, I.: Nanoparticles: Properties, applications and toxicities. Arabian Journal of Chemistry, 2019; 12(7), 908–931. doi:10.1016/j.arabjc.2017.05.011.
[6] Sen, M.: Nanocomposite Materials. In: Sen, M., editor: Nanotechnology and the Environment. London, UK: IntechOpen, 2020; pp. 106–118.
[7] Kebede, M. A.; Imae, T.: Low-Dimensional Nanomaterials. In: Ariga, K.; Aono, M., editors: Advanced Supramolecular Nanoarchitectonics. Amsterdam, Netherlands: Elsevier, 2019; pp. 3–16.
[8] Tiwari, J. N.; Tiwari, R. N.; Kim, K. S.: Zero-dimensional, one-dimensional, two-dimensional and three-dimensional nanostructured materials for advanced electrochemical energy devices. Progress in Materials Science, 2012; 57(4), 724–803. doi:10.1016/j.pmatsci.2011.08.003.
[9] Liu, C.; Chen, M.; Zhou, D.; Wu, D.; Yu, W.: Effect of filler shape on the thermal conductivity of thermal functional composites. Journal of Nanomaterials, 2017; 2017, 1–15. doi:10.1155/2017/6375135.
[10] Ko, F. K.; Wang, Y.: Introduction to Nanofiber Materials. Warrendale, PA, Cambridge: MRS Materials Research Society; Cambridge University Press, 2014.
[11] Akpan, E. I.; Shen, X.; Wetzel, B.; Friedrich, K.: Design and Synthesis of Polymer Nanocomposites. In: Pielichowski, K.; Maijka, T.M., editors: Polymer Composites with Functionalized Nanoparticles. Amsterdam, Netherlands: Elsevier, 2019; pp. 47–83.
[12] Anu Mary Ealia, S.; Saravanakumar, M. P.: A review on the classification, characterisation, synthesis of nanoparticles and their application. IOP Conference Series: Materials Science and Engineering, 2017; 263, 32019. doi:10.1088/1757-899x/263/3/032019.

[13] Gough, C. R.; Rivera-Galletti, A.; Cowan, D. A.; La Salas-de Cruz, D.; Hu, X.: Protein and polysaccharide-based fiber materials generated from ionic liquids: A review. Molecules, 2020; 25(15). doi:10.3390/molecules25153362
[14] Kroto, H. W.; Heath, J. R.; O'Brien, S. C.; Curl, R. F.; Smalley, R. E.: C60: Buckminsterfullerene. Nature, 1985; 318(6042), 162–163. doi:10.1038/318162a0.
[15] Iijima, S.: Helical microtubules of graphitic carbon. Nature, 1991; 354(6348), 56–58. doi:10.1038/354056a0.
[16] Bouchard, J.; Cayla, A.; Lutz, V.; Campagne, C.; Devaux, E.: Electrical and mechanical properties of phenoxy/multiwalled carbon nanotubes multifilament yarn processed by melt spinning. Textile Research Journal, 2012; 82(20), 2106–2115. doi:10.1177/0040517512450760.
[17] Harik, V.: Mechanics of Carbon Nanotubes: Fundamentals, Modeling and Safety. Saint Louis: Elsevier Science & Technology, 2018.
[18] Donnet, J.-B.; Bansal, R. C.; Wang, M.-J.: Carbon Black. Boca Raton, FL, USA: CRC Press, 2018.
[19] Wang, W.; Shen, C.; Li, S.; Min, J.; Yi, C.: Mechanical properties of single layer graphene nanoribbons through bending experimental simulations. AIP Advances, 2014; 4(3), 31333. doi:10.1063/1.4868625.
[20] Pierson, H. O.: Handbook of Carbon, Graphite, Diamond, and Fullerenes: Properties, Processing, and Applications. Park Ridge, N.J.: Noyes Publications, 1993.
[21] Kumar Jagadeesan, A.; Thangavelu, K.; Dhananjeyan, V.: Carbon Nanotubes: Synthesis, Properties and Applications. In: Pham, P.; Goel, P.; Kumar, S.; Yadav, K., editors: Twenty-first Century Surface Science – A Handbook. London, UK: IntechOpen, 2020; pp. 27–47.
[22] Tiwari, S. K.; Sahoo, S.; Wang, N.; Huczko, A.: Graphene research and their outputs: Status and prospect. Journal of Science: Advanced Materials and Devices, 2020; 5(1), 10–29. doi:10.1016/j.jsamd.2020.01.006.
[23] Neves, A. I. S.; Bointon, T. H.; Melo, L. V.; et al.: Transparent conductive graphene textile fibers. Scientific Reports, 2015; 5, 9866. doi:10.1038/srep09866.
[24] Bointon, T. H.; Khrapach, I.; Yakimova, R.; Shytov, A. V.; Craciun, M. F.; Russo, S.: Approaching magnetic ordering in graphene materials by FeCl3 intercalation. Nano Letters, 2014; 14(4), 1751–1755. doi:10.1021/nl4040779.
[25] Hirsch, A.; Vostrowsky, O.: Functionalization of Carbon Nanotubes. In: Schlüter, A.D., editor: Functional Molecular Nanostructures. Berlin, Germany: Springer, 2005; pp. 193–237.
[26] Punetha, V. D.; Rana, S.; Yoo, H. J.; et al.: Functionalization of carbon nanomaterials for advanced polymer nanocomposites: A comparison study between CNT and graphene. Progress in Polymer Science, 2017; 67, 1–47. doi:10.1016/j.progpolymsci.2016.12.010.
[27] Mody, V. V.; Siwale, R.; Singh, A.; Mody, H. R.: Introduction to metallic nanoparticles. Journal of Pharmacy & Bioallied Sciences, 2010; 2(4), 282–289. doi:10.4103/0975-7406.72127.
[28] Iravani, S.; Korbekandi, H.; Mirmohammadi, S. V.; Zolfaghari, B.: Synthesis of silver nanoparticles: Chemical, physical and biological methods. Research in Pharmaceutical Sciences, 2014; 9(6), 385–406.
[29] Yeh, Y.-C.; Creran, B.; Rotello, V. M.: Gold nanoparticles: Preparation, properties, and applications in bionanotechnology. Nanoscale, 2012; 4(6), 1871–1880. doi:10.1039/c1nr11188d.
[30] Sardar, R.; Funston, A. M.; Mulvaney, P.; Murray, R. W.: Gold nanoparticles: Past, present, and future. Langmuir, 2009; 25(24), 13840–13851. doi:10.1021/la9019475.
[31] Tang, B.; Lin, X.; Zou, F.; et al.: In situ synthesis of gold nanoparticles on cotton fabric for multifunctional applications. Cellulose, 2017; 24(10), 4547–4560. doi:10.1007/s10570-017-1413-8.
[32] Radetić, M.: Functionalization of textile materials with TiO2 nanoparticles. Journal of Photochemistry and Photobiology C: Photochemistry Reviews, 2013; 16, 62–76. doi:10.1016/j.jphotochemrev.2013.04.002.

[33] Haider, A. J.; Jameel, Z. N.; Al-Hussaini, I. H.: Review on: Titanium dioxide applications. Energy Procedia, 2019; 157, 17–29. doi:10.1016/j.egypro.2018.11.159.
[34] Nyamukamba, P.; Okoh, O.; Mungondori, H.; Taziwa, R.; Zinya, S.: Synthetic Methods for Titanium Dioxide Nanoparticles: A Review. In: Yang, D., editor: Titanium Dioxide – Material for a Sustainable Environment. London, UK: InTech, 2018; pp. 151–176.
[35] Asmat-Campos, D.; Delfín-Narciso, D.; Juárez-Cortijo, L.: Textiles functionalized with ZnO nanoparticles obtained by chemical and green synthesis protocols: Evaluation of the type of textile and resistance to UV radiation. Fibers, 2021; 9(2), 10. doi:10.3390/fib9020010.
[36] Haque, M. J.; Bellah, M. M.; Hassan, M. R.; Rahman, S.: Synthesis of ZnO nanoparticles by two different methods & comparison of their structural, antibacterial, photocatalytic and optical properties. Nano Ex, 2020; 1(1), 10007. doi:10.1088/2632-959X/ab7a43.
[37] Verbič, A.; Gorjanc, M.; Simončič, B.: Zinc oxide for functional textile coatings: Recent advances. Coatings, 2019; 9(9), 550. doi:10.3390/coatings9090550.
[38] Uğur, Ş. S.; Sarıışık, M.; Aktaş, A. H.: Nano-Al2O3 multilayer film deposition on cotton fabrics by layer-by-layer deposition method. Materials Research Bulletin, 2011; 46(8), 1202–1206. doi:10.1016/j.materresbull.2011.04.005.
[39] Xiao, X.; Chen, F.; Wei, Q.; Wu, N.: Surface modification of polyester nonwoven fabrics by Al2O3 sol–gel coating. Journal of Coatings Technology and Research, 2009; 6(4), 537–541. doi:10.1007/s11998-008-9157-x.
[40] Vasantharaj, S.; Sathiyavimal, S.; Saravanan, M.; et al.: Synthesis of ecofriendly copper oxide nanoparticles for fabrication over textile fabrics: Characterization of antibacterial activity and dye degradation potential. Journal of Photochemistry and Photobiology. B, Biology, 2019; 191, 143–149. doi:10.1016/j.jphotobiol.2018.12.026.
[41] Giannelis, E. P.: Polymer layered silicate nanocomposites. Advanced Materials, 1996; 8(1), 29–35. doi:10.1002/adma.19960080104.
[42] Alexandre, M.; Dubois, P.: Polymer-layered silicate nanocomposites: Preparation, properties and uses of a new class of materials. Materials Science and Engineering: R: Reports, 2000; 28(1–2), 1–63. doi:10.1016/s0927-796x(00)00012-7.
[43] Massaro, M.; Cavallaro, G.; Lazzara, G.; Riela, S.: Covalently Modified Nanoclays: Synthesis, Properties and Applications. In: Cavallaro, G.; Pasbakhsh, P.; Fakhrullin, R.F., editors: Clay Nanoparticles. Amsterdam, Netherlands: Elsevier, 2020; pp. 305–333.
[44] Solarski, S.; Mahjoubi, F.; Ferreira, M.; et al.: (Plasticized) Polylactide/clay nanocomposite textile: Thermal, mechanical, shrinkage and fire properties. Journal of Materials Science, 2007; 42(13), 5105–5117. doi:10.1007/s10853-006-0911-0.
[45] Kundu, C. K.; Li, Z.; Song, L.; Hu, Y.: An overview of fire retardant treatments for synthetic textiles: From traditional approaches to recent applications. European Polymer Journal, 2020; 137, 109911. doi:10.1016/j.eurpolymj.2020.109911.
[46] Ramesh, P.; Prasad, B. D.; Narayana, K. L.: Effect of MMT clay on mechanical, thermal and barrier properties of treated aloevera fiber/ PLA-hybrid biocomposites. Silicon, 2020; 12(7), 1751–1760. doi:10.1007/s12633-019-00275-6.
[47] Bee, S.-L.; Abdullah, M. A. A.; Bee, S.-T.; Sin, L. T.; Rahmat, A. R.: Polymer nanocomposites based on silylated-montmorillonite: A review. Progress in Polymer Science, 2018; 85, 57–82. doi:10.1016/j.progpolymsci.2018.07.003.
[48] Sinha Ray, S.; Okamoto, M.: Polymer/layered silicate nanocomposites: A review from preparation to processing. Progress in Polymer Science, 2003; 28(11), 1539–1641. doi:10.1016/j.progpolymsci.2003.08.002.
[49] Gokce, Y.; Aktas, Z.; Capar, G.; Kutlu, E.; Anis, P.: Improved antibacterial property of cotton fabrics coated with waste sericin/silver nanocomposite. Materials Chemistry and Physics, 2020; 254, 123508. doi:10.1016/j.matchemphys.2020.123508.

[50] Souzandeh, H.; Johnson, K. S.; Wang, Y.; Bhamidipaty, K.; Zhong, W.-H.: Soy-protein-based nanofabrics for highly efficient and multifunctional air filtration. ACS Applied Materials & Interfaces, 2016; 8(31), 20023–20031. doi:10.1021/acsami.6b05339.

[51] Ji, J.-Y.; Lively, B.; Zhong, W.-H.: Soy protein-assisted dispersion of carbon nanotubes in a polymer matrix. Mat Express, 2012; 2(1), 76–82. doi:10.1166/mex.2012.1055.

[52] Wu, M.-C.; Chan, S.-H.; Lin, T.-H.: Fabrication and photocatalytic performance of electrospun PVA/silk/ TiO2 nanocomposite textile. Functional Materials Letters, 2015; 8(03), 1540013. doi:10.1142/s1793604715400135.

[53] Belkhir, K.; Pillon, C.; Cayla, A.; Campagne, C.: Antibacterial textile based on hydrolyzed milk casein. Materials (Basel), 2021; 14(2). doi:10.3390/ma14020251

[54] Ghosh, T.; Dhar, P.; Katiyar, V.: 2. Nanocellulose: Extraction and Fabrication Methodologies. In: Katiyar, V.; Dhar, P., editors: Cellulose Nanocrystals. Berlin, Germany: De Gruyter, 2020; pp. 23–48.

[55] Phanthong, P.; Reubroycharoen, P.; Hao, X.; Xu, G.; Abudula, A.; Guan, G.: Nanocellulose: Extraction and application. Carbon Resources Conversion, 2018; 1(1), 32–43. doi:10.1016/j.crcon.2018.05.004.

[56] Mondal, S.: Review on nanocellulose polymer nanocomposites. Polymer-Plastics Technology and Engineering, 2018; 57(13), 1377–1391. doi:10.1080/03602559.2017.1381253.

[57] Panchal, P.; Ogunsona, E.; Mekonnen, T.: Trends in advanced functional material applications of nanocellulose. Processes, 2019; 7(1), 10. doi:10.3390/pr7010010.

[58] Kim, J.-H.; Shim, B. S.; Kim, H. S.; et al.: Review of nanocellulose for sustainable future materials. International Journal of Precision Engineering and Manufacturing-Green Technology, 2015; 2(2), 197–213. doi:10.1007/s40684-015-0024-9.

[59] Huang, S.; Tao, R.; Ismail, A.; Wang, Y.: Cellulose nanocrystals derived from textile waste through acid hydrolysis and oxidation as reinforcing agent of soy protein film. Polymers (Basel), 2020; 12(4). doi:10.3390/polym12040958

[60] Saake, B.; Lehnen, R.: Lignin. In: Ullman, F., editor: Ullmann's Encyclopedia of Industrial Chemistry. Weinheim, Germany: Wiley-VCH Verlag GmbH & Co. KGaA, 2000; pp. 13961–13976.

[61] Tribot, A.; Amer, G.; Abdou Alio, M.; et al.: Wood-lignin: Supply, extraction processes and use as bio-based material. European Polymer Journal, 2019; 112, 228–240. doi:10.1016/j.eurpolymj.2019.01.007.

[62] Laurichesse, S.; Avérous, L.: Chemical modification of lignins: Towards biobased polymers. Progress in Polymer Science, 2014; 39(7), 1266–1290. doi:10.1016/j.progpolymsci.2013.11.004.

[63] Timell, T. E.; Lin, S. Y.; Dence, C. W.; editors: Methods in Lignin Chemistry. Berlin, Heidelberg: Springer Berlin Heidelberg, 1992.

[64] Salaün, F.: Microencapsulation as an Effective Tool for the Design of Functional Textiles. In: Advances in Textile Engineering. Las Vegas: Open Access eBooks, 2017. https://openaccessebooks.com/advances-in-textile-engineering/microencapsulation-as-an-effective-tool-for-the-design-of-functional-textiles.pdf

[65] Salaün, F.: Microencapsulation Technology for Smart Textile Coatings. In: Hu, J., editor: Active Coatings for Smart Textiles. Amsterdam, Netherlands: Elsevier, 2016; pp. 179–220.

[66] Salaün, F.; Creach, G.; Rault, F.; Almeras, X.: Thermo-physical properties of polypropylene fibers containing a microencapsulated flame retardant. Polymers for Advanced Technologies, 2013; 24(2), 236–248. doi:10.1002/pat.3076.

[67] Fredi, G.; Bruenig, H.; Vogel, R.; Scheffler, C.: Melt-spun polypropylene filaments containing paraffin microcapsules for multifunctional hybrid yarns and smart thermoregulating thermoplastic composites. Express Polymer Letters, 2019; 13(12), 1071–1087. doi:10.3144/expresspolymlett.2019.93.
[68] Salaün, F.; Creach, G.; Rault, F.; Giraud, S.: Microencapsulation of bisphenol-A bis (diphenyl phosphate) and influence of particle loading on thermal and fire properties of polypropylene and polyethylene terephtalate. Polymer Degradation and Stability, 2013; 98(12), 2663–2671. doi:10.1016/j.polymdegradstab.2013.09.030.
[69] Li, W.; Ma, Y.-J.; Tang, X.-F.; et al.: Composition and characterization of thermoregulated fiber containing acrylic-based copolymer Microencapsulated Phase-Change Materials (MicroPCMs). Industrial & Engineering Chemistry Research, 2014; 53(13), 5413–5420. doi:10.1021/ie404174a.
[70] Salaün, F.; Vroman, I.: Influence of core materials on thermal properties of melamine–formaldehyde microcapsules. European Polymer Journal, 2008; 44(3), 849–860. doi:10.1016/j.eurpolymj.2007.11.018.
[71] Salaün, F.; Devaux, E.; Bourbigot, S.; Rumeau, P.: Influence of process parameters on microcapsules loaded with n-hexadecane prepared by in situ polymerization. Chemical Engineering Journal, 2009; 155(1–2), 457–465. doi:10.1016/j.cej.2009.07.018.
[72] Huang, X.; Zhu, C.; Lin, Y.; Fang, G.: Thermal properties and applications of microencapsulated PCM for thermal energy storage: A review. Applied Thermal Engineering, 2019; 147, 841–855. doi:10.1016/j.applthermaleng.2018.11.007.
[73] Arshady, R.: Suspension, emulsion, and dispersion polymerization: A methodological survey. Colloid and Polymer Science, 1992; 270(8), 717–732. doi:10.1007/bf00776142.
[74] Gupta, B.; Revagade, N.; Hilborn, J.: Poly(lactic acid) fiber: An overview. Progress in Polymer Science, 2007; 32(4), 455–482. doi:10.1016/j.progpolymsci.2007.01.005.
[75] Fedullo, N.; Sorlier, E.; Sclavons, M.; Bailly, C.; Lefebvre, J.-M.; Devaux, J.: Polymer-based nanocomposites: Overview, applications and perspectives. Progress in Organic Coatings, 2007; 58(2–3), 87–95. doi:10.1016/j.porgcoat.2006.09.028.
[76] Utracki, L. A.: Polymeric nanocomposites: Compounding and performance. Journal of Nanoscience and Nanotechnology, 2008; 8(4), 1582–1596. doi:10.1166/jnn.2008.18225.
[77] Herrero, M.; Asensio, M.; Núñez, K.; Merino, J. C.; Pastor, J. M.: Morphological, thermal, and mechanical behavior of polyamide11/sepiolite bio-nanocomposites prepared by melt compounding and in situ polymerization. Polymer Composites, 2019; 40(S1). doi:10.1002/pc.24962
[78] Veit, D.: Fibers-History, Production, Properties, Market. Springer, Switzerland, 2022, ISBN 978-3-031-15308-2
[79] Pötschke, P.; Brünig, H.; Janke, A.; Fischer, D.; Jehnichen, D.: Orientation of multiwalled carbon nanotubes in composites with polycarbonate by melt spinning. Polymer, 2005; 46 (23), 10355–10363. doi:10.1016/j.polymer.2005.07.106.
[80] Son, S. M.; Lee, J.-E.; Jeon, J.; et al.: Preparation of high-performance polyethersulfone/cellulose nanocrystal nanocomposite fibers via dry-jet wet spinning. Macromolecular Research, 2021; 29(1), 33–39. doi:10.1007/s13233-021-9001-z.
[81] Clemons, C.; Sabo, R.: A review of wet compounding of cellulose nanocomposites. Polymers (Basel), 2021; 13(6). doi:10.3390/polym13060911
[82] Hu, X.; Li, J.; Bai, Y.: Fabrication of high strength graphene/regenerated silk fibroin composite fibers by wet spinning. Materials Letters, 2017; 194, 224–226. doi:10.1016/j.matlet.2017.02.057.

[83] Liu, J.; Zhang, R.; Ci, M.; Sui, S.; Zhu, P.: Sodium alginate/cellulose nanocrystal fibers with enhanced mechanical strength prepared by wet spinning. Journal of Engineered Fibers and Fabrics, 2019; 14, 155892501984755. doi:10.1177/1558925019847553.

[84] Zhang, Y.; Hu, J.: Robust effects of graphene oxide on polyurethane/tourmaline nanocomposite fiber. Polymers (Basel), 2020; 13(1). doi:10.3390/polym13010016

[85] Lee, J.-E.; Shin, Y.-E.; Lee, G.-H.; Kim, J.; Ko, H.; Chae, H. G.: Polyvinylidene fluoride (PVDF)/ cellulose nanocrystal (CNC) nanocomposite fiber and triboelectric textile sensors. Composites Part B: Engineering, 2021; 223, 109098. doi:10.1016/j.compositesb.2021.109098.

[86] Kausar, A.: Polyacrylonitrile-based nanocomposite fibers: A review of current developments. Journal of Plastic Film & Sheeting, 2019; 35(3), 295–316. doi:10.1177/8756087919828151.

[87] Imura, Y.; Hogan, R. M. C.; Jaffe, M.: Dry Spinning of Synthetic Polymer Fibers. In: Zhang, D., editor: Advances in Filament Yarn Spinning of Textiles and Polymers. Amsterdam, Netherlands: Elsevier, 2014; pp. 187–202.

[88] Yu, M.; Dong, R.-H.; Yan, X.; et al.: Recent advances in needleless electrospinning of ultrathin fibers: From academia to industrial production. Macromolecular Materials and Engineering, 2017; 302(7), 1700002. doi:10.1002/mame.201700002.

[89] John, M. J.; Anandjiwala, R.; Oksman, K.; Mathew, A. P.: Melt-spun polylactic acid fibers: Effect of cellulose nanowhiskers on processing and properties. Journal of Applied Polymer Science, 2013; 127(1), 274–281. doi:10.1002/app.37884.

[90] Kim, H. A.; Kim, S. J.: Wear comfort properties of ZrC/Al 2 O 3 /graphite-embedded, heat-storage woven fabrics for garments. Textile Research Journal, 2019; 89(8), 1394–1407. doi:10.1177/0040517518770681.

[91] Leung, T.-K.; Lin, J.-M.; Chien, H.-S.; Day, T.-C.: Biological effects of melt spinning fabrics composed of 1% bioceramic material. Textile Research Journal, 2012; 82(11), 1121–1130. doi:10.1177/0040517512439917.

[92] Lin, J.-H.; Huang, C.-L.; Lin, Z.-I.; Lou, C.-W.: Far-infrared emissive polypropylene/wood flour wood plastic composites: Manufacturing technique and property evaluations. Journal of Composite Materials, 2016; 50(15), 2099–2109. doi:10.1177/0021998315602137.

[93] Norouzi, M.; Zare, Y.; Kiany, P.: Nanoparticles as effective flame retardants for natural and synthetic textile polymers: Application, mechanism, and optimization. Polymer Reviews, 2015; 55(3), 531–560. doi:10.1080/15583724.2014.980427.

[94] Rault, F.; Giraud, S.; Salaün, F.: Flame Retardant/Resistant Based Nanocomposites in Textile. In: Visakh, P. M.; Arao, Y., editors: Flame Retardants. Cham: Springer International Publishing, 2015; pp. 131–165.

[95] Minet, J.; Cayla, A.; Campagne, C.: Lignin as Sustainable Antimicrobial Fillers to Develop PET Multifilaments by Melting Process. In: Sand, A.; Zaki, E., editors: Organic Polymers. London, UK: IntechOpen, 2020; pp. 119–134.

[96] Zallen, R.: The Physics of Amorphous Solids. New York: Wiley, 1983.

[97] Kirkpatrick, S.: Percolation and conduction. Reviews of Modern Physics, 1973; 45(4), 574–588. doi:10.1103/RevModPhys.45.574.

[98] Lund, A.; van der Velden, N. M.; Persson, N.-K.; Hamedi, M. M.; Müller, C.: Electrically conducting fibres for e-textiles: An open playground for conjugated polymers and carbon nanomaterials. Materials Science and Engineering: R: Reports, 2018; 126, 1–29. doi:10.1016/ j.mser.2018.03.001.

Boris Mahltig

3 Finishing of textile materials

Keywords: nanoparticles, microparticle, effect pigments, microcapsules, dendrimers, basalt short fibers, transparent iron oxide, silver particles, photoactive, washing stability, fiber/dye interaction, cyclodextrin, particle surface, silica sol, nanosol, sol–gel technology, titania, self-cleaning, UV protection, aroma therapy, color pigments, calixarenes, crown ethers, surface active, antistatic, EMI-shielding, fire retardant, flame retardant, antimicrobial, controlled release

3.1 Introduction

Finishing of fiber materials means general treatment of fabrics or yarns mostly by using wet chemical methods. Wet chemical finishing processes are good comparable with dyeing processes, where a dye stuff is applied using a dye bath to a textile material. As part of a liquid recipe used in finishing processes, also particles can be applied onto textiles to gain new and advantageous properties. This application of recipes containing particles onto textiles could be seen as a kind of particle finishing. This type of particle application as finishing agent is different from the use of particles added as additives during fiber production by spinning. If particles are used as additives in spin processes, they are mainly inside the body of the fiber. In case of finishing, the particles are placed on the fiber surface or in areas near the fiber surface. For this, one main disadvantage of particle finishing in comparison to spin additives is a lower washing and abrasion stability of the gained effect. In contrast, one main advantage of finishing processes is the high flexibility during the complete textile production process because it is done at the end of the textile chain. A textile or garment can be prepared completely and after all the wished functional property can be added by a finally applied finishing agent.

Acknowledgments: The author would like to thank Mr. Thomas Heistermann (Hochschule Niederrhein) specially for very good and long-time cooperation in the finishing lab. Furthermore many thanks are owed to several master students and bachelor students such as B. Akbas, D. Darko, L. Greiler, H. Miao, J. Pallmann, C. Ruffen, E. Schüll, K. Topp, J. Zhang and others which are not mentioned personally. For permanent help with antimicrobial investigations and fluorescent measurements many thanks are owed to Prof. Dr. Hajo Haase (TU Berlin, Germany), M. Wendt, L. Wu.

Boris Mahltig, Hochschule Niederrhein, Faculty of Textile and Clothing Technology, Webschulstr. 31, 41065, Mönchengladbach, Germany, e-mail: boris.mahltig@hs-niederrhein.de

https://doi.org/10.1515/9783110670776-003

3.2 Types of particles in finishing processes

Particles applied in finishing processes can be used for different purposes (see Figure 3.1). First, particles themselves can exhibit a certain functional property, so if they are applied they can act as a functional agent itself (self-functional). The functional property of the particles is transferred to the textile to create a functional textile material. A good example is here a silver nanoparticle, which is antibacterial. By the addition of silver nanoparticles to textiles, this functional antimicrobial property is transferred to the textile material and an antimicrobial textile is created [1, 2]. Also, titanium dioxide particles are self-functional regarding their UV-absorptive properties. The application of those titania particles on textiles can be used for the production of UV-protective textiles [3, 4]. Second, the particles can act as a carrier for another component, which contain a functional property (carrier of functional property). Examples are here a nano-clay particle which releases a functional agent or inorganic particles carrying fluoroalkyl side-chains responsible for oleophobic and soil-repellent properties [5, 6]. Third, the particles are used as a coating forming agent, as it is done in the sol–gel process [7, 8]. Here, a coating is formed by coagulation of particles deposited on the fibers or textiles. This coating is in the end responsible for new properties of the treated textile material.

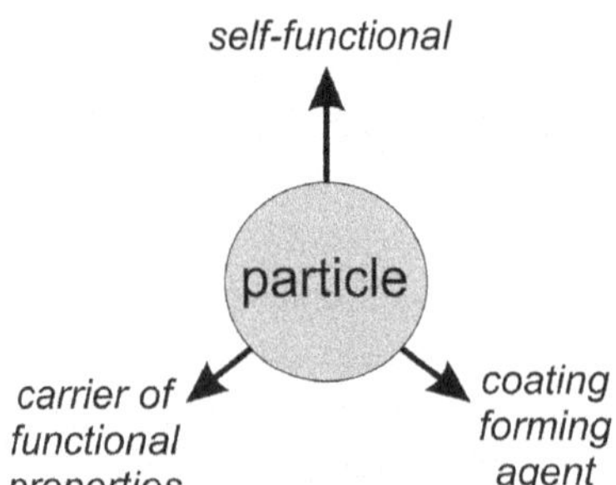

Figure 3.1: Overview on purposes of particles used in textile finishing.

Beside the classification of particles according to their purpose, it can be useful to categorize the particles according to their constitution. The particles used for application on textiles can be categorized according to size, shape and composition. Typical categories according to size and shape used for particles in textile treatment are summarized in Table 3.1. Particles are often distinguished according to their size into nanoparticles or microparticles. As rough estimation, particles with diameters smaller than 100 nm are named as nanoparticles, while larger particles are named as microparticles. Particles can have a regular isotropic shape. However, there are also very characteristic anisotropic particles, as for example, effect pigments. Also, small short fibers can be defined as anisotropic particles. Microcapsules can be understood as microscaled core–shell particles, which contain a functional compound in their core. In a broad range also dendrimers, which are star-like polymers can be named as

Table 3.1: Overview on main particle categories according to particle size and shape.

Particle category	Typical characteristic	Typical materials	Logo	references
Nanoparticles	Nanoscaled size with diameters from 1 nm to 100 nm	Nanosilver, titania, silica nanosols, carbon quantum dots		[7, 8, 10–13]
Microparticles	Microscaled size with diameters from 0.1 to 50 µm	Silver, titania, iron oxide, copper oxide		[11, 14–16]
Effect pigments	Anisotropic plain pigment	Metal effect pigments, pearlescent effect pigments		[9, 17]
Microcapsules	Microscaled hollow spheres; core–shell particle	Synthetic and natural polymers		[18, 19]
Short fibers	Short filaments with lengths in microscale; anisotropic	Short basalt fibers		[20]
Dendrimers	Hyperbranched polymer; star-like polymer	Synthetic polymers		[21–24]

polymer-based particles. The chemical composition of a particle is in fact the most important parameter determining the particle properties. A good example to explain this issue is the comparison of silver particles with silica particles. Silver as metal is conductive and applications with silver particles can lead to conductive materials. Silica in comparison is in any case nonconductive. However, the conductive properties of silver particles can be modified by size and shape of silver particles. Silver particles with anisotropic shape, for example, as effect pigments lead as coating additive finally to strong electrical conductivity [9].

The size and shape also have an influence on the properties of the particles. Additionally, these parameters can also influence the application process and the washing and abrasion properties of the finally realized product. Smaller particles often can be more easily dispersed in a coating recipe. Also, the abrasion stability of a final coating containing smaller particles is enhanced compared to the stability of coatings containing larger particles.

To give some pictorial examples for different categories mentioned in Table 3.1, Figures 3.2 3.4 exhibit SEM images of some typical particles used for textile coating. Figure 3.2 shows anisotropic short fibers from basalt material, which can be used as coating additive to reach UV-protective and infrared reflective properties [20]. Figure 3.3 shows a coating containing metallic effect pigments of the so-called cornflake type [25]. Such metallic effect pigments are traditionally used to gain metal-like prints on textiles. Additionally, other functional properties can be reached by using them as, for example, radiation protection, electrical conductivity or antimicrobial properties [9, 17, 25].

The application of a typical color pigment is shown in Figure 3.4 with an iron oxide pigment on a coated fabric. Iron oxides can occur in different coloration as

Figure 3.2: SEM image of basalt short fibers with 13 µm thickness. These basalt short fibers are suitable additives for textile coatings.

Figure 3.3: SEM image of copper-based effect pigments on a cotton substrate. The effect pigments exhibit the typical cornflake structure. These pigments are ideal coating additives to realize optical effects, antimicrobial properties and conductivity on textile materials.

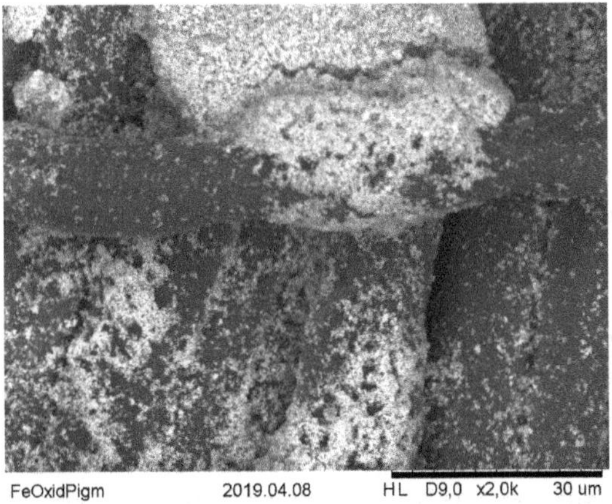

Figure 3.4: SEM image of iron oxide red pigment applied in a coating on a cotton fabric to realize coloration and UV-protective properties.

black, brown, red or yellow [26]. This coloration is determined by the type and chemical composition of the containing iron oxide. There are also transparent iron oxides of almost same composition but with significant smaller particle sizes. As a result of the smaller size a higher transparency of the pigments is achieved [27]. Such transparent iron oxide pigments are typically used for the treatment of wooden materials to obtain an improved light stability and special color effects.

3.3 Adhesion of pigments to textile substrates

The adhesion of pigments or particles in general to textile surfaces is the key parameter, which determines if a good rubbing fastness and a good washing fastness of the particle application is reached. For most applications, a good rubbing and washing stability is an absolute requirement without which an application or even a commercialization is not possible. Consequently, not only the functional property is important which is reached by addition of the particle on the textile but also it is important that the particles stay on the textile in case of usage, so a long-term usage is guaranteed.

There are different ways to gain a good adhesion between particle and fiber surface, and this is related to the issue which type of particle is applied to which type of fiber. The surface of a certain fiber contains different functional groups which are responsible for different adhesive effects connecting the applied particle to the fiber material.

Roughly different types for adhesion can be distinguished. Particles can exhibit a kind of self-adhesion to the fiber material. If a fiber and a particle fit together, the particle shows a certain attraction to the fiber. This type of self-adhesion is mainly found for smaller particles in the nanoscale. This attraction can be described with a certain fiber/particle relation (see also schematic drawing in Figure 3.5). It is in a certain way analogous to fiber/dye interactions, traditionally describing which type of dye is useful for which type of fiber [28].

Nevertheless, most particles are not self-adhesive to fiber surfaces, so they need a kind of agent supporting the connectivity. Mostly used are binder systems, which are polymers acting as a kind of glue fixing the particles on the fiber surface. Often used binder systems are based on acrylates or polyurethane.

Polyamide fibers and protein-based fibers (e.g., wool or silk) exhibit a positive electrical net-charge caused by the protonation of containing amino groups $-NH_2 \rightarrow -NH_3^+$ in acidic surrounding [28]. For this, negatively charged particles exhibit a certain attraction to these types of fibers, while positively charged particles are repulsed. In contrast, most types of acrylic fibers are negatively charged, due to a certain content of anionic sulphonate groups $-SO_3^-$ in the polymer structure [28]. For this, positively charged particles are usually attracted to acrylic fibers, while negative particles are repulsed (Figure 3.5). Polymer-based particles can also be equipped with anchor groups which allow the formation of covalent bonds to functional groups on the fiber surface. Often used are anchor groups which bond to the hydroxy groups on cotton fibers or viscose fibers.

Cotton fibers are from a chemical point of view cellulose-based fibers. Cotton fibers are quite hydrophilic because of the huge amounts of hydroxy groups $-OH$ which are present in the cellulose molecular structure. At these hydroxy groups hydrophilic compounds can be bonded by hydrogen bridges. Also, the formation of covalent chemical bonds to other molecules is possible by chemical reaction. Examples

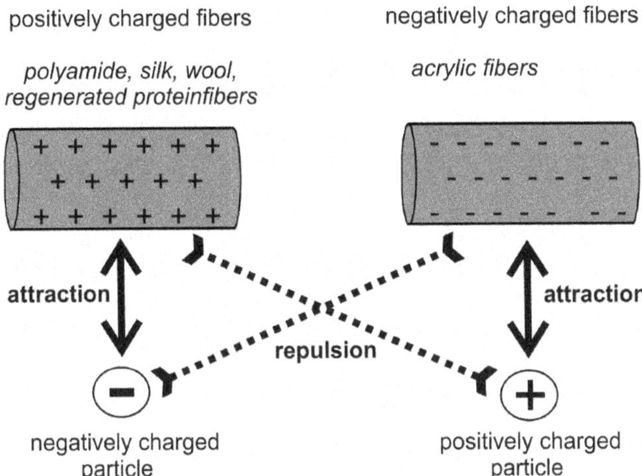

Figure 3.5: Schematic drawing of fiber particle relation based on electrical charges.

are here cyclodextrin or dendrimers modified with chlorotriazine groups [29, 30]. These groups act as reactive anchor, which fixes the polymer particle covalently to the cotton fiber surface. The reaction type is described as nucleophilic substitution. Such anchoring is in principle possible with each fiber containing functional groups suitable for this type of chemical reaction. Mainly these are cellulose-based fibers as cotton, viscose or lyocell. However, also protein-based fibers like silk, wool or regenerated protein fibers can react with reactive anchor groups. As synthetic fibers, fibers based on polyvinylalcohol and their derivatives can also be put in this category.

The most used fiber material across the world is polyester PET. The PET fiber is strongly hydrophobic and offers no functional groups on its surface on which a finishing agent can bond to be fixed. Usually these PET fibers are dyed using thermal processes [28]. PET exhibits a glass transition temperature of around 80 to 85 °C [31–33]. If PET is heated up in application processes to process temperatures of around 120 °C, hydrophobic compounds (as, e.g., disperse dyes) can penetrate into the polymer structure of the PET fibers [34]. After the application process, the fiber is cooled down and the penetrating hydrophobic compounds are physically embedded in the PET fibers. This procedure is principally useable to fix particles on PET fibers. If particles are equipped with large hydrophobic groups, these groups can penetrate the PET surface and act by this as a kind of anchor group for fixation of the attached particles.

3.4 Particle surfaces

Beside size, shape and composition a further parameter important for the particle properties is the surface of these particles. The surface is especially important for smaller particles because they exhibit a higher surface area compared to their volume.

The influence of the particle surface can be categorized into different types (see Figure 3.6). The particle surface can influence the properties and the functions of the particle itself. On the other hand the surface is important for application processes because it can influence the adhesion to the fiber or the stability in the coating recipe.

The relation of the function of a particle surface is clear especially, if particles with functionalized surface are used to modify textile surfaces. A good example in this field are silica sol particles which are modified with hydrophobic alkyl chains or fluoroalkyl groups at the surface to introduce water- or oil-repellent properties [35]. Another example is copper-based effect pigment equipped with a silver coating to improve their conductivity [17]. In contrast to this type of functionalization, a surface modification of a particle is used to protect the core of the particle. Metallic particles are often coated to protect them against corrosion by reaction with oxygen from air [9, 36, 37].

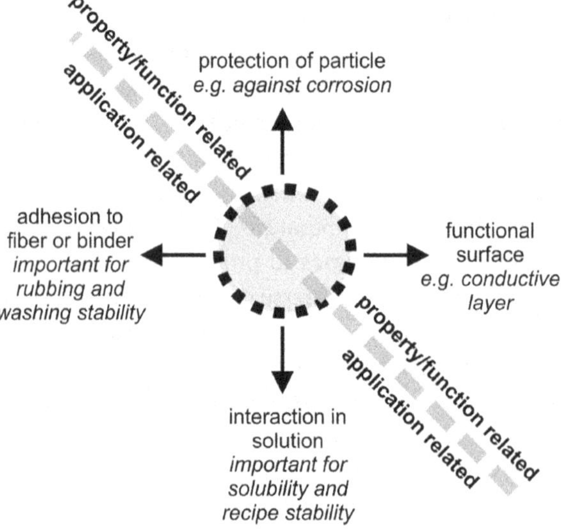

Figure 3.6: Particle surfaces with related properties.

3.5 Sol–gel technology

The sol–gel technology is one special process for particle production and their application as coating forming agent to realize functional textiles [7, 38]. Often sol–gel processes are referred to the buzz word "nanosol." This term "nanosol" is related to the fact that liquid coating agents are used, which contain inorganic particles in nanoscale, meaning that the particle diameters are in the range of 5–80 nm.

The development of sol–gel technology applied on textile substrates started roughly in the year 1999 with a patent invented by Textor et al. In the following decades there was a real boom of inventions and research activities in this field. From these activities also commercial products resulted, as, for example, the isys products supported by the CHT (Tübingen, Germany). However, nowadays only few commercially successful products are left, while the research activities are still on a high level.

The sol–gel process starts usually from a metal-organic precursor which is hydrolyzed under acidic or alkaline conditions (see Figure 3.7). The resulting product of the hydrolysis is condensed to metal oxide particles which are metastable in the liquid surrounding. This solution is often also named as nanosol and can be used as liquid coating agent for the treatment of textiles. The liquid coating agent can be applied by dipping or spraying onto textile substrates. During the coating and following drying process, the solvent evaporates and the metal oxide particles agglomerate to a three-dimensional network which builds up the matrix of the coating. Only in few cases, the metal oxide coating is alone suitable for modification. Examples in this field are silica coatings used to improve the abrasion resistance of a textile or titania coatings for photoactive applications. The abrasion resistance of glass fiber fabrics can be significantly improved by the application of silica-based sol–gel coatings [7, 39]. Photoactive titania coatings can be used for manifold different applications on textiles. They can be used to eliminate dyes from waste water or as a kind of self-cleaning application [40–45]. Titania-containing sol–gel coatings can also be used to realize textile fabrics with UV-protective properties [46]. Nevertheless, in most cases, the sol–gel coatings are modified with additives to introduce the functional properties.

Here, the sol–gel coating is modified with an additive which is responsible for a specific functional property. The sol–gel coating bonds this additive and fixes it to the textile surface, and by this the function is finally transferred to the textile. A prominent example in this area is the modification of simple silica sols by the addition of fluorinated alkyltrialkoxysilanes [35, 47, 48]. The fluorinated alkyl group is responsible for water- and oil-repellent properties. By hydrolysis and condensation of the alkoxy groups, this additive is covalently bonded to the silica sol matrix. The application of this recipe onto textile materials leads to textiles which are water and oil-repellent.

As it is shown in Figure 3.7, there are mainly two points in the sol–gel process when the functional additives can be added – before the hydrolysis starts or to the already prepared sol. If the additives are added before the hydrolysis, they can take part in hydrolysis and condensation processes and can be by this covalently bonded to the formed sol particles. This type of modification is often also named as chemical modification because the additives are chemically bonded to the sol–gel coating. If the additives are added to the sol-recipe and do not form covalent bonds to the sol particles, they are in a certain way embedded into the sol–gel matrix formed during the coating and the drying process. This kind of physical embedding is also named as physical modification of a sol–gel coating. Typical for physical modification is the embedding of silver particles or copper oxide particles in a silica coating used for the preparation of antibacterial sol–gel coatings on textiles [49–51]. Also, the embedding of dye molecules into sol–gel coating is possible [52].

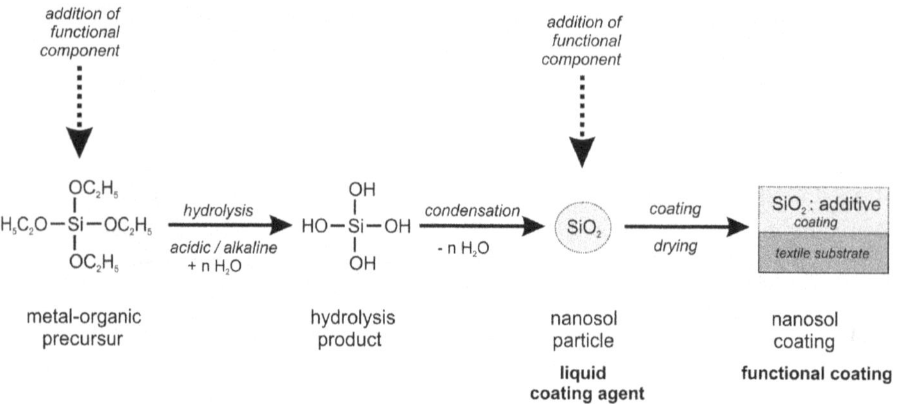

Figure 3.7: Schematic overview on the sol–gel process on the example of a silica-based nanosol.

By the use of sol–gel coatings, textile materials can be functionalized in a broad range [7]. Pure and modified silica coatings can be used to increase the abrasion fastness of textile surfaces. Titania coatings lead to UV-protective properties. This UV-protective effect can be enhanced by simultaneous embedding of fluorescent whitening agents, which are also organic UV-absorbers [53]. Further, titania of anatase type can also introduce photoactive properties. Such photoactive-coated textiles can be used as self-cleaning materials or for water and air cleaning purposes. The bonding of hydrophobic groups like alkyl groups to the silica sol is useful to realize water-repellent textiles. Quite useful for this purpose are alkyl silane compounds containing the long alkyl chain hexadecylakyl [35, 54, 55]. Also used for hydrophobic modification of silica sols are polymeric additives like polydimethylsiloxane [56, 57]. The modification with perfluoro alkyl groups leads beside to water repellent also to oil-repellent properties and is used to prepare soil-repellent textiles [58, 59]. One special feature of sol–gel coatings for water-repellent effects is that they cannot only introduce hydrophobic functional groups to textile surfaces but they can also increase a microscopical roughness of the textile surface. This microroughness can further decrease the wettability of a hydrophobic textile and by this a so-called superhydrophobicity can be realized on the textile [60–62].

Sol–gel coatings can also act as carrier for active substances, which are, for example, antimicrobial compounds that are released from these coatings over a longer period of time [63]. Beside inorganic materials like silver or copper oxide particles, also organic antimicrobial agents can be deposited by sol–gel method on textiles [49]. Other active substances which are embedded into sol–gel coatings are fragrances or corrosion inhibiting substances [64]. The embedding of fragrances can be useful for textiles used in aroma therapy or as wellness textiles.

3.6 Dyes and color pigments

Many researchers publish their use of particles for textile treatment with the statement that this is especially new and innovative. Especially in the decade from 2000 to 2010 the use of nanoparticles or nanotechnology in general was a trend word. Nevertheless, it should be kept in mind that the use of particles for textile treatment is in some fields very traditional. One field is the use of disperse dyes and color pigments for dyeing and printing processes [65]. Here, the main aimed property is the coloration of the textile. For advanced applications too, the modification of optical properties in the range of UV radiation or near infrared radiation is aimed for the use of color pigments [9, 66]. Other special color properties are gained by particles or pigments with fluorescent or phosphorescent properties [67].

3.7 Polymeric particles

Commonly particles are understood as the agglomerate of water-insoluble compounds from inorganic or organic nature. Typical inorganic particles are, for example, silica-based sol–gel particles or silver nanoparticles. A typical particle from organic material is a pigment dye which can be built up by a water-insoluble azo dye.

Beside this, particles can also be built up by polymeric structures. The simplest example is here probable a latex particle. For textile applications, the commonly used particles are microcapsules. Microcapsules are particles with sizes in micrometer scale, which are built up by a polymer shell and a core which contain a functional component. This functional component can be, for example, a fragrance or an active substance working as insect-repellent. In case of rubbing a textile surface containing the microcapsules, the polymer shell of the capsules breaks and by this the active substance is released [18, 19].

For other applications, microcapsules are mechanically stable and contain paraffin wax instead of volatile fragrances. These encapsulates paraffin wax melts at a certain temperature and by this uptakes a certain amount of surrounding heat energy. The uptaken heat is bonded as the energy needed to support the phase change from solid to liquid. The task of the microcapsules is here to keep the molten wax at the district place. These types of materials are often also named as phase-change materials, which can be used to implement temperature regulation processes to textile materials [68, 69].

Apart from classical particles, also single polymer structures could be counted in a certain way to a special group of particles used for the treatment of textile materials. A special polymeric material in this field are dendrimers. Dendrimers can be described as star-like polymers with a structure starting from a center point. From this center the dendrimer is structured with linking elements, as it is schematic shown in Figure 3.8. Dendrimers can be applied on textiles as carrier of functional properties. Dendrimers exhibit a high amount of functional end groups appearing in a high surface density on the surface of this star-like polymer (see Figure 3.8). For this, with the dendrimer high amounts of functional groups can be introduced onto a textile surface. This issue is, for example, used to introduce hydrophobic properties on textiles [21].

A second very interesting feature of dendrimers is that in contrast to the high density of groups on the dendrimer surface, the inner part of the dendrimers contains open areas, which can also be named as free spaces (Figure 3.8). In these areas metal particles as, for example, silver particles can be produced in situ and this silver/dendrimer arrangement can be applied onto textiles as a kind of antibacterial agent [70]. Also, fragrances or other active substances can be incorporated into these free areas and used for controlled release applications [71, 72].

Other compounds which can be used in textile functionalization are of oligomeric structure. Prominent examples here are cyclodextrins and calixarenes (Figures 3.9 and 3.10). These compounds are built up by a limited number of repeating units which are arranged in a cycled structure. Because the number of containing repeating units is small compared to typical polymers, these materials should be better named as oligomers instead of polymers. Cyclodextrins are carbohydrates built up by connected sugar molecules bonded together in larger cycles (Figure 3.9) [73]. A typical cyclodextrin is β-cyclodextrin which is built up by seven connected sugar molecules. In the hole in the center of the cyclodextrin certain molecules of the right size can be embedded. Applications are the embedding of fragrance molecules, which can be released from the cyclodextrin over a larger period of time [73–75]. Possible applications are found in aroma therapy [74, 76]. Also, empty cyclodextrins can uptake unwished bad-smelling substances from surrounding media and through this a kind of "stay fresh" or "odor-control" effect can be gained [77]. Cyclodextrins can be

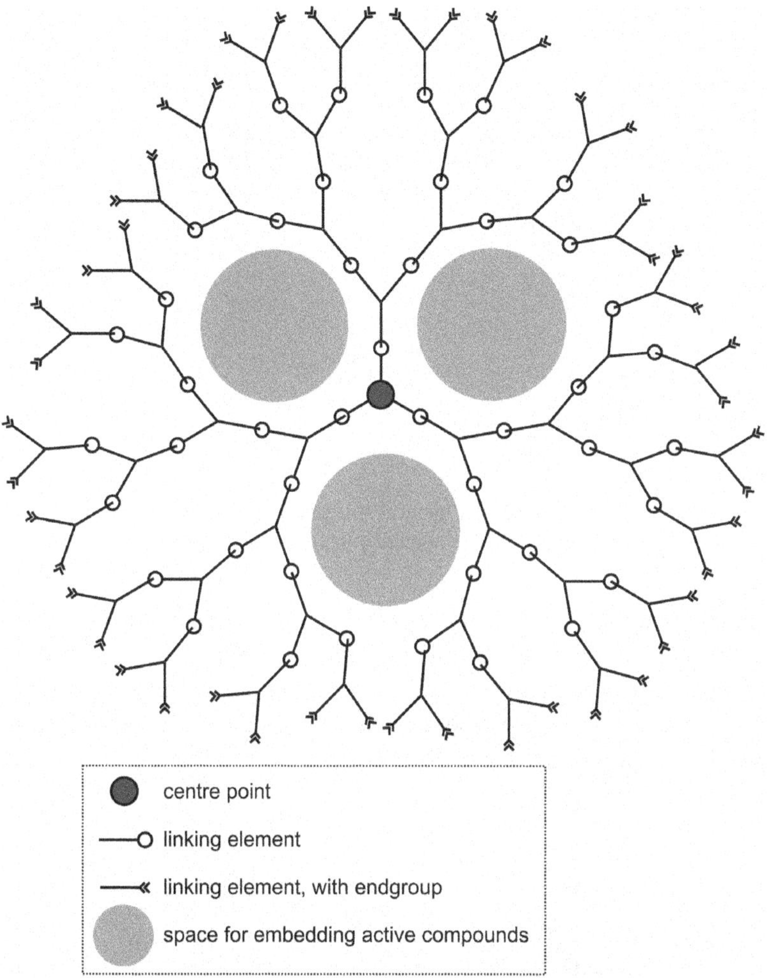

Figure 3.8: Schematic drawing of a dendrimer with four generations. Especially marked are the functional end groups on the dendrimer surface and the areas inside the dendrimer structure, there active compounds can be embedded.

equipped with reactive anchor groups, so they can be covalently bonded to cotton fabrics with the result of high wash fastness [78].

Calixarenes are cycled structures which can be used for the same purposes in textile finishing as cyclodextrins [79]. However, calixarene are built up by different repeating units and of aromatic nature (Figure 3.10). A special application is related to the calixarene property to bond special metal ions. By this, a textile filter material is realized, which especially binds uranium (VI) ions [80].

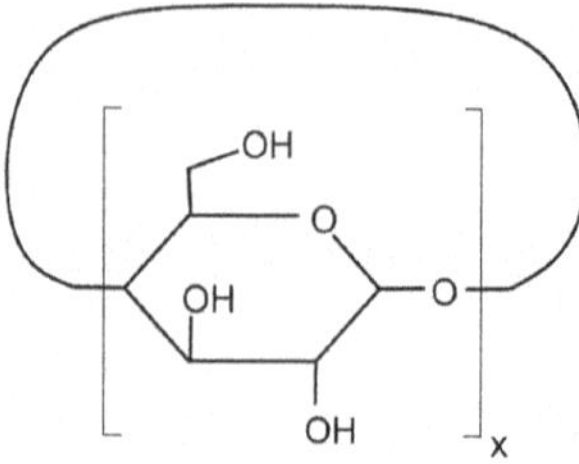

Figure 3.9: Chemical structure of cyclodextrins which are cycled oligomers. The number repeating units x is related to the type of cyclodextrin, for example, for alpha-cyclodextrin $X = 6$.

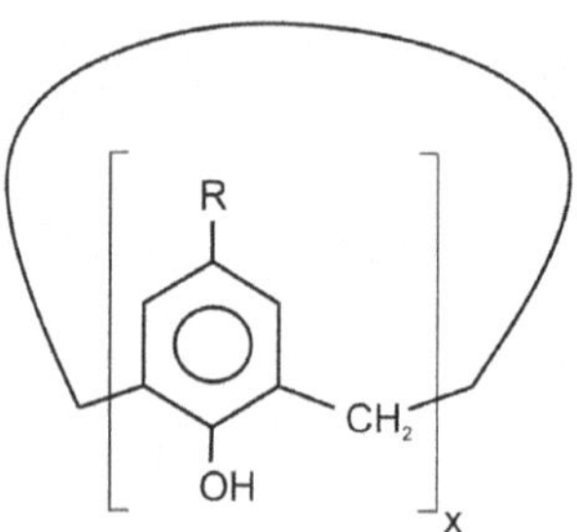

Figure 3.10: Chemical structure of calixarene as cycled oligomer. The number of repeating unit is $X > 4$.

Other simple cycled molecules useful for embedding of compounds are crown ethers [79, 81]. As shown in Figure 3.11, crown ethers are cycled ether compounds built up by ethylene ether units. The number of connected ether units determines the size of the crown ether. The replacement of one oxygen in the crown ether by an amino group leads to an aza-crown ether (Figure 3.11).

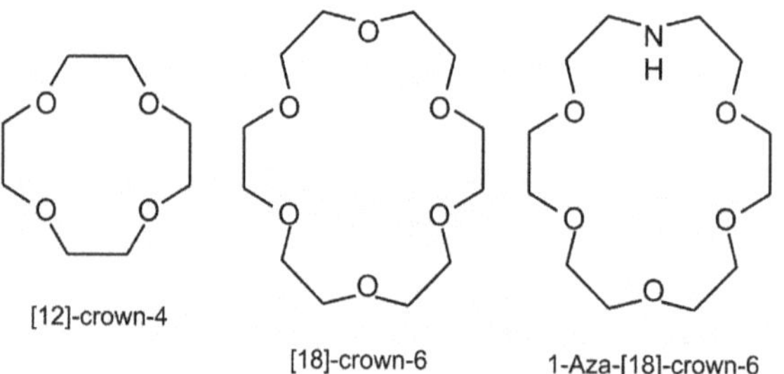

Figure 3.11: Chemical structures of crown ether of different sizes and the structure of one aza-crown ether.

3.8 Functional properties

In the previous sections, different particles are discussed mainly from the material point of view or as part of a preparation process with the sol–gel technology. In contrast to this, this section on functional properties presents mainly the different functional properties which can be realized by particle application. The addition of functional properties to textile materials is realized by particles which are the carrier of this function. These functions are mostly determined by the chemical composition of the particle and modified by the shape, size and surface of the particle. A good example for the influence of the particle size gives the comparison of conventional iron oxide color pigments with transparent iron oxide pigments. The transparent iron oxide pigments contain a certain coloration depending on the chemical type of the iron oxide. However, due to their smaller particle size, their transparency is high and because of a low reflection the color intensity is lower [27]. Figure 3.12 gives a broad overview on main functions, which are not related to dyeing or other optical properties. Here, the functions are shown in different categories and under each category several subpoints are mentioned. In some cases, there is a connection between subpoints of different categories, and this is illustrated in Figure 3.12 with dotted lines.

The functional category "surface active" is related to properties, which are depending on the interaction of the textile surfaces to a surrounding medium, mainly these are liquids or soils. To this category clearly belong hydrophobic, oleophobic and hydrophilic modifications because here the interaction of textiles to water or oil-based liquids is influenced. In a certain way, antistatic properties also belong to this category because a hydrophilic modification can lead to a certain moisture uptake

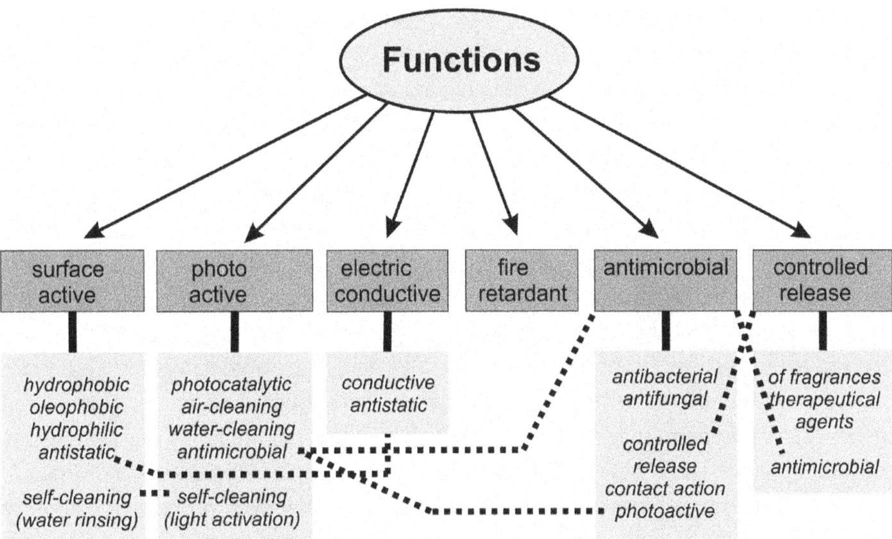

Figure 3.12: Overview on functions which can be reached on textiles by application of particles. The functions are categorized in seven main categories. The relations between different subcategories are marked by dotted lines. Coloration and other optical properties are not mentioned as functional properties in this image.

from surrounding atmosphere. By this uptaken water, the electrical surface resistance of the textile material is decreased, so antistatic properties can be introduced to the textile [82]. Also, the term “self-cleaning” is related to a surface-activated material. A hydrophobic material containing a certain roughness can exhibit self-cleaning properties, meaning that during the rinsing with water, dirt is removed easily from this surface. Such surfaces are for marketing reason often named as lotus-like surface [83, 84]. The term “self-cleaning” is also used for some photoactive materials. If these photoactive materials are illuminated with light, they can catalyze an oxidation process, which can decompose soils or dirt present on a textile surface [41]. The category “photoactive” is generally related to materials which can be activated by light. Often these are photocatalytic processes, which can be used for air or water cleaning applications [41, 44]. Also, antimicrobial effects can be reached by photoactive textile surfaces, for example, the growth of bacteria can be decreased by the photoactive titania modification of anatase applied as sol–gel material on textile fabrics [85]. The category “electric conductive” is related to particles which can increase the conductivity of a textile material, these can be, for example, metallic-based effect pigments or carbon-based materials applied as coating additives on textile fabrics [9, 17]. If the reached effect is weak, only antistatic properties are achieved by these products. Textiles with high conductivity often show the ability to shield against radiowave and microwave. This effect is also named EMI shielding [17]. Fire-retardant functions are related to the depression of a fire and the creation of self-extinguishing textile materials. Particles used for this function are able to disturb the chemical reactions supporting the fire. Examples in this field are pigments of organic compounds containing bromine and chlorine. During the exposition to heat, these substances decomposes and set free halogen radicals, which can interact with other radical components in the fire and by this stop the fire supporting reactions [86]. Other particle-based coatings contain amino- and phosphorous compounds for flame-retardant purposes. These coatings decompose under heat and form a protective char layer on the coated textile surface. This formed layer hinders the pyrolysis of the burning material and by this the extinguishing of the fire is possible [87]. The functional category “antimicrobial” is probably the most investigated function for textiles. This category is related to any type of material, which is able to limit the growth of microbes or even destroy them completely. Often here is distinguished between antibacterial and antifungal, if especially bacteria or fungi are influenced by the material. Most prominent examples are here applications with silver nanoparticles, copper oxide particles or metallic effect pigments [7, 9, 49]. Roughly, three different modes of action can be distinguished for antimicrobial textiles – controlled release of the antimicrobial agent, permanent fixed agents working in contact to the bacteria and photoactive processes. Antimicrobial photoactive textiles are strongly investigated and many papers are published [88–90]. However, most commercially available antimicrobial textiles are related to the controlled release of an antimicrobial substance or the contact action. The category “controlled release” is related to textiles which release an active substance over longer durations and in small amounts. The type of component, which is released, determines the final function of the textile. The release of antimicrobial agents into the surrounding medium stands for antimicrobial textiles, and this is probably the most used application. Other applications use the release of fragrances, for example, for textiles in aroma therapy or as wellness clothes [64].

3.9 Final statements

There are manifold different types of particles which can be used in finishing processes to give textile materials new and functional properties. Almost each type of function can be realized by application of particular materials. By view on the particles and their use, often one important issue influencing the particle properties is underestimated; this is the surface of the particles. The particle surface determines the particle stability, the behavior in the finishing recipe, the adhesion to the textile surface and also the functional effect gained by the particle. It should also be mentioned that many functional properties, which are realized by use of particles, can also be realized by the use of functional agents based on chemical molecules or ions. The use of particular agents is in some cases advantageous but not in all. For this, the producer of functional textiles has to select carefully the best finishing agent to reach the wished functional properties. This finishing agent can contain particles but this is not a must in any case.

3.10 Future developments

To realize a valid prognose for future developments in the use of particles for textile finishing is of course difficult because this prognose is depending on current developments and cannot see radical new developments in material science leading to absolutely new applications. For current developments there is always the danger that new particular materials are coming under legal restriction because of health risk concern.

Beside that concerns, the two main types of materials have a high potential to lead to new applications – these are carbon particles and encapsulated materials. Small carbon particles – also named as carbon quantum dots – are from the chemical point of very simple composition and the preparation is as well not difficult. In contrast, many different and new properties are proposed for these carbon particles, as, for example, color effects, fluorescent effects or antibacterial effects. If the technical preparation of these particles is possible in large amounts and high quality, they have a great chance to replace many other chemicals which are actually used for the same purposes.

In comparison to the carbon particles, the encapsulated materials are not made from one chemical composition. An interesting point is that the functional material is inside the capsule and the capsule supports the anchor function onto the textile surface and determines the washing and rubbing stability of the functionalization. A well-developed encapsulating technique could therefore offer a functionalization for different purposes which is determined by the content of the capsules. Especially is the field of flame-retardant finishing this could be a strong progress because here the washing stability is often inferior or reached by formaldehyde emitting components.

References

[1] Dubas, S. T.; Kumlangdudsana, P.; Potiyaraj, P.: Layer-by-layer deposition of antimicrobial silver nanoparticles on textile fibers. Colloids and Surfaces A, 2006; 289, 105–109.

[2] Radetic, M.: Functionalization of textile materials with silver nanoparticles. Journal of Materials Science, 2013; 48, 95–107.

[3] Paul, R.; Bautista, L.; De la Varga, M.; Botet, J. M.; Casals, E.; Puntes, V.; Marsal, F.: Nano-cotton fabrics with high ultraviolet protection. Textile Research Journal, 2010; 80, 454–462.

[4] Montazer, M.; Pakdel, E.; Moghadam, M. B.: Nano titanium dioxide on wool keratin as UV absorber stabilized by butane tetra carboxylic acid (BTCA): A statistical prospect. Fibers and Polymers, 2010; 11, 967–975.

[5] Ghosh, A.: Nano-clay particle as textile coating. International Journal of Engineering & Technology, 2011; 11, 40–43.

[6] Satoh, K.; Nakazumi, H.; Morita, M.: Novel fluorinated inorganic-organic finishing materials for nylon carpeting. Textile Research Journal, 2004; 74, 1079–1084.

[7] Mahltig, B.; Textor, T.: Nanosols and Textiles. Singapore: World Scientific, 2008.

[8] Mahltig, B.; Haufe, H.; Böttcher, H.: Functionalization of textiles by inorganic sol-gel coatings. Journal of Materials Chemistry, 2005; 15, 4385–4398.

[9] Mahltig, B.; Zhang, J.; Wu, L.; Darko, D.; Wendt, M.; Lempa, E.; Rabe, M.; Haase, H.: Effect pigments for textile coating – A review on the broad range of advantageous functionalization. Journal of Coatings Technology and Research, 2017; 14, 35–55.

[10] Mahltig, B.; Böttcher, H.: Veredlung von Textilien durch Nanosol-Beschichtungen. Melliand Textilberichte, 2002; 83, 251–253.

[11] Sojka-Ledakowicz, J.; Lewartowska, J.; Kudzin, M.; Jesionowski, T.; Siwińska-Stefańska, K.; Krysztafkiewicz, A.: Modification of textile materials with micro-and nano-structural metal oxides. Fibres & Textiles in Eastern Europe, 2008; 70, 112–116.

[12] Dastjerdi, R.; Montazer, M.; Shahsavan, S.: A new method to stabilize nanoparticles on textile surfaces. Colloids and Surfaces A, 2009; 345, 202–210.

[13] Greiler, L. C.; Haase, H.; Mahltig, B.: Microwave assisted conversion of an amino acid into a fluorescent solution. Acta Chimica Slovenica, 2018; 65, 865–874.

[14] Turalija, M.; Merschak, P.; Redl, B.; Griesser, U.; Duelli, H.; Bechtold, T.: Copper (i) oxide microparticles–synthesis and antimicrobial finishing of textiles. Journal of Materials Chemistry B, 2015; 3, 5886–5892.

[15] Kuhr, M.; Aibibu, D.; Cherif, C.: Targeted partial finishing of barrier textiles with microparticles, and their effects on barrier properties and comfort. Journal of Industrial Textiles, 2016; 45, 853–878.

[16] Salaün, F.; Vroman, I.; Elmajid, I.: A novel approach to synthesize and to fix microparticles on cotton fabric. Chemical Engineering Journal, 2012; 213, 78–87.
[17] Topp, K.; Haase, H.; Degen, C.; Illing, G.; Mahltig, B.: Coatings with metallic effect pigments for antimicrobial and conductive coating of textiles with electromagnetic shielding properties. Journal of Coatings Technology and Research, 2014; 11, 943–957.
[18] Azizi, N.; Chevalier, Y.; Majdoub, M.: Isosorbide-based microcapsules for cosmeto-textiles. Industrial Crops and Products, 2014; 52, 150–157.
[19] Rodrigues, S. N.; Martins, I. M.; Fernandes, I. P.; Gomes, P. B.; Mata, V. G.; Barreiro, M. F.; Rodrigues, A. E.: Scentfashion®: Microencapsulated perfumes for textile application. Chemical Engineering Journal, 2009; 149, 463–472.
[20] Ruffen, C.: Mahltig, B.: Basalt fibers as functional additives in coating of textiles. Journal of Coatings Technology and Research, 2021; 18, 271–281.
[21] GmbH, R.: (Geretsried, Germany), Bionic-Finish ECO, https://www.rudolf.de/technologien/bionic-finishreco/, last accessed: 15.03.2020
[22] Namligoz, E. S.; Bahtiyari, M. I.; Hosaf, E.; Coban, S.: Performance comparison of new (dendrimer, nanoproduct) and conventional water, oil and stain repellents. Fibres & Textiles in Eastern Europe, 2009; 17, 76–81.
[23] Mahltig, B.; Tatlises, B.; Fahmi, A.; Haase, H.: Dendrimer stabilized silver particles for the antimicrobial finishing of textiles. Journal of the Textile Institute, 2013; 104, 1042–1048.
[24] Akbari, S.: Kozłowski, R. M.: A review of application of amine-terminated dendritic materials in textile engineering. Journal of the Textile Institute, 2019; 110, 460–467.
[25] Wißling, P.: Metallic Effect Pigments: Fundamentals and Applications. Hannover: Vincentz Network GmbH, 2006.
[26] Seilnacht, T.: Pigmente und Bindemittel, 2nd ed. Bern: Seilnacht Verlag & Atelier, 2018.
[27] Sreeram, K. J.; Indumathy, R.; Rajaram, A.; Nair, B. U.; Ramasami, T.: Template synthesis of highly crystalline and monodisperse iron oxide pigments of nanosize. Materials Research Bulletin, 2006; 41, 1875–1881.
[28] Clark, M.: Handbook of Textile and Industrial Dyeing. Cambridge: Woodhead Publishing, 2011.
[29] Cabrales, L.; Abidi, N.; Hammond, A.; Hamood, A.: Cotton fabric functionalization with cyclodextrins. Journal of Materials and Environmental Science, 2012; 3, 561–574.
[30] Blotny, G.: Recent applications of 2,4,6-trichloro-1,3,5-triazine and its derivatives in organic synthesis. Tetrahedron, 2006; 62, 9507–9522.
[31] Thompson, A. B.; Woods, D. W.: The transitions of polyethylene terephthalate. Transactions of the Faraday Society, 1956; 52, 1383–1397.
[32] Loy, W.: Chemiefasern Für Technische Textilprodukte. Frankfurt: Deutscher Fachverlag, 2008.
[33] Michaels, A. S.; Vieth, W. R.; Barrie, J. A.: Solution of gases in polyethylene terephthalate. Journal of Applied Physics, 1963; 34, 1–12.
[34] De Clerck, K.; van Oostveldt, P.; Rahier, H.; van Mele, B.; Westbroek, P.; Kiekens, P.: Dye diffusion studies in PET fibres by confocal laser scanning microscopy and the interrelation with the glass transition. Polymer, 2004; 45, 4105–4112.
[35] Mahltig, B.; Böttcher, H.: Modified silica sol coatings for water-repellent textiles. Journal of Sol-Gel Science and Technology, 2003; 27, 43–52.
[36] Wang, D.; Bierwagen, G. P.: Sol–gel coatings on metals for corrosion protection. Progress in Organic Coatings, 2009; 64, 327–338.
[37] Kiehl, A.; Brendel, H.: Corrosion inhibited metal pigments. Macromolecular Symposia, 2002; 187, 109–120.
[38] Amberg-Schwab, S.: Spezifische Funktionalisierung von Chemiefasern durch neue Beschichtungsmaterialien. Technische Textilien, 2003; 46, 137–140.

[39] Textor, T.; Bahners, T.; Schollmeyer, E.: Organically modified ceramics for coating textile materials. Progress in Colloid and Polymer Science, 2001; 117, 76–79.

[40] Qi, K.; Daoud, W. A.; Xin, J. H.; Mak, C. L.; Tang, W.; Cheung, W. P.: Self-cleaning cotton. Journal of Materials Chemistry, 2006; 16, 4567–4574.

[41] Bozzi, A.; Yuranova, T.; Guasaquillo, I.; Laub, D.; Kiwi, J.: Self-cleaning of modified cotton textiles by TiO_2 at low temperatures under daylight irradiation. Journal of Photochem. & Photobiol. A: Chemistry, 2005; 174, 156–164.

[42] Liuxue, Z.; Peng, L.; Zhixing, S.: Photocatalysis anatase thin film coated PAN fibers prepared at low temperature. Materials Chemistry and Physics, 2006; 98, 111–115.

[43] Mahltig, B.; Gutmann, E.; Meyer, D. C.; Reibold, M.; Dresler, B.; Günther, K.; Faßler, D.; Böttcher, H.: Solvothermal preparation of metalized titania sols for photocatalytic and antimicrobial coatings. Journal of Materials Chemistry, 2007; 17, 2367–2374.

[44] Böttcher, H.; Mahltig, B.; Sarsour, J.; Stegmaier, T.: Qualitative investigations of the photocatalytic dye destruction by TiO_2-coated polyester fabrics. Journal of Sol-Gel Science and Technology, 2010; 55, 177–185.

[45] Mahltig, B.; Gutmann, E.; Meyer, D. C.: Solvothermal preparation of nanocrystalline anatase containing TiO_2 and TiO_2/SiO_2 coating agents for application of photocatalytic treatments. Materials Chemistry and Physics, 2011; 127, 285–291.

[46] Abidi, N.; Hequet, E.; Tarimala, S.; Dai, L. L.: Cotton fabric surface modification for improved UV radiation protection using sol-gel process. Journal of Applied Polymer Science, 2007; 104, 111–117.

[47] Takenori, T.; Hiroyuki, N.; Yoshie, K.; Yuji, O.; Kazufumi, O.: Development of a water- and oil-repellent treatment for silk and cotton fabrics with fluoralkyl-trimethoxysilane. Journal of Textile Engineering, 2009; 55, 13–21.

[48] Textor, T.; Mahltig, B.: Nanosols for preparation of antistatic coatings simultaneously yielding water and oil repellent properties for textile treatment. Materials Technology, 2010; 25, 74–80.

[49] Mahltig, B.; Fiedler, D.; Böttcher, H.: Antimicrobial sol-gel coatings. Journal of Sol-Gel Science and Technology, 2004; 32, 219–222.

[50] Mahltig, B.; Gutmann, E.; Reibold, M.; Meyer, D. C.; Böttcher, H.: Synthesis of Ag and Ag/SiO_2 sols by solvothermal method and their bactericidal activity. Journal of Sol-Gel Science and Technology, 2009; 51, 204–214.

[51] Mahltig, B.; Fiedler, D.; Fischer, A.; Simon, P.: Antimicrobial coatings on textiles–modification of sol–gel layers with organic and inorganic biocides. Journal of Sol-Gel Science and Technology, 2010; 55, 269–277.

[52] Mahltig, B.; Textor, T.: Combination of silica sol and dyes on textiles. Journal of Sol-Gel Science and Technology, 2006; 39, 111–118.

[53] Xu, P.; Wang, W.; Chen, S.-L.; UV Blocking Treatment of Cotton Fabrics by Titanium Hydrosol. AATCC Review, 2005; 28–31.

[54] Pipatchanchai, T.; Srikulkit, K.: Hydrophobicity modification of woven cotton fabric by hydrophobic fumed silica coating. Journal of Sol-Gel Science and Technology, 2007; 44, 119–123.

[55] Daoud, W. A.; Xin, J. H.; Tao, X.: Synthesis and characterization of hydrophobic silica nanocomposites. Applied Surface Science, 2006; 252, 5368–5371.

[56] Fir, M.; Vince, J.; Surca Vuk, A.; Vilcnik, A.; Jovanovski, V.; Mali, G.; Orel, B.; Simoncic, B.: Functionalisation of cotton with hydrophobic urea/polydimethylsiloxane sol-gel hybrid. Acta Chimica Slovenica, 2007; 54, 144–148.

[57] Vince, J.; Orel, B.; Vilcnik, A.; Fir, M.; Surca Vuk, A.; Jovanovski, V.; Simoncic, B.: Structural and water-repellent properties of urea/poly(dimethylsiloxane) sol-gel hybrid and its bonding to cotton fabric. Langmuir, 2006; 22, 6489–6497.
[58] Yu, M.; Gu, G.; Meng, W.-D.; Qing, F.-L.: Superhydrophobic cotton fabric coating based on a complex layer of silica nanoparticles and perfluorooctylated quaternary ammonium silane coupling agent. Applied Surface Science, 2007; 253, 3669–3673.
[59] Yeh, J.-T.; Chen, C.-L.; Huang, K.-S.: Preparation and application of fluorocarbon polymer/ SiO_2 hybrid materials, Part 2: Water and oil repellent processing for cotton fabrics by sol-gel method. Journal of Applied Polymer Science, 2006; 103, 3019–3024.
[60] Favret, E.; Löthman, P.: RIMAPS image analysis of biological and technical non-wettable surfaces. Microscopy and Analysis, 2007; 21, 7–9.
[61] Hoefnagels, H. F.; Wu, D.; de With, G.; Ming, W.: Biomimetic superhydrophobic and highly oleophobic cotton textiles. Langmuir, 2007; 23, 13158–13163.
[62] Gao, Q.; Zhu, Q.; Guo, Y.; Yang, C. Q.: Formation of highly hydrophobic surfaces on cotton and polyester fabrics using silica sol nanoparticles and nonfluorinated alkylsilane. Industrial & Engineering Chemistry Research, 2009; 48, 9797–9803.
[63] Haufe, H.; Thron, A.; Fiedler, D.; Mahltig, B.; Böttcher, H.: Biocidal nanosol coatings. Surface Coatings International Part B: Coatings Transactions, 2005; 88, 55–60.
[64] Haufe, H.; Muschter, K.; Siegert, J.; Böttcher, H.: Bioactive textiles by sol–gel immobilised natural active agents. Journal of Sol-Gel Science and Technology, 2008; 45, 97–101.
[65] Christie, R. M.: Colour Chemistry, Cambridge. Royal Society of Chemistry, 2001.
[66] CHT Beitlich GmbH (Tübingen, Germany), Calor Plus – warm white, https://www.cht.com/cht/web.nsf/id/pa_promo_calorplus_de.html, last accessed: 04.05.2020.
[67] Khattab, T. A.; Rehan, M.; Hamouda, T.: Smart textile framework: Photochromic and fluorescent cellulosic fabric printed by strontium aluminate pigment. Carbohydrate Polymers, 2018; 195, 143–152.
[68] Sánchez, P.; Sánchez-Fernandez, M. V.; Romero, A.; Rodríguez, J. F.; Sánchez-Silva, L.: Development of thermo-regulating textiles using paraffin wax microcapsules. Thermochimica Acta, 2010; 49, 16–21.
[69] Oliveira, F. R.; Fernandes, M.; Carneiro, N.; Pedro Souto, A.: Functionalization of wool fabric with phase-change materials microcapsules after plasma surface modification. Journal of Applied Polymer Science, 2013; 128, 2638–2647.
[70] Mahltig, B.; Cheval, N.; Astachov, V.; Malkoch, M.; Montanez, M. I.; Haase, H.; Fahmi, A.: Hydroxyl functional polyester dendrimers as stabilizing agent for preparation of colloid silver particles – A study in respect to antimicrobial properties and toxicity against human cells. Colloid and Polymer Science, 2012; 290, 1413–1421.
[71] Frérot, E.; Herbal, K.; Herrmann, A.: Controlled stepwise release of fragrance alcohols from dendrimer-based 2-carbamoylbenzoates by neighbouring group participation. European Journal of Organic Chemistry, 2003; 2003, 967–971.
[72] Asadi Fard, P.; Shakoorjavan, S.; Akbari, S.: The relationship between odour intensity and antibacterial durability of encapsulated thyme essential oil by PPI dendrimer on cotton fabrics. The Journal of the Textile Institute, 2018; 109, 832–841.
[73] Bhaskara-Amrit, U. R.; Agrawal, P. B.; Warmoeskerken, M. M. C. G.: Applications of β-cyclodextrins in textiles. AUTEX Research Journal, 2011; 11, 94–101.
[74] Wang, C. X.; Chen, S. L.: Fragrance-release property of β-cyclodextrin inclusion compounds and their application in aromatherapy. Journal of Industrial Textiles, 2005; 34, 157–166.
[75] Martel, B.; Morcellet, M.; Ruffin, D.; Vinet, F.; Weltrowski, L.: Capture and controlled release of fragrances by CD finished textiles. Journal of Inclusion Phenomena and Macrocyclic Chemistry, 2002; 44, 439–442.

[76] Sricharussin, W.; Sopajaree, C.; Maneerung, T.; Sangsuriya, N.: Modification of cotton fabrics with β-cyclodextrin derivative for aroma finishing. The Journal of the Textile Institute, 2009; 100, 682–687.
[77] Buschmann, H. J.; Knittel, D.; Schollmeyer, E.: New textile applications of cyclodextrins. Journal of Inclusion Phenomena and Macrocyclic Chemistry, 2001; 40, 169–172.
[78] Chao-Xia, W.; Shui-Lin, C.: Anchoring β-cyclodextrin to retain fragrances on cotton by means of heterobifunctional reactive dyes. Coloration Technology, 2004; 120, 14–18.
[79] Knittel, D.; Schollmeyer, E.: Technologies for a new century. Surface modification of fibres. Journal of the Textile Institute, 2000; 91, 151–165.
[80] Schmeide, K.; Heise, K. H.; Bernhard, G.; Keil, D.; Jansen, K.; Praschak, D.: Uranium (VI) separation from aqueous solution by calix [6] arene modified textiles. Journal of Radioanalytical and Nuclear Chemistry, 2004; 261, 61–67.
[81] Ten Breteler, M. R.; Nierstrasz, V. A.; Warmoeskerken, M. M. C. G.: Textile slowrelease systems with medical applications. AUTEX Research Journal, 2002; 2, 175–189.
[82] Textor, T.; Mahltig, B.: A sol-gel-based surface treatment for preparation of water repellent antistatic textiles. Applied Surface Science, 2010; 256, 1668–1674.
[83] Ramaratnam, K.; Tsyalkovsky, V.; Klep, V.; Luzinov, I.: Ultrahydrophobic textile surface via decorating fibers with monolayer of reactive nanoparticles and non-fluorinated polymer. Chemical Communications, 2007; 43, 4510–4512.
[84] Rossbach, V.; Patanathabutr, P.; Wichitwechkarn, J.: Copying and manipulating nature: Innovation for textile materials. Fibers and Polymers, 2003; 4, 8–14.
[85] Mahltig, B.; Haufe, H.: Biozidhaltige Nanosole zur Veredlung von weichen und temperaturempfindlichen Materialien. Farbe & Lack, 2010; 116(3), 27–30.
[86] Mischutin, V.: Caliban® F/RP®-44 a brominated flame retardant for textiles. Journal of Coated Fabrics, 1977; 6, 226–233.
[87] Kappes, R. S.; Urbainczyk, T.; Artz, U.; Textor, T.; Gutmann, J. S.: Flame retardants based on amino silanes and phenylphosphonic acid. Polymer Degradation and Stability, 2016; 129, 168–179.
[88] Ibanescu Busila, M.; Musat, V.; Textor, T.; Badilita, V.; Mahltig, B.: Photocatalytic and antimicrobial Ag/ZnO nanocomposites for functionalization of textile fabrics. Journal of Alloys and Compounds, 2014; 610, 244–249.
[89] Rehan, M.; Hartwig, A.; Ott, M.; Gätjen, L.; Wilken, R.: Enhancement of photocatalytic self-cleaning activity and antimicrobial properties of poly (ethylene terephthalate) fabrics. Surface & Coatings Technology, 2013; 219, 50–58.
[90] Rodriguez, C.; Di Cara, A.; Renaud, F. N. R.; Freney, J.; Horvais, N.; Borel, R.; Puzenata, E.; Guillard, C.: Antibacterial effects of photocatalytic textiles for footwear application. Catalysis Today, 2014; 230, 41–46.

Aurélie Cayla*, Eric Devaux, Fabien Salaün, Jeanette Ortega, Thomas Gries

4 Microfibers and nanofibers

Keywords: functionalized synthetic fibers, electrospun nanofibers, nanocomposite fibers, polymer nanocomposites, properties and applications

4.1 Introduction

Textile fibers can be classified according to their composition in two large families which are the natural fibers and the chemical fibers (man-made fibers). The latter has come into being in 1885 when Hilaire Bernigaud de Chardonnet submitted the first patent on artificial silk made from cellulose nitrate in a mixture of alcohol and ether. Then in 1892, Charles Cross, Edward Bevan and Clayton Beadler patented the viscose process (solubilization of cotton in sodium hydroxide and carbon sulphide). This was the beginning of the marketing of artificial fibers. Then following the discoveries and the understanding of the existence of long chains called macromolecules, the first synthetic fiber of polyamide 6.6 by Wallace Carothers (DuPont de Nemours) called Nylon is born in 1935. This was followed by the development of the different types of synthetic fibers that we know today: polyester, acrylic, polyolefin, but also technical fibers such as aramid or ultra-high molecular weight polyethylene (UHMWPE) fibers. Fiber consumption has more than doubled over the past 20 years, and synthetic fibers account for well over half of global fiber production (Figure 4.1). The global fiber production was around 111 million tons in 2019 and in 2020 it dropped slightly due to the COVID-19 pandemic.

Synthetic fibers due to the great modularity of the polymers that compose them and the different techniques of implementation (spinning) allow to cover a very wide range of applications via the multiple properties they confer to textiles. But in order to improve the intrinsic properties of polymers or to bring multifunctionality, the polymers blend and the addition of particles (especially at the nanoscale) allows to widen the range. After reviewing the different ranges of synthetic fibers that can be obtained, the improvement of fiber properties via the addition of particles will be detailed according to the different properties targeted.

*Corresponding author: Aurélie Cayla, Univ. Lille ENSAIT, Gemtex - Laboratoire de Génie et Matériau Textiles, F-59000 Lille, France, e-mail: aurelie.cayla@ensait.fr
Eric Devaux, Fabien Salaün, Univ. Lille ENSAIT, GEMTEX–Laboratoire de Génie et Matériaux Textiles, F-59000 Lille, France
Jeanette Ortega, Thomas Gries, RWTH Aachen University, Institut für Textiltechnik, D-52062 Aachen, Germany

https://doi.org/10.1515/9783110670776-004

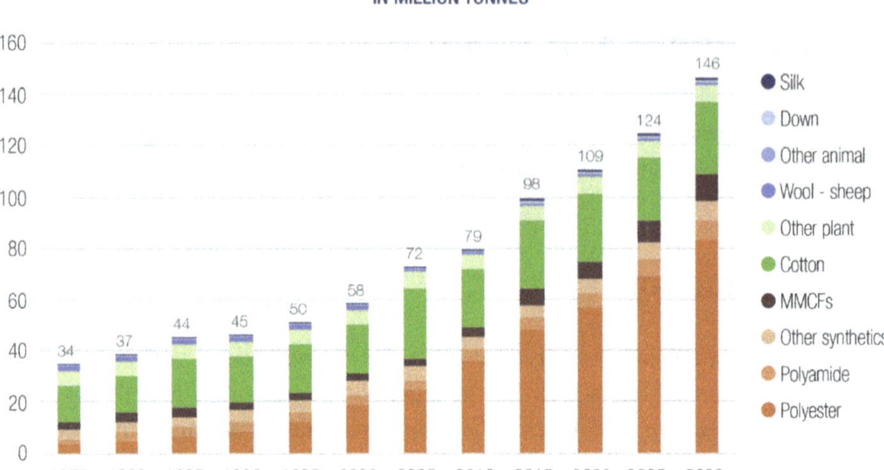

Figure 4.1: Global fiber production in million tonnes according the chemical nature (https://textileexchange.org/preferred-fiber-and-materials-market-report/, access 17/12/2021).

4.2 Functionalized synthetic fibers

4.2.1 Functional microfibers

Synthetic fibers have very different properties due to the chemical nature of the polymer and their formulation, but also the process used. In particular, the spinning process (melt or solvent) allows to obtain a multitude of filament counts ranging from several hundred microns for monofilaments (as for filaments intended for 3D printing) to a few nanometers (webs obtained by electrospinning) through microfibers. For the latter, there is currently no internationally recognized standard definition (ISO). A microfiber can be defined as a very fine fiber with a count of less than 1 dtex (diameter of fewer than 10 µm). Generally, they are made of polyester, polyamide, acrylic, modal, lyocell and viscose in the range of 0.5–1.2 dtex [1]. One of the conventional methods of fabricating microfibers is bicomponent (segmented cross section) melt-spinning step followed by dissolution in sodium hydroxide, but it has a high environmental impact [2]. Another method is the electrospinning method but with a low output. As a matter of fact, there is another approach to realize this goal. For the biphasic fibers from melt spinning, if the sacrificial phase is the matrix phase instead of the dispersed phase, the microfibers can be expected to be obtained. The introduction of nanofillers can be utilized not only for the binary blend but also for the porous materials. The porous materials have a very high specific area to embed the nanofillers. Some researchers adopted post-treatment method to endow the porous materials with additional properties. Furthermore, a permanent effect is hard to guarantee. Naturally, a single-pot strategy was proposed by Yan et al. [3], nanoparticles are incorporated into the polymer blends via melt compounding and located at the interface, the porous structure with surface-embedded nanofillers is fabricated after the selective extraction of the sacrificial phase. Silica nanoparticles with different surface chemistries were incorporated into polypropylene-70 (PP_{70}) and polyvinyl alcohol (PV_{30}) blends aimed at manufacturing surface functionalized porous PP fibers via melt spinning technology. The spinnability of the fibers with 1 wt.% of silica nanoparticles remains outstanding. Silica nanoparticles (SiR972) maintain a suitable localization during the fiber stage due to the thermodynamic equilibrium. The biphasic fibers provide a high specific interface area as well as a good mechanical property, upon which silica nanoparticles have no significant enhancement. The dominant localization of SiR972 at the biphasic interface has limited negative influence in the aggregation structure of the fibers as well as their mechanical properties. In addition, the increment of draw ratio generally has a positive impact on the specific interface area, PVA accessibility, PP crystallinity and mechanical properties of the biphasic fibers. This study demonstrates the feasibility of fabricating the surface-funtionalized porous fibers with silica nanoparticles. The PVA accessibility of PP_{70}-PVA_{30}-SiR972 with a theoretical drawing ration of about 3 reaches as high as 88.4%, and the tenacity of the fibers reaches to 16.2 cN/Tex and retained as 13.9 cN/Tex after selective extraction [4]. Apart from the homogenously modified particles, they also make efforts to introduce the Janus particles into the melt-spinning systems. They localize at the biphasic interface and significantly enhance the polymer compatibility, influencing the morphology of the blends. The obtained melt-spun yarns from PP_{70}-PVA_{30} are mechanically improved, of which the Young's modulus is significantly increased with a sharp 57% increment. It proves that the introduction of particles even has a potential to enhance rather than weaken the mechanical properties of fibers [5]. Other studies show the possibility of elaborating nanoporous polyethylene microfibers for thermal comfort applications [6]. Furthermore, microfibers excel due to their high surface area in medical applications [7].

4.2.2 Functional nanofibers

When the diameter decreases further we go from microfibers or nanofibers which is a fiber with a thickness or diameter of only a few nanometers. Obtaining fine fibers with a diameter of less than 2 µm is a difficult challenge. Indeed, the decrease in fiber diameter is limited in conventional manufacturing processes such as melt-spinning. Indeed, decreasing the diameter of a filament is possible by decreasing the material flow rate and increasing the draw. However, the stability of the flow is no longer guaranteed in the first case and the breakage of the filaments during the drawing stage limits the conditions of the processing of fine fibers [8]. The different ways of processing fine fibers are briefly presented in the following.

The electrostatic spinning technology more commonly known as electrospinning is a technique patented by Formhals [9] in 1934. This method has received renewed attention in the last two decades because of its ability to generate fiber diameters of the order of a hundred nanometers [10]. The polymer is solubilized in a solvent then injected in a syringe. A high voltage electric field (typically 5 to 30 kV) is applied to the syringe and to the surface of the solution. If this electric field is strong enough, nanofibers are then generated by the electrically charged jet coming out of the syringe. The solvent evaporates with the heat input during the experiment and the nanofibers are ejected and collected on a flat or rotating roller collector.

The production of nonwoven webs by the meltblown process is a solution for the production of fine fibers (currently used in protective masks) in the form of a web. It allows the production of fine fibers in large quantities without the use of solvents. After melt extrusion of the polymer, the material is passed through the spinning pack by a heated volumetric pump. At the die exit, a blowing system generates a jet of air and the filaments are drawn under heat. The resulting microfibers are then cooled to room temperature and collected on a collector [11]. The diameter of the obtained fibers is considerably reduced and can reach 2 µm; however the diameter distribution can vary along the length of the nonwoven web product.

Two centrifugal spinning processes exist: the "ForceSpinning®" technology manufactured by Fiberio Incorporation Technology and the Melt Spun Nanofiber Process from Dupont®. The spinning principle is based on centrifugal force. This spinning device includes a centrifugal spinning element that rotates around a spinning axis. The centrifugal force ensures the formation of the veil of nanofibers with an equivalent diameter of 250 nm [12]. The technology developed by Dupont also allows the raw material to be processed by melt and the fibers constituting the veils have a diameter of around 500 nm.

Today, these nanofibers have multiple applications in various sectors: biomedical, filtration, composites, personal protective equipment, optics or electronics, and so on. Recent reviews show the diversity of possible uses especially when the fibers are nanocomposites, as flexible electrodes, as composite nanomaterials in automotive and aerospace manufacturing [13].

4.2.3 Influence of filament cross sections and multicomponent filaments

Fibers produced by melt spinning have mostly a circular cross section, but depending on the application, the modification of the cross-sectional shape (spinneret hole) can allow to obtain additional functionalities [14]. For example (Figure 4.2), realization of cross sections with very high bending angles (trilobal fibers) allows to

Figure 4.2: Examples of several cross sections of polylactic acid (PLA) filaments by melt spinning process.

modify the optical properties of the fibers to increase the specific surface (antibacterial properties) or membrane gas separation processes with hollow fiber [15].

Whether it is the geometric shape of the filament section or the complexity of bi- or tri-component filaments that allow different polymers to be positioned side by side in complex and defined morphologies. The bi/tricomponent fibers are fibers that have been extruded then spun on the same spinneret with two or three polymers, with a cross-sectional shape specific to the intended use of the fiber's intended use. Among the structures obtained, the most common morphologies are the following: side-by-side, core/sheath, pie-wedge or islands at sea. This complex form modified performance of initial filament as mechanical stability or conductivity of core can be a desired performance in contrast with roughness in sheath surface [16]. Talbourdet et al. [17] developed a tricomponent filament: two layers (core and sheath) of conductive polymer composite (CPC) electrodes and the intermediate layer consist of piezoelectric polyvinylidene fluoride (PVDF). The conductive polymers act as an external and internat electrode, making it possible to carry the PVDF piezoelectric response when it is subjected to mechanical stresses. CPC constituting the electrodes layers were chosen according to their characteristic temperatures, such as the melting point and the crystallization temperature [17].

However, it is difficult to control all the implementation parameters in order to obtain the desired morphology. Indeed, both the intrinsic parameters of the polymers used (such as different molar masses of PP/polyamide 6 (PA6) in Figure 4.3), and the process parameters, play a key role on the stability, shape and quality of the interface [18].

Figure 4.3: Influence of molecular weight on morphologies of bicomponent fibers.

4.3 Improvement of fiber properties using nanocomposites

Composite (including nanocomposites) or multiphase materials can be specially tailored based on the selection of the single components. For example, in the case of fiber-reinforced plastic, one material, the fiber, is responsible for the tensile strength, and the other material, plastic, is responsible for the structure and transferring the loads. Such composites can be utilized as structures, where pure fiber or pure plastic is not suitable. The same concept is applied to particle modification of textiles or fibers. Such particles used in the textile industry are on the micro or nanoscale. Although these particles cannot be seen by the naked eye, the resulting functionalities in application are observable. Such example functionalities due to the addition of particles in fibers are shown in Table 4.1 and will be further discussed in the following section.

4.3.1 Hydrophobic and self-cleaning

It is often said that the best inspiration for new developments often stems from nature through bionics or biologically inspired engineering. Some common examples of bionic inventions are Velcro, many robotics and medical adhesives. In the field of textile engineering, this can be seen in the hydrophobic and self-cleaning surfaces. The biological basis behind these innovations is the microscopic structure of the Lotus leaf [19]. It was observed that water droplets did not spread on the surface of such leaves and additionally that contaminating particles are removed from the surface when water droplets roll off. These insights are applied to textiles be modification of the fiber or textiles structure.

One example of a commercially available water-repellent textile is from the American company Nanotex. The surface of cotton fibers is modified by incorporating

Table 4.1: Particles and their added functionalities.

	Hydrophobic	Self-cleaning	Antibacterial	UV-protective	Antistatic	Wrinkle resistant	Electro-conductive	Flame-retardant	Increased durability	Controlled release
Nanoclay				X				X		X
Carbon Nanotubes	X	X			X	X	X	X	X	
Carbon black					X		X		X	
Graphite nanofibers	X				X		X			
TiO_2	X	X	X	X	X	X		X		
ZnO	X	X	X	X	X	X		X		
MgO			X	X			X		X	
Al_2O_3			X	X			X		X	
SiO_2	X	X	X	X		X		X	X	X
Au			X							
Ag	X	X	X			X				
Fe_2O_3			X	X	X					
NiO			X						X	
CuO			X	X						

nanowhiskers onto the surface which provide a three-dimensional structure similar to the Lotus leaf [20, 21]. These whiskers and the cavities between them are responsible for the water-repellent behavior, subsequently allowing contamination to be easily carried away when the water rolls off the fabric.

Other examples of the hydrophobic and self-cleaning properties of textiles can be seen in the products from the Swiss Company Schoeller. Here, the surface is impregnated with an additive, again forming a structure on the textile resulting in a similar effect as seen in the Lotus leaf [20]. The development of these modified textiles is extremely relevant in the field of clothing in order to reduce the need to wash items, home textiles, to eliminate or reduce staining, and technical textiles in the outdoor sector, such as tents or sails.

4.3.2 Influence of additives on the mechanical properties

Carbonaceous particles have been regularly studied to improve the mechanical properties of fibers. In order to increase the mechanical properties of polypropylene filaments, carbon nanotubes have been repeatedly incorporated into this polymer matrix. Marcincin et al. [22] have developed a spin-draw process for these composites. First melt-blended by extrusion and recovered as pellets, the nanocomposites are introduced into the melt spinning machine. Filaments loaded between 0.02 and 0.3 wt% were produced. For one of the polypropylenes studied, a 0.1% carbon nanotubes (CNT) concentration increases the Young's modulus value by 4%, while 0.02% CNT is sufficient to increase the toughness by 5%. Beyond these concentrations, the mechanical properties of the filaments decrease by the creation of a large number of creations of a large number of defects within the polymer. To limit the agglomeration of particles in polypropylene (PP), Kearns and Shambaugh [23] have studied in PP/CNT blends in the solvent route (decalin) prior to melt spinning. The mechanical reinforcement of polycarbonate (PC) by the incorporation of carbon nanotubes has been studied. Fornes et al. [24] have made monofilaments loaded with 1.3 and 5% CNT by mass. The characteristics of the mixture containing 5% CNT do not allow the application of such a large stretch. The Young's modulus and stress at break are progressively increased with increasing filler content in the filaments while the elongation at break decreases. With 5% CNT, the modulus increases from 1.82 to 3.12 GPa, the stress from 43 to 64 MPa and the elongation from 40 to 47%. This increase in the mechanical properties of PC with high filler contents can be explained by a good dispersion of the nanotubes within the polymer and a preferential orientation of the nanotubes in the direction of stretching.

Additives of natural origin are also used to increase the mechanical properties of the fibers. Indeed, the addition of cellulose particles (nanocellulose, nanowhiskers or microcrystalline cellulose) can increase tenacity (textile strength) or Young's modulus (mechanical property measuring the stiffness) [25] but also can combine other properties like fire retardant (FR) [26].

4.3.3 Thermal comfort with functionalized fibers

The mid-infrared thermal radiation emitted by the human body corresponds to the wavelengths of 7–14 µm. One strategy adopted is to make textiles transparent to dissipate as much heat as possible. For example, Xiao et al. [27] designed an infrared-transparent nanofiber membrane using PA6 nanofibers containing silica microspores, which decreases the temperature by 0.4–1.7 °C compared to traditional materials. Various types of nanoparticles including transition metals (e.g., Ag, Ti and Al), inorganic or organic compounds (e.g., TiO_2, Fe_2CO_3, antimony-doped tin oxide and azo pigments) have been used as reflective materials in order to manage the control of the mid-IR thermal radiation [28]. Cai et al. [29] reported a new strategy utilizing inorganic nanoparticles as a coloring component for scalable brightly colored, infrared-transparent textiles. Inorganic pigment nanoparticles, such as Prussian Blue, iron oxide and silicon were compounded into the polyethylene matrix forms a uniform composite before melt spinning. The obtained textiles have a passive cooling effect of 1.6–1.8 °C. The production of thermoregulating fibers containing encapsulated phase change material has expanded in recent years for a wide range of applications, including clothing, footwear, blankets, carpets and mattresses but also higher value-added applications such as sportswear, spacesuits and automotive and aerospace textiles [30].

For the three last decades, extensive research has been carried out by many researchers' groups to load phase change microcapsules capsules into synthetic fibers using dry/wet-spinning [31] or melt spinning routes [32]. Thus, microcapsules were incorporated into viscose rayon and acrylic, polyacrylonitrile-vinylidene chloride (PAN/VDC) fibers up to 40% in weight, for a latent heat storage capacity ranging from 1 to 44 J/g, by wet spinning. The use of the process of melt spinning is limited by the size (less than 10 µm) and the thermal stability of the microcapsule. Nevertheless, Gao et al. [33] succeeded in obtaining PAN fibers with 5–25% of microcapsules for a latent heat reaching 25 J/g.

4.3.4 Flame-retardant properties

In certain sectors like home textiles, automotive and aerospace there are very high standards and requirements regarding the flammability of the materials used. This has been the motivation to develop FR textiles which fulfil these requirements. Such anti-flammable properties can be achieved either through the incorporation or coating of nanomaterials. Example materials for used for these methods are:
- Nanoclays
- Carbon nanotubes
- Metal oxides: titanium dioxide (TiO_2) and zinc oxide (ZnO)

In the case of nanoclays incorporated into polymer fibers, the nanoparticles migrate to the surface of the polymer in the presence of a flame. This migration forms a barrier for the oxygen, limiting the spread of the flame to the inside of the polymer and stifling the flame. Similarly, the carbon nanotubes for a char layer in the material, blocking the transfer of heat to the inside of the polymer fiber, delaying thermal degradation. Additionally, the metal oxide nanoparticles act as a heat barrier while also absorbing active species, such as free radicals [34].

The direct incorporation of FRs in the molten polymer during extrusion seems to be the simplest way to produce the FR fiber with an advantage to vary the types of FR additive and their concentration in polymer bulk. Once the additive is introduced into the polymer matrix by extrusion, melt spinning processes it. However, compatibility problems with several FR additives occur at the high temperatures used during melt spinning of polyamide (PA), polyester (PES) and PP [35]. The presence of relatively high FR concentrations (>20 wt%) necessary to confer FR properties creates a spinning fluid compatibility problem but also causes a reduction in fiber tensile strength and other essential textile properties. This may prevent the successful incorporation of FR during the process; as a consequence, only a few fire-retarded fibers are commercially available.

Trevira® CS (Trevira GmbH, formerly Hoechst) is one of the most successful commercially available FR PES fibers. The flame retardancy has been achieved in commercial polyethylene terephthalate (PET) fiber by directly incorporating reactive comonomeric organophosphorus units in the main chain [36, 37]. Another commercially utilized P-containing comonomer is the 9,10-dihydro-9-oxa-10-phosphaphenanthrenyl-10-oxide (DOPO). PET fibers containing this FR comonomer are commercially available from Toyobo under the trade name HEIM® and have limiting oxygen intake (LOI) ranging up to 28 vol% [37]. For polyamides, Bourbigot et al. [38] developed multifilament based on PA6 and Cloisite® 30B at a loading of 5 wt%. The fabrics produced from the loaded multifilament showed a reduction of 40% in the peak heat release rate (PHRR) under a heat flux of 50 kW/m^2 compared to the fabric from unfilled PA6. The same additives were studied by Solarski et al. [39] in a polylactide matrix. With 4 wt% of Cloisite® 30B, the authors observed improved fire retardancy for knitted structure. Vargas et al. [40] reported that a small amount of 0.75 wt% of sepiolite improved the fire performance of polypropylene multifilament. Textiles structure showed decreases in peak intensity of the heat release rate (PHRR) and total heat release (THR) compared to the pure PP textile. In addition, the FR additives for PP and their potential suitability for use in fiber application have been extensively reviewed. The same nanocomposite approach was adopted by Shanmuganathan et al. [41] to develop the polyamide 6/organo-montmorillonite nanoclay (PA6/OMMT) system. Nanocomposites with 8 wt% OMMT offered better spinnability, and the fibers have good physical properties to be knitted into fabrics. It was found that the FR effect depends on fabric geometry and the test conditions. Char forming kinetics was relatively slow to protect the material from flame spread in the horizontal flame spread test, and hence, the nanocomposite fabric does not show a significant difference in flame spread behavior compared to PA6 fabrics. However, under radiant heat conditions, the fibers melt and enable the formation of a continuous charred surface at very early stages of burning, resulting in improved flame retardancy properties. Fabric tightness factor plays a crucial role in favoring the formation of char, and with a higher tightness factor, enhanced flame retardancy was achieved. The material from flame spread in horizontal flame spread test and hence, the nanocomposite fabric does not show a significant difference in flame spread behavior compared to PA6 fabrics. However, under radiant heat conditions, the fibers melt and enable the formation of a continuous charred surface at very early stages of burning, resulting in improved flame retardancy properties. The fabric tightness factor plays a crucial role in favoring the formation of char, and a higher tightness factor allows the enhancement of flame retardancy properties.

During the last years, lignin (LL) has been employed to improve the thermal stability [42, 43] and fire retardancy of thermoplastic polymers [44, 45]. LL thermally decomposes over a broad temperature range because various aromatic functional groups have different thermal stability. Furthermore, LL is also able to generate a high amount (ranging in between 30 and 50 wt%) of char residue during its degradation in the temperature ranges of 500–700 °C under inert atmosphere, which allows the reduction of the heat release rate of the polymeric material during degradation steps [46, 47]. Nevertheless, in most cases, the unique properties are not sufficient to bring FR properties to fibers. Thus, intumescent FR additives based on phosphorus and nitrogen elements are added to the mixture [48].

The char forming ability of LL was exploited for designing FR formulation by Mandlekar et al. [48–51]. LL were combined with commercially available phosphinate FR. In the preliminary study, low sulphonate kraft (LS) was combined with zinc phosphinate (ZnP). Various formulations of ternary blends, including LS with ZnP, were prepared by extrusion to optimize the ratio to evaluate FR behavior. Thermogravimetric analysis (TGA) showed that incorporating lignin in the ternary mixtures increased the thermal stability of PA, promoting the formation of a stable char residue at the end of the experiments. Flammability tests showed that the addition of sulfonated lignin (LS) improved the FR properties due to the char formation, and most enhanced results were achieved when 10 wt% LS and ZnP were combined. In addition, cone calorimetry results showed that the interactions between lignin and ZnP promoted a significant reduction of PHRR and THR. This finding was attributed to the formation of a protective char layer. The most enhanced results were found when combined 10 wt% of LS and ZnP.

Moreover, increasing lignin content also effectively reduced the carbon monoxide (CO) and carbon dioxide-carbon mono oxide (CO_2/CO) yield compared to unfilled PA. They also used low-cost industrial lignins, that is, lignosulphonate and Kraft lignin, as carbon sources in combination with ZnP and aluminum phosphinate (AlP). Dispersion of lignin and FR in PA assessed by SEM showed that ternary blends of

ZnP and LL show uniform dispersion in PA matrix. However, LL and AlP blends have agglomerates of about 10 µm. TGA showed that incorporating lignin in ternary blends increased the thermal stability of PA, promoting the formation of a stable char residue at the end of the tests. The addition of LL favored the higher char formation due to the presence of sulphonate groups. The char residue obtained from ZnP blends was higher than that of AlP containing ternary blends, showing the interaction between ZnP and lignin favor the formation of a stable char residue. Flammability tests showed that the combination of LL and ZnP effectively improves FR properties by reducing total combustion time, self-extinction, and the V-1 rating was achieved for PA_{80}-LL_7-ZnP_{13} PA_{80}-LL_{10}-ZnP_{10} blends. Conversely, the combination of lignin with AlP did not improve the FR properties, all blends showed V-2 rating. Cone calorimetry results showed the interactions between lignin and phosphinate promoted reduction of PHRR and THR. In particular, the best FR performance was achieved by combining LL with AlP (i.e., PA_{80}-LL_{10}-AlP_{10}), resulting in a substantial reduction of PHRR (−74%). The presence of lignin in ternary blends reduced the CO and CO_2/CO yield compared to PA. Mainly, PA_{80}-LL_{10}-ZnP_{10} blend shows minimum CO release without affecting fire performance. The morphology of char residue showed that the formation of a compact char layer is primarily responsible for the improved FR properties.

Industrial LL and phosphinate FR combinations were used to develop multifilaments using the melt spinning process. All the LL and ZnP-based blends were successfully transformed into multifilaments. However, the spinning of AlP and LL mixture was not spinnable and showed the irregular flow of material with certain breakage of filaments. Multifilaments obtained from DL and ZnP have a comparable mean diameter to PA multifilaments (about 60 µm).

In contrast, LL and ZnP blends showed a higher mean diameter, about 70–76 µm, with a variation of 27% concerning PA. Mechanical testing results showed that PA-DL-ZnP filaments have higher tensile strength than PA-LL-ZnP filaments. Fire testing of PA-DL-ZnP fabric showed that the combination of kraft lignin (DL) and ZnP has little influence on HRR and PHRR reduction. However, the presence of lignin significantly reduced the THR and maximum rate of heat emission due to the char residue formation showing the charring effect.

Cayla et al. [45] have improved the FR properties of polylactic acid (PLA) with the addition of kraft lignin and ammonium polyphosphate (AP), selected as carbon and acid sources, respectively. The spinnability, thermal behavior and fire retardancy properties of PLA composites containing lignin and/or AP were studied by analyzing the variation of melt flow index (MFI) values thermal stability by TGA, UL-94 and cone calorimeter experiments. The incorporation of lignin led to a decrease in the MFI value, whereas it increased with the presence of AP, it was found due to the creation of free volume in the PLA matrix and the aggregation lignin particles. During the degradation steps, AP degraded the macromolecular PLA chains and modified the interactions between all the compounds. From the results, only five blends have been spinnable to obtain multifilaments, that is, PLA with 5 wt% of AP; PLA with 5, 10 and

20 wt% of LK and the ternary blend PLA_{90}-LK_{05}-AP_{05}. The thermal stability of the composites is slightly enhanced with the addition of AP, and the residue at 500 °C increases with lignin due to the charring capacity of this compound. The FR properties of PLA knitted composite structures (PLA_{95}-AP_{05}, PLA_{95}-LK_{05}, PLA_{80}-LK_{20} and PLA_{90}-LK_{05}-AP_{05}) were investigated by a cone calorimeter. PLA-LK-AP did not delay the ignition time but led to a significant reduction in the heat release rate due to char formation. Therefore, 5 wt% of LK and 5 wt% of AP were sufficient to obtain an efficient FR effect in PLA fabric.

The introduction of microcapsules containing FR actives into fibers by melt-spinning has also been studied by our research group in recent years [52, 53]. Analyses of the thermal and mechanical behavior of the obtained PP fibers have shown an improvement of these properties, influenced not only by the chemical composition of the microcapsule membrane (melamine-formaldehyde or silica), but also by the matrix/microcapsule interaction. Nevertheless, this process limits the loading rate of the microcapsules, which does not allow the use of these materials for specific applications.

4.3.5 Healthcare applications

Healthcare textiles include all technical textiles whose purpose is to provide benefits to human health. They have different properties depending on the final application [54, 55].

4.3.5.1 Antimicrobial properties

The main property of antimicrobial textiles is inhibiting the growth of microorganisms onto their surface since traditional textiles tend to be a good substrate for bacterial growth [56]. Antimicrobial textiles are produced by combining the fibrous material with an intrinsic antimicrobial agent. Combatting the growth of microbes such as bacteria and fungus on textiles is of large importance for sports clothing, home textiles and geotextiles. The direct incorporation of antimicrobial particles reduces the need to spray or treat surfaces with harmful chemicals, which end up in the water system through washing or rain, for clothing and geotextiles, respectively. Nanoparticles are especially effective for antimicrobial applications due to the high surface area of the particles which can come in contact with the microorganisms. Two kinds of additives which are mainly used in these functional textiles are [57, 58]:
- Metals: silver or gold
- Metal oxides: ZnO, TiO_2 or Cu_2O

These can be either antibiotic drugs, inorganic materials such as zinc titanium and silver oxides, or polymers like chitosan [59, 60]. The antimicrobial activity of these materials is because they own a positive surface charge that disrupts the microbial cell wall causing the bacterium death. Gawish et al. [61] produced polypropylene/Ag composite fibers, the antibacterial efficacy was evaluated by the percentage reduction of S. aureus and E. coli growth. The use of silver as an antimicrobial agent has been used since ancient times especially for wound or burn treatment or even water purification. Although the mechanism of action is not fully understood, it is thought that the silver nanoparticles release silver ions, which disrupt the DNA replication of the microbes, resulting in eventual death [58]. Metal oxides function through a different principle to fight against microbes: photocatalytic action. As the names suggests, light is needed to activate the antimicrobial functionality of the particle. The incident light changes the charge carriers at the surface of the material which react with surrounding oxygen to create radicals which destroy the microbe membrane [62]. Because this mechanism targets the membrane itself rather than the DNA of the microbe, a resistance cannot be developed. Additionally, this is effective to various microbes and bacteria, reducing the need to modify the particles for different microbe types [63].

Other studies have been conducted to introduce (ZnO) into synthetic fibers whether in PA6 [64] or PLA [65]. Furthermore, LL also seems to be a good candidate to replace silver nanoparticles in PET as observed by Minet et al. [66].

Electrospinning process may also to be used to produce nanofiber mats with incorporated metals because they can serve as antimicrobials. Through this route, it is possible to obtain nanofibers with unique properties, including high surface-to-volume ratio, very low weight, porous and flexibility nature. Thus, antibacterial nanofiber can be achieved by encapsulation/entrapment of several types of antibacterial agents, which allows the controlled release of antibacterial agents. Furthermore, this kind of entrapment of the antimicrobial protects them from temperature, oxidation, degradation and ambient conditions. Among the most studied agents, we find copper, gold, silver, titanium dioxide, iron and zinc oxide, which have been integrated in matrices of polyvinylpyrrolidone, cellulose acetate, Nylon-6, polyethylene-oxide (PEO), poly-D,L-lactide or poly(ε-caprolactone) (PCL), poly(3-hydroxyalkanoate) s and poly(vinyl alcohol) [67, 68].

4.3.5.2 Insect repellent

The application of repellents on the surface of fabrics has many limitations, that is

1. due to their volatile nature, their efficacy is limited to a short time;
2. and their durability and abrasion resistances are low when applied on the surface of fabrics;
3. in terms of toxicity, repellents such as permethrin and chili oil cause skin and eye irritation upon direct contact.

Encapsulation not only allows control of the release of the active but also limits the contact with the skin, thus reducing skin irritation and/or allergic reactions. Thus, direct microencapsulation of these repellents into nanofibers during emulsion electrospinning or integration of already microencapsulated repellents into nanofibers are solutions to these problems. Incorporating these solutions by electrospinning allows obtaining functional textiles with an optimum repellent power linked to the large specific contact surface, which favors optimal evaporation of the repellent when the capsules are broken.

Coeta et al. [69] have used the electrospinning method to optimize the incorporation of microcapsules containing *p*-menthane-3, 8-diol in PVA nanofibers. The incorporation of microcapsules in the nanofibers gives long-term functionality, and it still needs active breakage of the microcapsules to release the functional additive.

4.3.5.3 Wound healing applications

Hybrid organic/inorganic nanofibers have the particularity to present improved mechanical properties and to have a prolonged or controlled long-term release profile of encapsulated substances. Thus, they represent a material of choice for drug delivery and tissue engineering applications [8].

Biodegradable and biocompatible synthetic polymers are widely used for electrospinning for wound healing applications. Spontaneous release of active ingredients is an undesirable phenomenon that mainly occurs when they are incorporated into nanofibers. The controlled release of these active ingredients can be achieved through nanoparticles. Thus, silica particles are chosen as drug carriers because of their high surface area and porous interior that can be used as a loading site for hydrophobic and hydrophilic drugs. The entrapment of DOX hydrochloride (anticancer drug) into silica nanoparticles by sol–gel route prior to their incorporations into PCL nanofibers was studied by Gohary et al. [70]. This system allowed a continuous release of DOX for several days and was achieved from PCL/PEO nanofibers. Lópex-Esparza et al. [71] realized produced PCL nanofibers with silver nanoparticles and studied their antimicrobial activity against Gram-positive and Gram-negative bacteria. The minimum diameter of the obtained nanofibers was 160 nm with 1–10 mM of silver nanoparticles (AgNP). Al-Omair [72] demonstrated using these AgNPs with electrospun PCL/polymethacrylic acid (PMMA) graft copolymer nanofibers. The morphology of the nanofibers, the content of AgNPs, the water uptake of the nanofibers and the antimicrobial efficacy were analyzed. The obtained nanofibers had an average diameter between 200 and 570 nm, which was not affected by their presence. The resulting nanofibers possessed good hydrophilic properties and excellent antibacterial properties against Gram-negative bacteria *E. coli* and *P. aeruginosa* and Gram-positive bacteria *Bacillus thuringiensis* and *Staphylococcus aureus* with clear inhibition zones of about 22 and 53 mm. Chen et al. [73] trapped gentamicin in mesoporous silica and then electrospun them into PCL. The gentamicin-loaded PCL nanofibers showed high resilience against *E. coli*, while maintaining their bioactivity.

The incorporation of inorganic nanomaterials into electrospun polymer fibers improves the biocompatibility and mechanical strength of scaffolds. Furthermore, the nanoparticles can promote the cell adhesion, proliferation, migration and differentiation for tissue engineering applications. Amongst these inorganic particles, the most used are CNT [74, 75, 76, 77], GO [78], laponite [79], gold [80], silver [81] and zeolites [82].

4.3.5.4 Ultraviolet protection

For many textiles which are exposed to constant sunlight, resistance to ultraviolet (UV) radiation is necessary. In clothing, this UV protection textile provides an extra layer of shielding to the wearer as sunscreen would do. For outdoor textiles, such as umbrellas and awnings, the UV protection helps to maintain the mechanical and optical properties of the textile itself. A wide variety of material classes can be employed in order to provide textiles with UV protection [83, 84].

- Organic: organic polymers containing aromatic structures, collagen or dyes such as poly(diethyleneglycol bis allyl carbonate)
- Inorganic: natural mica, metal oxides, such as TiO_2, ZnO and aluminium oxide (Al_2O_3), or transition metal oxides such as zirconium dioxide (ZrO_2) or iron(III) oxide (FE_2O_3)
- Metallic: aluminium, copper, gold or silver.

These particles can either deflect, reflect, absorb or scatter the incident UV radiation protecting the bulk textile [83]. Since the particles are interacting with the external light and radiation source, they are often incorporated onto a final textile product as a coating in order to maximize this interaction with the surroundings.

4.3.6 Enhanced electrical properties

The addition of electrically conductive fillers (carbon, metal, intrinsic conductive polymer) inside isolated fibrous matrices in order to make the conductive yarns is not a new idea [85]. The main problem is that beyond a certain percentage of fillers (usually 5 wt%), the fibers are very difficult to melt process and lose their mechanical properties because fillers are placed in the interstices of the macromolecular chains. However, the idea is very simple, the preparation of a masterbatch with a polymer and conductive fillers at extrusion allows to obtain filled pellets. If the proportion of particles inside is sufficient (upon percolation threshold ϕc, Figure 4.4) this will allow to create, after the spinning operation, a path for the electrons to

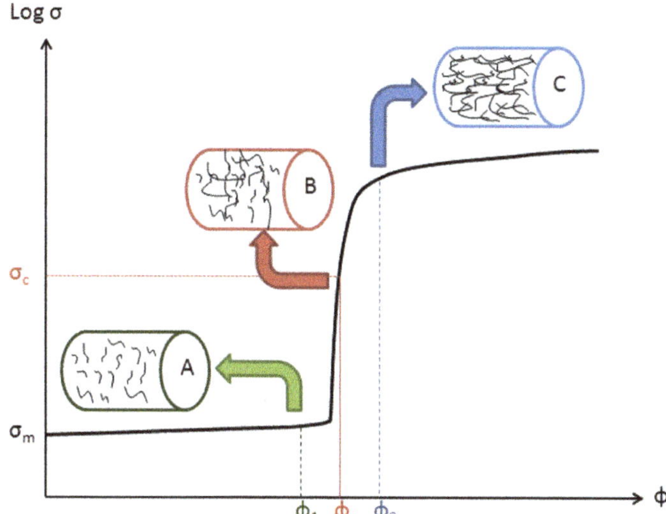

Figure 4.4: Representative electrical percolation curve upon insertion of conductive charges into a polymer with σ the electrical conductivity of the material, σ_m the electrical conductivity of the polymer matrix, ϕ the amount of fillers. σ_c and ϕ_c are the critical electrical conductivity and filler content, respectively, at which electrical percolation develops in the polymer (percolation threshold).

pass through the fiber. Moreover, the orientation given by the macromolecular chains during drawn step allows to intensify this phenomenon [86].

Studies showed good results of electrical conductivity, but unfortunately, the mechanical properties of these filaments made them very difficult to use in textile structures [87]. Some solution to reduce the impact on viscosity and to find a balance between electrical conductivity and processability is to use immiscible polymer blends [88] or to use different kinds of conductive fillers with several aspect ratio to have synergistic effect [89]. Today, with the help of structures such as bicomponent or tricomponent fibers core/sheath it is possible to integrate on the core of the fiber a masterbatch with a conductive filler allowing electrical conduction, while maintaining the mechanical properties thanks to the virgin polymer inside.

The intrinsic conductivity of the filaments guarantees the durability of the treatment and allows for a variety of applications, as sensors [90, 91], antennas [92], heating textiles [93], electromagnetic interference shielding [94] and 3D printing filaments [95].

The modification of electrical properties makes it possible to develop antistatic textiles. Static build up in textiles not only leads to unwanted cling of clothing, but, more dangerously, can also cause unexpected shocks and ignition of flammable vapor and dust. This static charge results from the separation of positive and negative charges in materials due to friction. One approach to reduce this is to increase the electrical conductivity of the textile. This conductivity allows for the charges to be carried away reducing the charge build up in one location. This can be done by simply incorporating thin metal wire into the textile structure. These wires, however, tend to result in a stiffer textile. By incorporating electrically conductive particles into the fiber material, the textile itself becomes conductive without the use of stiff metal wires. Example materials for this added functionality are [96]:

- Metal: silver, copper or gold
- Metal oxide: tin(II) oxide (SnO_2), ZnO or Antimony(III) oxide (Sb_2O_3)
- Carbon-based: carbon nanotubes or carbon black

Because the static charge is dissipated through the textile, the functional particles cannot act alone, but instead need to form a conductive network, so-called the percolation network, through the material. The formation of a suitable network depends on the size and shape of the particle, the dispersion and the concentration.

4.3.7 Energy harvesting

Piezoelectric textiles made from polymer fibers have several advantages over traditional ceramic piezoelectric materials, including flexibility and comfort, and can be more effectively used in wearable piezoelectric devices. In piezoelectric materials, micro and nanofibers improve the alignment of the dipoles, which results in a higher piezoelectric potential. A high piezoelectric coefficient is responsible for the broad study of the piezoelectric properties of PVDF and copolymers. In medical devices, PLA with biodegradable and biocompatible properties has excellent potential. PAN nanofibers prepared by electrospinning have piezoelectric properties that are comparable to PVDF. In addition, adding nanoparticles can significantly enhance the piezoelectric properties of nanofibers. Currently, smart textiles with micro and nanostructures are widely used in nanogenerators, sensors and energy storage.

The materials used in applications based on piezoelectric phenomena are mainly ceramics (PZT – lead zirconate titanate, KNN – $NaxK1\text{-}xNbO_3$, AlN-aluminum nitride, ZnO – zinc oxide, $BaTiO_3$ – barium titanate) or polymeric materials (PVDF, poly(vinylidene fluoride-trifluroethylene), PAN, PA11 and PLA. In the case of piezoelectric fiber development, it is necessary to distinguish between fibers with diameters in the nanometer range and those with diameters up to several tens of microns. Most piezoelectric nanofibers are produced by electrospinning, even if it is possible to prepare nanofibers from other spinning processes such as centrifugal spinning. Studies on the development of piezoelectric polymer fibers used in traditional textile structures (woven or knitted fabrics) are based on melt spinning. Although the literature mentions some research on the development of piezoelectric mono or multifilaments based on PP [97], PA11 [98] or PLA [99], research teams working on the development of piezoelectric filaments have focused on PVDF because of its interesting trade-off in terms of piezoelectric properties, cost and industrial-scale availability for mass applications of this polymer. Thus, many research groups in Europe, such as the Institute for Materials Research and Innovation laboratory at the University of Bolton (UK), the Department of Materials and Manufacturing Technology at Chalmers University of Technology (Sweden) [100], the ITA at RWTH Aachen University (Germany) [101], the University of Minho (Portugal) [102] and the GEMTEX laboratory (France) [103, 104], have studied the production of piezoelectric PVDF fibers. In each case, the aims were to determine the influence of spinning parameters, especially the influence of the drawing ratio, on the crystal structure of PVDF to optimize the content of the polar phase (mainly the β-phase). Apart from modifying the process parameters, the researchers tried to optimize the piezoelectric character of PVDF by adding fillers such as nanoparticles.

The development of composite piezoelectric materials by dispersing nanoscale ceramics in a polymer matrix is a solution to overcome the brittleness of ceramic materials and their poor piezoelectric properties [105]. The piezoelectric properties of PVDF can be improved by introducing nanofillers, which serve as nucleation and crystallization sites to improve the formation of the β-phase and the orientation of the crystals contributes to a better piezoelectric coefficient [106]. Two types of fillers are mainly used, that is, conductive fillers (graphene, CNT, carbon black, silver) that enhance nucleation and facilitate the transfer of generated charges and (ii) nonconductive fillers (BT, KNN, ZnO, MgO, TiO_2) promoting nucleation and dipole alignment.

4.4 Conclusion

The use of chemical fibers and more particularly synthetic fibers shows a very large diversity of applications in particular by the multiple properties that they can confer to textiles. Their functionalization can also be achieved via different techniques, whether chemical, via the judicious selection of the nature of the thermoplastic polymer and the additives incorporated but also brought by the choice of process, as by the shape of the section. All these elements and tools available do not allow to reach a simple functionality but to combine multiple. The notions of sustainability are now a major issue, which via the transformation of polymer with controlled degradation allow for the future to imagine functional fibers adapted to the life of a defined use.

References

[1] Mukhopadhyay, S.; Ramakrishnan, G.: Microfibres. Textile Progress, 2008; 40(1), 1–86. doi:10.1080/00405160801942585.

[2] Ohkoshi, Y. Shinshu University, Japan: 18 – Melt Spinning and Other Techniques for the Production of Nanofibres and Microfibres. In: Eichhorn, S. J.; Hearle, J. W. S.; Jaffe, M.; Kikutani, T., editors: Woodhead Publishing Series in Textiles (Cambridge), Handbook of Textile Fibre Structure. Woodhead Publishing, 2009; vol. 1, pp. 484–490. ISBN 9781845693800, https://doi.org/10.1533/9781845696504.2.484

[3] Yan, X.; Cayla, A.; Salaün, F.; Devaux, E.; Liu, P.; Huang, T.: A green method to fabricate porous polypropylene fibers: Development toward textile products and mechanical evaluation. Textile Research Journal, 2019; 90(5–6), 547–560. doi:10.1177/0040517519871944.

[4] Yan, X.; Cayla, A.; Salaün, F.; Devaux, E.; Liu, P.; Mao, J.; Huang, T.: Porous fibers surface decorated with nanofillers: From melt-spun PP/PVA blend fibers with silica nanoparticles. Journal of Applied Polymer Science, 2019; 137(11), 48470. doi:10.1002/app.48470.

[5] Yan, X.; Cayla, A.; Devaux, E.; Otazaghine, B.; Salaün, F.: Polypropylene/poly(vinyl alcohol) blends compatibilized with kaolinite Janus hybrid particles and their transformation into fibers. Industrial & Engineering Chemistry Research, 2019; 58(25), 10931–10940. doi:10.1021/acs.iecr.9b01990.

[6] Peng, Y.; Chen, J.; Song, A. Y., et al.: Nanoporous polyethylene microfibres for large-scale radiative cooling fabric. Nature Sustainability, 2018; 1, 105–112. https://doi.org/10.1038/s41893-018-0023-2

[7] Erika, A.; Baltušnikaitė-Guzaitienė, J.; Juškaitė, V.; Žilius, M.; Briedis, V.; Stanys, S.: Formation and characterization of melt-spun polypropylene fibers with propolis for medical applications. The Journal of the Textile Institute, 2018; 109(2), 278–284. doi:10.1080/00405000.2017.1341295.

[8] Montefusco, A. F.: The use of Nonwovens in air filtration. Filtration and Separation, 2005; 42, 30–31. doi:10.1016/S0015-1882(05)00446-5.

[9] Formhals, A.: US Patent, 1934 1,975,504. https://patents.google.com/patent/US1975504A/en

[10] Afshari, M.; Kotek, R.; Tonelli, A. E.: Producing Polyamide Nanofibers by Electrospinning. In: Brown, P. J.; Stevens, K. (Clemson University, USA), editors: Nanofibers and Nanotechnology in Textiles. Woodhead Publishing in Textiles (Cambridge, UK), 2007.

[11] Irwin M. Hutten, Chapter 4 - Raw Materials for Nonwoven Filter Media, Editor(s): Irwin M. Hutten, Handbook of Nonwoven Filter Media (Second Edition), Butterworth-Heinemann, 2016, Pages 158–275, ISBN 9780080983011, https://doi.org/10.1016/B978-0-08-098301-1.00004-6.

[12] Raghavan, B.; Soto, H.; Lozano, K.: Fabrication of melt spun polypropylene nanofibers by forcespinning. Journal of Engineered Fibers and Fabrics, 2013; 8, 9.

[13] Aruchamy, K.; Ashesh, M.; Nataraj, S. K.: Electrospun nanofibers, nanocomposites and characterization of art: Insight on establishing fibers as product. Nano-Structures & Nano-Objects, 2018; 16, 45–58. ISSN 2352-507X, https://doi.org/10.1016/j.nanoso.2018.03.013

[14] Hufenus, R.; Yan, Y.; Dauner, M.; Kikutani, T.: Melt-spun fibers for textile applications. Materials, 2020; 13, 4298. https://doi.org/10.3390/ma13194298

[15] Pelzer, M.; Vad, T.; Becker, A.; Gries, T.; Markova, S.; Teplyakov, V.: Melt spinning and characterization of hollow fibers from poly(4-methyl-1-pentene). Journal of Applied Polymer Science, 2020; 138, e49630. https://doi.org/10.1002/app.49630

[16] Naeimirad, M.; Zadhoush, A.; Kotek, R.; Esmaeely Neisiany, R.; Nouri Khorasani, S.; Ramakrishna, S.: Recent advances in core/shell bicomponent fibers and nanofibers: A review. Journal of Applied Polymer Science, 2018; 135, 46265. doi:10.1002/app.46265.

[17] Talbourdet, A.; et al.: Development of mono-component and tri-component fibres 100% polymer based piezoelectric PVDF to harvest energy IOP Conference Series: Materials Science and Engineering, 2017; 254, 072026. doi:10.1088/1757-899X/254/7/072026.
[18] Ayad, E.; Cayla, A.; Rault, F.; Gonthier, A.; Campagne, C.; Devaux, E.: Effect of viscosity ratio of two immiscible polymers on morphology in bicomponent melt spinning fibers. Advances in Polymer Technology, 2018; 37, 1134–1141. https://doi.org/10.1002/adv.21772
[19] Barthlott, W.; Neinhuis, C.: Purity of the sacred lotus, or escape from contamination in biological surfaces. Planta, Springer Verlag, 1997; 202, 1–8. https://doi.org/10.1007/s004250050096
[20] Das, S.: Application of nanotechnology in textile finishing. Textile Future, 2019. https://textile-future.com/archives/20042
[21] Wong, Y. W. H.; Yuen, C. W. M.; Leung, M. Y. S.; Ku, S. K. A.; Lam, H. L. I.: Selected applications of nanotechnology in textiles. AUTEX Research Journal, 2006; 6(1).
[22] Marcincin, A.; Hricova, M.; Marcincin, K.; Legén, J.; Ujhelyiova, A.; Bonduel, D.; Claes, M.: Conférence Lille3000 Futurotextiles Proceedings, 2006, Lille, pp. 135–146.
[23] Kearns, J. C.; Shambaugh, R. L. Polypropylene fibers reinforced with carbon nanotubes: Journal of Applied Polymer Science, 2002; 86, 2079–2084.
[24] Fornes, T. D.; Baur, J. W.; Sabba, Y.; Thomas, E. L. Morphology and properties of melt-spun polycarbonate fibers containing single- and multi-wall carbon nanotubes: Polymer, 2006; 47, 1704–1714.
[25] Clemons, C.;. Nanocellulose in spun continuous fibers: A review and future outlook. Journal of Renewable Materials, October 2016; 4(5), 327–339(13). https://doi.org/10.7569/JRM.2016.634112
[26] Aouat, T.; Kaci, M.; Devaux, É.; Campagne, C.; Cayla, A.; Dumazert, L.; Lopez-Cuesta, J. M.: Morphological, mechanical and thermal characterization of poly(lactic acid)/cellulose multifilament yarns prepared by melt spinning. Advances in Polymer Technology, 2018; 37(4), 1193–1205.
[27] Xiao, R.; Hou, C.; Yang, W.; Su, Y.; Li, Y.; Zhang, Q.; Gao, P.; Wang, H.: Infrared-radiation-enhanced nanofiber membrane for sky radiative cooling of the human body. ACS Applied Materials & Interfaces, 2019; 11(47), 44673–44681. doi:10.1021/acsami.9b13933.
[28] Peng, Y.; Cui, Y.: Advanced textiles for personal thermal management and energy. Joule, 2020; 4(4), 724–742. doi:10.1016/j.joule.2020.02.011.
[29] Cai, L.; Peng, Y.; Xu, J.; Zhou, C.; Zhou, C.; Wu, P.; Lin, D.; Fan, S.; Cui, Y.: Temperature regulation in colored infrared-transparent polyethylene textiles. Joule, 2019; 3(6), 1478–1486. doi:10.1016/j.joule.2019.03.015.
[30] Hassan, A.; Laghari, M. S.; Rashid, Y.: Micro-encapsulated phase change materials: A review of encapsulation, safety and thermal characteristics. Sustainability, 2016; 8(10). doi:10.3390/su8101046.
[31] Bryant, Y. G.; Colvin, D. P.: Fibre and reversible enhanced thermal storage properties and fabric made there from. Triangle Research and Development Corporation (Raleigh, NC), 1988.
[32] Hartmann, M. H.; Magil, M. C.: Melt Spinnable Concentrate Pellets Having Enhanced Reversible Thermal Properties. Outlast Technologies, Patent US20050035482A1 (United States) Inc, 2004.
[33] Gao, X.-Y.; Han, N.; Zhang, X.-X.; Wan-yong, Y.: Melt-processable acrylonitrile–methyl acrylate copolymers and melt-spun fibers containing MicroPCMs. Journal of Materials Science, 2009; 44(21), 5877–5884. doi:10.1007/s10853-009-3830-z.
[34] Norouzi, M.; Zare, Y.; Kiany, P.: Nanoparticles as effictive flame retardants for natural and synthetic textile polymers: Application, mechanism, and optimization. Polymer Reviews, 2015; 55(3), 531–560. https://doi.org/10.1080/15583724.2014.980427

[35] A Richard Horrocks, 4 - Textiles, Editor(s): A.R. Horrocks, D. Price, Fire Retardant Materials, Woodhead Publishing, 2001, Pages 128–181, ISBN 9781855734197, https://doi.org/10.1533/9781855737464.128.

[36] Bourbigot, S.: Flame Retardancy of Textiles: New Approaches. In: Advances in Fire Retardant Materials New York: Elsevier, 2008; pp. 9–40.

[37] Weil, E. D.; Levchik, S. V.: Flame retardants in commercial use or development for textiles. Journal of Fire Sciences, 2008; 26(3), 243–281. doi:10.1177/0734904108089485.

[38] Bourbigot, S.; Flambard, X.; Poutch, F.: Study of the thermal degradation of high performance fibres – Application to polybenzazole and p-aramid fibres. Polymer Degradation and Stability, 2001; 74(2), 283–290. doi:10.1016/s0141-3910(01)00159-8.

[39] Solarski, S.; Mahjoubi, F.; Ferreira, M.; Devaux, E.; Bachelet, P.; Bourbigot, S.; Delobel, R.; Coszach, P.; Murariu, M.; Da Silva Ferreira, A.; Alexandre, M.; Degee, P.; Dubois, P.: (Plasticized) polylactide/clay nanocomposite textile: Thermal, mechanical, shrinkage and fire properties. Journal of Materials Science, 2007; 42(13), 5105–5117. doi:10.1007/s10853-006-0911-0.

[40] Vargas, A. F.; Orozco, V. H.; Rault, F.; Giraud, S.; Devaux, E.; López, B. L.: Influence of fiber-like nanofillers on the rheological, mechanical, thermal and fire properties of polypropylene: An application to multifilament yarn. Composites. Part A, Applied Science and Manufacturing, 2010; 41(12), 1797–1806. doi:10.1016/j.compositesa.2010.08.018.

[41] Shanmuganathan, K.; Deodhar, S.; Dembsey, N. A.; Fan, Q.; Patra, P. K.: Condensed-phase flame retardation in nylon 6-layered silicate nanocomposites: Films, fibers, and fabrics. Polymer Engineering & Science, 2008; 48(4), 662–675. doi:10.1002/pen.20993.

[42] Sallem-Idrissi, N.; Sclavons, M.; Debecker, D. P.; Devaux, J.: Miscible raw lignin/nylon 6 blends: Thermal and mechanical performances. Journal of Applied Polymer Science, 2016; 133(6), n/a–n/a. doi:10.1002/app.42963.

[43] Fernandes, D. M.; Winkler Hechenleitner, A. A.; Job, A. E.; Radovanocic, E.; Gómez Pineda, E. A.: Thermal and photochemical stability of poly(vinyl alcohol)/modified lignin blends. Polymer Degradation and Stability, 2006; 91(5), 1192–1201. doi:10.1016/j.polymdegradstab.2005.05.024.

[44] Réti, C.; Casetta, M.; Duquesne, S.; Bourbigot, S.; Delobel, R.: Flammability properties of intumescent PLA including starch and lignin. Polymers for Advanced Technologies, 2008; 19(6), 628–635. doi:10.1002/pat.1130.

[45] Cayla, A.; Rault, F.; Giraud, S.; Salaun, F.; Fierro, V.; Celzard, A.: PLA with intumescent system containing lignin and ammonium polyphosphate for flame retardant textile. Polymers (Basel), 2016; 8(9). doi:10.3390/polym8090331.

[46] Li, B.; Zhang, X.; Su., R.: An investigation of thermal degradation and charring of larch lignin in the condensed phase: The effects of boric acid, guanyl urea phosphate, ammonium dihydrogen phosphate and ammonium polyphosphate. Polymer Degradation and Stability, 2002; 75(1), 35–44. doi:10.1016/s0141-3910(01)00202-6.

[47] Mandlekar, N.; Cayla, A.; Rault, F.; Giraud, S.; Salaün, F.; Malucelli, G.; Guan, J.-P.: An Overview on the Use of Lignin and Its Derivatives in Fire Retardant Polymer Systems. In: Poletto, M., editor: Lignin. Rijeka, Croatia: Intechopen, 2018; pp. 207–231.

[48] Mandlekar, N.; Malucelli, G.; Cayla, A.; Rault, F.; Giraud, S.; Salaun, F.; Guan, J. P.: Fire retardant action of zinc phosphinate and polyamide 11 blend containing lignin as a carbon source. Polymer Degradation and Stability, 2018; 153, 63–74. doi:10.1016/j.polymdegradstab.2018.04.019.

[49] Mandlekar, N.; Cayla, A.; Rault, F.; Giraud, S.; Salaun, F.; Guan, J.: Valorization of industrial lignin as biobased carbon source in fire retardant system for polyamide 11 blends. Polymers (Basel), 2019; 11(1), 1–18. doi:10.3390/polym11010180.

[50] Mandlekar, N.; Cayla, A.; Rault, F.; Giraud, S.; Salaun, F.; Guan, J.: Development of novel polyamide 11 multifilaments and fabric structures based on industrial lignin and zinc phosphinate as flame retardants. Molecules, 2020; 25, 21. doi:10.3390/molecules25214963.
[51] Mandlekar, N.; Cayla, A.; Rault, F.; Giraud, S.; Salaun, F.; Malucelli, G.; Guan, J.: Thermal stability and fire retardant properties of polyamide 11 microcomposites containing different lignins. Industrial & Engineering Chemistry Research, 2017; 56(46), 13704–13714. doi:10.1021/acs.iecr.7b03085.
[52] Salaün, F.; Creach, G.; Rault, F.; Almeras, X.: Thermo-physical properties of polypropylene fibers containing a microencapsulated flame retardant. Polymers for Advanced Technologies, 2013; 24(2), 236–248. doi:10.1002/pat.3076.
[53] Salaün, F.; Lewandowski, M.; Vroman, I.; Bedek, G.; Bourbigot, S.: Development and characterisation of flame-retardant fibres from isotactic polypropylene melt-compounded with melamine-formaldehyde microcapsules. Polymer Degradation and Stability, 2011; 96(1), 131–143. doi:10.1016/j.polymdegradstab.2010.10.009.
[54] Rtimi, S.; Giannakis, S.; Pulgarin, C.: Self-sterilizing sputtered films for applications in hospital facilities. Molecules, 2017; 22, 1074. https://doi.org/10.3390/molecules22071074
[55] Wang, W.; Hui, P. C. L.; Kan, C.-W.: Functionalized textile based therapy for the treatment of atopic dermatitis. Coatings, 2017; 7, 82. https://doi.org/10.3390/coatings7060082
[56] Morais, D. S.; Guedes, R. M.; Lopes, M. A.: Antimicrobial approaches for textiles: From research to market. Materials (Basel), 2016; 9(6). doi:10.3390/ma9060498.
[57] EMPA; TVS Textilverband Schweiz: Nano textiles, Gundlagen und Leitprinzipien zur effizienten Entwicklung nachhaltiger Nanotextilien. Empa Grafikgruppe und Hausdrukerei, 2011.
[58] Sun, G.: Antimicrobial Textiles. Cambridge: Woodhead Publishing Limited, 2016.
[59] Ul-Islam, S.; Butola, B. S.: Nanomaterials in the Wet Processing of Textiles. Wiley, Scivener publishing (Berverly) 2018.
[60] Parisi, O. I.; Scrivano, L.; Sinicropi, M. S.; Puoci, F.: Polymeric nanoparticle constructs as devices for antibacterial therapy. Current Opinion in Pharmacology, 2017; 36, 72–77. doi:10.1016/j.coph.2017.08.004.
[61] Gawish, S. M.; Avci, H.; Ramadan, A. M.; Mosleh, S.; Monticello, R.; Breidt, F.; Kotek, R.: Properties of antibacterial polypropylene/nanometal composite fibers. Journal of Biomaterials Science. Polymer Edition, 2012; 23(1–4), 43–61. doi:10.1163/092050610X541944.
[62] Ganguly, P.; Byrne, C.; Breen, A.; Pillai, S. C.: Antimicrobial activity of photocatalysts: Fundamentals, mechanisms, kinetics and recent advances. Applied Catalysis. B, Environmental, 2018; (225), 51–75. https://doi.org/10.1016/j.apcatb.2017.11.018
[63] Kubacka, A.; Diez, M. S.; Rojo, D.; Bargiela, R.; Ciordia, S.; Zapico, I.; Albar, J. P.; Barbas, C.; Martins Dos Santo, V. A. P.; Fernandez-Garcia, M.; Ferrer, M.: Understanding the antimicrobial mechanism of TiO_2-based nanocomposite films in a pathogenic bacterium. Scientific Reports, 2014 (2015); 4, 1–9. https://doi.org/10.1038/srep04134
[64] Dural Erem, A.; Ozcan, G.; Skrifvars, M.: Antibacterial activity of PA6/ZnO nanocomposite fibers. Textile Research Journal, 2011; 81(16), 1638–1646. doi:10.1177/0040517511407380.
[65] Doumbia, A. S.; Vezin, H.; Ferreira, M.; Campagne, C.; Devaux, E.: Studies of polylactide/zinc oxide nanocomposites: Influence of surface treatment on zinc oxide antibacterial activities in textile nanocomposites. Journal of Applied Polymer Science, 2015; 132(17). doi:10.1002/app.41776.
[66] Minet, J.; Cayla, A.; Campagne, C.: Lignin as Sustainable Antimicrobial Fillers to Develop PET Multifilaments by Melting Process. In: Sand, A.; Zaki, E., editors: Organic Polymers. IntechOpen London, 2020.

[67] Mohammadi, M. A.; Rostami, M.; Beikzadeh, S.; Raeisi, M.; Tabibiazar, M.; Yousefi, M.: Electrospun nanofibers as advanced antibacterial platforms: A review of recent studies. International Journal of Pharmaceutical Sciences and Research, 2019; 10(2). doi:10.13040/ijpsr.0975-8232.10(2).463-73.

[68] Rodriguez-Tobias, H.; Morales, G.; Grande, D.: Comprehensive review on electrospinning techniques as versatile approaches toward antimicrobial biopolymeric composite fibers. Materials Science & Engineering. C, Materials for Biological Applications, 2019; 101, 306–322. doi:10.1016/j.msec.2019.03.099.

[69] Ciera, L.; Beladjal, L.; Van Landuyt, L.; Menger, D.; Holdinga, M.; Mertens, J.; Van Langenhove, L.; De Clerk, K.; Gheysens, T.: Electrospinning repellents in polyvinyl alcohol-nanofibres for obtaining mosquito-repelling fabrics. Royal Society Open Science, 2019; 6(8), 182139. doi:10.1098/rsos.182139.

[70] El Gohary, M. I.; Abd El Hady, B. M.; Al Saeed, A. A.; Tolba, E.; El Rashedi, A. M. I.; Saleh, S.: Electrospinning of doxorubicin loaded silica/poly(ε-caprolactone) hybrid fiber mats for sustained drug release. Advances in Natural Sciences: Nanoscience and Nanotechnology, 2018; 9(2). doi:10.1088/2043-6254/aab999.

[71] López-Esparza, J.; Espinosa-Cristóbal, L. F.; Donohue-Cornejo, A.; Reyes-López, S. Y.: Antimicrobial activity of silver nanoparticles in polycaprolactone nanofibers against gram-positive and gram-negative bacteria. Industrial & Engineering Chemistry Research, 2016; 55(49), 12532–12538. doi:10.1021/acs.iecr.6b02300.

[72] Al-Omair, M.: Synthesis of antibacterial silver–poly(ε-caprolactone)-methacrylic acid graft copolymer nanofibers and their evaluation as potential wound dressing. Polymers, 2015; 7(8), 1464–1475. doi:10.3390/polym7081464.

[73] Chen, X.; Xu, C.; He, H.: Electrospinning of silica nanoparticles-entrapped nanofibers for sustained gentamicin release. Biochemical and Biophysical Research Communications, 2019; 516(4), 1085–1089. doi:10.1016/j.bbrc.2019.06.163.

[74] Qi, R.-L.; Tian, X.-J.; Guo, R.; Luo, Y.; Shen, M.-W.; Yu, J.-Y.; Shi, X.-Y.: Controlled release of doxorubicin from electrospun MWCNTs/PLGA hybrid nanofibers. Chinese Journal of Polymer Science, 2016; 34(9), 1047–1059. doi:10.1007/s10118-016-1827-z.

[75] Liao, H.; Qi, R.; Shen, M.; Cao, X.; Guo, R.; Zhang, Y.; Shi, X.: Improved cellular response on multiwalled carbon nanotube-incorporated electrospun polyvinyl alcohol/chitosan nanofibrous scaffolds. Colloids and Surfaces. B, Biointerfaces, 2011; 84(2), 528–535. doi:10.1016/j.colsurfb.2011.02.010.

[76] Luo, Y.; Wang, S.; Shen, M.; Qi, R.; Fang, Y.; Guo, R.; Cai, H.; Cao, X.; Tomas, H.; Zhu, M.; Shi, X.: Carbon nanotube-incorporated multilayered cellulose acetate nanofibers for tissue engineering applications. Carbohydrate Polymers, 2013; 91(1), 419–427. doi:10.1016/j.carbpol.2012.08.069.

[77] Hasanzadeh, E.; Ebrahimi-Barough, S.; Mirzaei, E.; Azami, M.; Tavangar, S. M.; Mahmoodi, N.; Basiri, A.; Ai, J.: Preparation of fibrin gel scaffolds containing MWCNT/PU nanofibers for neural tissue engineering. Journal of Biomedical Materials Research. Part A, 2019; 107(4), 802–814. doi:10.1002/jbm.a.36596.

[78] Luo, Y.; Shen, H.; Fang, Y.; Cao, Y.; Huang, J.; Zhang, M.; Dai, J.; Shi, X.; Zhang, Z.: Enhanced proliferation and osteogenic differentiation of mesenchymal stem cells on graphene oxide-incorporated electrospun poly(lactic-co-glycolic acid) nanofibrous mats. ACS Applied Materials & Interfaces, 2015; 7(11), 6331–6339. doi:10.1021/acsami.5b00862.

[79] Wang, S.; Zheng, F.; Huang, Y.; Fang, Y.; Shen, M.; Zhu, M.; Shi, X.: Encapsulation of amoxicillin within laponite-doped poly(lactic-co-glycolic acid) nanofibers: Preparation, characterization, and antibacterial activity. ACS Applied Materials & Interfaces, 2012; 4(11), 6393–6401. doi:10.1021/am302130b.

[80] Baranes, K.; Shevach, M.; Shefi, O.; Dvir, T.: Gold nanoparticle-decorated scaffolds promote neuronal differentiation and maturation. Nano Letters, 2016; 16(5), 2916–2920. doi:10.1021/acs.nanolett.5b04033.

[81] Son, W. K.; Youk, J. H.; Park, W. H.: Antimicrobial cellulose acetate nanofibers containing silver nanoparticles. Carbohydrate Polymers, 2006; 65(4), 430–434. doi:10.1016/j.carbpol.2006.01.037.

[82] Davarpanah Jazi, R.; Rafienia, M.; Salehi Rozve, H.; Karamian, E.; Sattary, M.: Fabrication and characterization of electrospun poly lactic-co-glycolic acid/zeolite nanocomposite scaffolds using bone tissue engineering. Journal of Bioactive and Compatible Polymers, 2017; 33(1), 63–78. doi:10.1177/0883911517707774.

[83] Edwards, S. D.; Edwards, K.; Parker, T. L.; Evans, J. M.: Ultraviolet ray (UV) blocking textile containing particles. US Patent 6 037 280, Date of Patent 14.03.2000.

[84] Hassan, B. S.; Islam, G. M. N.; Haque, A. N. M. A.: Applications of nanotechnology in textiles: A review. Advanced Research in Textile Engineering, 2019; 4(2), 1038.

[85] Lund, A.; van der Velden, N. M.; Persson, N.-K.; Hamedi, M. M.; Müller, C.: Electrically conducting fibres for e-textiles: An open playground for conjugated polymers and carbon nanomaterials. Materials Science and Engineering: R: Reports, 2018; 126, 1–29. Pages 1-29, ISSN 0927-796X, https://doi.org/10.1016/j.mser.2018.03.001

[86] Marischal, L.; Cayla, A.; Lemort, G.; Campagne, C.; Devaux, É.: Influence of melt spinning parameters on electrical conductivity of carbon fillers filled polyamide 12 composites. Synthetic Metals, 2018; 245, 51–60.

[87] Bouchard, J.; Cayla, A.; Lutz, V.; Devaux, É.; Campagne, C.: Electrical and mechanical properties of phenoxy/multiwalled carbon nanotubes multifilament yarn processed by melt spinning. Textile Research Journal, 2012; 82(20), 2116–2125.

[88] Cayla, A.; Campagne, C.; Rochery, M.; Devaux, É.: Melt spinning of carbon nanotubes-based polymeric blends: Electrical, mechanical and thermal properties. Synthetic Metals, 2012; 162, 759–767.

[89] Javadi Toghchi, M.; Campagne, C.; Cayla, A.; Bruniaux, P.; Loghin, C.; Cristian, I.; Chen, Y.: Electrical conductivity enhancement of hybrid PA6,6 composite containing multiwall carbon nanotube and carbon black for shielding effectiveness application in textile. Synthetic Metals, 2019; 251, 75–84.

[90] Rentenberger, R.; Cayla, A.; Villmow, T.; Jehnichen, C.; Campagne, C.; Rochery, M.; Devaux, É.; Pötschke, P.: Multifilament fibres of poly(lactide)/poly(ε-caprolactone) blends with multiwalled carbon nanotubes as sensor materials for ethyl acetate and acetone. Sensors and Actuators B, 2011; 160, 22–31.

[91] Cochrane, C.; Cayla, A.: Polymer-based Resistive Sensors for Smart Textiles. In: Kirstein, T., editor: Multidisciplinary Know-how for Smart-textiles Developers. Chapter 5 Cambridge, UK: Woodhead Publishing, 2013.ISBN: 978-0-85709-342-4.

[92] Krifa, M.: Electrically conductive textile materials – Application in flexible sensors and antennas. Textiles, 2021; 1, 239–257. https://doi.org/10.3390/textiles1020012

[93] Marischal, L.; Cayla, A.; Lemort, G.; Campagne, C.; Devaux, E.: Selection of immiscible polymers blends filled with carbon nanotubes for heating applications. Polymers, 2019; 11(11), 1827–1843.

[94] Maity, S.; Chatterjee, A.: Conductive polymer-based electro-conductive textile composites for electromagnetic interference shielding: A review. Journal of Industrial Textiles, 2018; 47(8), 2228–2252. doi:10.1177/1528083716670310.

[95] Eutionnat-Diffo, P. A.; Cayla, A.; Chen, Y.; Guan, J.; Nierstrasz, V.; Campagne, C.: Development of flexible and conductive immiscible thermoplastic/elastomer monofilament for smart textiles applications using 3D printing. Polymers, 2020; 12(10), 2300.

[96] Zhang, X.: Antistatic and Conductive Textiles. In: Pan, N.; Sun, G.: Functional Textiles for Improved Performance, Protection and Health. Cambridge: Woodhead Publishing Limited, 2011; pp. 27–44.
[97] Bayramol, D. V.: Polarize Edilmiş İzotaktik Polipropilen Monofilamentlerinin Voltaj Çıktıları Üzerine Çekim Oranının Etkisinin İncelenmesi. Tekstil Ve Mühendis, 2016; 23(103), 166–171. doi:10.7216/1300759920162310301.
[98] Vatansever, D. R.; Hadimani, L.; Shah, T.; Siores, E.: Yeşil Enerji Tekstil Uygulamaları İçin Piezoelektrik Monofilament Eldesi. Tekstil Ve Mühendis, 2012; 1–5. doi:10.7216/130075992012198501.
[99] Oh, H. J.; Kim, D. K.; Choi, Y. C.; Lim, S. J.; Jeong, J. B.; Ko, J. H.; Hahm, W. G.; Kim, S. W.; Lee, Y.; Kim, H.; Yeang, B. J.: Fabrication of piezoelectric poly(L-lactic acid)/BaTiO3 fibre by the melt-spinning process. Scientific Reports, 2020; 10(1), 16339. doi:10.1038/s41598-020-73261-3.
[100] Lund, A.; Gustafsson, C.; Bertilsson, H.; Rychwalski, R. W.: Enhancement of β phase crystals formation with the use of nanofillers in PVDF films and fibres. Composites Science and Technology, 2011; 71(2), 222–229. doi:10.1016/j.compscitech.2010.11.014.
[101] Walter, S.; Steinmann, W.; Schütte, J.; Seide, G.; Gries, T.; Roth, G.; Wierach, P.; Sinapius, M.: Characterisation of piezoelectric PVDF monofilaments. Materials Technology, 2013; 26(3), 140–145. doi:10.1179/175355511x13007211258962.
[102] Ferreira, A.; Costa, P.; Carvalho, H.; Nobrega, J. M.; Sencadas, V.; Lanceros-Mendez, S.: Extrusion of poly(vinylidene fluoride) filaments: Effect of the processing conditions and conductive inner core on the electroactive phase content and mechanical properties. Journal of Polymer Research, 2011; 18(6), 1653–1658. doi:10.1007/s10965-011-9570-1.
[103] Boudriaux, M.; Rault, F.; Cochrane, C.; Lemort, G.; Campagne, C.; Devaux, E.; Courtois, C.: Crystalline forms of PVDF fiber filled with clay components along processing steps. Journal of Applied Polymer Science, 2016; 133(14), n/a–n/a. doi:10.1002/app.43244.
[104] Talbourdet, A.; Rault, F.; Lemort, G.; Cochrane, C.; Devaux, E.; Campagne, C.: 3D interlock design 100% PVDF piezoelectric to improve energy harvesting. Smart Materials and Structures, 2018; 27(7). doi:10.1088/1361-665X/aab865.
[105] Sezer, N.; Koç., M.: A comprehensive review on the state-of-the-art of piezoelectric energy harvesting. Nano Energy, 2021; 80. doi:10.1016/j.nanoen.2020.105567.
[106] Ghosh, S. K.; Sinha, T. K.; Mahanty, B.; Mandal, D.: Self-poled efficient flexible "ferroelectretic" nanogenerator: A new class of piezoelectric energy harvester. Energy Technology, 2015; 3(12), 1190–1197. doi:10.1002/ente.201500167.

III Applications in technical textiles

Avinash P. Manian*, Thomas Bechtold, Tung Pham

5 Sports textiles/apparel

Keywords: micro- and nanoparticles, color, antimicrobial, hydrophobicity, smart textiles

5.1 Introduction

A wide number of possible applications for micro and nanoparticles in sports textiles and apparel as well as in many other industrial sectors has been proposed in the scientific literature, but the number of resulting commercial technologies and products is rather limited. Before introducing a new technological solution into commercial production, a wide range of conditions have to be fulfilled to enable companies to implement a new process and place a new product on the global market. Products enabled by micro- or nanoparticles are often based on a sophisticated chemical and physical background, which makes introduction in large-scale production challenging. Table 5.1 summarizes a selection of aspects that need consideration, with particular relevance to the application of micro- and nanoparticles, in the transfer of scientific laboratory experiments into commercial products.

Table 5.1: Representative aspects of technical and economic barriers in the transfer of a new micro or nanoparticle-based process from lab-scale to commercial textile production.

Aspect	Critical factor	Example of barrier
Raw material	Availability	Synthesis only in lab scale
	Safety	Information about hazards
Processes	Solvent-based	Environmental concerns Hazards during production
	Chemical balances and by-products	Emissions, effluents
Handling of chemicals	Working conditions	Dust formation

*Corresponding author: Avinash P. Manian, Research Institute of Textile Chemistry and Textile Physics, University of Innsbruck, Hoechsterstrasse 73, 6850 Dornbirn, Austria, e-mail: avinash.manian@uibk.ac.at
Thomas Bechtold, Tung Pham, Research Institute of Textile Chemistry and Textile Physics, University of Innsbruck, Hoechsterstrasse 73, 6850 Dornbirn, Austria

https://doi.org/10.1515/9783110670776-005

Table 5.1 (continued)

Aspect	Critical factor	Example of barrier
Equipment	Special machinery Low output	High investment required Production costs
Product	Level of performance	Not competitive, not durable
User	High content of unexplored chemicals	Health concerns, allergic potential, irritation
Disposal	Separation and recycling	Removal of additives

The synthesis of micro- or nanoparticle-based materials in large quantities is a critical step in the scale-up of processes. Complex synthetic pathways can easily be applied on a laboratory scale for the synthesis of smaller amounts of a functional product. However, a consideration of the necessary unit operations (e.g., mass and heat transfer, and solid and liquid transport processes) and of the equipment to accomplish them are prerequisites for process scale-up. As an example the synthesis of Cu_2O microparticles for antimicrobial finishing of synthetic fabric has been simplified to a series of precipitation/settling steps to achieve transfer from lab-amounts to production volumes without expensive investments [1, 2].

At the same time, new technical raw materials will have to pass careful safety considerations to ensure that any risks during production, application and use can be minimized. Examples for legal regulations which can lead to critical barriers during introduction of new products are the regulations for biocidal substances [3] as well as the special consideration of nanomaterial in the regulations for medical products [4]. To be implemented into textile chemical production plants a new process has to fulfil all existing regulations for textile production, for example, with regard to safe working conditions (toxicity, chemical hazards) and to environmental safety (e.g., wastewater and volatile emissions), all of which limit the use of organic solvent-based processes [5].

In an ideal case, existing machinery can be employed for the initial scale-up trials, which lowers the barrier to deliver marketable products without high investments. A clear balance between production costs, process efficiency and benefit in product performance must be available at an early stage of the development to justify expenses and risks with a substantial gain in product performance. As long as durability of an improved product does not match with consumer expectations, the product will not obtain a positive management decision for scale-up. Besides these mainly technical and commercial aspects, any combination of new material on textiles will also have to consider the postconsumer phase. New additives must not interfere with measures to recycle, separate, reuse or dispose textile wastes. The high number of technical and economic barriers explains the substantial reduction in number of successful technical approaches to utilize micro- and nanoparticle-based concepts. Thus, in order to pass through the "Valley of Death for innovations" on their way toward the commercial launch of a new product, much more efforts are required than just defining possible routes for new innovations [6].

The high competition on the textile markets has led to a substantial cost-pressure on standard products. As a result, innovation in low-price markets is rather stagnant and incremental. Often innovations cannot be justified by a possible gain in achievable market volume or prices. When production costs determine the position of a product; in many cases, incremental improvements are part of a continuous stepwise product innovation. From an analysis of the innovation demand in the field of textile products for an ageing society (garment, bedding, hygiene), the mechanism of innovation for such products has been analyzed [7]. Radical disruptive innovations can be observed in markets, which require specialized product performance, outstanding properties or new technical approaches, for example, products for aerospace or medical applications. Often, innovations diffuse from such sectors into markets characterized by high volumes and low prices. A typical example is the introduction of new material concepts, for example, high-performance fibers, from aerospace or medical products, to safety applications and then into the equipment for high performance sports equipment. The diffusion of successful material concepts into larger markets, for example, working clothes thus often proceeds via products developed for sports garment and equipment.

In many cases, innovation is derived from related innovative products developed for markets, which initially allow higher prices and better recognition for a product innovation. A considerable number of innovations proposed in the scientific literature disappear during the assessment for possible scale-up, process and cost analysis. The number of textiles, which have been enabled by micro or nanoparticles at consumer level, thus remains limited. In this chapter the focus is on applications, which:

- have achieved already substantial production volume,
- represent a remarkable technical concept or
- exhibit attributes promising larger production volumes in the near future.

Also rather traditional techniques will be mentioned because of their chemical background as micro /nanoparticle-based processes.

5.2 Color and structure

A traditional application of particle-based technology is found in dyeing procedures with the use of pigment dyes and vat dyes (a class of dyes that are applied in a reduction-oxidation process; the name "vat" dye stems from the historic tradition of performing these procedures in vats or barrels), sulfur dyes and indigo for denim production.

In vat dyeing, careful finishing of the synthesized filter cake of the pigment is required to achieve stable dispersion in the aqueous dyebath to avoid pigment filtration in yarn dyeing and dark spots due to bigger particles. Also, a well-controlled reduction reaction is supported by a uniform particle size distribution. Representative values for the maximum particle diameter of the finished pigment have been reported to be below 0.12 µm [8, 9]. During application, a reduction of the dye leads to dissolution and sorption on the polymer matrix of the cellulose fiber. Dyestuff immobilization is achieved by reoxidation of the leuco-form into the insoluble vat dye. The cellulose fiber matrix prevents the formation of higher aggregates; thus the vat dye pigment in the dyed substrate exhibits substantially smaller particle size, which has been estimated from X-ray diffraction studies of dimensions in the range of 50–100 nm. This explains the substantial change in color as well as the increase in color depth which is observed between the application of the finely divided finished vat dye pigment and the color of the dyed material [8]. The re-aggregation of the oxidized dye and possible recrystallization leads to changes in particle size and explains the color change and stabilization of shade during hot after-treatment of dyed material at the boil in the soaping step.

In a similar manner, a substantial reduction in particle size occurs on dyeing of cellulose fibers with indigo dye. While the raw indigo vat dye exhibits particle size distribution between 1 and 40 µm, a substantial reduction in particle dimensions results during the reduction and reoxidation step in dyeing, which can be estimated to be below 1 µm [10, 11]. A well-known phenomenon in indigo dyeing is the unwanted red shade of the dyed yarn, which is explained with the deposition of larger particles at the surface of the dyed yarn, for example, at high indigo concentrations in the dyebath. As a result, the reddish shade is accompanied by reduced rub fastness of the dyeing. Due to the small particle size, this unwanted effect cannot be removed with an intensive wash and requires a reduction of the particles to the soluble leuco form or oxidative destruction of the particles. A similar effect called bronzing is observed in dyeing with sulfur dyes, when larger dyestuff particles are deposited at the surface of the material [12].

Reduction of particle size of the oxidized vat dye through appropriate ball milling in combination with ultrasonic treatment and addition of surfactants leads to generation of nanosized vat dye particles with diameter below 20 nm, which also could be applied for dyeing and printing experiments without the use of reducing agents [13]. Also the classical azoic dyes base on the formation of insoluble dyestuff particles, which are formed in situ by coupling of a diazo compound with an activated aromatic (naphthol component). The average particle diameter is below 1 µm. The high fastness of these particle-based colorants is based on their insolubility in aqueous solutions due to the absence of any solubilizing groups in the dye molecule. A general disadvantage of these dyes is their tendency to exhibit limited rub fastness, as weakly bound deposited dye particles can be removed from the fiber surface during abrasive treatment.

Another widely used application of microparticles is found in the use of pigment dyes in dyeing and screen printing [14]. Organic or inorganic colorants are milled to a particle size below 0.5 µm. These particle-based colorants require binder systems, for example, polyacrylate- or polyurethane-based polymers to attach the particles to the fiber surface. Substantially smaller average particle size is used in ink-jet printing of organic pigments, where nanoscaled pigments with diameter in the dimension of 50–200 nm are used [15, 16].

Nanoscaled and microscaled TiO_2 particles are used as additives in spinning of dull fibers. The primary effect of these particles is to achieve a uniform, nondirectional light scattering and to reduce the high gloss of many man-made fibers. While often the rutile modification is used preferentially, the photocatalytic activity of the anatase form has been used to design self-cleaning textiles. Unfortunately, two general drawbacks have to be considered: first, the effect is dependent on the irradiation; thus production of oxidants is dependent on the access of UV light; and second, the formation of reactive oxygen species (ROS) is not only selective for the oxidation of contaminants but also contributes to fiber degradation and color destruction.

Particle-based colorants are also incorporated into man-made fibers during the fiber-spinning process. As an example, color master batches that contain the finely divided pigment are added to the pristine polymer during fiber production by melt extrusion. Such processes are common for all standard synthetic fibers; however it can also be adapted to regenerated cellulose fibers. An example of the use of vat dye containing pulp has been proposed as color master batch for spun-dyed lyocell-fiber production [17].

Besides coloration through light absorbing dyes and pigments, color can also be developed through structural effects on the fiber surface. Chromatic dispersion, interference, diffraction and photonic crystals can lead to the development of colored surfaces [18]. Structural colors are of interest due to their potential to generate colors of high luminance and high saturation. In many cases, the observed color is angle-dependent, thus leading to an iridescent effect. Relevant techniques to generate structural color can be based on ink-jet printing, electrospinning, micro-nanostructuring and thin-film deposition via sol–gel deposition, chemical vapour deposition, physical vapour deposition and photonic crystal self-assembly [18]. On textiles, for example, polyester or silk fibers, structural colors have been developed by deposition and self-assembly of colloidal microspheres. Particle size of microspheres ranges from 50 to 400 nm. An example for color development by the formation of nanoscaled structures by the electroless deposition of Cu-based particles on sol–gel-modified fibers is also reported in the literature. The development of colored surfaces in this case is believed to occur via localized surface plasmon resonance [19]. To achieve technical relevance, these techniques have to be transferred on textile substrates with the use of efficient and productive processes, and the durability of the structures must be sufficient to resist abrasive

Table 5.2: A summary of particles applied in coloration.

Colorant	Modes of application and size effects
Vat dyes	Dye powders are milled to less than 0.12 µm particle size to ensure uniform dispersion and reduction in the dyebath. The dyestuff molecules form aggregates in the substrate, and a size of 50–100 nm is generally found optimal for the required color shades and depths. Alternatively, oxidized vat dyes may be employed that require no reducing agents in the course of dyeing. Here the oxidized dyestuff is milled to particles of size less than 20 nm.
Indigo	Indigo is perhaps the most widely recognized and used dyestuff from the vat category. In dyebaths, the particle sizes range between 1 and 40 µm, while in substrates, aggregate of less than 1 µm is generally found optimal. The aggregate size in substrates affects the shade as well as crock fastness (i.e. fastness to rubbing).
Pigment dyes	For use in dyeing or screen printing, the organic or inorganic colorants are milled to a particle size of less than 0.5 µm. For ink-jet printing, particle sizes of 50–200 nm are required.
Azoic colorants	These are two component dyestuff systems, where substrates are first impregnated with one component (the naphthol), and then treated with a second component (the base) in a following step. The azo coupling reaction that ensues yields dyestuff aggregates of less than 1 µm size.
Structural coloration	This concept utilizes the color generation from light interaction with micro and nanosized particles (not themselves colored) due to chromatic dispersion, interference, diffraction and photonic effects. The reported sizes range between 50 and 400 nm.

forces during use and maintenance, for example, laundry. A summary of the particles and their applications discussed above is available in Table 5.2.

5.3 Printing of functional microparticles

Printing techniques with binder systems allow the fixation of particulate material on the surface of a textile structure. By this method, specific functional properties can be developed for a certain application. Printing techniques permit one-sided application of an intended effect, for example, light reflection and also allow the formation of patterned structures, for example, stripes or labels. Representative examples are

Printing of highly reflective patterns for safety purposes and special optic appearance [20]. A representative example of materials used are mica platelets with thickness of a few micrometers and a mean particle size of 20 μm [21]. Special effects appear when the platelets are coated with thin layers of other material, for example, TiO_2, SiO_2 or SnO_2. In sports textiles, stripes and patterns of highly reflective patterns are used to increase the visibility of athletes at conditions of twilight. A similar approach is used in working clothes and clothing for rescue forces.

Printing of light or temperature sensitive dyes, fluorescent or phosphorescent dyes. As an example, a bi-component fiber with a PP-based core that contains fluorescent particles and a sheet layer, for example, from PES or PA has been recommended for working cloths, interior textiles and carpets [22].
Printing of conductive pastes/inks permits introduction of electrically conductive structures. Printing of IR-reflective coatings allows adjustment of the reflectance of textile surfaces for infrared radiation [23, 24]. Modification of the thermal insulation properties and improvement of wear comfort through deposition of IR reflecting pigments is of growing interest.

5.4 Microparticles and antimicrobial effects

A considerable number of nanoparticle-based material has been tested for antimicrobial properties, among them ZnO, Ag, Cu, Fe_3O_4, Al_2O_3, TiO_2, SiO_2 and chitosan [25]. Substantial differences have been observed with regard to the minimum inhibitory concentration as well as with regard to the mechanism of action as shown in Table 5.3. Low concentrations are observed for materials, which lead to an inhibition through formation of ROS, cause membrane disruption or interfere with DNA replication. Representative examples include ZnO, Ag and Cu. Comparably higher concentrations are required for Fe_3O_4 and TiO_2, which indicates that possibly also other mechanisms such as flocculation and adsorption contribute in addition to the effects mentioned above. Flocculation has been reported as a significant contribution to the antimicrobial effects of Al_2O_3, SiO_2 and chitosan. Besides particle size, positive zeta potential has also been discussed as relevant factor determining antimicrobial properties of the particles. The positive charge promotes the interaction with the cell membranes, membrane disruption and also supports flocculation effects [25].

An example for the combination of improved UV-protective properties and antibacterial function has been reported by use of a combination of nanostructural ZnO (60 nm) and TiO_2 (460 nm) on polyester and cotton textiles [26]. A particular mechanism for antimicrobial activity is observed with TiO_2 where visible light-induced effects lead to the formation of hydrogen peroxide. An increase in performance can be achieved through doping with metal ions, for example, Cu, Mn and Fe. The formed peroxide decomposes with the formation of ROS, which then lead to the high antibacterial activity. The presence of metal ions that catalyze the decomposition of hydrogen peroxide also contributes to the antimicrobial effect. Among the three naturally

Table 5.3: Particles and their antimicrobial efficacy [25].

Material	Particle size (nm)	Minimum inhibitory concentration (mg/L)
ZnO	12–60	40–80
Ag	12–20	0.3–75
Cu	100	30
Fe_3O_4	8–9	2,000–3,000
TiO_2	17	500
Al_2O_3	60	20–500
SiO_2	20	20
Chitosan	40	8

occurring forms of TiO_2: rutile, anatase and brookite; the anatase modification exhibits the highest photochemical activity [27]. Representative dimensions of such nanoparticles are between 10 and 50 nm.

A representative example for particle-based antimicrobial functionality which has been successfully transferred on larger scale bases on the use of insoluble Cu(I) compounds is Cu_2O. The low solubility product of Cu_2O of 2×10^{-15} makes the particles almost insoluble in water; however slow oxidative corrosion of the particles leads to a local release of Cu^{2+} ions, which then produce an antimicrobial effect in the immediate surrounding of the particulate material. The red particles exhibit an average diameter between 1 and 2 µm and are bound to native or synthetic fibers through conventional pigment binder system [1]. Through the preceding slow oxidation step, a retarded release of Cu^{2+} ions from the surface of the material is observed and antimicrobial effects are located to the surface of the textile structure, where the concentration of Cu-ions increases above the minimal inhibitory concentration.

The small particle size of the microparticles is a requirement for introduction into technical scale processes as settling of the small particles is slow enough to permit application in continuous pad/dry/cure operations on conventional machinery. As an example, Figure 5.1 shows the light photomicrographs of preformed micro Cu particles (a) and the Cu microparticles on fibers after the finishing step. The use of Cu-based microparticles for antimicrobial finishing of microfibr-based cleaning wipes has passed the level of pilot-scale production.

Tests for antimicrobial activity demonstrated a substantial reduction in viability for the standard test strains, *Staphylococcus aureus* and *Klebsiella pneumoniae* at a concentration about 150 mg Cu/kg of material. Usually the formation of the Cu_2O particles proceeds via reduction of Cu^{2+} salts or corresponding complexes in alkaline solution (Figure 5.2). Depending on the chemicals used and the experimental conditions applied, different shapes of the particles are achieved. When the reduction of Cu^{2+} is initiated by a thermal step on alkali-activated cellulose, the Cu_2O crystals can exhibit the form of microneedles with typical dimensions of 0.13 µm diameter and 1.6 µm length [28]. Gentle conditions permit a slow growth of the particles on cellulose fibers,

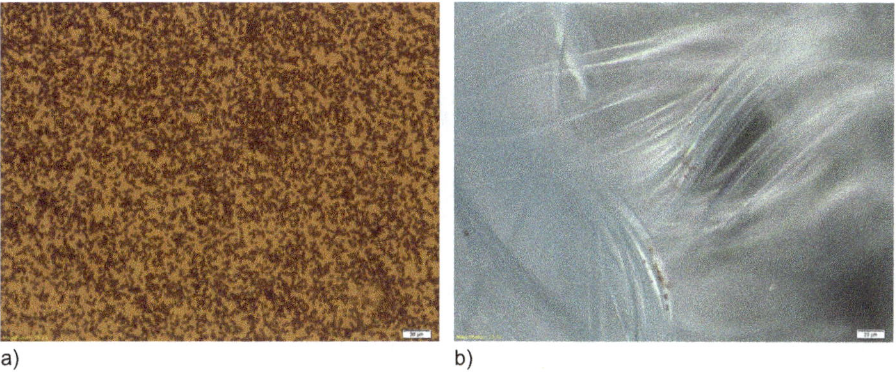

a) b)

Figure 5.1: Photomicrographs of Cu particles: (a) Light microscopy of particles and (b) Light microscopy of finished textile material.

which are then bound through Van der Waals forces. For practical application, however, higher particle density and improved attachment with the use of polymer binder systems are required to improve mechanical resistance [2].

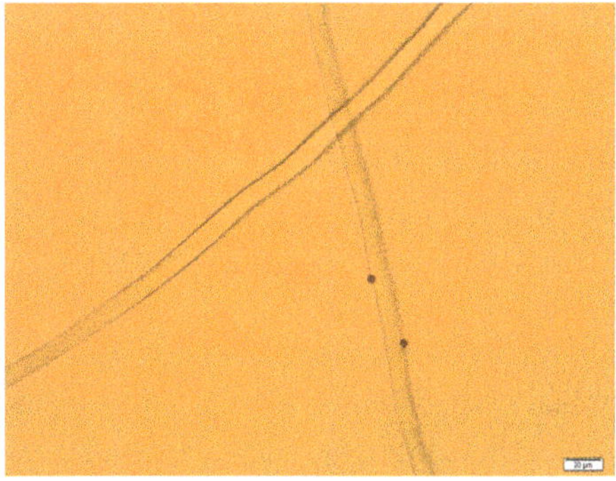

Figure 5.2: In situ formation of Cu_2O microparticles on cellulose fibers through reduction of Cu^{2+} ions.

The overall antimicrobial activity of a particle-based antimicrobial agent thus will be dependent on a number of conditions that determine the release of the active substance into the near environment. The particle size and form will determine the rate of dissolution, and in case of the Cu(I)-based system, also the rate of corrosion and oxidation into the more soluble Cu^{2+} form. Thus, besides the concentration of active antimicrobial agent in case of a particle-based system, the conditions of preparation of the particles and coating of the textile substrate will influence the overall activity of the agent. Sufficient permanency of the effect over a number of wash cycles will be important to achieve a controlled release. The deposited particles form a depot of antimicrobial agent, and rapid dissolution of the substance will lead to high activity during a short initial phase in use; however a rapid decrease in effect with time will be the consequence. The integration of Cu-based antimicrobial substances already has reached pilot-scale production for cleaning and disinfection wipes [29].

5.5 Silver

The most common application of silver particle technology in sports clothing is in the area of antimicrobial finishes. The antimicrobial effect of silver stems from both the metal and ions (Ag^+) released from the metal. The bulk metal adsorbs oxygen atoms, which in the presence of water can directly cause oxidative damage to microbial cells or indirectly through stable peroxide radicals formed from oxidation of other dissolved organic matter [30]. The antimicrobial action of Ag^+ is attributed to the inactivation of microbial enzymes, interference in cell propagation, damage to cell membrane function and the inducement of ROS [31–34]. Metal particles of nanometer dimensions, apart from releasing Ag^+, exert antimicrobial activity through the inducement of ROS in microbial cells [35] and have also been observed to penetrate the cell membranes [36, 37].

Another emerging area of application is in UV protection [38–42] as silver particles have been shown to protect human skin cells (keratinocytes) from the damaging effects of UVB radiation and are reported to function even better than zinc oxide and titanium dioxide particles. However, the focus in this section will be on their application as antimicrobial agents.

The aim of applying antimicrobial agents on clothing articles, such as for sports, is to protect the textile material against microbial degradation and to improve wearer comfort by preventing malodour development. This is demonstrated by the stated aims of commercially available technologies and products, selected examples of which are cited here [43–47].

The interaction between clothing and microbes is a function of the ability of microbes to adhere onto fiber surfaces (i.e. biofilm formation), fiber hydrophilicity, levels of unbound surface water and polymer susceptibility to hydrolysis by microbial enzymes [48, 49]. Natural fiber and regenerated cellulosics (e.g., cotton, wool and viscose rayon) are hydrophilic and exhibit rough surfaces that promote microbial adhesion, and the polymer structures are susceptible to hydrolysis by microbial enzymes. Thus, natural and regenerated fibers are prone to microbial degradation, which leads to strength loss and color changes. Animal hair fibers are found to be more susceptible to bacteria, while cellulosics more to fungi [50, 51]. Synthetic fibers (e.g., polyester and nylon) are hydrophobic and exhibit smooth surfaces that hinder microbial adhesion, and their polymer structures are not susceptible to enzymatic hydrolysis and therefore are not prone to microbial degradation. However, the poor moisture sorption propensities of synthetics allow for perspiration to pool in interstitial spaces, and the residues are susceptible to microbial colonization. The volatile by-products of microbial metabolism – carboxylic acids, amines and aldehydes – give rise to malodour [49, 52]. The general perception that malodour emanation is greater from synthetics as compared to natural and regenerated fibers is thought to result from differences in the microbial species enriched on one fiber type versus the other, and the ability of different fiber types to adsorb the volatile malodorous compounds [53–55].

The application of silver on textiles may be achieved by incorporation of the metal particles in manufactured fibers and filaments [56–59]. However, the antimicrobial functionality of silver requires the presence of water, and thus the efficacy of particle-incorporated products will depend on the degree to which the particles are accessible to moisture [57, 60, 61]. An alternative is to incorporate ion-exchange resin particles into manufactured fibers and then to load silver ions onto the fibers [62]. Silver particles may also be applied topically, as part of chemical finishing operations in textile processing. Two modes are discussed in the literature, one where textiles are treated with solutions of silver salts (e.g., nitrate, acetate and citrate) and the sorbed Ag^+ are reduced to metal particulates on the textile surface, and the other where textiles are treated with dispersions of silver metal particles [60, 63]. Reducing agents to convert Ag^+ into metal particles and stabilizing agents to prevent aggregation of the generated particles may be employed in both modes. The reducing agents commonly reported are $NaBH_4$, hydrazine, polysaccharides (e.g., chitosan and hydroxpropyl starch), glucose, plant extracts and fungi, and the stabilizing agents commonly reported are trisodium citrate and water-soluble polymers such as polyacrylates and polysaccharides (e.g., chitosan) [60, 63, 64]. Options are also discussed to avoid use of any chemical reducing/stabilizing agents such as employing sonolytic reduction or to use the textile material itself as both reducing and stabilizing agents [56, 65, 66].

A key issue with topical application of silver particles on clothing is their durability, that is, the duration over which effective antimicrobial activity is maintained and the impact of common maintenance operations, such as laundry. The common way reported for improving wash durability includes surface modifications of the textile surface to improve particle adhesion such as with plasma or other oxidative treatments; immobilization of particles with binders (e.g., polyacrylates and acrylate-based co-polymers), cross-linking agents and cross-linkable polymers; sol–gel finishing; and the use of bifunctional polymers that form covalent linkages with both the textile and the particles [64, 67–71]. But it should be noted that silver is a "leaching" type of agent [60], where the antimicrobial efficacy is proportional to the release of Ag^+ or particles into the immediate environment. Thus, improvement of the durability may have a negative influence on effectiveness. For illustration, wool fibers exhibit high binding capacity for Ag^+ and silver particles attributed to interactions with the disulphide groups and the nitrogen of amino acids [69, 72], and as a result, silver particle treatments of wool result in lower antimicrobial efficacy as compared to other fiber types [73].

Despite the long history of silver being used by humans (e.g., jewelry and utensils), its application on consumer products such as clothing, especially in the form of nanoparticles, has provoked investigations on their potential for causing harm either to humans or the environment. There are concerns on the possible effects of the sizes and shape of the nanoparticles as well as the chemical agents employed in their synthesis [63, 74, 75]. A transmission of nanoparticles through the skin is considered unlikely [76, 77], but the risk of their inhalation, especially from clothing with poor binding of particles, remains a matter of concern since silver nanoparticles exhibit the potential to cross the blood–brain barrier and cause neurotoxicity [75, 77–79]. There are concerns over their impact on flora and fauna, including to beneficial microorganisms such as nitrifying bacteria [75, 80, 81], but the potential for such environmental damage is unclear, as much remains unknown on the fate of nanoparticles in wastewater systems and the effect of environmental variables on their reactivity. The emergence of microbial strains exhibiting resistance to silver is also regarded as a potential threat that merits continuing investigation [61].

5.6 Microencapsulation

Microencapsulation describes the process where active substances (liquid, solid) are enveloped in polymeric shells, and the resulting particles have dimensions in the range of 1–1,000 µm [82]. The active substances are separate from the polymer matrix in microcapsules, and in this, they differ from "microparticles," which are defined to be polymeric particles of micrometric dimensions where the active substances are intermixed with the polymer matrix [83, 84]. A related, but separate category, is that of cyclodextrins, which are cyclic oligosaccharides employed to form host–guest inclusion complexes with active substance molecules [85–87]. The focus of discussion here is only on microencapsulation.

There are advantages to the application of encapsulated active substances on textiles as compared to their direct application [50, 82, 84, 88–91]:

- The encapsulation protects substances sensitive to the atmosphere (e.g., temperature, moisture, light and air oxygen) and hinders the evaporative loss of volatile substances. This improves the shelf life of products.
- Liquid substances may be turned into free-flowing powders and substances sensitive to textile processing conditions (e.g., acidity and alkalinity) may be protected. This improves the handling and processability of active substances.
- It enables a controlled and sustained delivery of functionality, the delivery may be targeted to a location or the particles may be engineered to deliver in

response to a trigger (e.g., pressure and temperature). Thus, desired levels of functionality may be achieved with lower amounts of active substances.

The concept of microencapsulation was first developed in the 1950s for carbonless copying paper [84], and from there its use has widened to pharmaceutics, laundry detergents, cosmetics, food formulations, coatings and so on. In clothing, microencapsulation technology is employed to incorporate antimicrobial agents, cosmetic agents, fragrances, insect repellents, polychromic colorants and phase change materials (PCMs) like paraffin in polystyrene microcapsules, as reflected in the descriptions of commercially available products [92–99]. An example of a commercial product is shown in Figure 5.3.

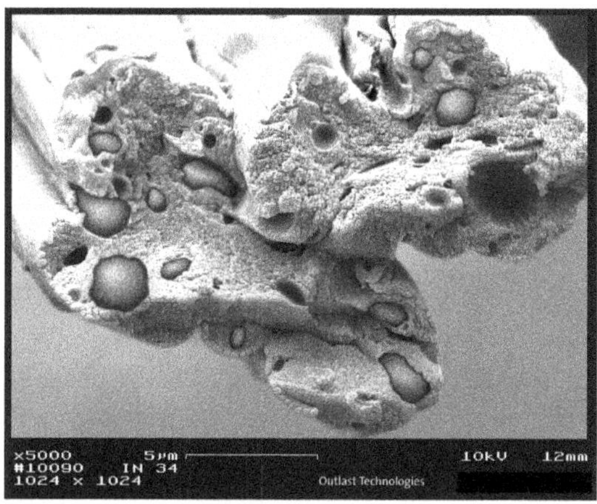

Figure 5.3: SEM micrograph of viscose fiber containing microencapsulated phase change materials. ©Outlast Technologies GmbH | www.outlast.com. Reproduced with permission.

The active substances employed for these purposes include plant extracts (essential oils and essences) for antimicrobial functionalities, fragrance, skin care or cosmetic applications, insect repellence [50, 88, 89, 100, 101]; inorganic salt hydrates and linear long chain hydrocarbons as PCMs for heat regulation (e.g., sodium sulfate decahydrate, manganese(II) nitrate hexahydrate, *n*-eicosane and *n*-octadecane) [102–104]; and synthetic organic substances for insect repellence (e.g., *N*,*N*-diethyl-*meta*-toluamide and permethrin). Some common methods of their microencapsulation for textile applications include the following:

Coacervation [82, 84, 86, 101, 105, 106]

Here, the active substance is dispersed/emulsified in a polymer solution, and then the polymer solubility is reduced so that it precipitates while enveloping the dispersed phase. If the active substance is dispersed in a solution of one polymer, that is termed "simple" coacervation, where the polymer solubility is reduced by adding electrolyte, changing the pH, adding a nonsolvent or reducing the temperature. If a second polymer is added to cause precipitation by interaction with the first polymer generally through electrostatic interactions, that is termed "complex" coacervation.

Examples of simple coacervation include the encapsulation in chitosan by the addition of NaOH or sodium tripolyphosphate, in sodium alginate by addition of calcium chloride and in gelatin by the addition of sodium sulfate. Examples of complex coacervation include the encapsulation in chitosan by mixing in gum arabic, silk fibroin or gelatin and in gelatin by mixing in gum arabic.

Interfacial polymerization [86, 88, 106]
The encapsulation proceeds through a polymerization reaction between two monomers, one of which is dissolved in the active substance. The mixture is dispersed/emulsified in a solution of the second monomer, and a polymerization reaction is induced between the monomers at the interface to trigger the encapsulation.

Examples include the encapsulation via reactions between isosorbide and methylene bis(phenyl isocyanate) monomers, acrylate and thiol monomers, succinyl chloride and 1,4 diaminobutane monomers, and acid chloride and an amine or alcohol monomers.

In situ polymerization [86, 88, 106]
The active substance is dispersed/emulsified in a solution of a polymer or a prepolymer, and then a cross-linking initiator or agent is added to the continuous phase to trigger the encapsulation. Examples include the encapsulation via formaldehyde-induced cross-linking of melamine, urea or cardanol.

A final step in the production of microcapsules is the hardening of capsule walls, which is usually accomplished by cross-linking with agents such as citric acid, glutaraldehyde or tripolyphosphate [50, 82, 88]. The application of microcapsules on textiles is accomplished by standard textile processing treatments such as padding, coating, exhaustion, screen printing, spraying and their fixation is achieved by techniques such as with a binder, cross-linking or plasma pretreatments [50, 82, 100]. The release of active substances is achieved by their diffusion through capsule walls or rupture or degradation of the walls through mechanical forces or change in the environment (e.g., temperature and pH) [50, 82, 100]. If photosensitive entities are incorporated in capsule walls, for example, carbon nanotubes or α-methyl stilbene and azobenzene groups, the release of active substances can be mediated by light irradiation [90, 107].

5.7 Superhydrophobicity

Nonwettability with water, or hydrophobicity, in sports clothing is desired to protect the wearer from weather elements and to enable their rapid drying. A relatively fresh concept that has emerged over recent years is that of "superhydrophobicity," which is linked to a growing interest in "biomimetics" or the desire to create artificial structures that mimic those in biologically occurring objects. A superhydrophobic surface is defined to be the one that exhibits a sessile water drop contact angle of greater than 150° and a contact angle hysteresis (i.e. difference between advancing and receding angles) of less than 10° [108–112]. In many instances, it is found easier to measure the sliding angle, shedding angle or tilting angle in place of contact angle hysteresis, and so they are often employed as substitute measures [110, 111].

The advantage cited of superhydrophobicity, especially for clothing, is self-cleaning which in principle reduces the need for laundry and thereby the resources and energy required for it [110]. The term self-cleaning describes the process wherein water drops rolling off a surface, collect and transport deposited dust and dirt particles with them. Termed the "lotus effect," it is named after the phenomenon occurring on lotus leaves which exhibit a water contact angle of 164° and contact angle hysteresis of 3° [111], that is attributed to the presence of hierarchical micro and nanometric protuberances (papillae), the spacing between them and a covering of hydrocarbon wax. Removing any one of these features, the wax or nanometric protuberances, is demonstrated significantly to impair their hydrophobic and self-cleaning behavior [110, 111].

Superhydrophobicity is modeled by a combination of the Wenzel and Cassie-Baxter modifications to the classical Young's model of wetting for an ideal surface [113, 114]. In essence, by introducing roughness (i.e. protuberances), the original hydrophilicity or hydrophobicity of surfaces becomes more intensified. And, the trapping of air at the solid–liquid interface, for example, in spaces between protuberances, increases contact angles and reduces wettability. Thus, a hydrophobic surface can be made more hydrophobic by increasing the roughness, and further, by increasing the degree of air entrapment at the solid–liquid interface.

To the best of the authors' knowledge, there are no commercial superhydrophobic clothing products available at the present time, and the information cited in this section are all from research literature. Further, it should be noted that measurements of self-cleaning are difficult. In addition to contact angles and contact angle hysteresis, the functionality is influenced by the propensity of surfaces for water "pinning," the kinetic energy of water drops, the pressure on water drops and the degree of adhesion of foreign particles on the solid surface [110, 111, 115]. Thus, in the research literature, it is commonly found that only water contact angles and measures of contact angle hysteresis are reported, and the self-cleaning functionality is not evaluated.

Generally, the treatments to impart superhydrophobicity to textiles involve the combination of applying a hydrophobicization agent and a process to introduce roughness. The roughness may be introduced by etching or by the deposition of particulate matter. However, mention must be made of work that shows sufficient degree of roughness for superhydrophobicity may be obtained by the engineering of textile construction parameters alone [110].

The hydrophobicization treatment generally involves the application of siloxane- or fluorocarbon-based agents, although it should be noted that perfluorooctanoic acids are disfavored due to their potential carcinogenicity [110]. The particle deposition for generation of roughness is achieved via the deposition or in situ generation of particulates from silver, zinc oxide, silica, titanium dioxide, carbon nanotubes, diamond-like carbon and so on with methods that include sol–gel treatments, immersion, coating, electrodeposition, electrospraying, layer-by-layer assembly, screen printing, chemical vapor deposition, pulsed laser deposition, dielectric barrier discharge and so on [56, 108–110, 116–128].

5.8 Conductive fibers through incorporation of carbon or metals

Two different groups of conductive materials should be recognized in the discussion of possible applications. Highly conductive material, for example, metallized fibers through deposition of copper in electroless coating [129] or the integration of conductive structures into the fabric (e.g., thin wires, carbon structures like graphene or metallized threads). Such materials can be used for the setup of electronic circuits for smart textiles applications, sensors and so on. The flexural rigidity, elasticity and density of metal wires are significantly higher than that of standard textile fibers and that can reduce wearer comfort. Hence they are blended (or mixed) with textile fibers to create hybrid yarns via techniques such as core spinning or wrap spinning. Other options include metal coating of textile fibers with elements such as tungsten and gold or the immobilization of conductive carbon such as carbon black or graphene The resistivity of these structures can be in the range of a few Ohms/centimeter [130]. Also printing of conductive inks containing electrically conductive particles, for example, silver or carbon permits formation of highly conductive structures, for example, for heating, electromagnetic shielding and sensor formation [131].

Medium level of conductivity is of interest in production of textile material with reduced static charge during use, which for example is of interest in production of sensitive electronics, for work in explosion-protected areas as well as for development of textile pressure sensors. Incorporation of conductive particles into the fiber is a widely used approach to obtain fibers with increased electrical conductivity, for example, achieved by incorporation of conductive organic polymers, carbon-based materials as well as metal particles.

A general restriction results from the incorporation of carbon and from the coating with conductive metal layers, which arises from the color of the modified material. For textile applications, usually a white material is the desirable basis, which allows coloration of the desired shade through dyeing and printing operations. The presence of an initial color limits this approach substantially. Thus, integration of conductive material has to consider this primary color or the conductive structures have to be positioned on inner layers of a garment.

5.9 Functionality through incorporation/deposition of particulates

For incorporation of functional additives into a certain fiber, the conditions applied during polymer shaping have to be considered. While for fibers from synthetic polymers, for example, polyester, polyamide or polypropylene the thermal stress has to be considered; chemical conditions applied during the viscose process or the NMMO-based lyocell production can limit the range of chemicals to be incorporated successfully.

In summer outdoor sports, for example, running, football, tennis and protection against UV irradiation from sunlight has gained rising attention to reduce risk of skin cancer. Reduction of UV transmittance can be achieved by selection of fiber material with low permeability, by use of denser material that exhibits a higher cover factor and by use of modified fibers or appropriate finishing operations. Incorporation of 0.1–1.5% nanoscaled pigments with high UV absorption in cellulose fibers permits the production of fibers with increased UV-protection factor [132]. Incorporation of 0.3 µm TiO_2 particles in synthetic fibers, for example, PES fibers leads to an increase in UVPF due to light scattering effects. Incorporation of silica into regenerated cellulose fibers leads to fibers with reduced flammability. The skeleton formed by SiO_2 hinders afterglow and flame propagation (Figure 5.4). Compared to other chemical additives or finishing concepts, the approach leads to fibers that do not contain a high load of chemicals and thus exhibit a low risk of irritation and will not release toxic decomposition products during burn. Thus, such fibers are promising for use in textiles, which come near to skin, for example, bedding textiles or underwear. An example from pilot-scale trials on a viscose product under development is shown in Figure 5.5.

A possible route to achieve increased thermal insulation capacity of textiles is the incorporation of particulate materials with low thermal conductivity into the fiber and textile structure. Besides the conventional layered approach used in outdoor garment, also incorporation of insulating material, for example, cork, in nonwoven structures has been shown to reduce thermal conductivity of materials which could be used for interior applications [133]. The effectiveness of incorporating insulating materials into

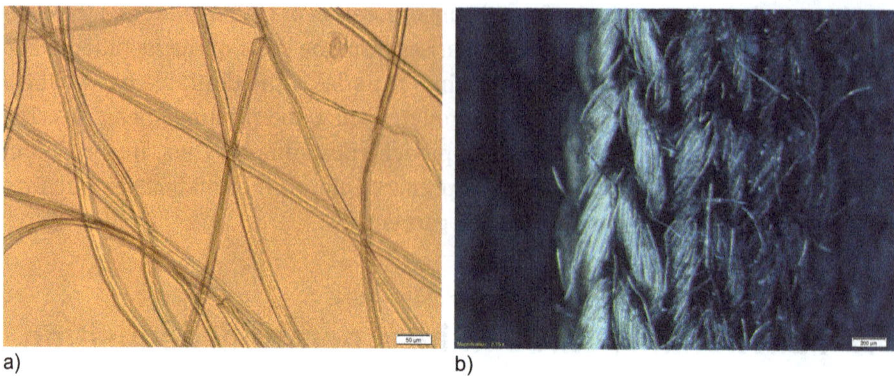

a) b)

Figure 5.4: (a) Pristine viscose fiber with incorporated silica and (b) residual white SiO_2 structure remaining after combustion of a blue-dyed knitted textile. After combustion, only the white skeleton consisting of SiO_2 can be seen in the original shape of the burnt fibers (b).

Figure 5.5: Photograph from a test on a SiO_2-incorporated viscose product under development, of ignitability from a smouldering cigarette (ISO 12952–1:2010).

fibers is limited compared to the insulation effects achieved by use of layers of bulky lightweight structures, for example, down, webs from crimped fibers and foams. However, incorporation of functional particles into fibers allows a modification of the reflectance properties for NIR irradiation and thus helps improve the wear comfort of garment.

5.10 Summary

In this chapter, specific technologies and innovations for the introduction of micro and nanoparticles in sports clothing and textiles for different functionalities are reviewed. We have attempted to maintain focus on those processes that are commercial, in that products are available, but have also discussed some technologies that are still under development. Some of the challenges specific to the textile-processing sector have been highlighted that need consideration when upscaling of laboratory-scale results to commercial-scale production. This has been an active area of research investigations for some time and promises to continue for much longer.

A myriad range of end applications are targeted with the introduction of micro and nanoparticles. They range from comfort (e.g., aroma-releasing microcapsules) to establishing electronic circuitry on a non-conductive substrate (e.g., for the integration of sensors), to reduce environmental impact of clothing use (e.g., by reducing the frequency of launderings), human health (e.g., by reducing UV transmission) and so on. It is therefore difficult to envisage the future scope of applications that may be targeted. However, with rising concerns over the trends in consumption and disposal of clothing there is increasing focus on "circularity," that is, recycling of fibers and polymers from waste clothing back into the manufacture of new textiles and clothing. For that to succeed, attention needs to be devoted on minimizing the use of any treatment or agent that will hinder recyclability. Already, discussions have emerged in the literature on circularity of "smart textiles," which are characterized by the integration of electronics and circuitry in textiles [134, 135]. Thus, a future trend that may be expected is efforts to design processes and technologies where micro and nanoparticles incorporated into textile substrates may easily be included in recycling technologies or be separated from the textile wastes.

References

[1] Turalija, M.; Merschak, P.; Redl, B.; Griesser, U.; Duelli, H.; Bechtold, T.: Copper(i)oxide microparticles-synthesis and antimicrobial finishing of textiles. Journal of Materials Chemistry B, 2015; 3(28). 10.1039/c5tb01049g.

[2] Emam, H. E.; Manian, A. P.; Široká, B.; Duelli, H.; Merschak, P.; Redl, B.; Bechtold, T.: Copper (I)oxide surface modified cellulose fibers – Synthesis, characterization and antimicrobial properties. Surface & Coatings Technology, 2014; 254. 10.1016/j.surfcoat.2014.06.036.

[3] European Parliament and Council: Concerning the making available on the market and use of biocidal products, in Regulation (EU) No 528/2012, European Union, Editor, 2012. http://data.europa.eu/eli/reg/2012/528/oj

[4] European Parliament and Council: Community code relating to medicinal products for human use. In: Directive 2001/83/EC, European Union, Editor, 2001. https://eur-lex.europa.eu/eli/dir/2001/83/oj

[5] Bundesministers für Land- und Forstwirtschaft Umwelt und Wasserwirtschaft: Verordnung des Bundesministers für Land- und Forstwirtschaft, Umwelt und Wasserwirtschaft über die Begrenzung von Abwasseremissionen aus der Textilveredelung und Behandlung (AEV Textilveredelung und -behandlung). In: BGBl. II Nr. 269/2003, 2003. https://www.ris.bka.gv.at/GeltendeFassung.wxe?Abfrage=Bundesnormen&Gesetzesnummer=20002744

[6] Hudson, J.; Khazragui, H. F.: Into the valley of death: Research to innovation. Elsevier Current Trends, 2013; 610–613. 10.1016/j.drudis.2013.01.012.

[7] Blaylock, A.; Constantin, F.; Ligabue, L.; Bocaletti, L.; Siroka, B.; Siroky, J.; Wright, T.; Bechtold, T.: Caregiver's vision of bedding textiles for elderly. Fashion and Textiles, 2015; 2(1), 6. 10.1186/s40691-015-0029-6.
[8] Valko, E. I.: Particle size in the vat dyeing of cellulose. Journal of the American Chemical Society, 1941; 63(5), 1433–1437. 10.1021/ja01850a083.
[9] Shim, W. S.; Lee, J. J.; Shamey, R.: An approach to the influence of particle size distribution of leuco vat dye converted by a reducing agent. Fibers and Polymers, 2006; 7(2), 164–168. 10.1007/BF02908261.
[10] Nicholson, S. K.; John, P.: The mechanism of bacterial indigo reduction. Applied Microbiology and Biotechnology, 2005; 68(1), 117–123. 10.1007/s00253-004-1839-4.
[11] Campos, R.; Kandelbauer, A.; Robra, K. H.; Cavaco-Paulo, A.; Gübitz, G. M.: Indigo degradation with purified laccases from Trametes hirsuta and Sclerotium rolfsii. Journal of Biotechnology, 2001; 89(2), 131–139. 10.1016/S0168-1656(01)00303-0.
[12] Rath, H.: Lehrbuch der Textilchemie. Berlin, Heidelberg: Springer-Verlag, 1963. 10.1007/978-3-662-00064-9.
[13] Hakeim, O. A.; Nassar, S. H.; Raghab, A. A.; Abdou, L. A. W.: An approach to the impact of nanoscale vat coloration of cotton on reducing agent account. Carbohydrate Polymers, 2013; 92(2), 1677–1684. 10.1016/j.carbpol.2012.11.035.
[14] Bechtold, T.; Pham, T.: Textile Chemistry. Berlin/Boston: Walter de Gruyter GmbH, 2019. 10.1515/9783110549898.
[15] Bermel, A. D.; Bugner, D. E.: Particle size effects in pigmented ink jet inks. Journal of Imaging Science and Technology, 1999; 43(4), 319–324.
[16] Daplyn, S.; Lin, L.: Evaluation of pigmented ink formulations for jet printing onto textile fabrics. Pigment and Resin Technology, 2003; 32(5), 307–318. 10.1108/03699420310497454.
[17] Manian, A. P.; Ruef, H.; Bechtold, T.: Spun-dyed lyocell. Dyes and Pigments, 2007; 74(3), 519–524. 10.1016/j.dyepig.2006.03.015.
[18] Huang, M.; Lu, S.-G.; Ren, Y.; Liang, J.; Lin, X.; Wang, X.: Structural coloration and its application to textiles: A review. The Journal of the Textile Institute, 2020; 111(5), 756–764. 10.1080/00405000.2019.1663623.
[19] Landsiedel, J.; Root, W.; Schramm, C.; Menzel, A.; Witzleben, S.; Bechtold, T.; Pham, T.: Tunable color and conductivity by electroless growth of Cu/Cu2Onanoparticles on sol-gel modified cellulose. Nano Research. 10.1007/s12274-020-2907-5.
[20] Malm, V.: Functional Textile Coatings Containing Flake-shaped Fillers: Investigations on selected optical and electrical properties, University of Borås, Sweden, 2018. https://www.diva-portal.org/smash/get/diva2:1183782/FULLTEXT01.pdf
[21] Malm, V.; Strååt, M.; Walkenström, P.: Effects of surface structure and substrate color on color differences in textile coatings containing effect pigments. Textile Research Journal, 2014; 84(2), 125–139. 10.1177/0040517513485626.
[22] Tsutsumi, H.; Takahira, A.: Light-storing conjugate fibers with durable light-storing properties comprising conjugate fibers having a polyolefin core containing light-storing particles and a sheath comprising fiber-forming polyamides or polyesters.
[23] Khajeh Mehrizi, M.; Mortazavi, S. M.; Mallakpour, S.; Bidoki, S.: The effect of nano- and micro-TiO2 particles on reflective behavior of printed cotton/nylon fabrics in Vis/NIR regions. Color Research and Application, 2012; 37. 10.1002/col.20675.
[24] Abbasipour, M.; Khajeh Mehrizi, M.: Investigation of changes of reflective behavior of cotton/polyester fabric by TiO 2 and carbon black nanoparticles. Scientia Iranica, 2012; 19(3), 954–957. 10.1016/j.scient.2012.04.016.

[25] Seil, J. T.; Webster, T. J.: Antimicrobial applications of nanotechnology: Methods and literature. International Journal of Nanomedicine, 2012; 7, 2767–2781. 10.2147/IJN.S24805.
[26] Sójka-Ledakowicz, J.; Lewartowska, J.; Kudzin, M.; Jesionowski, T.; Siwińska-Stefańska, K.; Krysztafkiewicz, A.: Modification of textile materials with micro-and nano-structural metal oxides. Fibres and Textiles in Eastern Europe, 2008; 16(5), 112–116.
[27] Liao, C.; Li, Y.; Tjong, S. C.: Visible-light active titanium dioxide nanomaterials with bactericidal properties. Nanomaterials, 2020; 10(1). 10.3390/nano10010124.
[28] Emam, H. E.; Ahmed, H. B.; Bechtold, T.: In-situ deposition of Cu2O micro-needles for biologically active textiles and their release properties. Carbohydrate Polymers, 2017; 165, 255–265. 10.1016/j.carbpol.2017.02.044.
[29] Bechtold, T.; Intemann, W.: Putz-, Wasch- und/oder Poliersubstrat bestehend aus textilen und/oder vlieseartigen und/oder schwammartigen Strukturen zur Reinigung von Oberflächen beliebiger Art mit antimikrobiellenEigenschaften, DE102015001669A1. 2016.
[30] Davies, R. L.; Etris, S. F.: The development and functions of silver in water purification and disease control. Catalysis Today, 1997; 36(1), 107–114. 10.1016/S0920-5861(96)00203-9.
[31] Batarseh, K. I.: Anomaly and correlation of killing in the therapeutic properties of silver (I) chelation with glutamic and tartaric acids. Journal of Antimicrobial Chemotherapy, 2004; 54(2), 546–548. 10.1093/jac/dkh349.
[32] Edwards-Jones, V.: The benefits of silver in hygiene, personal care and healthcare. Letters in Applied Microbiology, 2009; 49(2), 147–152. 10.1111/j.1472-765X.2009.02648.x.
[33] Park, H.-J.; Kim, J. Y.; Kim, J.; Lee, J.-H.; Hahn, J.-S.; Gu, M. B.; Yoon, J.: Silver-ion-mediated reactive oxygen species generation affecting bactericidal activity. Water Research, 2009; 43(4), 1027–1032. 10.1016/j.watres.2008.12.002.
[34] Percival, S. L.; Bowler, P. G.; Russell, D.: Bacterial resistance to silver in wound care. Journal of Hospital Infection, 2005; 60(1), 1–7. 10.1016/j.jhin.2004.11.014.
[35] Flores-López, L. Z.; Espinoza-Gómez, H.; Somanathan, R.: Silver nanoparticles: Electron transfer, reactive oxygen species, oxidative stress, beneficial and toxicological effects. Mini review. Journal of Applied Toxicology, 2019; 39(1), 16–26. 10.1002/jat.3654.
[36] Morones, J. R.; Elechiguerra, J. L.; Camacho, A.; Holt, K.; Kouri, J. B.; Ramírez, J. T.; Yacaman, M. J.: The bactericidal effect of silver nanoparticles. Nanotechnology, 2005; 16(10), 2346–2353. 10.1088/0957-4484/16/10/059.
[37] Zhang, L.; Wu, L.; Si, Y.; Shu, K.: Size-dependent cytotoxicity of silver nanoparticles to Azotobacter vinelandii: Growth inhibition, cell injury, oxidative stress and internalization. PLOS ONE, 2018; 13(12), e0209020. 10.1371/journal.pone.0209020.
[38] Arora, S.; Tyagi, N.; Bhardwaj, A.; Rusu, L.; Palanki, R.; Vig, K.; Singh, S. R.; Singh, A. P.; Palanki, S.; Miller, M. E.; Carter, J. E.; Singh, S.: Silver nanoparticles protect human keratinocytes against UVB radiation-induced DNA damage and apoptosis: Potential for prevention of skin carcinogenesis. Nanomedicine: Nanotechnology, Biology and Medicine, 2015; 11(5), 1265–1275. 10.1016/j.nano.2015.02.024.
[39] Palanki, R.; Arora, S.; Tyagi, N.; Rusu, L.; Singh, A. P.; Palanki, S.; Carter, J. E.; Singh, S.: Size is an essential parameter in governing the UVB-protective efficacy of silver nanoparticles in human keratinocytes. BMC Cancer, 2015; 15(1), 636. 10.1186/s12885-015-1644-8.
[40] Tyagi, N.; Srivastava, S. K.; Arora, S.; Omar, Y.; Ijaz, Z. M.; Al-Ghadhban, A.; Deshmukh, S. K.; Carter, J. E.; Singh, A. P.; Singh, S.: Comparative analysis of the relative potential of silver, Zinc-oxide and titanium-dioxide nanoparticles against UVB-induced DNA damage for the prevention of skin carcinogenesis. Cancer Letters, 2016; 383(1), 53–61. 10.1016/j.canlet.2016.09.026.
[41] Li, R.; Yang, J.; Xiang, C.; Song, G.: Assessment of thermal comfort of nanosilver-treated functional sportswear fabrics using a dynamic thermal model with human/clothing/

environmental factors. Textile Research Journal, 2018; 88(4), 413–425. 10.1177/0040517516679147.

[42] Ouadil, B.; Cherkaoui, O.; Safi, M.; Zahouily, M.: Surface modification of knit polyester fabric for mechanical, electrical and UV protection properties by coating with graphene oxide, graphene and graphene/silver nanocomposites. Applied Surface Science, 2017; 414, 292–302. 10.1016/j.apsusc.2017.04.068.

[43] DuPont and Dow: How Silvadur™ Antimicrobial Technology Works. [cited 2020, 29 April]. http://silvadur.dupont.com/how-it-works/

[44] Thomson Research Associates Inc.: Silpure Silver Antimicrobial Fabric Treatment. [cited 2020, 29 April]. https://ultra-fresh.com/silpure-silver-antimicrobial-fabric-treatment/

[45] Sciessent LLC: Smarter Antimicrobial Technology. [cited 2020, 29 April]. https://www.sciessent.com/wp-content/uploads/2019/07/Sciessent-Agion-Product-Sheet-R4-sm.pdf

[46] F Group Nano: How SmartSilver Works. [cited 2020, 29 April]. http://www.smartsilver.com/how

[47] Switzerland, S.: active>silver™. [cited 2020 07 May]. https://www.schoeller-textiles.com/en/technologies/activesilver

[48] Sun, G.: 17 – Antimicrobial Finishes for Improving the Durability and Longevity of Fabric Structures. In: Sun, G., editor: Antimicrobial Textiles. Cambridge, England: Woodhead Publishing, 2016; pp. 319–336. 10.1016/B978-0-08-100576-7.00017-1.

[49] Teufel, L.; Schuster, K. C.; Merschak, P.; Bechtold, T.; Redl, B.: Development of a fast and reliable method for the assessment of microbial colonization and growth on textiles by DNA quantification. Journal of Molecular Microbiology and Biotechnology, 2008; 14(4), 193–200. 10.1159/000108657.

[50] Yip, J.; Luk, M. Y. A.: 3 – Microencapsulation Technologies for Antimicrobial Textiles. In: Sun, G., editor: Antimicrobial Textiles. Cambridge, England: Woodhead Publishing, 2016; pp. 19–46. 10.1016/B978-0-08-100576-7.00003-1.

[51] Hearle, J. W. S.; Lomas, B.; Cooke, W. D.: 41 – Introduction. In: Hearle, J. W. S.; Lomas, B.; Cooke, W. D., editors: Atlas of Fibre Fracture and Damage to Textiles, 2nd ed. Boca Raton, USA: Woodhead Publishing Ltd, 1998; pp. 377–381. 10.1533/9781845691271.8.377.

[52] Thilagavathi, G.; Viju, S.: 16 – Antimicrobials for Protective Clothing. In: Sun, G., editor: Antimicrobial Textiles. Cambridge, England: Woodhead Publishing, 2016; pp. 305–317. 10.1016/B978-0-08-100576-7.00016-X.

[53] Callewaert, C.; De Maeseneire, E.; Kerckhof, F.-M.; Verliefde, A.; Van de Wiele, T.; Boon, N.: Microbial odor profile of polyester and cotton clothes after a fitness session. Applied and Environmental Microbiology, 2014; 80(21), 6611–6619. 10.1128/aem.01422-14.

[54] Abdul-Bari, M. M.; McQueen, R. H.; Nguyen, H.; Wismer, W. V.; de la Mata, A. P.; Harynuk, J. J.: Synthetic clothing and the problem with odor: Comparison of nylon and polyester fabrics. Clothing and Textiles Research Journal, 2018; 36(4), 251–266. 10.1177/0887302x18772099.

[55] McQueen, R. H.; Laing, R. M.; Brooks, H. J. L.; Niven, B. E.: Odor intensity in apparel fabrics and the link with bacterial populations. Textile Research Journal, 2007; 77(7), 449–456. 10.1177/0040517507074816.

[56] Rivero, P. J.; Urrutia, A.; Goicoechea, J.; Arregui, F. J.: Nanomaterials for functional textiles and fibers. Nanoscale Research Letters, 2015; 10(1), 1–22. 10.1186/s11671-015-1195-6.

[57] Morais, D. S.; Guedes, R. M.; Lopes, M. A.: Antimicrobial approaches for textiles: From research to market. Materials, 2016; 9(6), 498. 10.3390/ma9060498.

[58] Wendler, F.; Meister, F.; Montigny, R.; Michael, W.: A new antimicrobial ALCERU® fibre with silver nanoparticles. FIBRES & TEXTILES in Eastern Europe, 2007; 64, 41–45.

[59] Chen, J. Y.; Sun, L.; Jiang, W.; Lynch, V. M.: Antimicrobial regenerated cellulose/nano-silver fiber without leaching. Journal of Bioactive and Compatible Polymers, 2015; 30(1), 17–33. 10.1177/0883911514556960.

[60] Simoncic, B.; Klemencic, D.: Preparation and performance of silver as an antimicrobial agent for textiles: A review. Textile Research Journal, 2016; 86(2), 210–223. 10.1177/0040517515586157.

[61] Maillard, J.-Y.; Hartemann, P.: Silver as an antimicrobial: Facts and gaps in knowledge. Critical Reviews in Microbiology, 2013; 39(4), 373–383. 10.3109/1040841X.2012.713323.

[62] Büttner, R.; Markwitz, H.; Knobelsdorf, C.: ALCERU® silver – A new ALCERU® Fibre with versatile application potential. Lenzinger Berichte, 2006; 85, 131–136.

[63] Gubala, V.; Johnston, L. J.; Liu, Z.; Krug, H.; Moore, C. J.; Ober, C. K.; Schwenk, M.; Vert, M.: Engineered nanomaterials and human health: Part 1. Preparation, functionalization and characterization (IUPAC Technical Report). Pure and Applied Chemistry, 2018; 90(8), 1283. 10.1515/pac-2017-0101.

[64] Radetic, M.: Functionalization of textile materials with silver nanoparticles. Journal of Materials Science, 2013; 48(1), 95–107. 10.1007/s10853-012-6677-7.

[65] Harifi, T.; Montazer, M.: A review on textile sonoprocessing: A special focus on sonosynthesis of nanomaterials on textile substrates. Ultrasonics Sonochemistry, 2015; 23, 1–10. 10.1016/j.ultsonch.2014.08.022.

[66] Emam, H. E.; Manian, A. P.; Široká, B.; Duelli, H.; Redl, B.; Pipal, A.; Bechtold, T.: Treatments to impart antimicrobial activity to clothing and household cellulosic-textiles – Why "Nano"-silver?. Journal of Cleaner Production, 2013; 39, 17–23. 10.1016/j.jclepro.2012.08.038.

[67] Xu, Q. B.; Xie, L. J.; Diao, H.; Li, F.; Zhang, Y. Y.; Fu, F. Y.; Liu, X. D.: Antibacterial cotton fabric with enhanced durability prepared using silver nanoparticles and carboxymethyl chitosan. Carbohydrate Polymers, 2017; 177, 187–193. 10.1016/j.carbpol.2017.08.129.

[68] Xu, Q.; Wu, Y.; Zhang, Y.; Fu, F.; Liu, X.: Durable antibacterial cotton modified by silver nanoparticles and chitosan derivative binder. Fibers and Polymers, 2016; 17(11), 1782–1789. 10.1007/s12221-016-6609-2.

[69] Syafiuddin, A.: Toward a comprehensive understanding of textiles functionalized with silver nanoparticles. Journal of the Chinese Chemical Society, 2019; 66(8), 793–814. 10.1002/jccs.201800474.

[70] Zhang, Y. Y.; Xu, Q. B.; Fu, F. Y.; Liu, X. D.: Durable antimicrobial cotton textiles modified with inorganic nanoparticles. Cellulose, 2016; 23(5), 2791–2808. 10.1007/s10570-016-1012-0.

[71] Shahid Ul, I.; Butola, B. S.; Mohammad, F.: Silver nanomaterials as future colorants and potential antimicrobial agents for natural and synthetic textile materials. RSC Advances, 2016; 6(50), 44232–44247. 10.1039/C6RA05799C.

[72] Masri, M. S.; Friedman, M.: Effect of chemical modification of wool on metal ion binding. Journal of Applied Polymer Science, 1974; 18(8), 2367–2377. 10.1002/app.1974.070180815.

[73] Denning, R.: Enhancing Wool Products Using Nanotechnology. In: Johnson, N. A. G.; Russell, I. M., editors: Advances in Wool Technology. Cambridge, England: Woodhead Publishing Ltd., 2009; pp. 248–264. 10.1533/9781845695460.2.248.

[74] Souza, L. R. R.; da Silva, V. S.; Franchi, L. P.; Jorge de Souza, T. A.: Toxic and Beneficial Potential of Silver Nanoparticles: The Two Sides of the Same Coin. In: Saquib, Q.; et al., editors: Cellular and Molecular Toxicology of Nanoparticles. Springer Cham, Switzerland, 2018; pp. 251–262. 10.1007/978-3-319-72041-8_15.

[75] Pulit-Prociak, J.; Stoklosa, K.; Banach, M.: Nanosilver products and toxicity. Environmental Chemistry Letters, 2015; 13(1), 59–68. 10.1007/s10311-014-0490-2.

[76] von Goetz, N.; Lorenz, C.; Windler, L.; Nowack, B.; Heuberger, M.; Hungerbuhler, K.: Migration of Ag- and TiO2-(nano)particles from textiles into artificial sweat under physical

stress: Experiments and exposure modeling. Environmental Science & Technology, 2013; 47(17), 9979–9987. 10.1021/es304329w.

[77] Gubala, V.; Johnston, L. J.; Krug, H. F.; Moore, C. J.; Ober, C. K.; Schwenk, M.; Vert, M.: Engineered nanomaterials and human health: Part 2. Applications and nanotoxicology (IUPAC Technical Report). Pure and Applied Chemistry, 2018; 90(8), 1325–1356. 10.1515/pac-2017-0102.

[78] Menzel, M.; Fittschen, U. E. A.: Total reflection X-ray fluorescence analysis of airborne silver nanoparticles from fabrics. Analytical Chemistry, 2014; 86(6), 3053–3059. 10.1021/ac404017u.

[79] Strużyńska, L.; Skalska, J.: Mechanisms Underlying Neurotoxicity of Silver Nanoparticles. In: Saquib, Q.; et al., editors: Cellular and Molecular Toxicology of Nanoparticles. Springer International Publishing Cham, 2018; pp. 227–250. 10.1007/978-3-319-72041-8_14.

[80] Donia, D. T.; Carbone, M.: Fate of the nanoparticles in environmental cycles. International Journal of Environmental Science and Technology, 2019; 16(1), 583–600. 10.1007/s13762-018-1960-z.

[81] Lai, R. W. S.; Yeung, K. W. Y.; Yung, M. M. N.; Djurišić, A. B.; Giesy, J. P.; Leung, K. M. Y.: Regulation of engineered nanomaterials: Current challenges, insights and future directions. Environmental Science and Pollution Research, 2018; 25(4), 3060–3077. 10.1007/s11356-017-9489-0.

[82] Massella, D.; Giraud, S.; Guan, J.; Ferri, A.; Salaun, F.: Textiles for health: A review of textile fabrics treated with chitosan microcapsules. Environmental Chemistry Letters, 2019; 17(4), 1787–1800. 10.1007/s10311-019-00913-w.

[83] Nordstierna, L.; Abdalla, A. A.; Nordin, M.; Nydén, M.: Comparison of release behaviour from microcapsules and microspheres. Progress in Organic Coatings, 2010; 69(1), 49–51. 10.1016/j.porgcoat.2010.05.003.

[84] Meirowitz, R.: Microencapsulation Technology for Coating and Lamination of Textiles. In: Smith, W. C., editor: Smart Textile Coatings and Laminates. Boca Raton, USA: Woodhead Publishing Ltd., 2010; pp. 125–154. 10.1533/9781845697785.2.125.

[85] Buschmann, H. J.; Dehabadi, V. A.; Wiegand, C.: Medical, Cosmetic and Odour Resistant Finishes for Textiles. In: Paul, R., editor: Functional Finishes for Textiles. Improving Comfort, Performance and Protection. Amsterdam, Netherlands: Woodhead Publishing Ltd., 2015; pp. 303–330. 10.1533/9780857098450.1.303.

[86] Nelson, G.: Microencapsulated Colorants for Technical Textile Application. In: Gulrajani, M. L., editor: Advances in the Dyeing and Finishing of Technical Textiles. Oxford, England: Woodhead Publishing Ltd., 2013; pp. 78–104. 10.1533/9780857097613.1.78.

[87] Persico, P.; Carfagna, C.: Cosmeto-textiles: State of the art and future perspectives. Advances in Science and Technology, 2013; 80, 39–46. 10.4028/www.scientific.net/AST.80.39.

[88] Bruyninckx, K.; Dusselier, M.: Sustainable chemistry considerations for the encapsulation of volatile compounds in laundry-type applications. ACS Sustainable Chemistry & Engineering, 2019; 7(9), 8041–8054. 10.1021/acssuschemeng.9b00677.

[89] Van Langenhove, L.; Paul, R.: Insect Repellent Finishes for Textiles. In: Paul, R., editor: Functional Finishes for Textiles. Improving Comfort, Performance and Protection. Amsterdam, Netherlands: Woodhead Publishing Ltd., 2015; pp. 333–360. 10.1533/9780857098450.2.333.

[90] Ilatje, C. P.; Gumi, T.; Valls, R. G.: Emerging application of vanillin microcapsules. Physical Sciences Reviews, 2015; 1(4), 77–95. 10.1515/9783110331998-006.

[91] Van Parys, M.: Smart Textiles Using Microencapsulation Technology. In: Ghosh, S. K., editor: Functional Coatings: By Polymer Microencapsulation. Weinheim, Germany: Wiley-VCH Verlag GmbH & Co. KGaA, 2006; pp. 221–258. 10.1002/3527608478.ch7.

[92] Insilico Co. Ltd.: Functional Microcapsules, Insilico, 2020. [cited 2020 08 May]. http://www.insilico.co.kr/source/microcapsule_brochure_eng.pdf
[93] Devan: Devan & Textiles that's Chemistry, 2020. [cited 2020 08 May]. http://devan.net/
[94] smartessences: Smart Fabrics: Applications, 2020. [cited 2020 08 May]. https://www.smartessences.com/smart-fabrics-applications/
[95] Outlast Technologies LLC: Technology, 2020. [cited 2020 08 May]. http://www.outlast.com/en/technology/
[96] Celessence Technologies: What is Celessence™ Technology? 2020. [cited 2020 08 May]. http://www.celessence.com/the-science/
[97] TANATEX Chemicals B.V.: Find the Product You Need, 2020. [cited 2020 08 May]. https://tanatexchemicals.com/products/?filter_product-process=cosmetic-finishes#mk-archive-products
[98] Lipotec SAU: QUIOSPHERES® Cosmetotextile Microcapsules, 2020. [cited 2020 08 May]. https://www.lipotec.com/en/products/quiospheres-reg-cosmetotextile-microcapsules/
[99] Schoeller Switzerland: Optimum Temperature Equalization, 2020. https://www.schoeller-textiles.com/en/technologies/schoeller-pcm
[100] Ghayempour, S.; Montazer, M.: Micro/nanoencapsulation of essential oils and fragrances: Focus on perfumed, antimicrobial, mosquito-repellent and medical textiles. Journal of Microencapsulation, 2016; 33(6), 497–510. 10.1080/02652048.2016.1216187.
[101] Xiao, Z.; Liu, W.; Zhu, G.; Zhou, R.; Niu, Y.: A review of the preparation and application of flavour and essential oils microcapsules based on complex coacervation technology. Journal of the Science of Food and Agriculture, 2014; 94(8), 1482–1494. 10.1002/jsfa.6491.
[102] He, F.: Study on manufacturing technology of phase change materials and smart thermo-regulated textiles. Advanced Materials Research, 2013; 821–822, 130–138. 10.4028/www.scientific.net/AMR.821-822.130.
[103] Mondal, S.: Phase change materials for smart textiles – An overview. Applied Thermal Engineering, 2008; 28(11–12), 1536–1550. 10.1016/j.applthermaleng.2007.08.009.
[104] Iamphaojeen, Y.; Manian, A. P.; Wright, T.; Caven, B.; Bechtold, T.; Siriphannon, P.: Polyelectrolyte-assisted immobilization of oil-based nanocapsules on cotton fabric. Australian Journal of Chemistry, 2016; 69(7), 811–816. 10.1071/CH15746.
[105] Timilsena, Y. P.; Akanbi, T. O.; Khalid, N.; Adhikari, B.; Barrow, C. J.: Complex coacervation: Principles, mechanisms and applications in microencapsulation. International Journal of Biological Macromolecules, 2019; 121, 1276–1286. 10.1016/j.ijbiomac.2018.10.144.
[106] Peng, H.; Zhang, D.; Ling, X.; Li, Y.; Wang, Y.; Yu, Q.; She, X.; Li, Y.; Ding, Y.: n-Alkanes phase change materials and their microencapsulation for thermal energy storage: A critical review. Energy & Fuels, 2018; 32(7), 7262–7293. 10.1021/acs.energyfuels.8b01347.
[107] Tylkowski, B.; Giamberini, M.; Underiner, T.: Photosensitive Microcapsules. In: Giamberini, M.; Prieto, S. F.; Tylkowski, B., editors: Microencapsulation. Berlin, Germany: Walter de Gruyter GmbH, 2015; pp. 1–18. 10.1515/9783110331998-003.
[108] Ramaratnam, K.; Iyer, S. K.; Kinnan, M. K.; Chumanov, G.; Brown, P. J.; Luzinov, I.: Ultrahydrophobic textiles using nanoparticles: Lotus approach. Journal of Engineered Fibers and Fabrics, 2008; 3(4), 1–14. 10.1177/155892500800300402.
[109] Duta, L.; Popescu, A. C.; Zgura, I.; Preda, N.; Mihailescu, I. N.: Wettability of Nanostructured Surfaces. In: Aliofkhazraei, M., editor: Wetting and Wettability. London, United Kingdom : IntechOpen Limited, 2015; pp. 207–252. 10.5772/60808.
[110] Park, S.; Kim, J.; Park, C. H.: Superhydrophobic textiles: Review of theoretical definitions, fabrication and functional evaluation. Journal of Engineered Fibers and Fabrics, 2015; 10(4), 1–18. 10.1177/155892501501000401.

[111] Nishimoto, S.; Bhushan, B.: Bioinspired self-cleaning surfaces with superhydrophobicity, superoleophobicity, and superhydrophilicity. RSC Advances, 2013; 3(3), 671–690. 10.1039/C2RA21260A.

[112] Gao, L.; McCarthy, T. J.: Teflon is hydrophilic. comments on definitions of hydrophobic, shear versus tensile hydrophobicity, and wettability characterization. Langmuir, 2008; 24(17), 9183–9188. 10.1021/la8014578.

[113] Seo, K.; Kim, M.; Kim, D. H.: Validity of the equations for the contact angle on real surfaces. Korea-Australia Rheology Journal, 2013; 25(3), 175–180. 10.1007/s13367-013-0018-5.

[114] Banerjee, S.: Simple derivation of Young, Wenzel and Cassie-Baxter equations and its interpretations. arXiv Condensed Matter, 2008. https://arxiv.org/abs/0808.1460

[115] Melki, S.; Biguenet, F.; Dupuis, D.: Hydrophobic properties of textile materials: Robustness of hydrophobicity. The Journal of the Textile Institute, 2019; 110(8), 1221–1228. 10.1080/00405000.2018.1553346.

[116] Zhu, T.; Li, S.; Huang, J.; Mihailiasa, M.; Lai, Y.: Rational design of multi-layered superhydrophobic coating on cotton fabrics for UV shielding, self-cleaning and oil-water separation. Materials & Design, 2017; 134, 342–351. 10.1016/j.matdes.2017.08.071.

[117] Tian, N.; Wei, J.; Li, Y.; Li, B.; Zhang, J.: Efficient scald-preventing enabled by robust polyester fabrics with hot water repellency and water impalement resistance. Journal of Colloid and Interface Science, 2020; 566, 69–78. 10.1016/j.jcis.2020.01.067.

[118] Li, G.; Mai, Z.; Shu, X.; Chen, D.; Liu, M.; Xu, W.: Superhydrophobic/superoleophilic cotton fabrics treated with hybrid coatings for oil/water separation. Advanced Composites and Hybrid Materials, 2019; 2(2), 254–265. 10.1007/s42114-019-00092-w.

[119] Xiong, M.; Ren, Z.; Liu, W.: Fabrication of UV-resistant and superhydrophobic surface on cotton fabric by functionalized polyethyleneimine/SiO2 via layer-by-layer assembly and dip-coating. Cellulose, 2019; 26(16), 8951–8962. 10.1007/s10570-019-02705-5.

[120] Sohbatzadeh, F.; Farhadi, M.; Shakerinasab, E.: A new DBD apparatus for super-hydrophobic coating deposition on cotton fabric. Surface & Coatings Technology, 2019; 374, 944–956. 10.1016/j.surfcoat.2019.06.086.

[121] Li, J.; Wu, X.; Jiang, P.; Li, L.; He, J.; Xu, W.; Li, W.: A facile method to fabricate durable super-hydrophobic cotton fabric. Journal of Vinyl & Additive Technology, 2020; 26(1), 3–9. 10.1002/vnl.21709.

[122] Ahmed, H. M.; Abdellatif, M. M.; Ibrahim, S.; Abdellatif, F. H. H.: Mini-emulsified copolymer/silica nanocomposite as effective binder and self-cleaning for textiles coating. Progress in Organic Coatings, 2019; 129, 52–58. 10.1016/j.porgcoat.2019.01.002.

[123] Sohbatzadeh, F.; Eshghabadi, M.; Mohsenpour, T.: Controllable synthesizing DLC nano structures as a super hydrophobic layer on cotton fabric using a low-cost ethanol electrospray-assisted atmospheric plasma jet. Nanotechnology, 2018; 29(26), 265603/1–265603/10. 10.1088/1361-6528/aabdae.

[124] Wang, H.; Li, W.; Li, Z.: Preparation of fluorinated PCL porous microspheres and a super-hydrophobic coating on fabrics via electrospraying. Nanoscale, 2018; 10(39), 18857–18868. 10.1039/c8nr05793a.

[125] Mai, Z.; Xiong, Z.; Shu, X.; Liu, X.; Zhang, H.; Yin, X.; Zhou, Y.; Liu, M.; Zhang, M.; Xu, W.; Chen, D.: Multifunctionalization of cotton fabrics with polyvinylsilsesquioxane/ZnO composite coatings. Carbohydrate Polymers, 2018; 199, 516–525. 10.1016/j.carbpol.2018.07.052.

[126] Zhou, C.; Li, Y.; Jin, X.; He, Y.; Xiao, C.; Wang, W.: Highly hydrophobic conductive polyester fabric based on homogeneous coating surface treatment. Polymer-Plastics Technology and Materials, 2019; 58(3), 246–254. 10.1080/03602559.2018.1466178.

[127] Guo, R.; Lan, J.; Pen, L.; Jiang, S.; Yan, W.: Microstructure and hydrophobic properties of silver nanoparticles on amino-functionalized cotton fabric. Materials Technology. Advanced Performance Materials, 2016; 31(3), 139–144. 10.1179/1753555715Y.0000000032.

[128] Nateghi, M. R.; Dehghan, S.; Shateri-Khalilabad, M.: A facile route for fabrication of conductive hydrophobic textile materials using N-octyl/N-perfluorohexyl substituted polypyrrole. International Journal of Polymeric Materials and Polymeric Biomaterials, 2013; 62(12), 648–652. 10.1080/00914037.2013.769167.

[129] Root, W.; Aguiló-Aguayo, N.; Pham, T.; Bechtold, T.: Conductive layers through electroless deposition of copper on woven cellulose lyocell fabrics. Surface & Coatings Technology, 2018; 348, 13–21. https://doi.org/10.1016/j.surfcoat.2018.05.033

[130] Pham, T.; Bechtold, T.: Conductive Fibers. In: Hu, J.; Kumar, B.; Lu, J., editors: Handbook of Fibrous Materials. Weinheim, Germany: Wiley-VCH Verlag GmbH & Co. KGaA., 2020; pp. 233–262. 10.1002/9783527342587.ch9.

[131] Root, W.; Wright, T.; Caven, B.; Bechtold, T.; Pham, T.: Flexible textile strain sensor based on copper-coated lyocell type cellulose fabric. Polymers (Basel), 2019; 11(5), 784. 10.3390/polym11050784.

[132] Clemens, B.; Andreas, G.; Dobson, P.; Kämpf, K.; Schuster, C.; Kroner, G.: Ultraviolet Protective Fabrics Based on Man-made Cellulosic Fibers. WO2010144925A1. 23.12.2010.

[133] Carvalho, R.; Fernandes, M.; Fangueiro, R.: The influence of cork on the thermal insulation properties of home textiles. Procedia Engineering, 2017; 200(December), 252–259. 10.1016/j.proeng.2017.07.036.

[134] Schischke, K.; Nissen, N. F.; Schneider-Ramelow, M.: Flexible, stretchable, conformal electronics, and smart textiles: Environmental life cycle considerations for emerging applications. MRS Communications, 2020; 10(1), 69–82. 10.1557/mrc.2019.157.

[135] Chae, Y.; Hinestroza, J.: Building circular economy for smart textiles, smart clothing, and future wearables. Materials Circular Economy, 2020; 2(1), 2. 10.1007/s42824-020-00002-2.

Mark Pätzel*, Kyra Scheerer, Thomas Gries

6 Applications in agricultural textiles

Technical textiles

Keywords: agricultural textiles, silicate particles, repulsive agents, insect repellent

The world population is projected to increase to 9 billion people in the next 20 years [8]. This directly results in an increased demand for nourishments. The global food security is already unsustainable and is further endangered by increasing droughts and insect plagues. If agricultural productivity remains constant, this means that the area under cultivation would have to cover about 80% of all arable land worldwide, i.e. seven billion hectares, by 2050 to ensure food security for these people without crop protection. The area includes primeval forests and inhabited regions. This extrapolation shows that in order to increase agricultural productivity urgent protective efforts must be made to combat weather conditions and insect plagues. This is the only way to ensure the world food security in 2050. Agricultural textiles can contribute to such protective measures. The application of agricultural textiles can improve the cultivation conditions and increase the crop yield. Especially the regional food crises caused by insect plagues such as the migratory locust in East Africa, and local droughts such as in Pakistan show the acute importance of food security [9]. That is why the sector of agricultural textiles is among those with the strongest growth predictions. The agrotextile market is estimated to grow by about 8–10% per annum in developing countries and by about 3.9% per year in developed countries [6, 11].

6.1 General overview

The use of agricultural textiles has several advantages. On the one hand, the use of pesticides and herbicides required to maximize yields is reduced, as is the amount of water needed to ripen the fruit. On the other hand, depending on the design, textile structure or material, agricultural textiles can offer protective functions against various environmental influences such as UV radiation or low temperatures. In addition, the use of agricultural textiles facilitates the harvest.

As fiber material, both synthetic and natural fibers are used for agricultural textiles. As a general rule, the fiber material is selected depending on the function and

***Corresponding author: Mark Pätzel,** Institut für Textiltechnik of RWTH Aachen University, Otto-Blumenthal-Straße 1, 52074 Aachen, Germany, e-mail: mark.paetzel@ita.rwth-aachen.de
Kyra Scheerer, Thomas Gries, Institut für Textiltechnik of RWTH Aachen University, Otto-Blumenthal-Straße 1, 52074 Aachen, Germany

https://doi.org/10.1515/9783110670776-006

application. Synthetic, man-made fibers have the advantage that the fiber properties can be adjusted to a certain extent. Natural fibers are selectively used in the appropriate application fields. Therefore, synthetic fibers have greater commercial utility due to the adjustable product properties and the wide range of applications [6].

The most commonly used fiber material is polypropylene (PP). This is achieved by various properties of the material, such as low weight (17–60 g/m^2), high elasticity and good transmission of sunlight (80–94%). The preferred natural fibers are jute, coir, wool, sisal and hemp. Natural fibers are used due to their biodegradability and are often used as mulch mats in the form of needle fleeces. Thus, they serve as a natural fertilizer. In addition, natural fibers such as wool, jute, coir, sisal, hemp and flax are used for their moisture retaining and wet strength properties. Synthetic fibers, such as nylon, polyester, polyethylene and, as already mentioned, PP, are used because of their high strength, lightweight, long life and space-saving storage and, above all, low price [6, 11, 12].

Applications of agricultural textiles can be divided into the areas of forestry, landscaping, agriculture and fishery. In the following, the focus will be on agricultural textiles, as they are mainly used for food production. The areas of plant protection in the agricultural sector are as follows:

- Protection at low temperatures and frost
- Protection against UV radiation
- Protection against hail
- Protection against grazing
- Protection against wind
- Protection against weeds

For hail protection, leno[1] fabrics made of monofilaments are used in the open field (Figure 6.1a). Nonwovens are often used for early harvesting and frost protection. The production is carried out by the melt-blown process[2] at low cost for use in the open field (Figure 6.1b). In warmer regions, shading materials such as sun protection nets to protect plants against excessive UV radiation are used. They are manufactured as knitted fabrics from multifilaments and fiber ribbons. Shading fabrics can be used in the greenhouse or in the open field (Figure 6.1c). Depending on the species of animals, various textiles made of mono- and multifilaments are used for protection against animal feeding. For example, a very open knitted fabric is used to protect against birds and a fine mesh knitted fabric against insects. The textile is used outdoors and in greenhouses (Figure 6.1d). Protection against wind is used to stabilize the plants. Depending on how much wind reduction is to be achieved,

1 Leno weave fabric (also called gauze weave or cross weave) is a weave in which two warp yarns are twisted around the weft yarns to provide a strong and durable yet sheer fabric.

2 Melt blowing is a conventional fabrication method of microfibers and nanofibers, where a polymer melt is extruded through small nozzles surrounded by high-speed blowing gas.

Table 6.1: Areas of plant protection in the agricultural sector.

Plant protection	Textile structure	Fiber structure	Field of application	In Figure 6.1
Against hail	Leno fabric	Monofilament	Open field	a
Against frost	Nonwoven	–	Open field	b
Against UVradiation	Knitted fabric	Multifilament, fiber ribbon	Greenhouse, open field	c
Against grazing	Woven fabric, knitted fabric	Multifilament, monofilament	Greenhouse, open field	d
Against wind	Woven fabric, knitted fabric	Fiber ribbon, monofilament	Greenhouse, open field	e
Against weeds	Woven fabric	Fiber ribbon	Greenhouse, open field	f

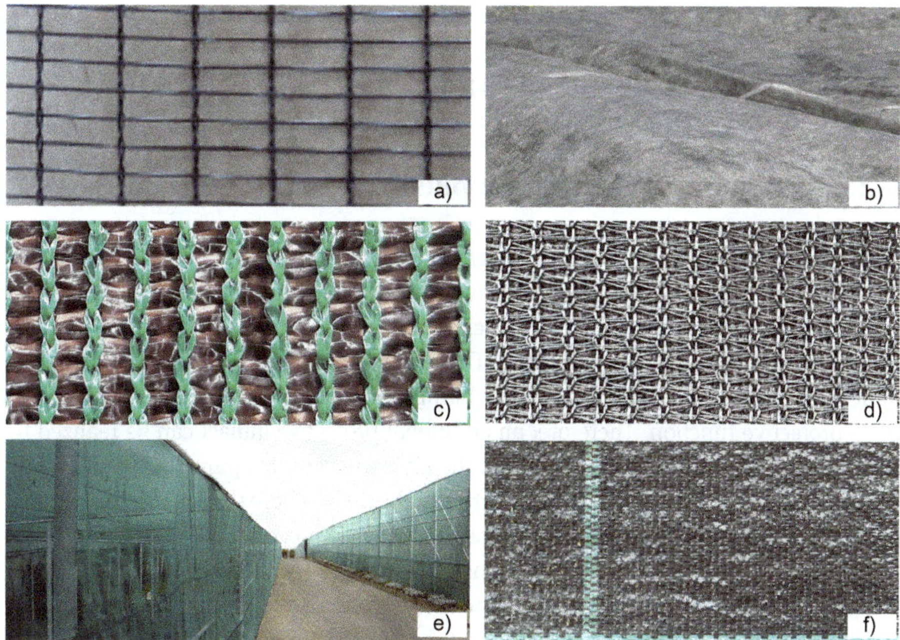

Figure 6.1: Examples of agricultural textiles, from top left to bottom right: (a) hail protection net, (b) early harvest nonwoven, (c) sun protection net, (d) insect protection, (e) wind protection and (f) soil weed protection [7].

dense fabrics or more open knitted fabrics made of monofilaments or fiber ribbons are used. Here too, there are applications in open spaces or in the greenhouse (Figure 6.1e). Finally, protective textiles against weeds are also produced. For soil protection, a dense fabric of ribbons is used, which allows water to run off but minimizes weed growth. In the greenhouse, white fabrics are used that reflect light back to the plant. Outdoor black fabrics are used because they allow less UV radiation to pass into the soil, thus inhibiting growth (Figure 6.1f).

6.2 Insect-repellent fabrics

In order to provide these textiles with an insect-repellent effect, they can be functionalized with different substrates. Insecticides can be dispersed from the textile itself or they can unfold their lethal effect by releasing pheromones in combination with traps in or on the textile. The application of repulsive substances is also possible. Another solution is coating with silicate particles to selectively integrate toxic properties on the textile. Two approaches are currently being pursued:
- Aversive approach with repulsive agents
- Dehydration approach with silicate particles

6.2.1 Aversive approach with repulsive agents

The mechanical protective function of agricultural textiles is advantageous and mostly desirable, but not for all applications that are sufficient. Often the individual filaments are damaged by environmental influences, such as installation and deinstallation, extreme weather conditions or browsing. The dysfunctional integrity of the agricultural structure causes a potential intrusion of harmful insects and thus a considerable loss of its protective function. Therefore, an additional protective impact can be realized by the insertion of repulsive substances into the filaments. For this purpose, several repulsive substances can be considered. Currently, besides ethereal oils from plants and herbs, synthetic substances are also in use. It should be noted that the respective repellents are most specifically effective against one or more insect species (Table 6.2). For example, the ethereal oil produced from nettles is particularly effective against aphids.

In addition to the specific selection of the required repulsives, there is another aspect to be considered, the long-term effect. A continuous release of the inserted repulsive substance over at least one season is necessary to ensure the additional protective function for the entire time period. For this purpose, the repulsives must be embedded in a closed structure from which they can diffuse out. Thus, the active ingredients are released very slowly, and a long-term effect is realized. One possibility for this continuous release is the use of biodegradable fibers. For this purpose, the repulsives are

Table 6.2: Herbal-based repulsive substances.

Herbal basis for the ethereal oil	Effective against	Herbal basis for the ethereal oil	Effective against
Basil	Mildew, white flies	Radish	Leek moth
Savory	Black aphids	Marigold	Nematodes, cabbage white
Nettle	Aphids	Rosemary	Cabbage white
Oats	Lice	*Salvia*	Cabbage white
Onion	Spider mites, carrot flies	*Celery*	Caterpillars
Chervil	Lice	*Tagetes*	Nematodes, white flies

compounded into the biodegradable polymer fibers and mixed with conventional polymer fibers. During the breakdown of biodegradable fibers, the compounded repulsives are continuously released. The stability of the agricultural textile is still maintained by the conventional fibers. For the production of melt-spun filaments with a repulsive agent, temperature resistance is a decisive limiting factor. The process temperatures vary depending on the polymer used. Tests have shown that nettle extract is thermally stable up to 25 °C, neem tree oil up to 220 °C and hemp plant extract up to 250 °C.

In order to investigate the effectiveness of the repulsive agent, in addition to field tests under real conditions, preliminary investigations on a smaller scale must be carried out. By these preliminary investigations it is possible to find out first tendencies with less time and money expenditure than with field tests. For these investigations of the aversive effect of individual repulsives on, for example, insects, so-called olfactometer tests are carried out (Figure 6.2).

For the examination, 30 insects are placed in the glass and their behavior is observed and noted. In the left, Y-arm is the repellent, and in the right Y-arm is the reference sample. Only the insects that are behind the evaluation boundary are counted. Due to the light irritation of most insects, the test environment should be darkened and before and after the use of a scent source, the olfactometer should be rinsed in hexane, acetone or dichloromethane [4].

2.2 Dehydration approach with silicate particles

In addition to cultivation in open fields, the use of greenhouses for the production of food is particularly common. The largest area of cultivation for the European market for greenhouse vegetables is the region around the Spanish city of Almeria.

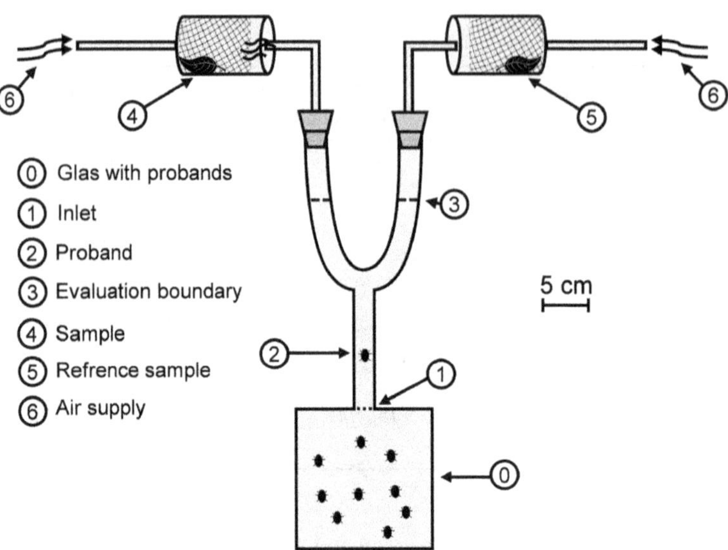

Figure 6.2: Schematic structure of an olfactometer.

Here, 3 million tons of vegetables are grown annually on an area of 36,000 hectares. This provides space for 40,000 greenhouses. In these greenhouses, 30% of the German vegetables grow on an annual average. In winter, however, the average is much higher, up to 50%. Most of the rest of our vegetables also come from greenhouses in other Mediterranean regions with similar growing conditions. In the greenhouses, primarily yield-optimized plants are grown in monoculture to increase productivity. The soil is covered with light-reflecting textiles to promote the ripening process. The warm and humid greenhouse climate not only stimulates the growth of the plants but also promotes the growth of parasites. Since greenhouses are equipped with ventilation openings to prevent mold, for example, windows and movable hatches, which take up approx. 30% of the greenhouse area, the invasion of parasites cannot be avoided. Through yield-optimized breeding, the plants increasingly lose their natural protection so that they have little chance of resisting the pests. In order to ensure that the integrity and appearance of the vegetables does not deteriorate, the increased parasite infestation is combated with the use of insecticides when growing these monocultures. In the cultivation of potatoes, for example, it is possible to double production by using pesticides. The pesticides initially act against the insects which however build up resistances with prolonged exposure. This leads to an increased application of the pesticides and requires a variation of the insecticide use. This results in a vicious circle of resistance build-up and the use of more and stronger toxins. This trend can be seen from the amount of insecticide used in Spain. For example, it increased by 10% in the period from 2013 to 2016.

An alternative is the approach of dehydrating the insects. For this purpose, the agricultural textile is provided with a coating of silicate particles. Silicate particles have the advantage that they are not toxic to humans or other mammalians and are also permitted for agricultural use [2]. The coated agricultural textile ensures that the harmful insects are killed but does not cause the insect to build up resistance. The surface structure of the amorphous silicate particles is highly porous. The porosity ensures a dehydration effect of the particle. In case of contact between the coated agricultural textile and an insect, the particles destroy the thin wax layer of the insect either by abrasion or adsorption or by a combination of both [10]. The wax layer normally surrounds the chitinous exoskeleton and has among other purposes the function to form a water barrier and to protect the insect from water loss. At the contact area, the protective function of the wax film is no longer given (Figure 6.3). The consequence is a water loss of the insect, which leads to dehydration and thus to death of the insect [1]. This method is therefore effective against many harmful arthropods, such as beetles, flies, moths, cockroaches, lice, mites and ants, which have a protective wax layer [5, 3].

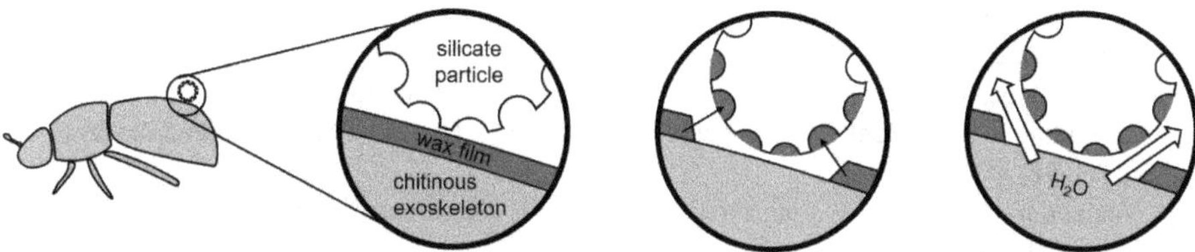

Figure 6.3: Mechanism of amorphous silicate particles [7].

6.3 Conclusion and outlook

In conclusion, there are two promising approaches to the use of particles in agriculture: firstly, the use of repulsive substances to repel insect pests; secondly, the use of highly porous silicate particles to dehydrate insect pests with a chitinous shell. Both approaches can be combined with a variety of agricultural textiles and reduce the need for the use of insecticides. As a result, those approaches can contribute to more environmentally friendly and higher yielding production of various agricultural goods. Two factors will play a decisive role, particularly as a result of the continuing increase in the world's population and the advancing climate crisis.

Research and development tasks for agrotextiles in the future will involve the study of recyclability, the study of polymer degradability and the increased use of biomaterials. These developments are not only important for the agricultural applications but also for aquaculture, horticulture and floriculture. This agrotextile market has a market volume of 13.0 billion USD € and a compound annual growth rate of 4.7% [13]. At the same time, it is necessary to further investigate and improve the knowledge gained on insect protection in order to provide more targeted and comprehensive protection of agricultural products. In parallel, the additivation of commercial polymers such as PP with particles to protect them from UV radiation, hail and weathering [14] is currently already in use and would have to be reexamined and redesigned as a result of the use of biomaterials.

References

[1] Adler, C.; Frielitz, C.; Günther, J.: Kieselgur gegen vorratsschädliche Insekten im Getreidelager Landbauforschung Völkenrode. Schwerpunkt: Pflanze Sonderheft, 2007; 314, p. 33.

[2] Beck, H.-J.: Sicherheitsdatenblatt (93/112/EG), Biobeck PA 910 (2010)

[3] Umwelt- & Gesundheitsschutz Stadt Zürich Silikat-Staub zur Bekämpfung von Hausschädlingen, 2012.

[4] Belz, E.; Kölliker, M.; Balmer, O.: Olfactory attractiveness of flowering plants to the parasitoid Microplitis mediator: Potential implications for biological control. BioControl, DOI:10.1007/s10526-012-9472-0, 2013, p. 166.

[5] Islam, M. S.; Rahman, M. M.: Diatomaceous earth-induced alterations in the reproductive attributes in the housefly *Musca domestica* L. (Diptera: Muscidae). Elixir Applied Zoology, 2016; 96, 41941–41944.

[6] Annapoorani, G.: Agro Textiles and Its Applications. Woodhead Publishing India in Textiles, WPI Publishing, 2018.

[7] Orth, M.: Innovative Agrartextilien Dissertationsvortrag. Aachen: Institut für Textiltechnik of RWTH Aachen, 2020.

[8] URL:https://de.statista.com/statistik/daten/studie/1717/umfrage/prognose-zur-entwicklung-der-weltbevoelkerung, access: 15 October 2020.

[9] URL: https://www.wfp.org/publications/2020-global-report-food-criseshttps://de.statista.com/statistik/daten/studie/1717/umfrage/prognose-zur-entwicklung-der-weltbevoelkerung/, access: 15 October 2020.

[10] Quarles, W.: Diatomaceous earth for pest control IPM practitioner, monitoring the field of pest management. The IPM Practitioner, Vol. XIV, No. 5/6, May/June, p.

[11] Kumar, R. S.: Textile is a boon agriculture. IRE Journals, 2017; Volume 1 Issue 2, ISSN: 2456-8880, p. 74. Studie von David Rigby Associates.

[12] Demšar, A.; Žnidarčič, D.; Gregor-Svetec, D.: Primerjava lastnosti polipropilenskih vlaken, namenjenih za izdelavo vrtnarskih vlaknovin. Acta agriculturae Slovenica, Volume 93, Issue 2, 2009, p. 211–217.

[13] URL: https://www.grandviewresearch.com/press-release/global-agro-textiles-market, access: 26 May 2021.

[14] URL: https://www.fortunebusinessinsights.com/agro-textiles-market-102963, access: 26 May 2021.

Kira Heins, Martin Scheurer*

7 Particle applications in technical textiles in construction

Keywords: membrane construction, textile facades, lightweight construction, textile reinforced concrete, fiber reinforcement bars, insulation, geotextiles, fire protection, fiber–matrix–bond strength, antimicrobial agent

One main field of application for technical textiles is the building and construction sector. In 2010, approximately 7% of the worldwide turnover in technical textiles was attributable to this "BuildTech" sector [1]. In the coming decades, architecture and the building industry will undergo lasting changes due to the necessary reduction of resource usage and construction costs as well as other ongoing megatrends such as urbanization. Innovative composite materials support the resulting demands of architects and constructors for greater flexibility and height of buildings and structures. Technical textiles in construction cover a wide range of applications with specific demands on the textile configuration. Particles can be used to adapt the properties of the textiles accordingly, as the examples shown in the following sections highlight.

7.1 Facade and membrane systems

One immediately visible application of textiles in construction is in textile-based membrane and facade systems. Both rely on the full exploitation of the flexibility and lightweight of textiles as well as their extensive design possibilities.

7.1.1 Membrane construction

Membrane systems are widely used in modern architecture, for example, for stadiums, sport arenas and exhibition tents (e.g., the Olympiastadion in Berlin). Their main advantages are the possibility to create structural spans of up to 60 m and more while weighing around 1% of conventional alternatives, as well as their flexibility

***Corresponding author: Martin Scheurer,** Institut für Textiltechnik of RWTH Aachen University, Otto-Blumenthal-Straße 1, 52074 Aachen, e-mail: martin.scheurer@ita.rwth-aachen.de
Kira Heins, Institut für Textiltechnik of RWTH Aachen University, Otto-Blumenthal-Straße 1, 52074 Aachen

https://doi.org/10.1515/9783110670776-007

and adjustability to different environments and requirements [2–4]. A membrane system consists of two components: the support structure and the membrane itself. The systems carry over their curved surface. The stronger the curvature, the lower the resulting membrane forces, and the more favorable the load-bearing behavior of the overall structure. To achieve a biaxial stress transfer via the membrane, it is prestressed on all sides [5].

A membrane can be a coated or uncoated fabric, a thin plastic film or even a metal sheet. One of the more commonly used materials is polyester in the form of a coated woven fabric. Alternatively, polyamide, polypropylene, polyethylene and glass are also used as membrane materials. To ensure long-term durability of the membrane, protection of the fabric against environmental stresses is necessary. These environmental stresses include UV radiation, abrasion, humidity, temperature and microorganisms [4]. The effect of these stresses is strongly dependent on the material used. To achieve protection and prevent material aging, the membranes are commonly coated using polymers such as polyvinylchloride (PVC), polytetrafluoroethylene (PTFE), polyvinylidene difluoride (PVDF) and silicone [4]. In addition, the coating allows the membranes to be welded. Polyester fabrics are usually finished with a PVC coating [4]. Acrylic lacquers or mixtures of acrylic and PVDF lacquers may be applied as an additional top coat. Another commonly used membrane material is glass fiber. The surface of glass fiber fabrics can either be coated or, in the case of coarser glass fabrics, sealed with a laminated film. A common coating material is PTFE. In the case of glass fabrics with PTFE coating, no significant aging processes can be observed even after many years [4].

7.1.2 Textile facade

A textile facade is the external vertical textile cladding of a building. The textiles are always integrated into a frame either made from aluminum, steel or wood. While the general task of the facade is the creation of an aesthetically appealing outside of the building, it can perform a variety of functions [3, 6]. The facade system should be chosen according to the architecture and function of the building to be clad. The building cladding takes on various tasks as an interface between the inside and outside world. Each urban situation requires an individual response to specific local conditions. For example, a facade can serve as protection against environmental influences and as sound and heat insulation. Overall, textile facades offer many advantages in the areas of design, energy efficiency and sustainability [6, 7]. A textile facade is, for example, used at Burj Al Arab (Dubai) with each two-layered PTFE glass-fiber membrane spanning 2,500 m^2 [3]. A simpler application of textile facades is shown in Figure 7.1, which depicts the textile facade of a commercial building in Bielefeld, Germany.

Figure 7.1: Textile facade by Serge Ferrari at Eggersmann Küchen Hiddenhausen, Bielefeld, Germany (picture by Jan Serode, ITA).

7.1.3 Particle applications in membrane systems and textile facades

Particles are used for various purposes in membrane systems and textile facades, always according to the specific requirements of the construction project. The most obvious application of particles in membrane systems and facades is in the form of pigments used for coloration, since a pleasing aesthetic is often required. However, particles can also assist in most of the other functions of membrane systems and facades, such as thermal or acoustic insulation. Particles can also improve longevity of the textiles through increased UV or abrasion resistance, and can also add other additional functions to the textile. The possibilities of particles in the area of application are illustrated in the following two examples.

Through the application of metal particles in the micrometer range, a German research project developed a low thermal emissivity coating, greatly improving thermal insulation of coated membranes while still retaining the possibility of coloring the membrane using conventional colors [8, 9].

One recent application of particles giving textile facades' additional functions is in an anti-NOX facade developed in Germany. Utilizing the photocatalytic effect of titanium dioxide, which is applied to the facade with particle sizes of 7 nm in a coating, the facade is able to reduce the amount of harmful nitrogen oxides in the air. The nitrogen oxides are bound to the facade surface under UV light (which is also present in sunlight) and oxidize into harmless salts which are washed away by rain. This application is particularly interesting in dense urban settings [10, 11].

7.2 Lightweight construction in civil engineering

Another application of technical textiles in civil engineering is lightweight constructions. The use of textiles in the form of reinforcement for polymeric or cementitious matrices enables a significant reduction of weight while retaining strength.

7.2.1 Fiber-reinforced plastics

Fiber-reinforced plastics (FRP) are composite materials consisting of high-performance fibers and a polymeric matrix. FRP are used in civil engineering in the form of lamellas for the external reinforcement and retrofitting of existing concrete structures, in the form of tension cables and in the form of reinforcing bars and textiles for the internal reinforcement of concrete elements. For reinforcement purposes, the fiber materials carbon and glass are mainly used. In addition to these two, basalt and aramid fibers are also suitable for some applications, but not as widely used in the market. While the fiber material provides tensile strength and stiffness, the polymeric matrix protects the fibers against UV radiation, moisture and chemicals. Two distinct types of matrix materials exist: thermoplastic and thermosetting materials. The latter are commonly used in civil engineering, especially unsaturated polyester resin, vinyl ester, phenolic or epoxy resin. They provide excellent load-bearing properties, are waterproof and provide thermal insulation [12–16]. Besides their application as tensile elements, FRP can also be utilized as construction materials for structures such as pedestrian and vehicle bridges, buildings, pipelines or in the offshore energy sector [13, 17]. The orthotropic (properties dependent upon orientation) behavior of the composite must be taken into account for all applications. In the direction of fibers, the mechanical properties of the fiber material are mainly exploited, while the properties of the polymeric matrix are used transversely [18, 19].

7.2.1.1 FRP lamellae

FRP lamellae have a high tensile strength and stiffness and have been used since the mid-1990s for the repair and subsequent reinforcement of concrete components [18, 19]. They are adhesively applied to the structure's surface in the form of sheets or strips by a groove filler-like epoxy paste or by cement grout. The fibers in a lamella are stretched and arranged unidirectionally [13, 15]. Figure 7.3 shows carbon fiber lamellae that are used to reinforce an existing structure.

Figure 7.2: Carbon fiber lamellae used to reinforce an existing structure [53].

7.2.1.2 GFRP – glass fiber reinforcement bars

FRP bars are used as a substitute for conventional steel reinforcement bars. Commonly used rebars consist of alkali-resistant glass fibers embedded in resin by the pultrusion process, whereby the reinforcing fibers are continuously impregnated with the liquid polymer resin, formed and then thermally cured. Between 60% and 70%, fiber volume contents are achieved depending on the matrix material. The glass fibers are oriented unidirectional in the rod. This results in high strength in the axial direction. In contrast, glass fiber reinforcement plastics (GFRP) are susceptible to transverse stress [19, 20]. One of their major disadvantages is the weak bond with the concrete matrix due to the smooth surface resulting from pultrusion. The bond between the concrete matrix and the bar surface is particularly important for the effectiveness of the reinforcement. This bond can be improved by structuring the surface of the bars. Most rebars are produced with a profiled surface, created by placing or milling ribbed structures. An alternative surface treatment is a particle coating of the surface, most commonly with sand particles (cf. Figure 7.3). The ribbed structure results in a significantly improved composite behavior, depending on the type and design of the profiling [16, 21, 22].

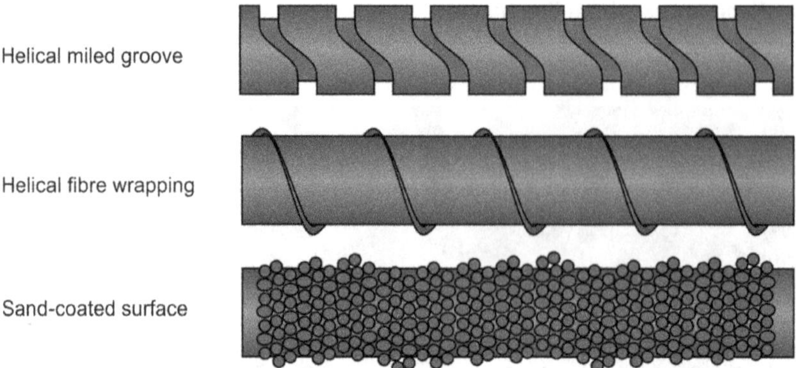

Figure 7.3: Types of surface treatments for GFRP bars.

7.2.1.3 GFRP profiles

FRP can be used as structural profiles, most commonly in bridges. The cross sections of profiles are based on those used in steel. However, some material-specific shape adaptations are possible. The profiles are mainly manufactured in the pultrusion process. The thickness of the components is limited by the heat generated during the curing of the thermoset polymer matrix. Again, the advantages of using FRP are the low weight, corrosion resistance, dimensional stability, electrical neutrality and resistance in aggressive environments [15]. When used in bridges, special attention must be paid to dynamic load cases due to the low mass and slimness of the elements [23]. Examples for FRP bridges within Europe are the West Mill FRP Road Bridge in England and the Aberfeldy Bridge in Scotland [24].

7.2.2 Reinforcement of cementitious matrices

In recent decades, the reinforcement of cement-bonded matrices with high-performance fiber materials has increasingly caught the interest of research and industry [25–27].

Steel-reinforced concrete is the most successful building material of the last century in terms of quantity used and is considered to be the almost perfect combination of two building materials [28]. The main advantages are the low cost of steel reinforcement and the extensive experience in handling and designing reinforced concrete structures. Disadvantages are the risk of steel corrosion due to water ingress (carbonation) and the minimum wall thickness (60–80 mm) required to protect against corrosion [28, 29].

The reinforcement of concrete with short and long fibers, mostly made of glass or asbestos, has been tested and is technically mature. When reinforcing with asbestos fibers, care must be taken to ensure that the corresponding surfaces are sealed. The main parameters are fiber volumes and orientations [28].

In composites, the concrete can only absorb compressive stresses, but not tensile forces. Therefore, concrete must be reinforced for applications where tensile or bending forces occur [28]. In textile-reinforced concrete, grid structures of high-performance filaments are embedded in a matrix of concrete. This reinforcement replaces the steel reinforcement and enables completely new architectural possibilities [26, 30]. The advantages of reinforcements made of fibers or textiles are the thin wall thickness, since the reinforcement does not have to be protected from corrosion by a thick layer of concrete, and the possibility of developing application-oriented textile structures can absorb the forces and moments occurring at precisely defined points [28, 31]. In addition, it offers good mechanical properties, less cracking compared to steel-reinforced concrete, better formability and generally better handling of the concrete parts (lower weight) [32, 33]. Figure 7.4 shows multiple layers of carbon textile as reinforcement in concrete.

Figure 7.4: Concrete reinforced with two layers of carbon textile.

The disadvantages of textile-reinforced concrete components include the currently still limited production methods available (high manual effort), the often still missing design and dimensioning bases and the associated often nonexistent approvals, as well as the relatively high material and manufacturing costs [34]. Great efforts are currently being made to bring corresponding manufacturing processes to market maturity.

Rovings made of continuous glass filaments are usually used for the reinforcement of concrete components. Two- and three-dimensional textiles made of combination yarns are also often used (core: glass yarn; sheath: polypropylene staple fibers). There is a similar variety as for the textiles used in concrete. Depending on the composition (binder, aggregate and water content), the adhesion between concrete and textile matrix is very different. However, it is decisive for the strength and cohesion of the component [30, 33, 35].

Regarding the discussion about a reduction of CO_2 emissions, structural elements with textile reinforcements offer considerable advantages, as considerably less concrete is required compared to steel-reinforced concrete (no danger of the reinforcement rusting). This means that up to 80% of concrete and thus the corresponding CO_2 can be saved. The integration of heat and sound insulation offers further advantages, especially for industrial buildings [29].

However, the recycling of textile-reinforced concrete has so far only been possible to a very limited extent. A separation with a high degree of purity of both textiles and concrete is possible as long as sufficient impregnation is available (epoxy resin coating), and a suitable comminution process (hammer mill) has been selected. As the textile is usually heavily damaged, current applications of the recycled short textile fibers are being developed and investigated [36].

Typical examples of applications for fiber-reinforced concrete components are facade elements and roof tiles, pipes, cable ducts, elements for thermal and electrical insulation and bank reinforcement, but also lightweight construction walls and bridges [37, 38].

7.2.3 Particle applications in lightweight construction in civil engineering

Concrete is the second most used material in the world, only surpassed by water [39]. Concrete consists at least of cement, water and aggregates, usually sand, but can be tailored using a wide range of additives. These additives are often particles such as pigments for coloration. More complex additives are also available, like expanded clay particles containing bacteria with concrete healing properties [40]. The use of particles, including nanoparticles, is therefore native to concrete production.

Properties of FRP elements can also be altered by the use of additives and particles. For example, inorganic materials with diameters of a few nanometers can be added to the polymer matrix material. These include clay, carbon nanotubes, graphene or materials such as SiO_2, ZnO, Al_2O_3 and TiO_2. Prerequisites for an efficient use of particles are the type and quantity as well as their uniform, homogeneous distribution in the matrix. The nanoparticles influence mechanical, thermal and electrical properties, resistance to environmental influences and fire behavior [14].

A common use of particles in textile lightweight construction components is the sand coating of GFRP rebars. Sand particles are added to a resin coating and applied after the rebar production. The efficiency of the treatment depends on the utilized resin as well as sand. The bond strength between the rebar and the concrete is mainly developed through an enhanced friction and interlocking of the components. An improvement of bond strength from 17% up to 58% can be realized by the sanded surface [20–22]. Similar approaches are also used for textiles for TRC (textile-reinforced concrete), where sand particles are added to the coating after impregnation of the textile [41].

Another approach to the coating of reinforcing textiles for TRC is a substitution of the polymer-based coatings commonly used with particle-based mineral coatings. The main advantages of mineral coatings are better heat resistance, which is relevant in case of fire, and better recyclability, since no polymer is added to the component. Initial trials involved mixtures of polyvinyl alcohol polymer and cement particles [42], but recent developments utilize purely mineral, particle-based coatings [43, 44].

7.3 Insulation materials

Insulation materials serve to reduce noise pollution and heat loss through the building envelope and prevent climate-related moisture damage. The most important properties of insulating materials are thermal conductivity and water vapor diffusion resistance. The thermal conductivity depends on the solids used, structure of the solids, porosity or density, conductivity of the solids, type and structure of the gas inclusions, gas pressure in the pores, radiation and surface properties of the structure, airtightness of the structure, moisture and temperature of the building material [45–47]. It is differentiated between applications in exterior, interior and core insulation. Inorganic fiber insulation materials such as glass and rock wool are classified as mineral wool. Their combined market shares in Germany amounts to approx. 55% [46]. A further material category for insulation materials is that of renewable raw materials such as wood, flax, coconut, cork, sheep wool or cellulose. In general, the sound insulation performance of rigid thermal insulation materials (such as rigid foam or wood wool lightweight building boards) is lower than that of fiber-based insulation materials [46, 47]. Various common types of insulation materials are shown in Figure 7.5.

7.3.1 Particle applications in insulation materials

Particles are mainly used in insulation materials to improve and adapt the existing properties. Examples are borates (boric salt) and ammonium sulfates. Stable borate salt compounds are used in insulation made of renewable raw materials. They serve to improve flame protection, fungus protection and insect protection. In one example, 8 mass% of borate salt is added to cotton insulation. The borate salts

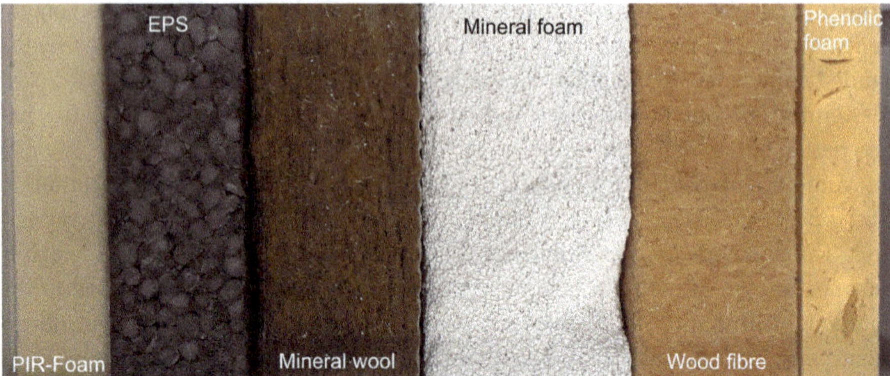

Figure 7.5: Various common insulation materials.

prevent combustion and show a sporocidal effect on five common types of fungal spores. These particles are broken down at 120–130 °C, forming a permanent wax layer on the fiber surface. Ammonium sulfates and phosphates are also added to the renewable resources as flame retardants [46, 47].

7.4 Geosynthetics

Geosynthetics or geotextiles are commonly used in or close to the ground, as depicted in Figure 7.6. They are usually applied to improve the physical and chemical properties of the soil in agriculture or around buildings and infrastructure [48, 49]. Geotextiles assume a variety of functions including separation and filtration, draining, reinforcement and erosion control. Depending on the function, material and textile architecture differs. Geotextiles are mainly nonwoven and woven structures. The textiles can be made from natural or synthetic fiber materials. The former includes jute, bamboo, cotton and coir for short-term application; the latter includes polymers such as polypropylene, polyester and polyethylene. One of the main issues is the biodegradability, especially of natural fibers [48, 50, 51]. Biodegradation can be caused by bacteria, microbes and fungi. Hence, the textiles are treated with antimicrobial agents of biocides in the form of particles or coating [51, 52].

7.4.1 Particle applications in geotextiles

Natural fibers show a high degree of biodegradation which can be slowed by particle application. Plant materials are more likely to be attacked by fungi, while bacteria tend to attack protein fibers. Most commonly used particle treatments are silver

Figure 7.6: Geotextile grids used for ground reinforcement [54].

particles at a size of 10 nm and quaternary ammonium compounds [51]. While synthetic materials show high chemical stability and hydrophobic surfaces, damages and dirt particles can serve as starting points of a microbial attack. Silver particles are also used as antimicrobial agents in synthetic fibers. They are applied to the surface during the finishing stage. Further metallic biocides are titanium dioxide, zinc oxide and copper oxide. Their functionality is based on oxidative stress that damages the microorganisms. Further protection can be ensured by applying UV blocking systems such as titanium dioxide or zinc oxide to the fiber surface [51, 52].

7.5 Summary and outlook

Particles are widely used to modify the properties of technical textiles to ensure their functionality in specific applications. The particles are most frequently used in the coating of textiles. Due to the large number of fields of application of textiles in the building and construction industry, the particles used and their tasks differ immensely. Their tasks range from protection against aggressive biogenic life forms to improving the bond between the textile and matrix. The use of textiles in construction will continue to increase due to their flexibility, variability and considerable savings potential. At the same time, further fields of application within the construction industry will be opened up for the use of technical textiles. Since every structure in every application makes special demands on the textile, the use of particles to functionalize and support the textiles will continue to be an important factor.

References

[1] Koch, A.: Multifunktionale Textilbetonsysteme Für Die Gebäudehülle. Düren: Shaker Verlag, 2018, zugl. Dissertation, RWTH Aachen University, Aachen.

[2] Hunkin, S.; Ling, J.; Severin, A.; Ackermann, P.; Moritz, K.; Alonso, A. R. L.; Gómez Plaza, D.; Menéndez Estrada, A.: Textile-based Architecture – Exploring the State-of-the Art. In: Hunkin, S.; Ling, J.; Severin, A., editors: Etfe-mfm. Belgium, 2017, 8. Nov. 2017.

[3] Cremers, J.: Tents, Sails, and Shelter: Innovations in Textile Architecture in Book: Toward Sustainable Communities & Buildings – A Reader for Professionals and Students. In: Land, W.; McClain, A., editors: Center for Sustainable Development. The University of Texas at Austin School of Architecture, 2009; pp. 153–160.

[4] Joao, L. S.; Carvalho, R.; Fangueiro, R.: A study on the durability properties of textile membranes for architectural purposes. Procedia Engineering, 2016; 155, 230–237, Elsevier B.V., Amsterdam, Netherlands: doi:10.1016/j.proeng.2016.08.024.

[5] Seidel, M.: Tensile Surface Structures. A Practical Guide to Cable and Membrane Construction: Materials, Design, Assembly and Erection. Berlin: Ernst & Sohn, Verlag für Architektur und technische Wissenschaften GmbH & Co. KG, 2009: doi:10.1002/9783433600269.

[6] Monitcelli, C.; Zanelli, A.; Campiolo, A.: Life cycle assessment of textile façades, beyond the current cladding systems; 2013; Conference: [RE]Thinking Lightweight Structures, TensiNet, Tensinet Symposium; Istanbul.

[7] Schmid, F.; Haase, W.; Sobek, W.; Veres, E.; Mehra, S.-R.; Sedlbauer, K.: Schallschutz und akustische Wirkweise bei mehrlagigen textilen Fassadensystemen. Bauphysik, 2014; 36(1), 1–10, Ernst & Sohn Verlag für Architektur und technische Wissenschaften GmbH & Co. KG, Berlin: doi:10.1002/bapi.20141000.

[8] Manara, J.; Lenhart-Rydzek, M.; Wolf, N.; Arduini-Schuster, M.: Niedrigemittierende Funktionsschichten auf Membranen zur Erhöhung der Energieeffizienz. Bautechnik, 2013; 90(4), 219–224, Ernst & SohnVerlag für Architektur und technische Wissenschaften GmbH & Co. KG, Berlin: doi:10.1002/bate.201300012.

[9] Manara, J.; Reidinger, M.; Rydzek, M.; Arduini-Schuster, M.: Polymer-based pigmented coatings on flexible substrates with spectrally selective characteristics to improve the thermal properties. Progress in Organic Coatings, 2011; 70(4), 199–204, Elsevier B.V., Amsterdam, Netherlands: doi:10.1016/j.porgcoat.2010.09.024.

[10] Serode, J.;: First anti-NOX textile facade in Hamburg, Germany, enables nitrogen oxide reduction, https://www.ita.rwth-aachen.de/cms/ITA/Das-Institut/Aktuelle-Meldungen/~fxalt/Weltweit-erste-Anti-NOX-Fassade-in-Hambu/?lidx=1, last accessed 20.05.2020

[11] https://www.ece.com/de/presse/pressemeldungen/details/?tx_news_pi1%5Bnews%5D=702&tx_news_pi1%5Bcontroller%5D=News&cHash=8ec2d2da159af5be1e4d3a52d15a8510, last accessed 20.05.2020

[12] Bedon, C.: Review on the use of FRP composites for façades and building skins. American Journal of Engineering and Applied Sciences; 9(3), 713–723: doi:10.3844/ajeassp.2016.713.723.

[13] Das, S. C.; Nizam, E. H.: Applications of Fibber Reinforced Polymer Composites (FRP) in civil engineering. International Journal of Advanced Structures and Geotechnical Engineering; 03(03), 299–309, BASHA RESEARCH CENTRE, 2014.

[14] Frigione, M.; Lettieri, M.: Durability issues and challenges for material advancements in FRP employed in the construction industry. Polymers, 2018; 10(3), 247, MDPI, Basel, Switzerland: doi:10.3390/polym10030247.

[15] Gudonis, E.; Timinskas, E.; Gribniak, V.; Kaklauskas, G.; Arnautov, A. K.; Tamulėnas, V.: FRP reinforcement for concrete structures: State-of-the-art review of application and design. Engineering Structures and Technologies, 2013; 5(4), 147–158, Vilnius Gediminas Technical University, Lithuania: doi:10.3846/2029882X.2014.889274.
[16] Pritschow, A.: Zum Verbundverhalten von CFK-Bewehrungsstäben in Bauteilen Aus Ultrahochfestem Beton, Dissertation. Stuttgart: Fakultät Bau- und Umweltingenieurwissenschaften der Universität, 2016.
[17] Shin, Y. H.; Yoong, Y. Y.; Hejazi, F.; Saifulnaz, M. R. R.; Review on pultruded FRP structural design for building construction; IOP Conf. Series: Sustainable Civil and Construction Engineering Conference; Earth and Environmental Science, 357, 2019: doi:10.1088/1755-1315/357/1/012006
[18] Liu, Y.; Zwingmann, B.; Schlaich, M.: Advantages of using CFRP cables in orthogonally loaded cable structures. AIMS Materials Science, 2016; 3(Issue3), 862–880, AIMS Press, Springfield, USA: doi:10.3934/matersci.2016.3.862.
[19] Schumann, A.; May, M.; Curbach, M.: Carbonstäbe im Bauwesen Teil 1: Grundlegende Materialcharakteristiken. Beton- Und Stahlbetonbau, 2018; 113(12), 868–876, Ernst & Sohn Verlag für Architektur und technische Wissenschaften GmbH & Co. KG, Berlin: doi:10.1002/best.201800077.
[20] Ju, M.; Park, G.; Lee, S.; Park, C.: Bond performance of GFRP and deformed steel hybrid bar with sand coating to concrete. Journal of Reinforced Plastics and Composites, 2017; 36(6), 464–475, Sage Publishing, London, UK: doi:10.1177/0731684416684209.
[21] Goraya, R. A.; Ahmed, K.; Tahir, M. A.: Effect of surface texture on bond strength of GFRP rebar in concrete. Mehran University Research Journal of Engineering & Technology, 2011; 30(No 1), 44–52, Meheran University, Jamshoro, Pakistan.
[22] Katz, A.: Bond mechanism of FRP rebars to concrete. Materials and Structures/Matériaux Et Constructions, 1999; 32, 761–768, Springer Nature, Springer-Verlag, Berlin-Heidelberg, Germany, December: doi:10.1007/BF02905073.
[23] Casalegno, C.; Russo, S.: Dynamic characterization of an all-FRP bridge. Mechanics of Composite Materials, 2017; 53(1), (Russian Original 53, 1, January-February, 2017), 17–30, Springer Nature, Springer-Verlag, Berlin-Heidelberg, Germany, March: doi:10.1007/s11029-017-9637-0.
[24] Li, Y.-F.; Badjie, S.; Chiu, Y.-T.; Chen, W. W.: Placing an FRP bridge in Taijiang national park and in virtual reality. Case Studies in Construction Materials, 2018; 8, 226–237, Elsevier B.V., Amsterdam, Netherlands: doi:10.1016/j.cscm.2018.02.005.
[25] Naaman, A. E.: Textile reinforced cement composites: Competitive status and research directions; Brameshuber, W., editors. International RILEM Conference on Material Science – MATSCI, Aachen 2010, Vol. 1, pp. 3–22, ICTRC, RILEM Publications SARL.
[26] Lorenz, E.; Schütze, E.; Schladitz, F.; Curbach, M.: Textilbeton – Grundlegende Untersuchungen im Überblick. Beton- Und Stahlbetonbau, 2013; 180(10), 711–722, Ernst & Sohn Verlag für Architektur und technische Wissenschaften GmbH & Co. KG, Berlin, October: doi:10.1002/best.201300041.
[27] Hegger, J.; Schneider, M.; Kulas, C.: Dimensioning of TRC with Application to Ventilated Façade systems; Brameshuber, W., editors. International RILEM Conference on Material Science MATSCI Aachen 2010, 1 ICTRC, pp 393–403, RILEM Publications SARL.
[28] Curbach, M.; Jesse, F.: Eigenschaften und Anwendungen von Textilbeton. Beton- Und Stahlbetonbau, 2009; 104(1), 9–16, Ernst & Sohn Verlag für Architektur und technische Wissenschaften GmbH & Co. KG, Berlin, Germany: doi:10.1002/best.200800653.

[29] Hajek, P.: Concrete structures for sustainability in a changing world. Procedia Engineering, 2017; 171, 207–214, Elsevier B.V., Amsterdam, Netherlands: doi:10.1016/j.proeng.2017.01.328.

[30] Butler, M.; Mechtcherine, V.; Hempel, S.: Durability of textile reinforced concrete made with AR glass fibre: Effect of the matrix composition. Materials and Structures, 2010; 43(10), 1351–1368, Springer Nature, Springer-Verlag, Berlin-Heidelberg, Germany, December: doi:10.1617/s11527-010-9586-8.

[31] Brückner, A.; Ortlepp, R.; Curbach, M.: Textile reinforced concrete for strengthening in bending and shear. Materials and Structures, 2006; 39(8), 741–748, Springer Nature, Springer-Verlag, Berlin-Heidelberg, Germany, October: doi:10.1617/s11527-005-9027-2.

[32] Hegger, J.; Kulas, C.; Horstmann, M.: Realization of TRC façades with impregnated AR-glass textiles. Key Engineering Materials, 2011; 466, 121–130, Trans Tech Publications Ltd, Baech, Switzerland: doi:10.4028/www.scientific.net/KEM.466.121.

[33] Colombo, I. G.; Magri, A.; Zani, G.; Colombo, M.; Di Pricso, M.: Erratum to: Textile reinforced concrete: Experimental investigation on design parameters. Materials and Structures, 2013; 46(11), 1953–1971, Springer Nature, Springer-Verlag, Berlin-Heidelberg, Germany, November: doi:10.1617/s11527-013-0023-7.

[34] Scheerer, S.; Michler, H.: Freie Formen mit Textilbeton. Supplement: Verstärken Mit Textilbeton, 2015; 110(1), 94–100, Beton- und Stahlbetonbau Spezial, Ernst & Sohn Verlag für Architektur und technische Wissenschaften GmbH & Co. KG, Berlin: doi:10.1002/best.201400113.

[35] Younes, A.; Seidel, A.; Rittner, S.; Cherif, C.; Thvroff, R.: Innovative textile Bewehrungen für hochbelastbare Betonbauteile; Supplement: Verstärken mit Textilbeton. Beton- Und Stahlbetonbau Spezial, 2015; 110(1), 16–21, Ernst & Sohn Verlag für Architektur und technische Wissenschaften GmbH & Co. KG, Berlin, January: doi:10.1002/best.201400101.

[36] Kimm, M.; Gerstein, N.; Schmitz, P.; Simons, M.; Gries, T.: On the separation and recycling behaviour of textile reinforced concrete: An experimental study. Materials and Structures, 2018; 51(122), 765, Springer-Verlag, Berlin-Heidelberg: doi:10.1617/s11527-018-1249-1.

[37] Brameshuber, W.; Hegger, J.; Gries, T.; Dilger, K.; Böhm, S.; Mott, R.; Voss, S.; Barlé, M.; Hartung, I.: Serielle Einzelfertigung (Stückfertigung) von Bauteilen aus textilbewehrtem Beton; Abschlussbericht zum AiF-Forschungsprojekt, 28.09.2007

[38] Peled, A.; Bentur, A.; Mobasher, B.: Textile Reinforced Concrete. Boca Raton, Florida, USA: CRC Press, 2007: doi:10.1201/9781315119151.

[39] Gagg, C. R.: Cement and concrete as an engineering material: An historic appraisal and case study analysis. Engineering Failure Analysis, 2014; 40, 114–140, Elsevier B.V., Amsterdam, Netherlands: doi:10.1016/j.engfailanal.2014.02.004.

[40] Lucas, S. S.; Moxham, C.; Tziviloglou, E.; Jonkers, H.: Study of self-healing properties in concrete with bacteria encapsulated in expanded clay. Science and Technology of Materials, 2018; 30(1), 93–98, Elsevier B.V., Amsterdam, Netherlands: doi:10.1016/j.stmat.2018.11.006.

[41] Donnini, J.; Corinaldesi, V.; Nanni, A.: Mechanical properties of FRCM using carbon fabrics with different coating treatments. Composites Part B: Engineering, 2016; 88, 220–228, Elsevier B.V., Amsterdam, Netherlands: doi:10.1016/j.compositesb.2015.11.012.

[42] Glowania, M.; Weichold, O.; Hojczyk, M.; Seide, G.; Gries, T.: Neue Beschichtungsverfahren Für PVA-Zement-Composite in Textilbewehrtem Beton. In: Curbach, M.; Jesse, F., editors: Textilbeton: Theorie Und Praxis; Tagungsband Zum 4.Kolloquium Zu Textilbewehrten Tragwerken (CTRS4) Und Zur 1, Anwendertagung, Dresden, 3. 6.-5.6.2009. Dresden: Techn. Univ., 2009; pp. 75–86.

[43] Nadiv, R.; Peled, A.; Mechtcherine, V.; Hempel, S.; Schroefl, C.: Micro- and nanoparticle mineral coating for enhanced properties of carbon multifilament yarn cement-based

composites. Composites Part B: Engineering, 2017; 111, 179–189, Elsevier B.V., Amsterdam, Netherlands: doi:10.1016/j.compositesb.2016.12.005.
[44] Schneider, K.; Michel, A.; Liebscher, M.; Mechtcherine, V.: Verbundverhalten mineralisch gebundener und polymergebundener Bewehrungsstrukturen aus Carbonfasern bei Temperaturen bis 500 °C. Beton- Und Stahlbetonbau, 2018; 113(12), 886–894, Ernst & Sohn Verlag für Architektur und technische Wissenschaften GmbH & Co. KG, Berlin: doi:10.1002/best.201800072.
[45] Brombacher, V.; Michel, F.; Krug, D.; Torres, M. U.; Niemz, P.: Untersuchungen zur Optimierung von Holzfaserdämmstoffen in Abhängigkeit von den Aufschlussbedingungen. Bauphysik 38(5); Ernst & Sohn Verlag für Architektur und technische Wissenschaften GmbH & Co. KG, Berlin, Germany, 2016: doi:10.1002/bapi.201610026.
[46] Fouad, N. A.: Bauphysikkalender 2019 – Energieeffizient; Kommentar DIN V 18599. Berlin, Germany: Wilhelm Ernst & Sohn, Verlag für Architektur und technische Wissenschaften GmbH & Co. KG, 2019: doi:10.1002/9783433609842.
[47] Hurtado, P. L.; Rouilly, A.; Vandenbossche, V.; Raynaud, C.: A review on the properties of cellulose fibre insulation. Building and Environment, 2016; 96, 170–177, Elsevier B.V., Amsterdam, Netherlands: doi:10.1016/j.buildenv.2015.09.031.
[48] Desai, A. N.; Kant, R.: 4 – Geotextiles Made from Natural Fibres. In: Koerner, R. M., editor: Geotextiles. Amsterdam, Netherlands: Woodhead Publishing, Woodhead Publishing, Elsevier B.V., 2016; 61–87: doi:10.1016/B978-0-08-100221-6.00004-8.
[49] Fung, Y. C.; Kaniraj, S. R.: Comparison of the behavior of fiber and mesh reinforced soils. Journal of Civil Engineering, Science and Technology, 2017; 8(2), 82–88, University of Malaysia, Sarawak: doi:10.33736/jcest.441.2017.
[50] Sumi, S.; Unnikrishnan, N.; Mathew, L.: Effect of antimicrobial agents on modification of coir. Procedia Technology, 2016; 24, 280–286, Elsevier B.V., Amsterdam, Netherlands: doi:10.1016/j.protcy.2016.05.037.
[51] Sun, G.: 17 – Antimicrobial Finishes for Improving the Durability and Longevity of Fabric Structures. In: Sun, G., editor: Woodhead Publishing Series in Textiles, Antimicrobial Textiles. Amsterdam, Netherlands: Woodhead Publishing, Elsevier B.V., 2016; pp. 319–336: doi:10.1016/B978-0-08-100576-7.00017-1.
[52] Choudhury, A. K. R.: 11 – Finishes for Protection against Microbial, Insect and UV Radiation. In: Choudhury, A. K. R., editor: Woodhead Publishing Series in Textiles, Principles of Textile Finishing. Amsterdam, Netherlands: Woodhead Publishing, Elsevier B.V., 2017; pp. 319–382: doi:10.1016/B978-0-08-100646-7.00011-4.
[53] https://www.sp-reinforcement.de/sites/default/files/field_product_image_list/cfk_lamelle_-_stahlbetonverstarkung_1.jpg, last accessed 28.07.2020
[54] https://www.huesker.de/fileadmin/Pictures/_processed_/a/4/csm_Environmental_Engineering_Sludge_lagoon_remediation_over_of_brown_coal_slurries_GE_80ec2e657e.jpg, last accessed 28.07.2020

Jamal Sarsour*, Thomas Stegmaier, Goetz Gresser

8 Ecotech textiles in environment protection and energy efficiency: current examples from research

Keywords: aerosol separation, fine dust separation, spacer textiles, fog catcher, water harvesting, waste water treatment, textile carrier materials for biomass, textile prefiltration, textile heat collector, heat storage in knitted structure, textile heat exchanger panel

8.1 Utilization of textiles in Ecotech

Technical textiles are a steadily growing market in high-wage countries like Germany. Textiles and products manufactured with a great deal of know-how have great potential to transform the traditional sector for clothing and home textiles into a high-tech industry. The research of the German Institutes for Textile and Fiber Research Denkendorf (DITF) plays a pioneering role in this process.

In the field of environmental technology, so-called ecotextiles are of great importance. They are used for soil sealing, erosion control, air purification, prevention of water pollution and water purification.

This chapter gives an overview of the development projects at the DITF Denkendorf, which were carried out with partners from textile production, plant and mechanical engineering, the chemical auxiliary industry and users of such specialized textiles. Selected developments on textile materials for dry and wet filtration, drinking water production and wastewater/exhaust air treatment, for regenerative energies (solar thermal energy) as well as biotechnological applications are presented.

***Corresponding author: Jamal Sarsour,** German Institutes of Textile and Fiber Research, Koerschtalstrasse 26, 73770 Denkendorf, Germany, e-mail: jamal.sarsour@ditf.de
Thomas Stegmaier, Goetz Gresser, German Institutes of Textile and Fiber Research, Koerschtalstrasse 26, 73770 Denkendorf, Germany

https://doi.org/10.1515/9783110670776-008

8.2 Examples of DITF projects

8.2.1 Textiles for exhaust air purification

8.2.1.1 Highly efficient coalescence separator based on novel three-dimensional, nanostructured filter media

Industrial manufacturing processes often require the separation of liquid particles from a gaseous medium. Such separation processes are required, for example, in machine tools in the separation of oil mist in the exhaust air. In this process, the misting and partial evaporation and condensation of cooling lubricants produces harmful submicron aerosols with a particle size of 100–500 nm.

If these aerosols enter the human body via the respiratory tract, they can trigger pathological reactions such as coughing, toxic-allergic phenomena, bronchitis or smoldering infections. The minimization of respiratory diseases in the metalworking industry cannot be achieved through the cooling lubricant, but only through protective measures. For example, exhaust systems directly at the machine tools have already led to significant improvements. However, suction without precipitation and the recirculation of the lubricating oil carries the oil mist to the outside and falls under emission protection.

In order to separate these health-endangering substances from the air, textile filter media have been used till now, which only inadequately separate small aerosols (<1 µm), cause a high differential pressure and thus a high energy consumption and which have to be replaced when saturated.

The aim of the research project was the development of innovative filter media for the separation of finest droplets (aerosols) from gases with minimal pressure loss. The idea for new filter principles originated in living nature: The new filter media [1] have wetting and – in close proximity – nonwetting surfaces for the liquid to be separated. The new coalescence separators should enable a synergy effect between novel filter materials, surface structuring and optimized filter architecture.

The model for the development was the desert grass *Stipagrostis sabulicola* – a biomimetic model organism for a directed separation of liquids from aerosols (Figure 8.1). The surface properties on the blade of grass result from morphological structures along the blade (Figure 8.2). These structures could be transferred to technical fibers (Figures 8.3–8.5). These fibers exhibited a similar behavior to the natural model and thus proved the successful transfer of bionic features to technical material developments.

Within the scope [2, 3] of the project, technical knitted fabrics were examined for their ability as highly efficient coalescence separators. For this purpose, the newly developed materials (pile fabrics, spacer textiles and quilted rib fabrics with inserted monofilaments) were compared with respect to their structure, their textile-physical parameters and their separation efficiency in the coolant mist. In order to filter out particles with a size of less than 1 µm, a structure was required which offers the inflowing particles a large surface area and is dense enough to retain the small particles. The knitted fabrics under investigation were able to continue to separate the incoming particles even when saturation was reached. The knitted fabrics thus have a major advantage over nonwovens, which have to be replaced when saturation is reached. This function could be achieved by the successful bionic transfer of the liquid discharge during filtration.

To further optimize the separation [4] behavior of the textile filter media, the materials used were tested in various combinations. The positive properties of the individual groups of materials tested were combined to achieve synergy effects.

In summary, it can be stated that a combination of three textile filter media at an inflow velocity of 1.0 m/s showed the best separation performance of the particles <1 µm. It was 94.2% at a pressure drop of 2,740 Pa. These were higher than with previously known filter knitted fabrics, which separate about 30%. With the knitted fabrics developed, the set separation target for healthy breathing air can be met.

Figure 8.1: Desert grass *Stipagrostis sabulicola*.

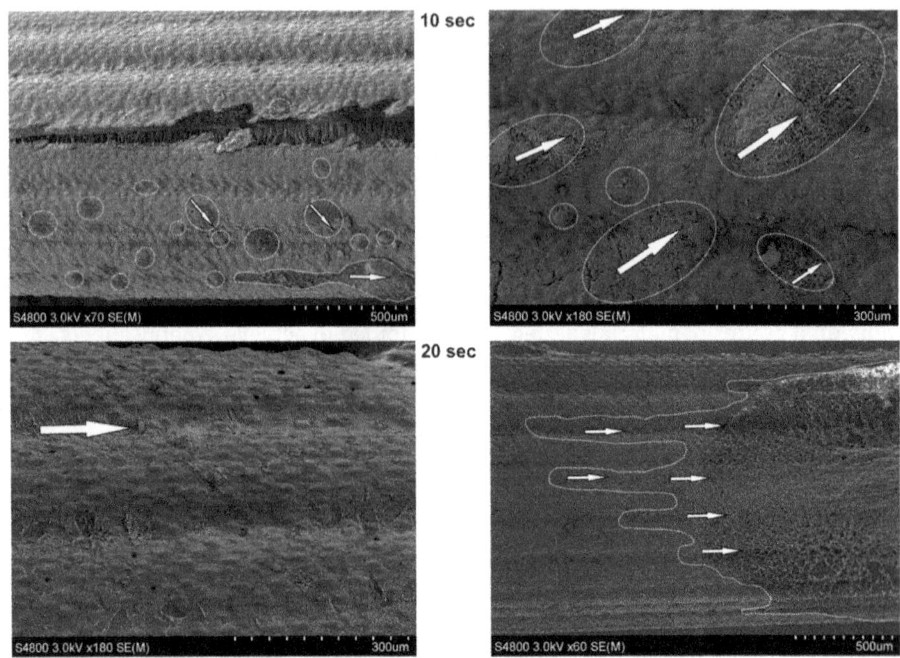

Figure 8.2: Cryo-REM investigations (Stanislav Gorb and Dagmar Voigt, MPI for Metals Research Stuttgart – University of Kiel).

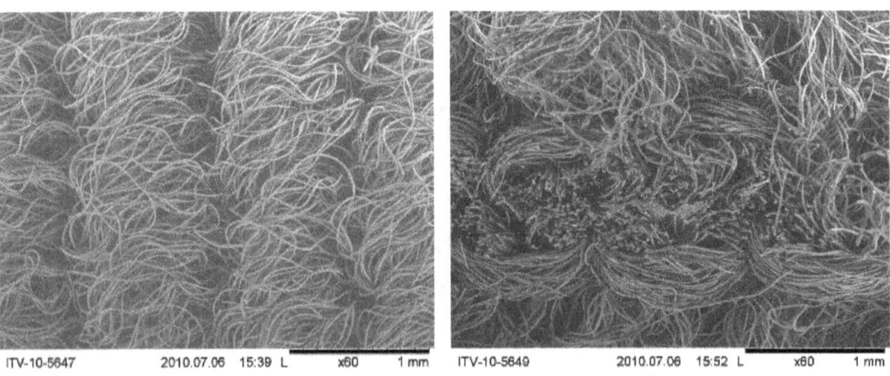

Figure 8.3: Front/rear and cross view of the textile filter material (pile fabric).

Figure 8.4: Front, back and cross view of the textile filter material (quilted rib knitted fabric).

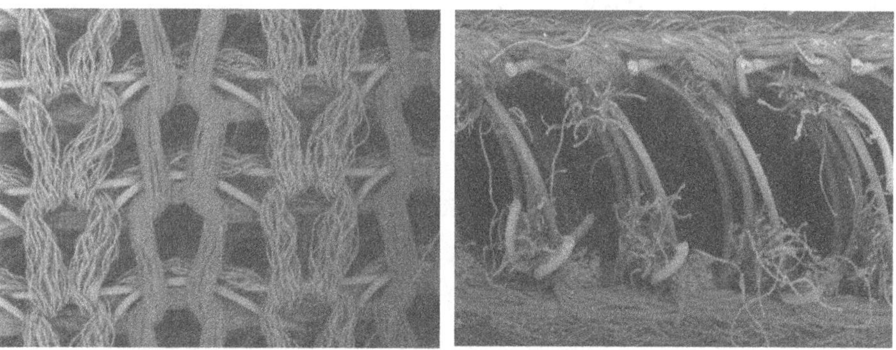

Figure 8.5: Front/rear and cross view of the textile filter material (spacer textile).

8.2.1.2 Three-dimensional textile aerosol separator system for flue gas desulfurization

The requirements for industrial emission control are constantly increasing. Besides the classical air quality parameters such as total dust, SO_x and NO_x, the focus is increasingly on the reduction of respirable fine dust and aerosols in the range of <10 µm. The relationships and dependencies between process parameters such as temperature and dust content in the individual process stages of flue gas cleaning and the formation of sulfur oxides and SO_3 aerosols are complex. In simple terms, SO_3 aerosols are mainly formed during the combustion of sulfur-containing fuels. In the course of flue gas cleaning, sulfur dioxide (SO_2) is oxidized to sulfur trioxide (SO_3) by catalytic processes such as catalytic denitrification (SCR). If the temperature during desulfurization falls below the dew point of sulfuric acid, further aerosols are formed by quenching the flue gas with the scrubbing solution. These aerosols are enriched with fine dust, gypsum and salts.

Operators of large furnaces in [5] power generation as well as of other thermal plants in the basic materials and chemical industry are faced with the task of improving the separation efficiency of their flue gas and exhaust air systems and making them more cost-effective.

There are only a few process engineering processes available that have the necessary separation efficiency of such substances. The retrofitting of fine separators is often costly, since large systems are required for this purpose. If the conventional processes are supplemented with additional cleaning stages, pressure losses occur, which in turn have to be compensated by higher blower capacities and thus higher energy consumption.

The commercially available separators, for example, in lamella form, are installed as multistage packages in vertical gas ducts. Their function is coupled to typical empty pipe speeds of 1–4 m/s. Due to the separation of aerosols and entrained insoluble salts such as gypsum, solids are deposited on the separation surfaces. The separators should therefore always be designed in combination with a cleaning system to prevent a reduction of the flow cross sections in the separation channels.

In the project [6, 7], chemically resistant fiber materials with varying production or finishing and coating parameters were processed into different textile spacer structures (Figure 8.6) and tested in a test rig at the DITF (Figures 8.7 and 8.8) with regard to their separation efficiency of finest liquid droplets.

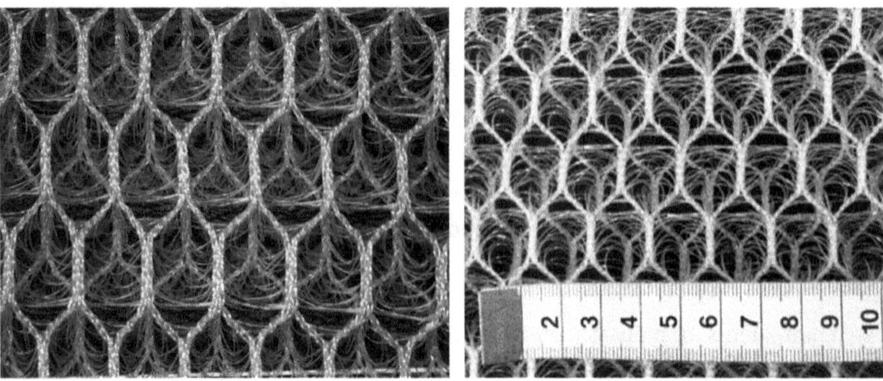

Figure 8.6: Three-dimensional spacer fabric honeycomb size (length/width) (25 mm/14 mm).

Figure 8.7: Mechanical fixation of 3D spacer fabric.

Figure 8.8: Installed filter system in test unit.

The results showed that spacer fabrics can exhibit a separation efficiency of up to 85% at a pressure drop of 1.0 mbar at high volume flows of 100,000 m^3/h depending on porosity and fiber fineness as well as the number of layers in a composite.

8.2.1.3 Textile-based fine-dust collector for separation of fine dust in urban areas [8, 9]

In 2005, a new EU limit value for fine dust particulate matter (PM 10 µm) was introduced; in 2015, a new limit value was used for particulate matter (PM 2,5 µm). However, since then, the limit values have been massively exceeded at many measuring stations in Germany. This mainly concerned the short-term limit value (24-h limit value) PM10, which may not exceed 50 µg/m^3 more than 35 times a year. Fine dust has detrimental health effects: The small particles can enter the human organism via respiration. There they act like a foreign body and cause inflammation.

Till now, there are no active air filter methods [9] for the outside air that are effective or can be implemented at reasonable cost.

The fine dust separation system developed within the project "Development of a textile-based fine dust catcher for the separation of fine dust in urban areas" offers a simple possibility to reduce fine dust emissions at low investment and maintenance costs. The textiles are sprinkled with washing liquid in the fine dust catcher system and strengthen the separating effect of the trickle film. It is also possible to introduce the washing liquid into the system via pressure nozzles.

Till now, the system has only been investigated with regard to its ability to separate fine dusts (PM10) with water. In principle, however, odors and, depending on the washing liquid, various gases can also be removed in this way.

Within the scope of the project, 3D knitted fabrics made of innovative and chemically resistant fibers with improved functionality due to the use of particles were developed. Thanks to the 3D bionically inspired mesh structure, it was possible to create the best conditions for bringing together and binding the dust particles to the water droplets (Figure 8.9). Particle separation is achieved by collision of the particles with water droplets, the droplet separation is defined on the 3D knitted fabric.

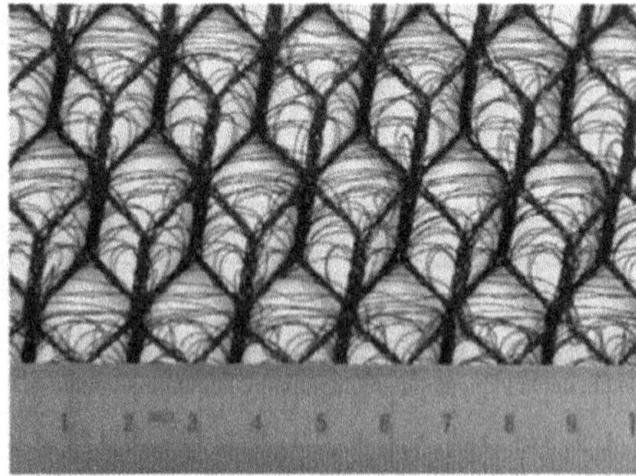

Figure 8.9: Three-dimensional knitted fabric with water drops.

To evaluate the effectiveness, 3D knitted fabrics sprinkled with water were tested in a road tunnel where the fine dust concentration is relatively high. The separation efficiency was approximately 40% for the fine dust fractions PM10.

At the end of the project, a demonstrator of the particulate trap system [8] was set up near the particulate hotspot "Am Neckartor" in Stuttgart, Germany (Figure 8.10). The measurements results were not yet available when this article went to press.

Figure 8.10: Fine dust trap system.

8.2.2 Textiles for wastewater treatment

8.2.2.1 Immobilization surfaces for biotechnological applications

In biological wastewater treatment, the biomass retention in the reactor is of crucial importance for the performance and economy of the system. For this reason, reactors with immobilized biomass are currently predominantly used. This immobilization can take place in form of a biofilm, settled on a carrier material. Requirements for such a material are:
- Large specific surface
- Good material exchange rates

- Avoidance of partial accumulation of biomass or waste products
- Material with hydrophilic and hydrophobic components for rapid biofilm formation

Previous experiences at the DITF for wastewater treatment with textile structures have shown that textile systems as carrier material optimally meet the requirements mentioned above [10, 11, 12].

Examples of this are the following work:

- AIF ZIM project "Development of retrofittable textile pre-filtration to reduce the cleaning frequency of water treatment plants" FKZ: ZF4060041SB7
  Subproject: Development of 3D textile growth carriers for microorganisms (01.01.2018–31.12.2019).
- AIF ZIM-Project "Development of a novel process with selective, textile, biodegradable 3D filter inserts for the treatment of precipitation run-off water of polluted motor vehicle traffic areas for subsequent infiltration into the ground water FKZ: ZF458170CM8 (01.09.2018–31.08.2020).
- Mr. Beryl Olila: Biological wastewater treatment using textile carrier materials; master thesis at the University of Applied Sciences Biberach at ITV-Denkendorf 2017

The properties of textile carriers are already being used in various applications, for example with nonwovens for denitrification of drinking water and for biomass retention in activated sludge tanks of municipal sewage treatment plants (Figures 8.11 and 8.12).

In biological wastewater treatment, textile materials offer the following advantages as carriers for biomass:

- The large surface area is ideal for colonization by microorganisms.
- The surface condition can be varied so that the colonization adheres well.
- The mesh size can be varied to ensure good flow and thus limitless fabric transport.

Another special feature is that the textile materials move in the water flow, but still form a solid and geometrically ordered stationary phase.

Textile materials as carriers for microbial growth are already being used successfully in some processes for the purification of waste water and surface water.

Figure 8.11: Textile support (fiber polymer is PP) in bioreactor. Thermophilic anaerobic fixed bed reactor (42 m3).

Figure 8.12: Textile carrier materials for biomass.

8.2.2.2 Textile prefiltration to reduce the cleaning frequency of water treatment plants

Biofouling is a common problem in the operation of membrane plants worldwide. It is particularly serious because it usually does not occur until 6–12 months after commissioning, when the structural measures are completed. Biofouling must be removed by additional intensive chemical cleaning processes, which has a negative effect on the lifetime of the membrane. This increases the operating costs of the membrane plant. In addition, environmentally harmful biocides often have to be used to prevent excessive pressure increases.

Higher energy consumption at higher inlet pressure, additional cleaning, a possibly earlier membrane change and biocide application mean an increase in operating costs, which can account for more than half of the total costs.

Traditional methods cannot control persistent biofouling [13], but reducing or even preventing biofouling growth in industrial processing plants is critical to performance and reliability. Traditional biocides control biofouling, but they are harmful to the environment and some promote corrosion of materials. Enzymatic approaches have proven effective in individual cases and can overcome the disadvantages of traditional biocides. However, they are usually uneconomical and too specific for routine biofouling control.

By using innovative spacer fabrics in a prefiltration stage, biofouling in a membrane plant can be prevented to a large extent (Figure 8.13).

Figure 8.13: Textile structure (3D spacer textile with structured porous polyester fibers).

By using textile carrier structures to immobilize the microorganisms [15, 16], a more resource-saving, energy-efficient and cost-effective method of controlling biofouling in water treatment can be achieved.

Such 3D spacer textiles for waste water treatment were tested by the DITF in the water treatment plant with ultra- and nanofiltration systems in a big swim arena.

Figure 8.14 shows the test results obtained over a period of 4 weeks after the start of the reactor. The degradation of the total organic carbon (TOC) of the bioreactor with the 3D spacer textiles exceeded 90% at a residence time of 6 h. The TOC

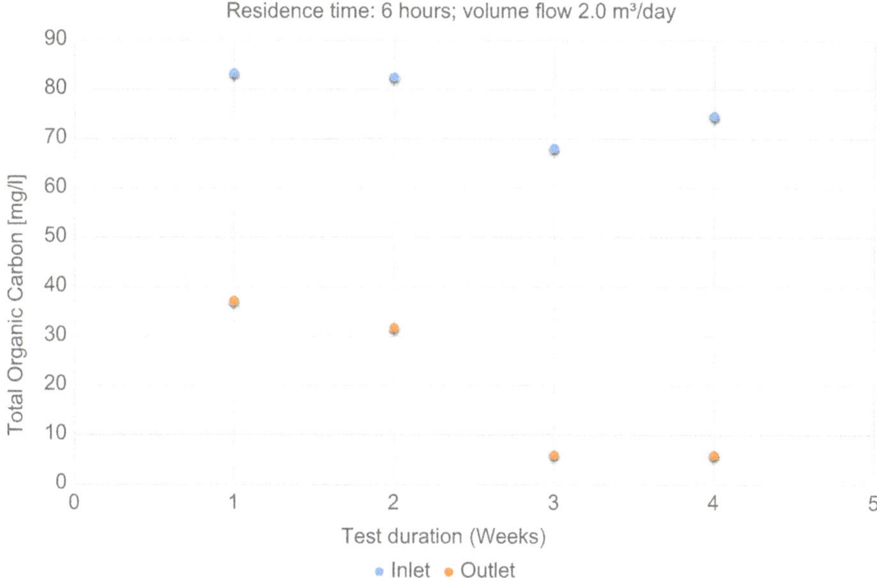

Figure 8.14: Results of cleaning swimming pool water with vertical flow textile modules (10 m^2) in fixed bed reactor (volume 0.5 m^3).

Figure 8.15: Left: Microorganism growth on 3D textile; right: after cleaning.

is one of the most important sum parameters for the assessment of the organic load of a water. The increase of biomass (TS: dry substance) on the textile was on average 567 mg/m^2 in laboratory tests.

The effluent of the bioreactor [13, 14] was examined microscopically for residual microorganisms and μ-plastic. The bacteria were counted under the optical microscope (Motic B3 Series) with objective 100. They were between 20,000 and 55,000 microorganisms in the inlet and 0 microorganisms in the outlet. In the samples taken over several weeks no microplastic was found in the effluent.

For annual cleaning, the flow velocity in the spacer textile is increased (see Figure 8.15). The excess sludge settled after filtration is disposed of via the wastewater treatment plant.

8.2.3 Textiles for water collection

8.2.3.1 Obtaining drinking water from fog

According to the WasserStiftung in Germany, 1.2 billion people do not have secure access to a spring, a well or a pipe system. In 2025, this figure will rise to 2.3 billion people. No raw material is as important as water; access to clean water is a human right. The fact that it is unequally distributed on earth causes conflicts that are intensified by the increasing desertification caused by climate change. The problem exists above all in developing countries. There, centralized water supply systems often cannot be implemented for technical and logistical reasons, or the connection of remote settlements, for example on islands or in high altitudes, is uneconomical.

In arid regions, living nature has developed fascinating methods to ensure the survival of plants (e.g., *Pinus* (pine), *Myrica arborea*, *Stipagrostis sabulicola*) and animals (such as the desert beetle *Onymacris unguicularis*) (Figure 8.16) by extracting water from the air. Particularly in extreme desert areas, astonishing, coordinated mechanisms are often at work to extract vital water from the atmosphere.

What nature uses so extremely successfully under extreme conditions has so far only been partially achieved by technology: highly efficient mist droplet separation with rapid discharge of the collected liquid.

This is where the research work of the DITF [17, 18] came in: the development of innovative, textile-based mist collectors using the third dimension (Figure 8.17). The aim of the development work was to transfer the plant functions for mist separation into textiles and technical applications for water extraction.

The biological models of vegetation in arid or semi-arid areas such as the Namib Desert in Namibia were analyzed in detail and textile structures were subsequently produced that exhibit similar properties in water absorption, conduction and release. The surface of the textile fibers was modified in such a way that they can optimally collect the mist from the air. To be able to use the "third dimension," structured textiles (spacer textiles) were produced. They offer the necessary high air permeability with low air resistance for storms with wind speeds of up to 100 km h^{-1}, are sufficiently UV-resistant (by using of microparticles to achieve improved functionality) even under extreme solar radiation and are characterized by high mist aerosol separation.

The developed textile fog collectors have been successfully used in basic tests in a fog chamber as well as in long-term field tests at fogging sites in the Namib desert (Namibia) and on Crete (Greece). The field tests complement the results from the successful basic tests with fog generators and certify the developed 3D textiles with a separation efficiency of over 85%.

The 3D textiles are available under [19] the brand name **Fog Ha**rvesting – **T**extile **i**nspired by **N**ature (FogHa-TiN). They were installed in the "CloudFisher" fog collector with which the WasserStiftung has been conducting long-term field research in the Anti-Atlas Mountains in Morocco since November 2013 (Figure 8.18). Mount Boutmezguida near the city of Sidi Ifni was an ideal test site due to its climatic location.

The results of the field tests show:

- The CloudFisher is the world's first series-produced fog catcher that can withstand wind speeds of up to 120 km/h.
- The 3D textile fabric – FogHa-TiN – is about four times more effective than the conventional material, which has been tested as the best to date. The newly developed textile structure achieved peak values of up to 66 L water/m^2 textile per day in the study in Morocco.
- It can also be installed quickly and easily by unskilled helpers
- The new fog catcher requires no energy and is extremely low maintenance.
- All materials used are food-safe.

The CloudFisher can supply hundreds of thousands of people with high-quality drinking water that meets the World Health Organization standard. It is used in mountainous and coastal regions around the world where rainfall is rare but fog is frequent and heavy.

Figure 8.16: Fog drinker beetle (*Onymacris unguicularis*) (Photo: Henschel, Gobabeb).

Figure 8.17: DITF fog catcher FogHa-TiN (photo: DITF).

Figure 8.18: The new dynamic fog catcher CloudFisher (photo: Aqualonis).

8.2.4 Textiles for energy generation

8.2.4.1 Energy-efficient textile construction with solar thermal use based on the polar bear skin

In times of climate protection, dwindling fossil energy sources and rising crude oil and gas prices, the further development of solar thermal energy, that is, the conversion of solar radiation into heat and its use, is an important social task.

Textile materials are becoming increasingly important in the construction industry. Characteristic of conventional textile structures such as roofs for stadiums, airports, train stations, air domes and tents are prestressed, self-supporting roof and wall surfaces made of coated fabrics or films. These single-layer membranes have no thermal resistance and are therefore not suitable for air-conditioned interiors.

The aim of the project was not only to create a textile building with air-conditioned [23] interiors, but also to store the energy generated in the process.

In the course of the project, "the Polar Bear Pavilion" – [20] an energy self-sufficient textile membrane construction with futuristic architecture – was created on the DITF premises (Figure 8.19).

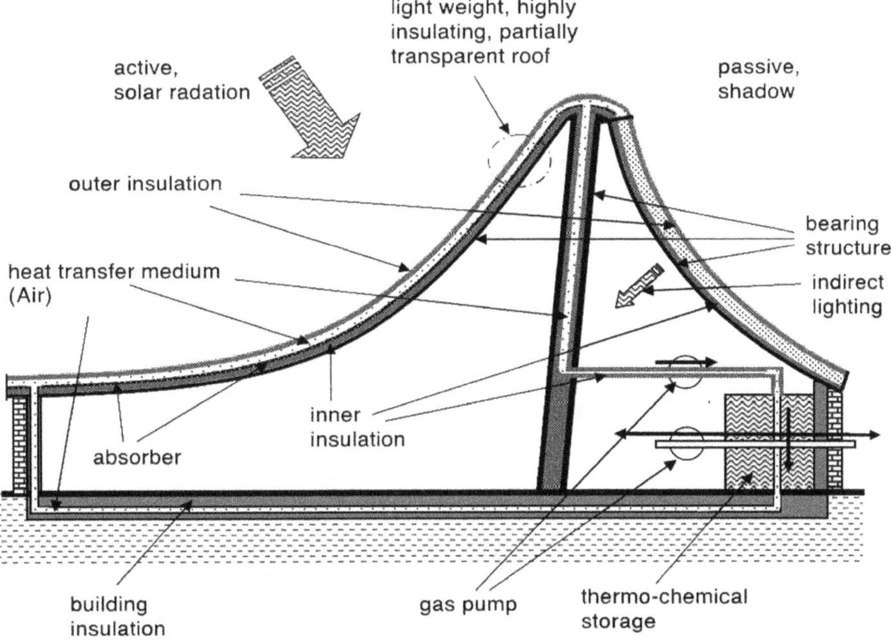

Figure 8.19: Structure of the Polar Bear Pavilion.

Innovative textile-based air solar collector

The development of the textile heat collector is modeled on the solar thermal functions of polar bear fur. The dense heat-insulating fur with colorless hairs transmits part of the sunlight to the black epidermis, which absorbs the sunlight. There is hardly any loss of heat to the outside; the solar radiation is effectively converted into heat energy. The textile shell of the Polar Bear Pavilion follows this principle: the incident sunlight hits a solar collector through which air flows. This flexible solar collector serves as an efficient energy heat exchanger:

- The bottom layer consists of a black-coated textile fabric (silicone-coated glass fabric). It forms the absorber that can withstand temperatures of up to 160 degrees.
- The air-bearing layer above is made of an open spacer fabric. The upper side is coated with silicone. This enables a high degree of transparency and provides flexible thermal insulation when combined. The upper side is permeable to visible light and the near infrared (NIR) of the solar spectrum, while the emission of long-wave light and convection are greatly reduced.
- An ethylene tetrafluoroethylene film with high transmission of visible light and the NIR is used as a cover.
- Additional translucent cover layers increase the thermal insulation and still allow sunlight to pass through.

The system allows unsupported spans of up to 10 m and curved, freely formable surfaces – a good basis for many architectural solutions. The Polar Bear Pavilion has five flexible panels with solar collectors on the south side. The warm air generated here is fed through the roof of the pavilion to heat the building (Figure 8.20).

Figure 8.20: The Polar Bear Pavilion, inspired by the solar thermal functions of polar bear fur.

Seasonal heat energy storage

The textile-based air solar collector is complemented by another innovation: a long-term heat energy storage system. The energy storage system can convert heat energy into physical-chemical energy and store it almost without loss (Figure 8.21). The patented development can compensate for day-night temperature changes and store sufficient heat in summer to heat the pavilion in winter. The storage medium consists of silica gel, which has an extraordinarily large inner surface area of approximately 600 m^2/g and can therefore absorb a great deal of moisture. During drying (desorption), the gel absorbs heat, which it releases again as soon as it is moistened (adsorption). During humidification, the air temperature increases by approximately 25 °C and is used for heating via a heat exchanger. The silica gel is suitable for desorption temperatures of 80–130 °C. This results in a maximum load of 25 wt% and a maximum value for the adsorption enthalpy of 0.174 kWh/kg. The granulate can handle 2,000 loading and unloading cycles. The energy density achieved so far in the storage facility is 133 kWh/m^3 of storage capacity, which is close to the theoretical maximum of 165 kWh/m^3.

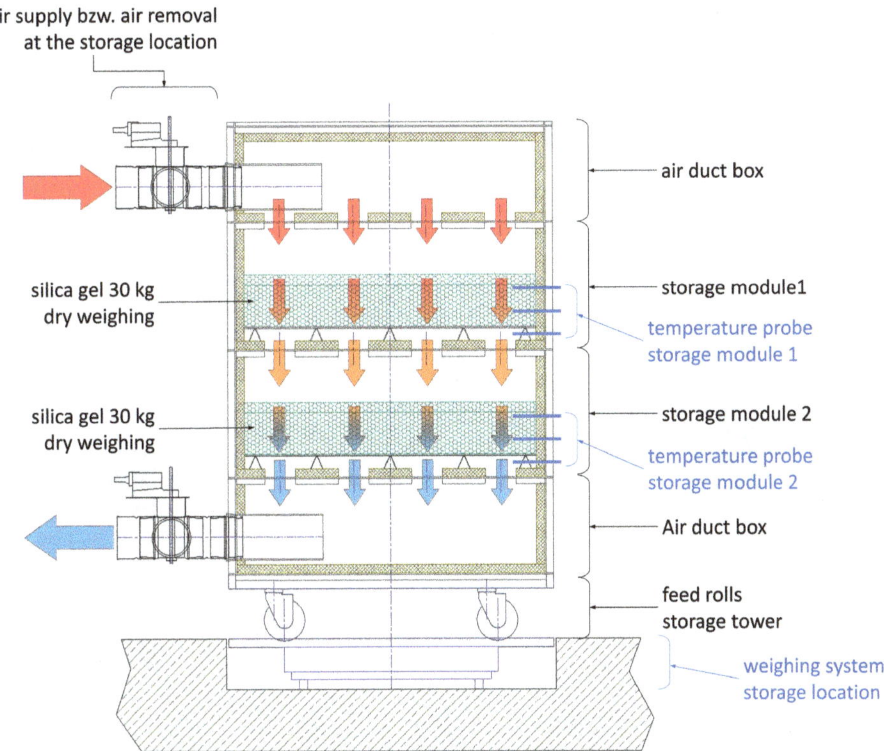

Figure 8.21: Principle of operation of the sorption accumulator (source: TAO).

Textile insulation for low energy standards

On the north side of the Polar Bear Pavilion the roof was insulated with polyester fleece, on the south side, under the solar collectors, with temperature-resistant mineral fiber felt. The building thus exceeds the low-energy standard and represents an unusually energy-efficient membrane construction.

The Polar Bear Pavilion, an energy self-sufficient textile membrane construction with futuristic architecture. The flexible solar collectors in the textile building envelope in combination with the patented heat storage system ensure optimal energy efficiency and make the pavilion energy-independent: On beautiful summer days in Germany, solar radiation can achieve an output of 800–1,000 W/m². The sun alone is sufficient to keep the pavilion warm in summer and winter.

The measured values were subjected to a daily, constant, automatic evaluation, which determines, for example, the maximum values of the solar radiation of the day, the various temperature measurements and volume flows of the collectors. According to the results obtained so far, an air temperature slightly above 150 °C could be reached (Figures 8.22 and 8.23).

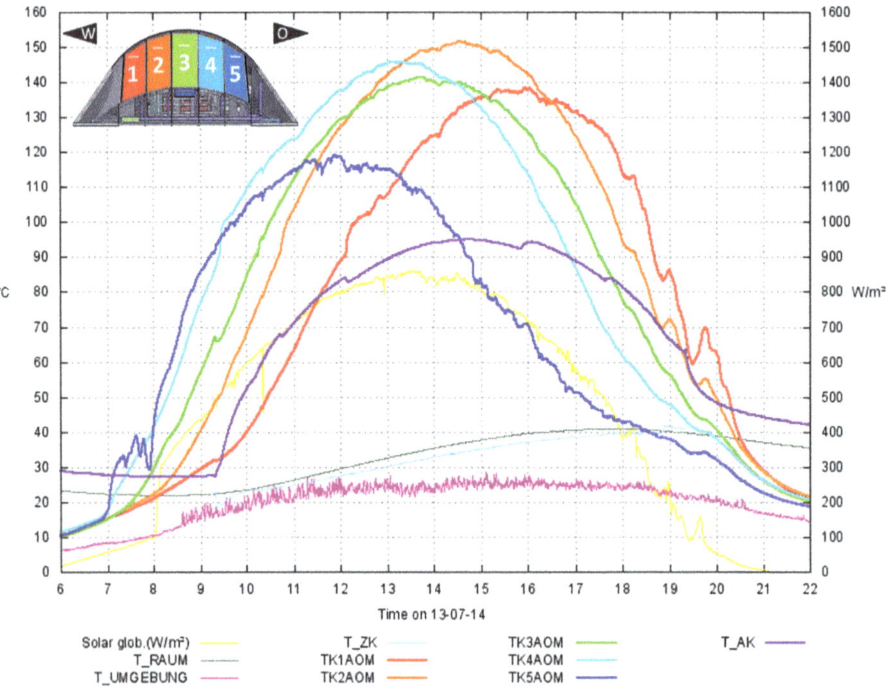

Figure 8.22: Graph of collector temperatures (TKiAOM) and irradiation.
Yellow line: solar global irradiation (W/m², scale right), all other lines: temperatures (°C, scale left), T_RAUM: room temperature, T_UMBIENT: ambient temperature, T_ZK: temperature supply air collectors, T_AK: temperature exhaust air collectors (total).

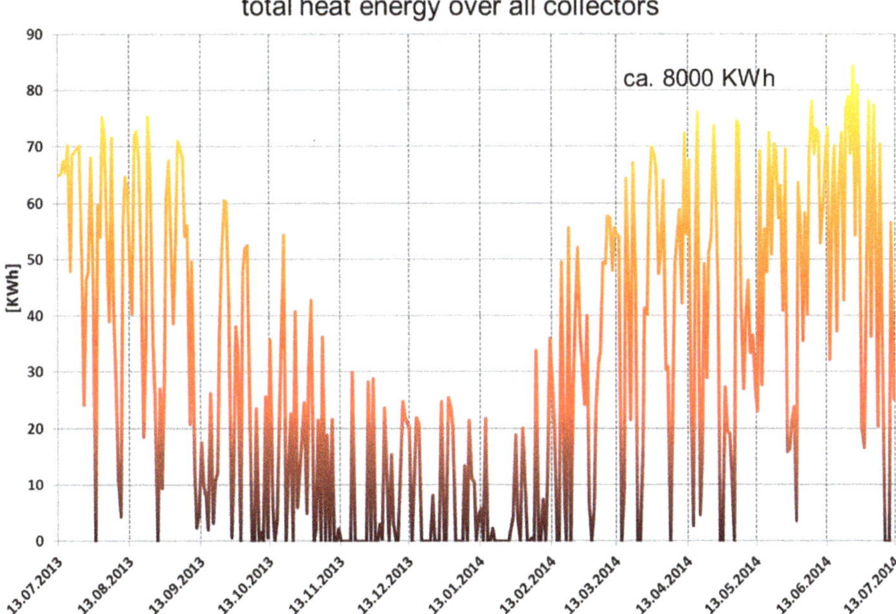

Figure 8.23: Thermal energy yield (in kWh) in the period July 2013 to July 2014.

Efficiencies/efficiencies

At a maximum air flow rate of 50.7 Nm^3/h in the solar collectors, an solar thermal efficiency of 51% was determined.

Regarding the energy storage, drying efficiencies of 70% to almost 100% were calculated from the operation of the first applications. This means that at least 70% of the supplied thermal energy was absorbed in the storage (charging efficiency).

The results give several impulses for textile construction: Lightweight, translucent buildings can have an energy balance that meets current and future requirements for resource-saving construction.

8.2.4.2 Textile-based collector with integrated latent heat storage for solar thermal energy use

In a further development, the aim was to develop a textile-based collector for solar thermal energy use with integrated heat storage for niche markets, which works in an environmentally friendly and resource-saving manner and pays for itself within a reasonable period of time. The main areas of application were the air conditioning of buildings and applications such as seawater desalination and drying processes.

Here, too, the solar-thermal functional principle of polar bear fur served as a model: Incoming sunlight hits the flexible, textile solar collector, through which warm air flows. The collector forms a direct unit with the heat storage tank. The connection of collector and heat storage tank and the underlying solar thermal utilization concept, which connects the building's outer skin directly to the storage tank, enables numerous new and individual architectural (or structural) solutions. The textile layered construction of the air collector and the latent heat storage tank is intended to optimize and significantly increase the heat energy loss compared to similar systems currently available on the market. The following research objectives were aimed at:

1. development of a layered construction for flexible textile air collectors with integrated textile latent heat storage, which is in principle industrially producible
2. optimization of the technical solution with regard to costs, buildability, applicability and with regard to ecological aspects of production, use and disposal
3. development of different demonstrators for testing
4. gaining basic knowledge and process fundamentals as a contribution to a more efficient use of renewable energies (turning away from fossil fuels)

In the development of a textile-based collector for solar thermal energy use with integrated heat storage, both economic and ecological goals were in the foreground. Applications for air conditioning of buildings, desalination of sea water and drying processes were investigated.

The test results and findings [21, 22] from the project are promising. Based on them, a textile collector with integrated phase change material (PCM) latent heat storage can be calculated and manufactured. The textile design allows a free design for flexible applications.

More than 60 wt% of PCM could be integrated into a knitted fabric, which proved to be an efficient heat accumulator in numerous practical tests. In a suitable design, the storage unit was also flexible.

A lightweight demonstrator was built at the DITF and tested under controlled conditions. The collector achieved an overall energy efficiency of up to 48% with the storage modules made of PCM knitted fabrics (Figure 8.24).

The flexible and robust design and the relatively large freedom of shape of the system allow easy installation. The risk of breakage is low and the solar collector is insensitive to damage such as storms with hail. The low investment and operating costs make the system very attractive for applications in buildings and processes that previously had no heat supply of their own (Figure 8.25) [24, 25].

Figure 8.24: Phase change material (PCM) heat storage in knitted structure.

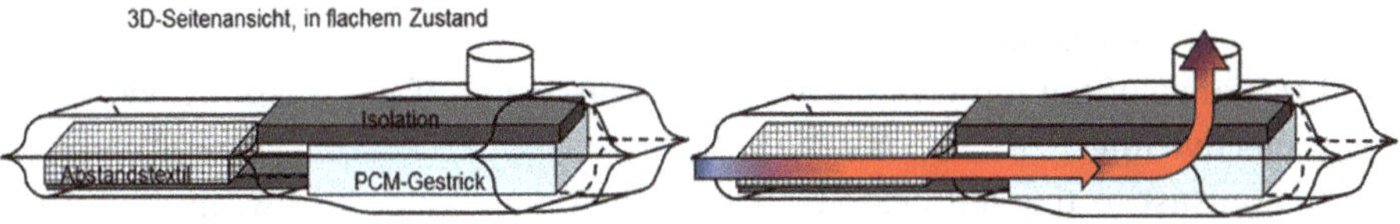

Figure 8.25: Demonstrator for the textile collector with integrated storage tank.

8.2.4.3 Textile-based heat exchanger panel as system solution for the transfer of fluid-bound heat potentials

Wastewater is a waste product that is hardly ever recycled. Depending on its use, the wastewater occurs at higher temperatures than drinking water and the strata and groundwater in the ground. Due to its temperature, wastewater represents a long-term safe, natural, regenerative and locally available energy source. The waste water, which is transported daily in large quantities through the sewage system, purified in sewage treatment plants and then discharged into the receiving water represents a material flow without economic value. In the process, approximately 5–6 billion m^3 of waste water are produced in Germany every year. If only about 3 K heat energy were to be extracted from the wastewater, a heat yield of about 20 TWh would be achieved. With this available heat energy, one could heat more than 5%

of all buildings in Germany in an environmentally friendly way. With today's heat pump technology, the use of waste water offers many economic and ecological advantages over conventional heating technologies.

Depending upon local conditions the primary energy consumption as well as the CO_2 – emission between 25% and 50% could be lowered with the today well-known techniques of the waste water heat use. The use of waste water heat contributes to the increase of the gross national product, while the added value of conventional heating systems with oil or gas takes place mainly outside the own national economy, consumes scarce resources and increases the dependence on the energy price on the world markets.

The aim of the [27, 28] research project was the development and testing of a textile heat exchanger panel (WTP) to recover energy from waste water. The panel is installed during the construction of sewers or retracted during [26] sewer rehabilitation. This is also possible with small cross-sections.

The heat exchanger panel should consist of a spacer fabric, the open sides of which are closed with foil or by direct coating. The spacer fabric creates channels through which the heat transfer medium flows and thereby releases the heat from the waste water. The knitted fabric is designed in such a way that optimum heat transfer is possible.

Properties of the heat exchanger panel:

- High thermal efficiency (heat transfer to the carrier medium) due to turbulent flow and large contact surface
- Resistance of the top layer to abrasive waste water, if necessary by additional protective liner
- In contrast to the stainless-steel components used up to now, no biofilm is formed, which can lead to heat gain losses of up to 40% and makes an interval flushing of the sewage pipe necessary
- Flow and return of the fluid in the heat exchanger panel can be integrated,
- Low pressure loss in the heat exchanger panel
- Installation in sewage pipes <DN 800 (the Uhrig system can only be used in large accessible dimensions), possible from DN 200, that is, close to the sewage generator with possibly increased sewage temperatures
- Can be used in connection with renovation procedures (liner procedure) and thus cost allocation
- The WTP inserted [29] between two liners is protected and does not represent a drain obstacle
- Easy installation in the sewer and attachment of connections and valves
- Flexible formability of the WTP
- considerably cheaper than currently used stainless steel heat exchangers (approximately 250 €/m compared to stainless steel components, depending on the dimension of the duct 750–900 €/m)

Structure of the textile heat exchanger panel

The textile heat exchanger panel consists of a spacer fabric, the open sides of which are closed with foil lamination or with a direct coating, Figure 8.26. The long sides of the 3D spacer fabric can be joined by gluing or welding. The foils form the waterproof layer and are connected to the spacer fabric by means of an adhesive layer. The special construction of the spacer fabric creates channels in the heat exchanger panel.

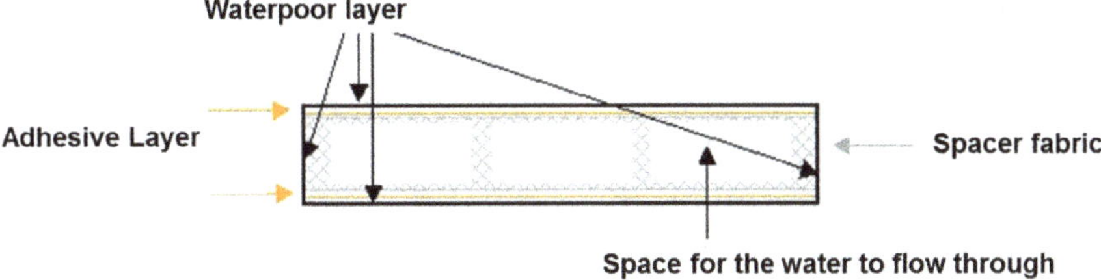

Figure 8.26: Heat exchanger panel structure.

The spacer threads are arranged one behind the other so that the water experiences as little resistance as possible when flowing through. The flow and return of the water in the heat exchanger panel (WTP) is made possible by separating the individual chambers in the spacer fabric, for example by ultrasonic welding.

Figure 8.27 shows the fluid flow from the heat pump into the WTP in blue color. The WTP is divided into three chambers by two ultrasonic welds; one flow and two return chambers. In this example, the return of the heated fluid takes place at the

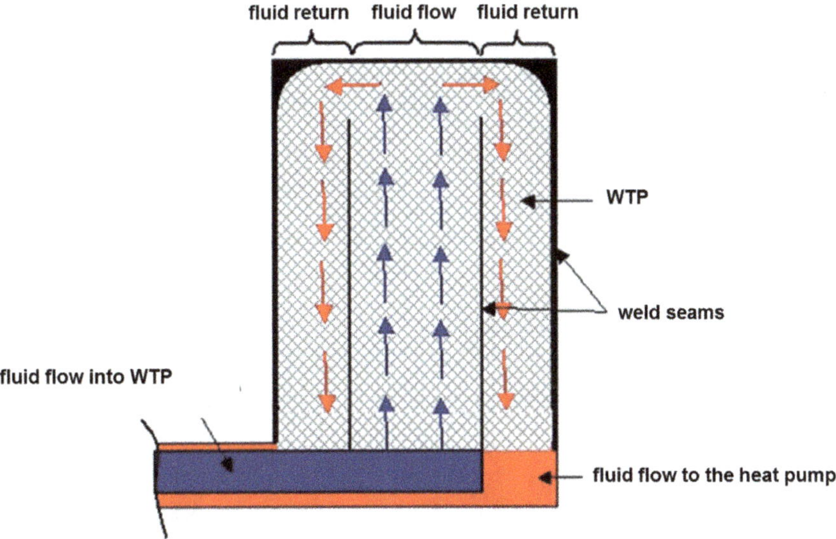

Figure 8.27: Top view of flow and return system heat exchanger panel WTP.

lateral chambers of the panel. The sides of the WTP can also be sealed with weld seams or laminated with a film. The return flow (red color) of the fluid is fed to the heat pump.

The WTP can be fixed in the sewer by gluing it to the bottom of the sewer pipe. Alternatively, the heat exchanger panel can be placed in the pipe together with a liner using a modified technology, whereupon the panel is firmly attached to the liner on the pipe bottom. If the upper foil layer, that is, the top layer of the heat exchanger panel, does not have the required waste water resistant properties, an additional resin-impregnated protective liner can be inserted, which is in direct contact with the waste water. The heat exchanger panel is flexible and can be transported in roll form. The dimensions of the heat exchanger can be freely adjusted depending on the pipe diameter, so that the panel can be adapted to the area of the pipe through which the wastewater flows.

Figure 8.28 illustrates the functional principle of wastewater heat recovery using a WTP. The WTP is placed in the bottom section of the sewer so that the wastewater flows around it. The wastewater is about 8–15 °C warm on an annual average. A heat exchange now takes place between the wastewater and the fluids in the heat exchanger system (here the WTP). The fluid in the WTP heats up and is fed to the heat pump. From the heat pump, the temperature is raised to a level of approximately 40–70 °C and can be used for heating purposes. A heat exchanger panel on textile basis can optimize the flow and thus enable a more effective heat transfer.

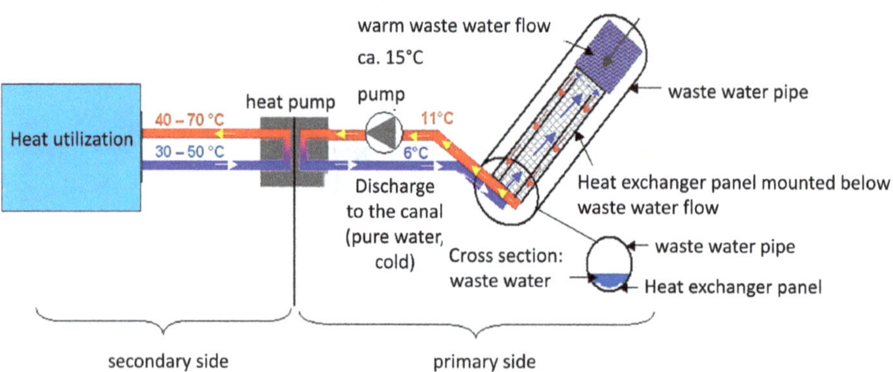

Figure 8.28: Principle of the heat exchange system.

The panel can be used in smaller sewers from DN 300 upwards without the need for further installations for its protection (Figure 8.29).

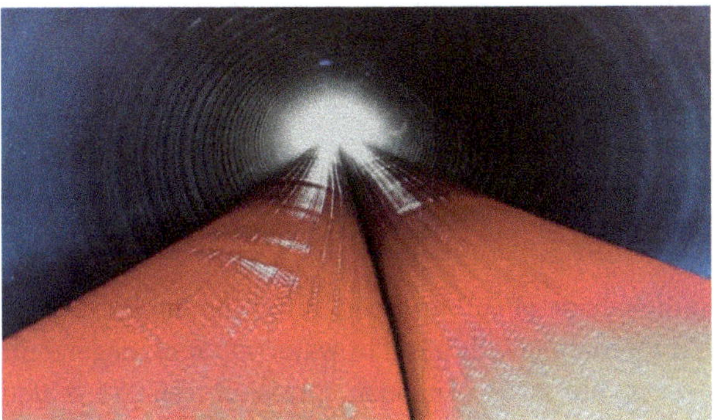

Figure 8.29: filled WTP in the pipe.

Results to date

Compared to the systems offered on the market, the textile heat exchanger panel transfers significantly higher amounts of energy to the fluid due to highly turbulent flows. In addition, the design for smaller pipe cross-sections makes them suitable for inner-city manifolds – a method that has not been available on the market until now.

Compared to the established processes, the WTP is significantly cheaper at 960 €/kW. The processes already available on the market are offered at prices ranging from 700 to 3,800 €/kW.

8.2.5 Outlook

Fiber-based materials for technical applications are growing rapidly worldwide. In Germany, more than 50% of the textiles produced are used in the field of technology. The trend is still increasing.

The current developments in our technology-oriented society require new, highly functional materials, whereby fiber-based materials contribute with a high level of benefit. The endeavor to stop climate change encourages the development of CO_2-neutral processes, largely foregoing fossil raw materials and increasing the share of renewable energies. Textile materials also contribute to the formation of microplastics in our environment. This must be avoided in the future. There is therefore a great need for new textile materials that are made from renewable raw materials or that are biodegradable. For this, enormous efforts are necessary to make natural resources accessible, also within the framework of the goals of the bioeconomy, by processing these raw materials into high-quality biopolymer semifinished products using the simplest possible low-energy processes. For materials that have to be resistant and are not allowed to simply biodegrade, recycling cycles must be set up and established. In the case of biodegradable products, which are mainly used outdoors, certain durability must still be achieved for the desired service life.

In the field of renewable energies, technical textiles are already making important contributions:

- Wind power rotors, the blade length of which has meanwhile exceeded 60 m, are technically not feasible without high modulus glass fiber and carbon fiber in composite materials
- In the fuel cell, porous textiles made of carbon fibers fulfill the important task of transporting protons in combination with water or water vapor management.
- In biogas plants, the gas covers are made of robust, corrosion resistant coated fabrics

New requirements arise from the technologies of hydrogen storage and semiflexible photovoltaics.

The use of textile carriers for microorganisms is already being used successfully in sewage treatment plants. New processes in bioprocess engineering require new, adapted carrier materials for bacteria, fungi and small organisms.

Associated with this are efforts to bring agriculture back to the cities, at least in part. Urban farming calls for fiber-based innovations in order to be able to cultivate plants without humus and soil. Textiles offer enormous advantages in terms of water transport via capillary forces. Fibers can be designed so that they can be used as sensors for moisture management.

For textiles in filter applications, new requests and requirements are constantly being brought up. In particular, thanks to the large surface area of the fibers, highly efficient materials can be developed that can selectively absorb ingredients. This can be used, for example, to absorb heavy metals from wastewater. Catalytic processes also use the extremely large surface area of fibers to effectively convert ingredients into gases and liquids.

This means that there is still a great need for the further development of textile materials in the future in order to protect our environment and the climate.

References

[1] DITF Denkendorf: Entwicklung energetisch hocheffizienter Koaleszenzabscheider auf Basis neuartiger dreidimensionaler, nanostrukturierter Filtermedien. BMBF Abschlussbericht 2012, Förderkennzeichen 01RB0706A.

[2] Roth-Nebelsick, A.; Ebner, M.; Miranda, T.; Gottschalk, V.; Voigt, D.; Gorb, S.; Stegmaier, T.; Sarsour, J.; Linke, M.; Konrad, W.: Leaf surface structures enable the endemic Namib desert grass Stipagrostis sabulicola to irrigate itself with fog water. Journal of the Royal Society Interface, Published:22 February 2012.

[3] Ebner, M.; Miranda, T.; Roth-Nebelsick, A.: Efficient fog harvesting by Stipagrostis sabulicola (Namib dune bushman grass). Journal of Arid Environments, 2011; 75, 524–531.

[4] Stegmaier, T.; Sarsour, J.; Linke, M.: Desert Plants Inspire High-tech Filtration Processes. In: Biomimetics International Industrial Convention. Berlin, 16–17.03.2011.

[5] DITF Denkendorf: Entwicklung eines textilen 3D Aerosolabscheidersystems zur Rauchgasentschwefelung. Abschlussbericht zum ZIM-Kooperationsprojekt, Förderkennzeichen ZF4060002CM5, 2018.

[6] Schwarz, J.: Bionische Abluftreinigung und -entfeuchtung der Prozessabluft industrieller Reini-gungsmaschinen mit textilen Filtern, Fachkolloquium – Faserbasierte Lösungen für Energie und Umwelt, ITV-Denkendorf, 20. Mai 2015.

[7] Kneer, A.: Numerische Untersuchung des Wärme- und Stoffübertragungsverhaltens von Abstandsgewirken mit Hilfe von Modellen mit aufgelösten Textilstrukturen, Fachkolloquium – Faserbasierte Lösungen für Energie und Umwelt, ITV-Denkendorf, 20. Mai 2015.

[8] DITF Denkendorf: Entwicklung eines textilbasierten Feinstaubfängers zur Abscheidung von Feinstäuben in Ballungszentren. Abschlussbericht zum ZIM-Kooperationsprojekt, Förderkennzeichen ZF4184802NK7, 2020.

[9] UBA, 2016: Gesundheitsrisiken der Bevölkerung in Deutschland durch Feinstaub https://www.umweltbundesamt.de/daten/umwelt-gesundheit/gesundheitsrisiken-der-bevoelkerung-in-deutschland

[10] DITF Denkendorf: Entwicklung und Erprobung einer modularen Anlagenlösung zur innerbetrieblichen Aufbereitung von Teilströmen hoch und/oder schwach belasteter Industrieabwässer bei gleichzeitiger Rückführung in den Brauchwasserkreislauf. Abschlussbericht zum ZIM-Kooperationsprojekt, Förderkennzeichen ZF4060013SA6, 2018.

[11] Sarsour, J.; Stegmaier, T.; Gresser, G.: Abstandstextilien in umwelttechnischen Systemen. FACHTEX Arbeitskreis Technische Textilien, Oberflächenmodifikation und Technische Textilien im Forschungskuratorium Textil e.V. Bayreuth, 01. 10. 2014.

[12] Sarsour, J.; Pfeiffer, R.: Entwicklung Einer Energiearmen Demonstrationsanlage Zur Abwasserreinigung Und Rückgewinnung von Brauchwasser. In: Innovative Ressourcennutzung Und Abtrennung von Schadstoffen. Ergebnisse Aus Dem Förderprogramm EFRE-Umwelttechnik. Stuttgart, 17.10.2014.

[13] DITF-Denkendorf: Entwicklung nachrüstbarer, textiler Vorfiltration zur Senkung der Reinigungshäufigkeit von Wasser-aufbereitungsanlagen. Abschlussbericht zum ZIM-Kooperationsprojekt, Förderkennzeichen ZF4060041SB7, 2021.

[14] Stoll, P.: Untersuchung Des Biofoulings Sowie der Reinigungsfähigkeit an Einer Umkehrosmose Flachzellentestanlage. Bachelorarbeit, 2016.

[15] Hans-Curt Flemming: Elements of an integrated anti-fouling strategy with emphasis on Monitoring. University of Duisburg-Essen, 2011.

[16] Vrouwenvelder, J. S.; Kruithof, J.; van Loosdrecht, M.: Biofouling of Spiral Wound Membrane Systems 2011, ISBN: 9781843393634.

[17] Stegmaier, T.; Sarsour, J.; Ewert, B.; Gresser, G. T.:Textile Lösungen für die Trinkwassergewinnung und die Abwasseraufbereitung Vortrag auf dem Webinar Perspektiven 2035: Sauberes Wasser – Ein Thema von globaler Bedeutung des Forschungskuratoriums Textil und des VDMA, 14. Dezember 2020.

[18] Stegmaier, T.; Aliabadi, M.; Arnim, V. V.; Kaya, C.; Sarsour, J.; Gresser, G.: Bionics/biomimetic in textile research. Journal of Textile Engineering and Fashion Technology, 2020; 2(1), 6–7.

[19] Sarsour, J.; Stegmaier, T.; Gresser, G. T.: 3D Textile Structures for Harvesting Water from Fog: Overwiew and Perspectives. In: Mittal, K. I.; Bahners, T., editors: Textile Finishing. Scrivener Publishing LLC, 2017; pp. 325–343.

[20] Stegmaier, T.; Sarsour, J.; Gresser, G. T.; Wagner, R.; Kröplin, B.; Kungl, P.; Kneer, A.; Arnold, W.; Bögner-Balz, H.: Energy-efficient textile building with transparent thermal insulation for the solar thermal use according to the model of the Polar Bear´s Fur. Innovative Energy & Research, 2018; 7(3), 1000218, ISSN: 2576-1463.

[21] Stegmaier, T.; Sarsour, J.; Gresser, G.:Textilbasierter Kollektor mit integriertem Latenwärmespeicher zur solarthermischen Energienutzung Proceedings, Global Fiber Congress, Dornbirn, 12–14. September 2018.

[22] Sarsour, J.; Stegmaier, T.; Gresser, G.:Textile-based collector with integrated latent heat storage for the use of solar thermal energy Lecture on the TechTextil, 11. Mai 2017.

[23] Stegmaier, T.; Sarsour, J.: Der Inneffizienz bei der Solarenergienutzung aufs Dach steigen. Technical Textiles Kettenwirk-Praxis, 2016; 03, 26–27.

[24] August, A.; Kneer, A.; Reiter, A.; Wirtz, M.; Sarsour, J.; Stegmaier, T.; Barbe, S.; Gresser, G. T.; Nester, B.: A Bionic approach for heat generation and latent heat storage inspired by the polar bear. Energy, 2019; 168, 1017–1030.

[25] Stegmaier, T.; Sarsour, J.; Gresser, G.: Textile based collector with integrated latent heat storage for solar thermal energy use. Proceedings, China-German Workshop for Modern Membrane Structure, October 20th–21st 2018, Shanghai, China.

[26] DITF Denkendorf: Entwicklung eines textilbasierten Wärmetauscherpaneels (WTP) als Systemlösung zur Übertragung fluid gebundener Wärmepotentiale. Schlussbericht zum BMWi-Verbundvorhaben 2019; Förderkennzeichen: 03ET1288 A-B-C-D-E-F.

[27] Rometsch, L.: Heat recovery from sewers, development of a catalog of requirements for sewage treatment plant and sewer network operators, short report, IKT Gelsenkirchen, January 2005.

[28] Christ, O.; Mitsdoerffer, R.: Regenerative Energie nutzen – Wärmequelle Abwasser; Wasserwirtschaft Wassertechnik 5/2008M6 – M11.

[29] FITR-Forschungsinstitut für Tief – und Rohrleitungsbau gemeinnützige GmbH: Wärmetauschermatte in Abwasserkanälen in Verbindung mit Innensanierung – Heatliner-, gefördert vom Bundesministerium für Wirtschaft und Technologie BMWi, 10/2006 – 03/2009.

Rahel Krause*, Carolin Schwager, Jan Jordan

9 Protective textiles

Keywords: PPE, personal protective equipment, protective clothing, risk categories, risk mitigation, particles for PPE, functional textiles, technical textiles, protection

9.1 Technical textiles

9.1.1 Protective textiles/personal protective equipment

Personal protective equipment (PPE) is the last barrier between a person and a hazard. In sports applications, the use of PPE is mostly voluntary whereas in occupational health employers are obliged to provide their employees with suitable PPE. The selection of PPE depends on the risk assessment of the working environment of the employee. In the following an overview on the most relevant and critical types of hazards and risks is introduced. The types of protection that are commonly used are described. An outlook on the current or potential future use of particle technology in the presented risk scenarios is provided. To obtain an understanding on the application of particle technology in PPE, the characterization and mitigation of risks will be introduced.

9.1.1.1 Types of risks

Risks to a person wearing a PPE ("user") generally originate from (a) external hazards from the user's environment, (b) a defective PPE or (c) a critical health status of the user itself as shown in Figure 9.1.

The European Chemicals Agency's "Introductory Guidance on the CLP Regulation" (CLP stands for classification, labeling and packaging of substances and mixtures) specifies in more detail potential physical and health hazards as listed in Table 9.1 [1].

The associated PPE, which protects against the different physical or health hazards, is divided into several categories according to the EU regulation on PPE, published in 2016. These categories (I, II and III) are described in more detail in the following Section 9.1.1.2 [3].

***Corresponding author: Rahel Krause,** Institut für Textiltechnik of RWTH Aachen University, Otto-Blumenthal-Strasse 1, 52074 Aachen, Germany, e-mail: rahel.krause@ita.rwth-aachen.de
Carolin Schwager, Jan Jordan, Institut für Textiltechnik of RWTH Aachen University, Otto-Blumenthal-Strasse 1, 52074 Aachen, Germany

https://doi.org/10.1515/9783110670776-009

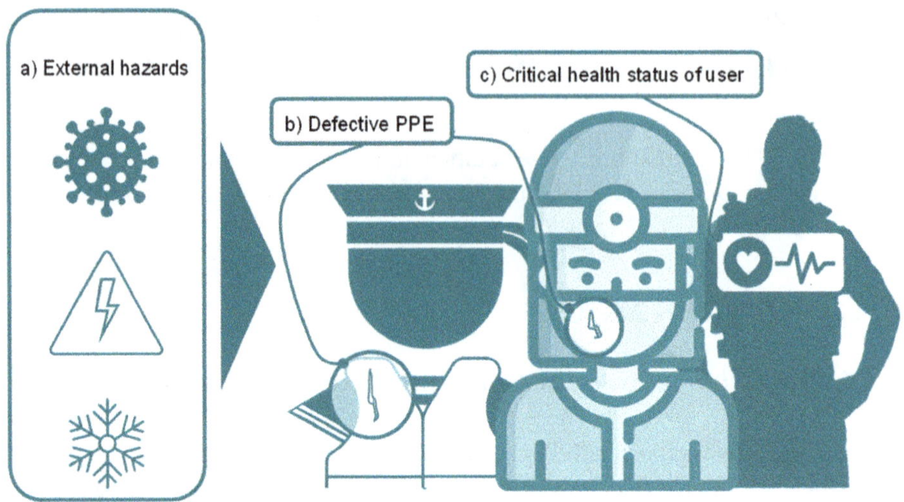

Figure 9.1: Origins of risks (own composition of images including work from: Clker-Free-Vector-Images, OpenClipart- Vectors, Mohamed Hassan, biohardlos92 (each on Pixabay)).

Table 9.1: Hazard class table according to the ECHA [2].

Physical hazards		Health hazards	
Hazard class	**Chapter in guidance**	**Hazard class**	**Chapter in guidance**
Explosives	2.1	Acute toxicity	3.1
Flammable gases	2.2	Skin corrosion/irritation	3.2
Flammable aerosols and aerosols	2.3	Eye damage/irritation	3.3
Oxidizing gases	2.4	Respiratory/skin sensitization	3.4
Gases under pressure	2.5	Mutagenicity	3.5
Flammable liquids	2.6	Carcinogenicity	3.6
Flammable solids	2.7	Toxic for reproduction	3.7
Self-reactive substance/mixture	2.8	Specific target organ toxicity	3.8
Pyrophoric liquids	2.9	(single exposure)	
Pyrophoric solids	2.10	Specific target organ toxicity	3.9
Self-heating substance/mixture	2.11	(repeated exposure)	
Water-reactive – emits flammable gases	2.12	Aspiration hazard	

Table 9.1 (continued)

Physical hazards		Health hazards	
Hazard class	**Chapter in guidance**	**Hazard class**	**Chapter in guidance**
Oxidizing liquids	2.13		
Oxidizing solids	2.14		
Organic peroxides	2.15		
Corrosive to metals	2.16		

9.1.1.2 Types of risk mitigation

The types of risk mitigation can be described with the risk priority number (RPN) generated in the failure mode and effects analysis. The RPN describes the criticality of a failure by evaluating the severity (S) of a failure, the probability (P) of the occurrence of a failure as well as the ease of detection of a failure (D). The RPN is defined as product of S, P and D: $RPN = S \cdot P \cdot D$ [4].

To reduce the criticality of a failure (i.e., health risk) in the occupational context, employers are obliged to implement all available measures to compensate, reduce or even neutralize any type of risk to the user. The first measures to be implemented target a complete neutralization of the risk. According to an updated risk assessment remaining risks are identified which cannot be compensated by organizational or structural measures implemented in the user's working environment. It is for these remaining risks that suitable PPE must be selected and provided to the user. In sports the responsibility of selecting and wearing suitable PPE is generally with the user.

Wearing PPE does not guarantee a complete safety from health risks. Referring to the origins of risks introduced in Section 9.1.1.1 the user's health is in danger if (a) the external hazards exceed an intensity for which the PPE is certified, (b) the PPE is defective without the user's awareness or (c) the user's health has reached a critical status.

To illustrate these aspects, selected critical situations of a firefighter are discussed. The external hazards can exceed an intensity for which the PPE is certified in case of a "flashover." A flashover occurs in closed rooms in which the combustible material and objects reach their ignition temperature nearly at the same time. If firefighters enter a room in such a situation, they are exposed to temperatures of above 1,000 °C. The PPE can withstand the resulting heat flux for only a few seconds. If the firefighter's way out of the zone of risk is then blocked by, for example, debris, the firefighter's life is in danger. Another external risk can be the exposure to a hazardous gaseous substance leaking from a container for which the fire fighters were not prepared and hence not equipped with the specific set of PPE.

A further critical situation occurs, if the PPE of the firefighter is defect without his or her awareness. One example is the firefighter's helmet which has absorbed an excessive amount of UV radiation over the use time due to sunlight. The helmet shell can hence deteriorate due to an increased brittleness of the material and fail in the moment of a sudden mechanical impact (e.g., from falling debris).

The third origin of risk lies within the firefighters themselves. In a rescue situation, the adrenaline level tends to be that high that the own exhaustion is not sensed. Dehydration due to excess transpiration as well as an overheated body temperature may cause nausea. If the team members are unable to find their collapsed colleague, this results in danger to life.

PPE is divided into three different categories according to the EU Regulation on PPE [3]:

I. Minimal risks: (a) superficial mechanical injury; (b) contact with cleaning materials of weak action or prolonged contact with water; (c) contact with hot surfaces not exceeding 50 °C; (d) damage to the eyes due to exposure to sunlight (other than during observation of the sun); (e) atmospheric conditions that are not of an extreme nature
II. Risks other than those listed in Categories I and III
III. Risks that may cause very serious consequences such as death or irreversible damage to health

The use of particles is particularly relevant for the PPE assigned to category 3. Therefore, the various risk types and the PPE of this Category III are explained in more detail below.

9.1.1.3 PPE of risk category III

This chapter presents the different types of risk and the associated PPE that protects the wearer from the risk. To protect the human body, PPE is worn close to the body. Therefore, a high percentage of PPE is made of textiles, which provide sufficient flexibility and comfort. Table 9.2 gives some examples of textile solutions for protection against the different types of risks within category III.

In the following chapter, the functions of particles in PPE is further described.

9.1.1.4 Functions of particles in PPE

Particles can be used in many different applications of PPE to enhance its protection level. One example for the usage of particles are stab-resistant waistcoats. Shear

Table 9.2: Risk and PPE overview for category III [3].

Risk	Exemplary textile solution within PPE
(a) Substances and mixtures which are hazardous to health	Barrier textiles with coatings or membranes, keeping the substances away from the wearer; filter textiles with small pore size (e.g., gloves, gowns, protective suits, respiratory mask)
(b) Atmospheres with oxygen deficiency	Oxygen masks supplying the user with breathable air, for example, with filter textiles to prevent the user from breathing polluted air
(c) Harmful biological agents	Barrier textiles with coatings or membranes, keeping the harmful biological agents away from the wearer; filter textiles with small pore size (e.g., gloves, gowns, protective suits, respiratory mask)
(d) Ionizing radiation	Barrier textiles preventing the penetration of radioactive dust, gases, liquids or mixtures (e.g., gloves, gowns, protective suits, respiratory mask)
(e) High-temperature environments the effects of which are comparable to those of an air temperature of at least 100 °C	Insulating textiles intended for brief use in high-temperature environments and which may be splashed by hot products (e.g., jackets, protective suits, gloves)
(f) Low-temperature environments the effects of which are comparable to those of an air temperature of – 50 °C or less	Insulating textiles with a coefficient of transmission of incident thermal flux as low as required under the foreseeable conditions of use (e.g., jackets, protective suits, gloves)
(g) Falling from a height	High-strength textiles preventing a collision between user and obstacle, including a shock-absorbent material to prevent injury (e.g., ropes, body harnesses, airbags)
(h) Electric shock and live working	Insulating textiles protecting the body against the effects of electric current or conductive textiles ensuring that there is no difference of potential between the user and the installations
(i) Drowning	Textiles enabling the user to return to the surface as quickly as possible and keeping the user afloat in a position which permits breathing (e.g., life jackets, inflated by gas)
(j) Cuts by hand-held chainsaws	Textiles providing sufficient resistance to abrasion, perforation and gashing to protect the user from superficial injuries (e.g., pants, suits, gloves)

Table 9.2 (continued)

Risk	Exemplary textile solution within PPE
(k) High-pressure jets	Textiles providing sufficient resistance to high-pressure jets to protect the user from superficial injuries (e.g., pants, suits, gloves)
(l) Bullet wounds or knife stabs	High-strength textiles protecting the user from superficial injuries and penetration into the body (e.g., protective vests, helmets)
(m) Harmful noise	Insulating textiles protecting the user from harmful noise (e.g., headphones)

thickening fluids are added to the fabric. Thus, textile layers can be reduced which means that the user´s wearing comfort increases [5].

The role that particle technology can play in risk mitigation using PPE will in the following be discussed in the exemplary fire fighters' scenario. The RPN as above introduced will be used to identify potentials of particle technology in PPE.

An external hazard such as the unexpected leakage of a gaseous hazardous material could be addressed by reducing the severity of hazardous material getting in contact with the user's PPE using neutralizing particles. The probability of a diffusion of the hazardous material into the PPE could be reduced by using particles that form a sealing layer on the inner layers of the user's PPE. The ease of detection of hazardous material in the proximity of the user can be enhanced by using chemochromic pigments which react in the presence of specific chemicals by changing their color.

The risk of a defective PPE (such as a helmet exposed to excessive UV radiation) can be mitigated by reducing the severity of a resulting failure using a shear-thickening or self-healing layer. The probability of occurrence of a failure of the helmet material can be reduced by applying a coating with reflective particles leading to a reduced absorption of UV radiation in the helmet and hence a longer lifetime of the PPE. The ease of detection of a defect in the protective function of the helmet can be increased by applying photochromic particles which indicate an excessive UV radiation by changing their color.

In case of a critical health status of the user, such as nausea due to overheating, particle technology based on cooling particles can be used for a reduction of severity by extending the time for a rescue mission. Such cooling particles can furthermore be used preventively to reduce the probability of occurrence of a nausea due to overheating. In case the user collapses and must be found by a rescue team within instants, luminescent particles could enhance the ease of his detection by making the surface of the firefighter's PPE more visible.

The types of risk mitigation and the identification of suitable particle technology using the factors of the RPN are summarized in Table 9.3 for the context of the presented firefighting scenarios.

Table 9.3: Types of risk mitigation and identification of suitable particle technology in the context of firefighters.

Origin of risk	Factor of RPN to be reduced	Suitable concept of particle technology
(a) External hazard (leakage of hazardous gaseous substance)	Severity (S)	Neutralizing particles
	Probability of occurrence (P)	Reactive particles for enhanced sealing
	Ease of detection (D)	Chemochromic pigments
(b) Defective PPE (helmet with exposed to UV radiation)	S	Self-healing effect
	P	Protective coating with reflective particles
	D	Photochromic pigments
(c) Critical health status (nausea due to overheating)	S	Cooling particles
	P	Cooling particles
	D	Luminescent pigments

Due to the variety of risks, causes for the risks as well as functionalities of particles, there is a multitude of possibilities to integrate particles into PPE in a meaningful way. Table 9.4 provides an overview of the functionalities of particles and lists examples of individual applications for particles within PPE. For categorization purposes, the following functionalities of particles have been classified:
- Light emission
- Color/visibility
- Conductivity
- Flame resistance/retardancy
- Substance degradation/disarming
- Isolation thermal/radiation/reduction of impact
- Isolation against substances
- Mechanical resistance
- Influencing surface tension

Table 9.4: Particles and their functions for PPE applications (PPE category III includes exclusively the risks that may cause very serious consequences such as death or irreversible damage to health).

Functions of particles in PPE	Particles	Examples	Type of risk												
			Substances and Mixtures hazardous to Health	**Atmospheres with Oxygen Deficiency**	**Harmful biological Agents**	**Ionizing Radiation**	**Warm Environment having Effects comparable to those of an Environment with an Air Temperature of 100 °C or more**	**Cold Environments having Effects comparable to those of an Environment with an Air Temperature of -50 °C or less**	**Falls from a Height**	**Electric Shock and Work on electrically Live Parts**	**Drowning**	**Injuries caused by Hand Chainsaws**	**High Pressure Jet**	**Injuries caused by Projectiles or Knife stabs**	**Harmful Noise**
			Harmful substances	**Hazard of breathless-ness and suffocation**	**Harmful substances**	**Damage to individual cells or organs**	**Harmful temperature**	**Harmful temperature**	**Mechanical impact**	**Cardiac Crisis; Hazard of breathless-ness and suffocation**	**Hazard of breathless-ness and suffocation**	**Mechanical impact**	**Mechanical impact**	**Mechanical impact**	**Damage to the hearing**
Light emission	Fluorescent phosphor particles	Indication of substance	x*	x*	x*	x*	x*	x*	x*	x*	x*	x*	x*	x*	x*
Color/ visibility	pigments, etc.	color, high visibility	x*	x*	x*	x*	x*	x*	x*	x*	x*	x*	x*	x*	x*
		Reflection, high visibility	x*	x*	x*	x*	x*	x*	x*	x*	x*	x*	x*	x*	x*

Conductivity	Carbo-nanotubes	Antistatic product protective clothing for the ESD area/high-frequency (HF) protective clothing, indication by forwarding sensor signals	X*	X*	X*	X*	X*	X*	X*	X* X	X*	X*	X*	X*	X*
	Graphene	See above	X*	X*	X*	X*	X*	X*	X*	X X*	X*	X*	X*	X*	X*
Flame resistancy/retardancy	Exfoliated graphite	Inflates and thereby reduce oxygen supply from flames	–	–	–	–	X	–	–	–	–	–	–	–	–
Substance degradation/disarm	Titanium dioxide	Uses solar energy – conversion into chemical energy – dissolve harmful substances (organic materials) in water and air	X	–	X	–	–	–	–	–	–	–	–	–	–
	Activated carbon		X	–	X	–	–	–	–	–	–	–	–	–	–
	Magnesium, silver, calcium, zinc, copper, etc.	Antimicrobial effect	X	–	X	–	–	–	–	–	–	–	–	–	–

(continued)

Table 9.4 (continued)

Functions of particles in PPE	Particles	Examples	Type of risk												
			Substances and Mixtures hazardous to Health	Atmospheres with Oxygen Deficiency	Harmful biological Agents	Ionizing Radiation	Warm Environment having Effects comparable to those of an Environment with an Air Temperature of 100 °C or more	Cold Environments having Effects comparable to those of an Environment with an Air Temperature of -50 °C or less	Falls from a Height	Electric Shock and Work on electrically Live Parts	Drowning	Injuries caused by Hand Chainsaws	High Pressure Jet	Injuries caused by Projectiles or Knife stabs	Harmful Noise
			Harmful substances	Hazard of breathless-ness and suffocation	Harmful substances	Damage to individual cells or organs	Harmful temperature	Harmful temperature	Mechanical impact	Cardiac Crisis; Hazard of breathless-ness and suffocation	Hazard of breathless-ness and suffocation	Mechanical impact	Mechanical impact	Mechanical impact	Damage to the hearing
Isolation thermal/ radiation/ reduction of impact	PCM, for example containing paraffin	Phases transition to equalize the temperature	–	–	–	–	X	X	–	–	–	–	–	–	–
	Magnesium oxide + silver (MgO-$AgNO_3$)	Slow down heat + oxygen transfer	–	–	–	–	X	X	–	–	–	–	–	–	–
	Sb_2O_3 + DBDPE	Solar reflective particles – protection against heat radiation (infrared, UV)	–	–	–	–	X	–	–	–	–	–	–	–	–

	Exfoliated graphite	Thermal isolation	–	–	–	–	X	–	–	–	–	–	–	–	–
	Polydopamine, AgNP, PDM	Nonwoven coated with various particles	–	–	–	X	–	–	–	–	–	–	–	–	–
	Aerogel	Out of different materials	–	–	–	–	X	X	–	–	–	–	–	–	X
Isolation against substances	PTFE + fluorinated alkyl	On aramid fibers – self-cleaning and protection against individual acids and bases	X	–	–	–	–	–	–	–	–	–	–	–	–
	Activated carbon	s.o.	X	–	X	–	–	–	–	–	–	–	–	–	–
Mechanical resistance	Silicon carbide	Cut protection	–	–	–	–	–	–	–	–	–	X	–	X	–
	Hard metal oxide particles + silica nanoparticles	Sheer thickening fluids	–	–	–	–	–	–	X	–	–	–	X	X	–
Influencing surface tension	PTFE + fluorinated alkyl	on aramid fibers – self-cleaning and protection against individual acids and bases	X	–	X	–	–	–	–	–	–	–	–	–	–

*Indirect effect: function of the particles is used for indication of risks or warning of user/surrounding people to minimize risk.

9.1.1.4.1 Particles to create an optical effect

On the one hand, light emitting/reflecting particles or color particles enhance the visibility of PPE, but on the other hand, they can warn against dangerous situations in combination with sensors or actors. Therefore, these particles increase the protection level of PPE. Light emitting/reflecting and color particles can be used wherever a high visibility of the personnel or an advanced warning system is reasonable, such as in firefighters´ PPE, first responders´ PPE or PPE against hazardous substances, see table 9.4.

Fluorescent particles

Lu et al. [6] described how paraffin waxes and fluorescent phosphor particles are encapsulated in a phase change material (PCM) and integrated into polyaniline fibers. The fluorescent phosphor ($SrAl_2O_4$: Eu^{2+}, Dy^{3+}) has a green afterglow of about 1 h after exposed to UV-light irradiation for 1 min. Due to easy afterglow recovery, the phosphor is applicable for cyclic utilization.

Luminescent particles

Electroluminescent layers, which emit light when an electrical voltage is applied, can be used to produce warning clothing, which work reliably even in absolute darkness. A textile with luminescent properties can be prepared as follows:

- optional application of a protective layer on a first surface of the textile substrate;
- applying a first transparent conductive layer to the first surface of the textile substrate or the protective layer;
- applying at least one light emitting layer to the transparent conductive layer;
- applying at least one dielectric layer to the light emitting layer;
- application of at least one second conductive layer to the dielectric layer;
- optionally, application of connecting elements to the conductive layers;
- electrically connecting the first transparent conductive layer and the second conductive layer with a voltage source.

As those steps are quite time expensive, a different method to produce electroluminescent textiles is described [7].

The electroluminescent textile consists of a textile substrate with an applied layer structure. This layer structure includes two electrically conductive layers, which are arranged at a certain distance from each other. A light emitting layer with luminescent pigments is arranged between these conductive layers. This light emitting layer has dielectric properties.

The luminescent properties are reached by means of luminescent particles. Suitable luminescent particles or pigments are zinc sulfide, cadmium sulfide, cadmium selenide or zinc telluride and gallium nitride doped with copper or manganese [7].

9.1.1.4.2 Conductivity

Particles with conductive properties are particularly relevant in the field of smart textiles or smart PPE. Due to the electrically conductive particles, electricity can be transmitted within the PPE. Thus, it is possible, to supply integrated sensors or other electric components. Examples of conductive particles are silver and copper particles, but also carbon modifications. Conductive carbon particles include, for example, carbon black, carbon nanotubes or graphene [26].

9.1.1.4.3 Flame resistancy/retardancy

In case of a fire *expandable graphite* expands to several hundred times of its original volume due to heat exposure. It forms an intumescent layer (volume of layer increases, density decreases) on the material surface which slows down the spread of fire and counteracts the most dangerous consequences of fire to humans, namely the formation of toxic gases and smoke. Expandable graphite is most suitable as flame retardant in films or coatings and can be used to protect against environments with an air temperature of 100 °C or more (see table 9.4 [8]).

9.1.1.4.4 Isolation/reduction of thermal or radiant impact

There are many different particles which decrease the impact of cold and warm environments or the impact of radiation (see table 9.4). That is why these particles are sufficient to use in applications like firefighter´s clothing, EMI protection or cold protection clothing. Some examples of particles will be described in the following.

Phase change materials

PCM can change its aggregate state from solid to liquid or the other way around (as shown in Figure 9.2). In textiles, PCMs are either half-solid or semi liquid at body temperature. If the body temperature increases, the half-solid part of the PCM changes into a liquid state and thus it absorbs a part of the additional body heat. In case of cold temperature, the semi liquid part of the PCMs change into a solid state and warms up the body [5, 9].

A product example to protect against cold temperature is "Outlast™" by Outlast Technologies, Inc., USA. Small microparticles of PCM are encapsulated in a polymer which can be produced into fibers, films or coatings [11].

Magnesium oxide and silver nitrate particles

Sbai et al. [12] described a method to increase flame-retardant properties of textiles by adding magnesium oxide (MgO) and silver nitrate ($AgNO_3$) particles into the polymer substrate of electrospun nanofibers. The combination of MgO and Ag forms a barrier which slows the heat and oxygen transfer. Electrospun nylon 6 fibers coated with MgO-$AgNO_3$ particles are applied on a cotton substrate. Afterwards, a horizontal and vertical burning test according to UL 94 standard was

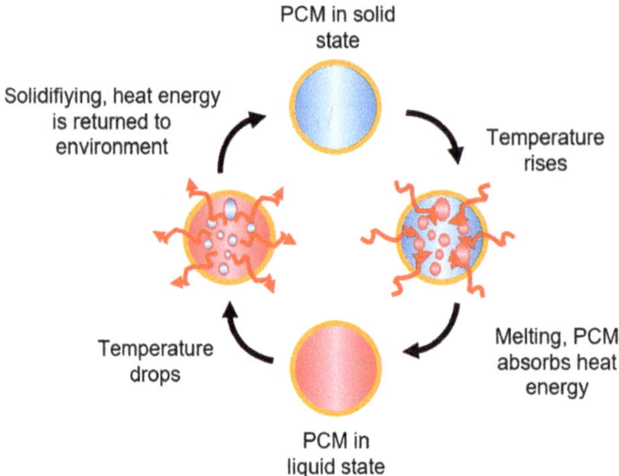

Figure 9.2: Behavior of PCM [10].

carried out. The best combustion rate of 1.56 mm/s was reached at 3 wt.% MgO and 0.25 wt.% $AgNO_3$.

Solar-reflective particles
Antimony oxides (Sb_2O_3) and decabromodiphenylethane (DBDPE) particles can be used as solar-reflective particles and flame retardants as well as an increase of water repellence. A reflectance of up to 86.1% in visible light region and up to 72.9% near infrared light region can be reached by adding 5 wt.% Sb_2O_3, 15 wt.% DBDPE and 1 wt.% of antidripping agent on a material surface. Compared to a polymer matrix these properties lead to a maximum temperature drop of 31.5 °C in indoor environment test and 14 °C in outdoor environment test [13].

Expandable graphite
As already mentioned in 1.1.4.4 expandable graphite can be used as a flame retardant but it can also be used as heat isolation. The product Flacavon DPL by Schill + Seilacher GmbH, Böblingen which consists of an ethylene vinyl acetate adhesive and expandable graphite can be used as a textile coating. Flacavon DPL has an application temperature of up to 130 °C [14].

Silver particles in combination with polydopamine
Lou et al. [15] introduced a method to produce a superhydrophobic and electromagnetic interference (EMI) shielding nonwoven fabric. A polypropylene (PP) nonwoven was coated with several layers. First, the PP nonwoven was impregnated with a polydopamine solution for 12 h and by 20 °C. As a second step, a silver trifluoroacetate (STA) solution was prepared. The nonwoven was dipped into it for 10 min and afterwards into a $N_2H_4 \cdot H_2O$ solution for reducing the STA to Ag nanoparticles. As a last step, a polydimethylsiloxane solution was applied onto the nonwoven and then

the coated textile dried in a drying oven at 80 °C for 1 h. An EMI shielding nonwoven with was produced.

Due to the very high conductivity of 81.2 S/cm and the porous structure, the nonwoven has an excellent EMI shielding performance with the peak shielding effectiveness of 71.2 dB and the specific shielding effectiveness of 270.7 dB cm^3/g in the X-band.

9.1.1.4.5 Neutralization of harmful substances/viruses/bacteria/fungi/vectors

Particles for degradation of harmful substances, viruses, bacteria, fungi or vectors can be used in PPE against "substances and mixtures hazardous to health" or against "harmful biological agents" (see table 9.4).

Metal oxide particles

Metal oxide (nano)particles have antimicrobial behavior. Some of them are listed in the following [12, 16]:

- Magnesium oxide (MgO)
- Calcium oxide (CaO)
- Zinc oxide (ZnO)
- Copper oxide (CuO)
- Cerium oxide (CeO_2)
- Silver nanoparticles (Ag)

Koper et al. [17] presented MgO and CaO nanoparticles . These nanoparticles are carrying active forms of halogens, for example, $MgO \cdot Cl_2$ and $MgO \cdot Br_2$. By means of these particles, about 90% of *Escherichia coli*, *Bacillus cereus* and *Bacillus globigii* are killed within several minutes. Other bacillus species can be decontaminated within hours.

Titanium dioxide particles

Using titanium dioxide (TiO_2) particles is a common method to degrade harmful organic materials in air and water by means of photocatalysis. But the degradation of harmful toxic chemicals is still a challenge. The TiO_2 particles have a high reaction efficiency if they are exposed to natural sunlight or mild UV light. Due to this behavior it is necessary to adjust an adequate activity without any degradation of the textile substrate. To achieve a great photocatalytic activity, a high surface area is elementary. To set the high surface area, TiO_2 has to be in its anatase crystalline structure [18].

In Lee et al. [18], negatively charged TiO_2 particles are directly coated onto electrospun fibers. By using layer-by-layer deposition, TiO_2 particles and positive charged polyhedral oligosilsesquioxane (POSS) molecules are applied onto the fibers. Thus, the cohesion between the particles and the fibers can be enhanced and a high surface area is reached. The coated electrospun fibers degrade harmful organic chemicals without harming the textile substrate.

Silica gel particles plus N-halamine siloxan

The functionalization of the surface of silica gel particles with *N*-halamine siloxan is described in [19]. It is possible to gain a covalent bonding of hydantoinylsiloxane (monomeric or polymeric form) and coat particles of silica gel. To enhance a biocidal behavior, the coated silica particles were chlorinated with aqueous solution of household bleaches. Thus, *S. aureud* and *E. coli* O157 were disarmed.

Silver glass particles

In the SuperFabric®-AmTex™, Superfabric, Oakdale, USA, the antimicrobial behavior is achieved by silver glass particles. Hard guard plates (to achieve mechanical resistance) are applied on the base fabric. Into these guard plates, small particles of silver glass are embedded. In this way, the fabric degrades viruses, bacteria, mold, mildew, fungi and algae [20].

9.1.1.4.6 Isolation against substances/viruses/bacteria/fungi/vectors

To enhance the protection of PPE against different substances, viruses, bacteria, fungi and vectors certain particles like polytetrafluoroethylene (PTFE) or activated carbon can be used as functionalization. Especially particles, which influence the surface tension of a textile or of PPE are relevant for the isolation of different substances.

In Yeerken et al. [21] PTFE microparticles in combination with fluorinated alkyl silanes are coated on aramid fabrics. Thus, a self-cleaning and superamphiphobic (see Figure 9.3) behavior can be set. The functionalized fabric was stressed with H_2SO_4 (40%) and NaOH (40%). A penetration of the acid and the alkali into the textile could not be detected within a period of 180 min. There was also no noticeable change in weight-loss rate after 100 h of exposure to common acids and alkalis.

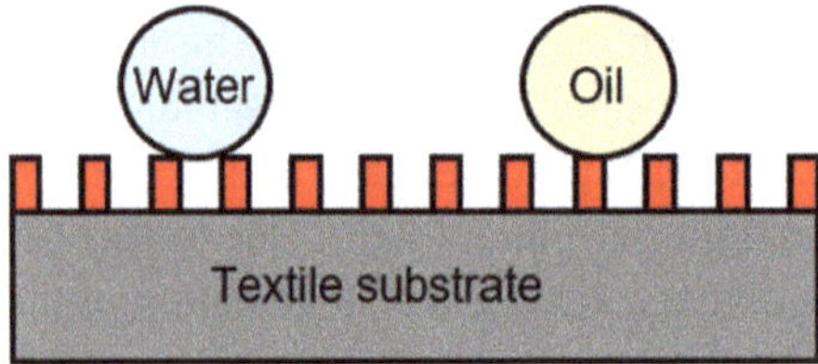

Figure 9.3: Superamphiphobic and self-cleaning behavior.

9.1.1.4.7 Mechanical resistance

To protect the users against projectiles, knife stabs, high pressure jets or chainsaws the mechanical resistance of PPE has to be increased. One method to improve the mechanical resistance is the use of particles with a shear thickening behavior or silicon carbide particles.

Shear thickening fluids
Using shear thickening fluid increases the protection performance of textiles against mechanical impact (e.g., bullets, knife stabs; see table 9.4). In Figure 9.4, the different behavior of a shear thickening fluid in comparison to a Newtonian and dilatant fluid can be seen. The shear thickening behavior is reversible which means that the disappearance of the impact force reduces the viscosity directly [5, 22].

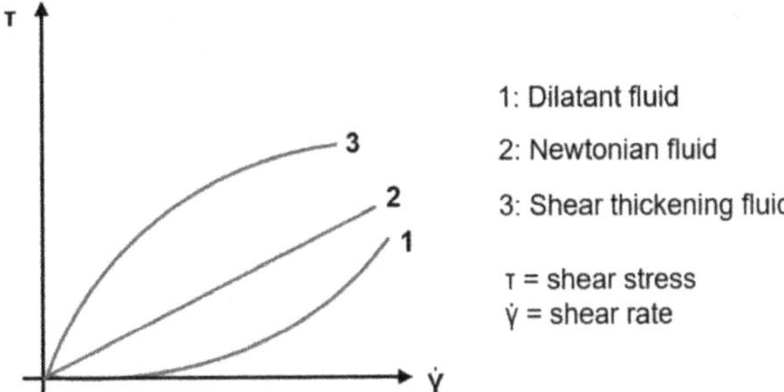

Figure 9.4: Shear stress–shear rate diagram of a shear thickening, Newtonian and dilatant fluid.

In Hassan et al. [22], a shear thickening fluid made of a combination of hard metal oxide particles and silica particles is invented. It is used to coat the textile layer of bullet-resistant body armor. In this way, it is possible to reduce the number of textile layers of the body armor which improves the wearing comfort of the user. The hard metal oxides are suspended in a liquid polymer which consists of silica particles dispersed in liquid polyethylene glycol and was exposed to an ultrasound irradiation. If a bullet hits the body armor the material changes from a fluid-like to a solid-like status. This behavior increases the protection level [9].

Another shear thickening fluid to improve the protection level of body armor is introduced in [23]. Silica particles are also dispersed in polyethylene glycol. Fabrics made of Spectra®, Honeywell International Inc., Morriston, USA, Kevlar® 802 F, DuPont de Nemours Inc., Wilmington, USA or Kevlar® 363 2S, DuPont de Nemours Inc., Wilmington, USA are coated with the shear thickening fluid. The energy absorption against ballistic impact was increased between 19–58%.

Silicon carbide particles
Silicon carbide particles can be used to enhance the protection against bullets, needles, knives or chainsaws. On the one hand. these particles blunt the knife point. and on the other hand. the mechanical impact can be reduced [11, 24].

Teijin Twaron GmbH, Wuppertal, Germany has invented the fabric Twaron® SRM. Silica carbide particles are dispersed in a polymer matrix and fixed on a *para*-Aramid fabric which results in a sealed surface. While the coating absorbs the impact of needles and bullets, the *para*-Aramid reduces the impact energy afterwards [24, 25].

9.1.1.5 Summary and outlook

Particle technology offers a wide range of potentials for the enhancement of safety in working and leisure environments. A number of these potentials are already widely in use, some require further research as well as supporting actions for industrial take-up and market entry.

The aspects of sustainability/circular economy, production technology on industrial scale, standardization and conformity assessment are the most prominent challenges for a comprehensive industrial take-up. Research and innovation activities will hence need to be focused on these fields. life cycle assessment as well as product environmental footprint will be relevant methods and evaluation methods to be used, once industrial value chains have been described. An engagement in standardization committees furthermore supports the exchange among a spectrum of stakeholder groups interested in a safe and sustainable use of particle technology in PPE.

References

[1] European Chemicals Agency: Leitlinie zur CLP-Verordnung. https://echa.europa.eu/de/guidance-documents/guidance-on-clp. Accessed 30 July 2021.

[2] European Chemicals Agency: Hazard Class Table. https://echa.europa.eu/de/support/mixture-classification/hazard-class-table. Accessed 30 July 2021.

[3] European Parliament and Council: Regulation (EU) 2016/425 of The European Parliament and of the Council of 9 March 2016 on personal protective equipment and repealing Council Directive 89/686/EEC, 2016.

[4] Kiran, D. R.: Total Quality Management. Elsevier, 2017; pp. 373–389. Oxford, United Kingdom.

[5] Cao, H.: Active Coatings for Smart Textiles. Elsevier, 2016; pp. 375–389. Sawston, United Kingdom.

[6] Lu, Y.; Xiao, X.; Liu, Y.; Wang, J.; Qi, S.; Huan, C.; Liu, H.; Zhu, Y.; Xu, G.: Achieving multifunctional smart textile with long afterglow and thermo-regulation via coaxial electrospinning. Journal of Alloys and Compounds, 2020; 812.

[7] Schneider, R.; Frick, S.; Brühl, H.: Elektrolumineszierendes Textil: Offenlegungsschrift, 2012. DE 10 2012 109 763 A1.

[8] Rathberger, K.: Dämmschichtbildner Blähgraphit – hochwirksamer mineralischer Flammschutz für Kunststoffe, 2012. https://www.openpr.de/news/657460/Daemmschichtbildner-Blaehgraphit-hochwirksamer-mineralischer-Flammschutz-fuer-Kunststoffe.html. Accessed 19 October 2020.

[9] Dolez, P. I.; Mlynarek, J.: Smart materials for personal protective equipment. Smart Textiles and Their Applications. Elsevier, 2016; pp. 497–517. Sawston, United Kingdom.

[10] Juarez, A.; Balart, R.; Ferrándiz, S.; Peydro, M.: Classification of phase change materials and his behaviour in SEBS / PCM blends; 5th Manufaturing Engineering Society International Conference; June 2013; Zaragoza, Spain. 2013.

[11] Holmes, D. A.; Horrocks, A. R.: Technical textiles for survival. Handbook of Technical Textiles. Elsevier, 2016; pp. 287–323. Sawston, United Kingdom.

[12] Sbai, S. J.; Boukhriss, A.; Majid, S.; Gmouh, S.: The recent advances in nanotechnologies for textile functionalization. Advances in Functional and Protective Textiles. Elsevier, 2020; pp. 531–568. Woodhead Publishing.

[13] Qi, Y.; Zhu, S.; Zhang, J.: The incorporation of modified Sb2O3 and DBDPE: A new member of high solar-reflective particles and their simultaneous application in next-generation multifunctional cool material with improved flame retardancy and lower wetting behaviour. Energy and Buildings, 2018; 172; pp. 47–56.

[14] Schill + Seilacher GmbH: Technical Textiles: Performance chemicals for technical textiles and the associated industry, 2020. https://www.schillseilacher.de/fileadmin/user_upload/Produkte/Chemikalien_fuer_technische_Textilien/Textil_04_2019_Web.pdf. Accessed 19 October 2020.

[15] Luo, J.; Wang, L.; Huang, X.; Li, B.; Guo, Z.; Song, X.; Lin, L.; Tang, L.-C.; Xue, H.; Gao, J.: Mechanically Durable, Highly Conductive, and Anticorrosive Composite Fabrics with Excellent Self-Cleaning Performance for High-Efficiency Electromagnetic Interference Shielding. ACS Applied Materials & Interfaces, 2019; 11(11). pp. 10883–10894.

[16] Turaga, U.; Kendall, R. J.; Singh, V.; Lalagiri, M.; Ramkumar, S. S.: Advances in materials for chemical, biological, radiological and nuclear (CBRN) protective clothing. Advances in Military Textiles and Personal Equipment. Elsevier, 2012; pp. 260–287. Sawston, United Kingdom

[17] Koper, O. B.; Klabunde, J. S.; Marchin, G. L.; Klabunde, K. J.; Stoimenov, P.; Bohra, L.: Nanoscale powders and formulations with biocidal activity toward spores and vegetative cells of bacillus species, viruses, and toxins. Current Microbiology, 2002; 44(1); pp. 49–55.

[18] Lee, J. A.; Krogman, K. C.; Ma, M.; Hill, R. M.; Hammond, P. T.; Rutledge, G. C.: Highly Reactive Multilayer-Assembled TiO 2 Coating on Electrospun Polymer Nanofibers. Advanced Materials, 2009; 21(12). pp. 1252–1256.

[19] Liang, J.; Owens, J. R.; Huang, T. S.; Worley, S. D.: Biocidal hydantoinylsiloxane polymers. IV. N-halamine siloxane-functionalized silica gel. Journal of Applied Polymer Science, 2006; 101(5). pp. 3448–3454.

[20] SuperFabric®: SuperFabric®-AmTex™: A New Line of Antimicrobial SuperFabric®, 2020. http://www.superfabric.com/pdf/hdm11731_catalog_v1_6_2_spread.pdf. Accessed 14 October 2020.

[21] Yeerken, T.; Yu, W.; Feng, J.; Xia, Q.; Liu, H.: Durable superamphiphobic aramid fabrics modified by PTFE and FAS for chemical protective clothing. Progress in Organic Coatings, 2019; 135; pp. 41–50.

[22] Hassan, T. A.; Rangari, V. K.; Jeelani, S.: Synthesis, processing and characterization of shear thickening fluid (STF) impregnated fabric composites. Materials Science and Engineering: A, 2010; 527(12) pp. 2892–2899.

[23] Arora, S.; Majumdar, A.; Butola, B. S.: Soft armour design by angular stacking of shear thickening fluid impregnated high-performance fabrics for quasi-isotropic ballistic response. Composite Structures, 2020; 233. pp. 111–720.

[24] Aramid, T.: Ballistic materials handbook; https://www.teijinaramid.com/wp-content/uploads/2019/11/TEIJ_Handbook_Ballistics_2019_DEF.pdf; pp.1–13; Accessed 25 October 2022 2019.

[25] Gadow, R.; von Niessen, K.: Lightweight Ballistic with Additional Stab Protection Made of Thermally Sprayed Ceramic and Cermet Coatings on Aramide Fabrics; International Journal of Applied Ceramic Technology, Volume 3, Issue 4 pp. 284–292. 2006.

[26] Kirschner, C.; Elektrisch leitfähige Silicone mit niedriger Viskosität; 2019; https://epub.uni-regensburg.de/41124/1/Dissertation_ClaudiaKirschner.pdf; Accessed 22 June 2021

Leonie Beek
10 Filtration

Keywords: filtration, nanofibers, membrane, nonwoven, electrospinning, slip-flow effect

10.1 Introduction

Filtration is an emerging field with an estimated growth of 6.9% in 5 years to a total volume of 34.7 billion € by 2025. Additional growth is projected because of higher legal restrictions, treatment of industrial waste and demand for a safe working environment [1].

Filtration can be classified by air and gas filtration and liquid filtration [2]. One of the greatest current challenges in filtration is the separation of fine particles. Particulate matter (PM) is produced anthropogenically by industry, traffic and households as well as naturally, for example, by wind erosion. Fine dust is classified according to the size of the PM:

- PM10: particle diameter <10 µm
- PM2.5: particle diameter <2.5 µm

The former is filtered from the air through the nasal hairs and mucous membranes in humans. PM2.5 class particles are respirable and can be transported up to 1,000 km by wind [3, 4].

Such particles can trigger a wide range of diseases such as asthma, allergies, silicosis and lung cancer. On average, PM reduces the life of every EU citizen by 12 years. This depends very much on where the person lives. In Germany, the WHO limit value of 10 µg/m^3 is exceeded by about 30%. In comparison, the average PM2.5 pollutions in Delhi, India, in 2019 were 98.6 µg/m^3 and in Dhaka, Bangladesh was 83.3 µg/m^3 [3, 5].

Furthermore, fine dust pollution is also crucial inside closed rooms. The stress indoors increases by use of laser printers, candles, cooking areas and even vacuum cleaners. Overall, 90% of humanity is affected by PM pollution and 11% of all deaths worldwide are due to fine dust pollution [6, 7]. The following measures are used to lessen fine dust [1]:

- Reduce sources
- Separate at the source
- Use protective equipment

Leonie Beek, ITA – Institut für Textiltechnik der RWTH Aachen University, Otto-Blumenthal-Str. 1, 52074 Aachen, Germany, e-mail: leonie.beek@ita.rwth-aachen.de

https://doi.org/10.1515/9783110670776-010

Not only air pollution but also water pollution is a worldwide problem. Clean water is essential for ensuring the health of the world's population. Despite its huge importance, there are threats against safe water management like overexploitation, climate change and pollution. For example, 2.2 billion people lack safely managed drinking water in 2017 and 80% of wastewater is not being treated or reused before flowing back into the environment [2, 8]. On the other hand, the water consumption per capita per year in the USA is at 1.2 Mio. L [9]. As a result, filtration and separation processes play an important role to ensure a better health of the world's population and our planet.

10.2 Basics of filtration and fine dust separation

Filtration is the separation of solid components from a liquid or gaseous medium. During this process, an accumulation of particles can form on the filter surface, which is called filter cake. Examples for the filtration of liquids are drinking water and wastewater treatment. Common examples of gas filtration are the purification of breathing air and exhaust gases.

Within these categories, a distinction is made between depth and surface filtration. In depth filtration no filter cake is formed because the particles are separated within the filter. Surface filtration, on the other hand, can be recognized by the formation of a filter cake, as the particles are separated on the filter's surface (Figure 10.1).

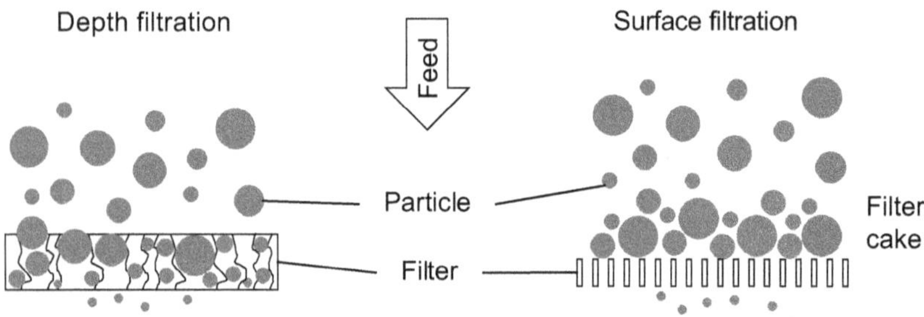

Figure 10.1: Principle of depth and surface filtration.

Textiles are often used as filter media because of their relatively low price (e.g., in comparison to membranes). Besides, the properties of the filter such as pore size and material can be adjusted very well concerning the application. Another advantage of textiles is that they can be produced in large quantities. Three different types of textiles are essentially used: woven fabrics, knitted fabrics and nonwovens (Table 10.1). Woven are mainly used for surface filtration and nonwovens for depth filtration. Knitted fabrics are applicable for both types of filtration. On one hand, an advantage of woven fabrics is that the filtration properties like porosity can be influenced very well whereas in knitted fabrics it is not controllable optimally. In nonwovens the filtration properties are influenced by fiber selection and weight per area. On the other hand, particles with a diameter less than 1 µm size can be separated in nonwovens. For woven and knitted fabrics, the minimum particle size is 2.5 µm. For woven as well as nonwoven filter media, natural and synthetic materials are being used. Materials for knitted filter media are restrained by their flexibility. Due to that, mainly metal and thermoplastic polymers are being used. The applications for nonwoven and woven media range widely, while knitted fabrics are only used in niche applications like mist/liquid filtration [10–12].

Table 10.1: Comparison of textile filters.

	Woven fabrics	Knitted fabrics	Nonwovens
Definition	Flat structure made of filaments crossed at right angles	Fabric based on stitches	Flat structure made of interconnected fibers
Filtration type	Surface filtration	Surface/depth filtration	Surface/depth filtration
Influence on filtration properties	Precisely controllable	Not optimally controllable	Controllable
Filtration particle size	>2.5 µm	>2.5 µm	<0.1 µm
Fiber material	Natural (e.g., wool, cotton, flax), synthetic (e.g., metal, glass, thermoplastic polymers)	Metal, thermoplastic polymers	Natural (e.g., wool, cotton, flax), synthetic (e.g., metal, glass, thermoplastic polymers)
Fiber size	>25 µm (monofilament)	>10 µm (monofilament)	>0.1 µm
Application	Diverse, mainly for cleanable/regenerable filter systems	Niche applications, for example, mist/liquid filtration	Diverse applications, mainly air and gas filtration

In particular, nonwovens are used particularly often as a filter because they show an outstanding surface to volume ratio. A distinction must be made between nonwoven and nonwoven fabrics. Nonwoven is generally used to describe intermediate products for waddings, felts and nonwoven fabrics. The nonwoven itself is a nonwoven bonded by adhesive or entanglement of the fibers [10].

Particle separation in nonwovens is based on three different mechanisms. The inertia effect can filter house dust, the barrier effect occurs during the filtration of fine dust and bacteria, and by diffusion, for example, viruses can be separated. Besides, there is also the sieve effect, through which pollen, for example, can be filtered, but this is not relevant for filtration by use of nonwoven (Figure 10.2) [13].

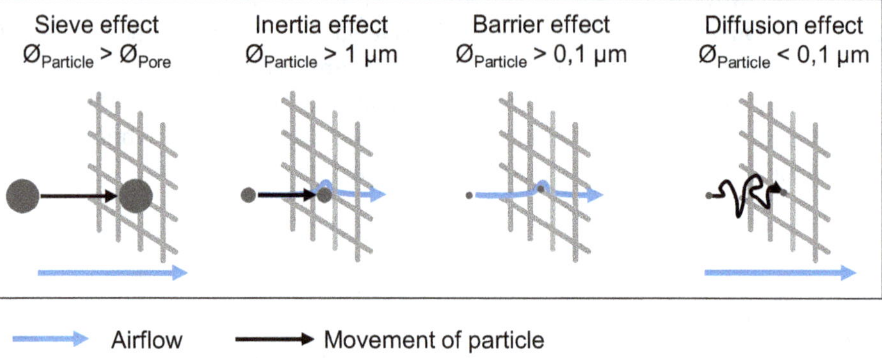

Figure 10.2: Four different physical effects on which filtration is based.

10.3 The challenge of fine dust filtration and the motivation for nanofibers

The effects described above mean that not all particles can be filtered consistently well (Figure 10.3). This so-called fractional effectiveness depends on the particle diameter and is expressed in varying degrees of filter efficiency. The filter efficiency is lowest at the most-penetrating particle size (MPPS): between particle sizes of 0.1–0.4 µm, the diffusion effect decreases, and the barrier effect increases. The problem here is that the diffusion effect decreases with increasing particle diameter, but the barrier effect is constant [13].

To filter particles in the MPPS area reliably, despite this challenge, a denser nonwoven structure can be used. As a result, more than 99% of the particles can be separated at the filter. One example for this kind of filter systems are high-efficiency particulate air filters. However, the limitation is that the pressure behind the filter medium drops sharply, which means that more energy must be used to keep the volume flow constant [13–15].

The pressure drop Δp can be determined by the difference in pressure upstream (p_{in}) and downstream (p_{out}) of the filter [10]:

$$\Delta p = p_{in} - p_{out}$$

The product of the pressure drop and the volume flow $\dot{V}$ determines the power P that is needed to guide the medium through the filter [10]:

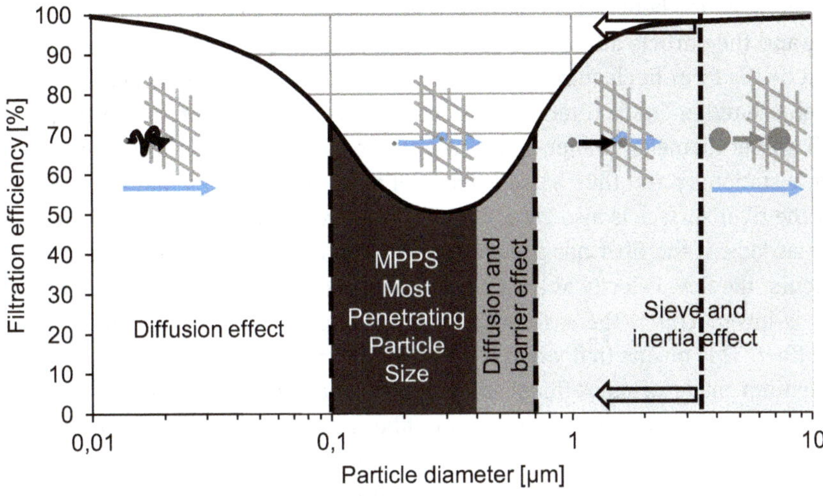

Figure 10.3: Dependence of filtration performance on particle diameter with main effects.

$$P = \dot{V} \cdot \Delta p$$

The Davies equation can also be used to design the filter. This is composed of the gas velocity u_0 of the dynamic viscosity η, the thickness of the filter h, the fiber diameter d_f and the packing density of the filter Φ [10]:

$$\Delta p = \frac{u_0 \cdot \eta \cdot h}{d_f^2}\left(64 \cdot \Phi^{1.5} + \left(1 + 56 \cdot \Phi^3\right)\right)$$

where Δp is the pressure loss, u_0 is gas velocity, η is gas viscosity, h is the thickness of the filter, d_f is fiber diameter and Φ is the packing density of the filter.

The formula shows that the fiber diameter is quadratically inversely proportional to the pressure loss. Following this the fiber diameter should be as large as possible if the pressure drop is to be kept as low as possible. Furthermore, the packing density should be as low as possible. This represents a conflict of objectives with the small fiber diameter and high packing density required for the filtration of fine particles.

To increase the separation in the MPPS area despite these conflicting objectives, the barrier effect must be examined closely. Here the center of mass passes the fiber, but the particle remains attached to the fiber due to its diameter. The effect can be quantified by the barrier effect parameter R, which is composed of the particle diameter ∅P and the fiber diameter $\varnothing_F$ [13]:

$$R = \frac{\varnothing_P}{\varnothing_F}$$

Since the fiber diameter acts inversely proportional to R, it must be as small as possible. Fine fibers are therefore preferred. With a particle size between 0.1 and

0.6 µm, the distance between the airflow and the fiber is greater than the particle diameter and the particle adheres to the fiber surface. This can be explained on the one hand by the form-fit clasping of the pores and stiffeners (mechanical adhesion). On the other hand, a dipole interaction (polarization theory) acts on the surface.

With a fiber diameter greater than 1 µm, the so-called nonslip effect is effective: the fiber is stationary and thus has a velocity of zero, which means that the flow velocity at the fiber surface is also zero. Particles that are in the airflow thus adhere to the fiber surface. If the fiber has a diameter between 65 nm and 1 µm, the slip-flow effect occurs: the flow velocity at the fiber surface is not zero, as the air resistance on the fiber is lower. This is the case because fewer molecules exchange the impulse with the fiber. This means that when a filter medium made of nanofibers (d_f < 1 µm) and a medium made of microfibers (d_f > 1 µm) of the same fiber length is passed through, the pressure drop through the nanofiber medium is lower. Due to the slip flow phenomena, the gas flows closer to the surface of a nanofiber than to that one of a microfiber (Figure 10.4). This means that the direct deposition of small particles in the gas flow is improved as more of these particles pass close enough to collide with the nanofiber than with the microfiber. Therefore, it makes sense to use nanofibers for the development of nonwovens for fine dust filtration [16].

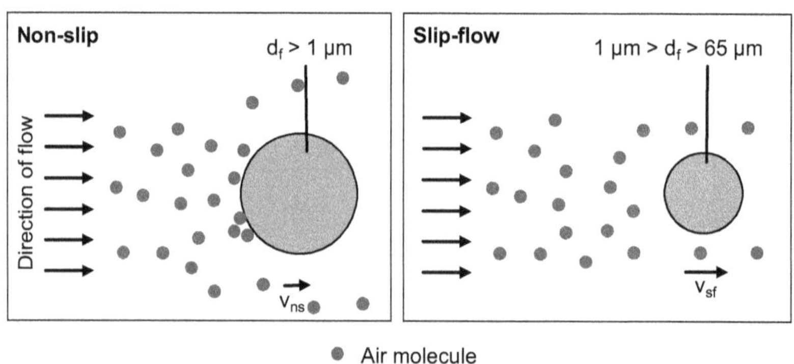

Figure 10.4: Scheme of nonslip effect and slip-flow effect.

10.4 Production of nanofibers

Fibers with a diameter <1 µm can be produced using bicomponent spinning, the melt-blown process and electrospinning (Table 10.2).

In bicomponent spinning, two polymers are spun into one fiber by melt spinning. By use of the matrix/fibril type, it is possible to solve one component or to split the two components after spinning. “Island-in-the-sea” is one type of matrix/fibrils

Table 10.2: Comparison of processes for the production of nanofibers.

	Bicomponent spinning	Melt-blown	Electrospinning
Fiber diameter	>0.7 μm	>0.5 μm	>0.01 μm
Fiber uniformity	High	Medium	High
Throughput	>100 kg/h	>100 kg/h	>100 g/h
Variety of materials	Low	Medium	High
Total rating	Low	Medium	High

fibers whereby many fibrils of one polymer are dispersed in the matrix of another polymer. The matrix is known as the "sea" and the fibrils are known as "islands". The matrix is a soluble material that is washed away by a suitable solvent after the spinning. What remains are bundles of thin parallel fibers, resulting in a fabric that is very soft and flexible. This technique allows the production of fibers with a diameter of between 0.7 μm and 3 μm. The main advantages here are the high fiber uniformity and a throughput of more than 100 kg h^{-1}. However, the choice of material is limited because of specific requirements regarding melting point and viscosity (Figure 10.5, Table 10.2) [17–20].

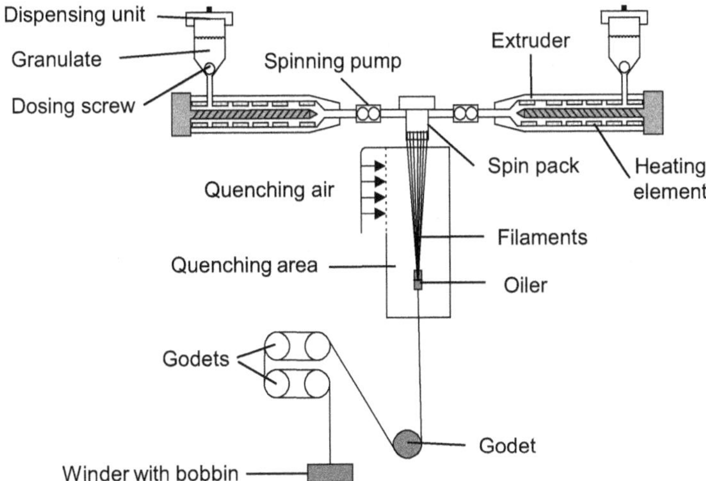

Figure 10.5: Schematic overview of the bicomponent melt spinning process.

In the melt-blown process, the polymer melt is drawn into fibers by compressed air. The fiber diameter is in the range between 0.5 and 20 μm and more than 100 kg/h can be converted. The disadvantages are that the fibers are not uniform, lumps can form and the choice of material is limited by the required melt viscosity (Figure 10.6 and Table 10.2) [17–19].

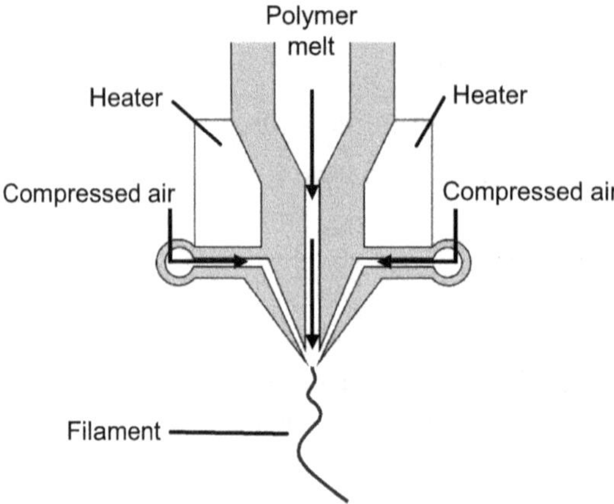

Figure 10.6: Schematic overview of the filament building process in melt-blown technology.

In electrospinning, a polymer is stretched within a high-voltage field. Solvent spinning currently outweighs melt spinning, as melt spinning does not produce a uniform fiber quality with a diameter <1 µm. Generally, the fiber diameter is between 0.01 µm and 10 µm with high fiber uniformity. A disadvantage is the low throughput of slightly more than 100 g h^{-1}. The choice of material is only limited by the requirements on the molecular weight (Figure 10.7 and Table 10.2) [17–19, 21, 22].

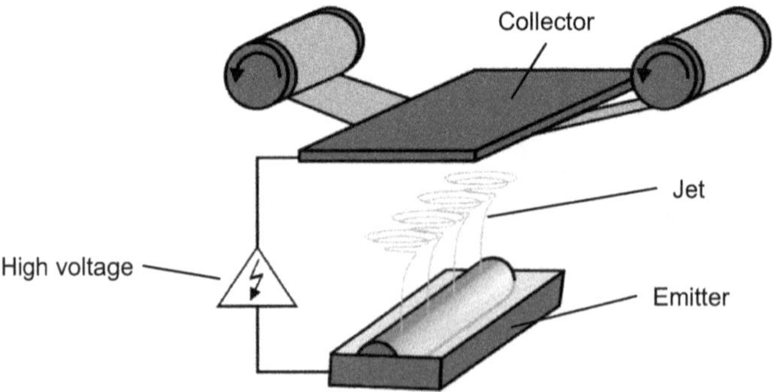

Figure 10.7: Schematic overview over the electrospinning process.

Nanofibers can also be applied as a **coating** to carrier materials without spinning directly onto the carrier. The advantage of the coating technique is that, in contrast to electrospinning, a higher depth of the nonwovens can be achieved. The nanofiber coating not only achieves a higher separation efficiency, but also enables the cleaning of the filter media. In some cases, the properties are so good that they are referred to as having self-cleaning properties. Backwashing does not lead to the detachment of the nanofiber coating because of good adhesion between the coating and the carrier [23].

Furthermore, coatings are highly relevant for filter media to improve functionality. For example, personal protection equipment like face masks can be coated with different materials, like copper oxide or silver nanoparticles, to inactivate Sars-Cov-2-viruses [24].

Particles can not only be deposited through the nanofibers, but certain particles are also implemented into the fibers to achieve additional functionalities. For example, silver nanoparticles are compounded into man-made fibers so that they have antimicrobial properties. This is particularly important for filter media in ventilation systems for hospitals, food production and in the pharmaceutical industry [25].

10.5 Consolidation of nanofiber-based nonwovens for fine dust filtration

The filtration performance as well as the pressure drop of the electrospun nonwoven is superior to conventional membranes. To guarantee good filtration properties with nanofibers, it is essential that the nonwoven layers cannot move relatively to each other. This movement can lead to pressed and damaged nanofibers. Therefore, the nonwoven layers must be consolidated after spinning. Needling, hydroentanglement and hot pressing can be used for this purpose [26, 27]. The aim is to bond the carrier nonwoven to the nanofiber layer without destroying the latter.

During needling, needles pierce the nonwoven, causing the fibers to interlock. The size of the needles must be adjusted to the fineness of the fibers. By use of needling, the nanofibers cannot be bonded to the layer of the carrier nonwoven and holes are created in the nanofiber layer (Figure 10.8). In hydroentanglement, the fibers are intertwined by water jets, whereby the water pressure must be adapted to the type of fiber. The nanofibers are inserted into deeper layers, which also creates holes (Figure 10.8). In hot pressing, the layers are bonded together by pressure and heat. An adhesive can be used optionally. This process allows the nanofiber layer to be bonded to the carrier nonwoven with or without adhesive without damaging it (Figure 10.8) [26, 27].

In general, the peeling force required to separate the nanofiber layer from the carrier web can be increased by hot pressing. The use of a polypropylene (PP)

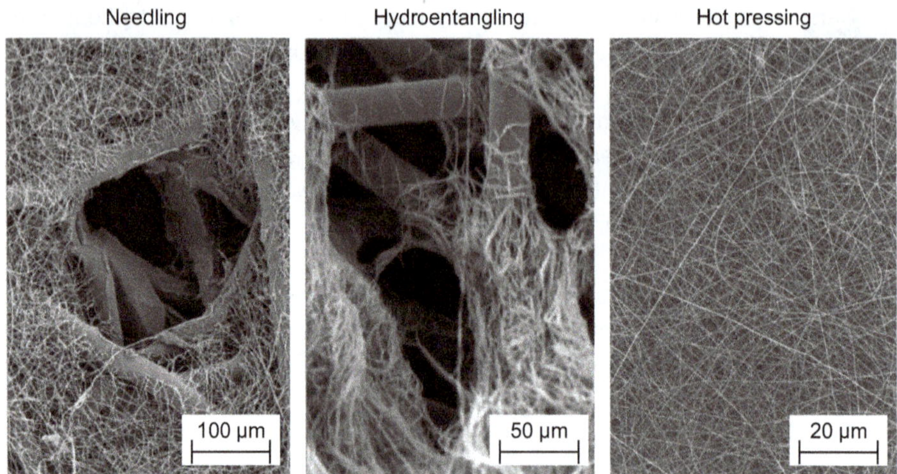

Figure 10.8: Microscopic images of the electrospun nonwoven on the carrier nonwoven bonded with different methods [28].

adhesive only increases the peeling force when using a polyethylene terephthalate (PET) carrier nonwoven. In this case the average peeling force is approximately 51 N/m. When examining a cotton carrier nonwoven and a PP carrier nonwoven, no increase can be measured (Figure 10.9) [28].

Based on these results, hot pressing with the use of a PET carrier nonwoven in combination with a PP adhesive is recommended to achieve a firm bond between the nanofiber layer and carrier nonwoven [28].

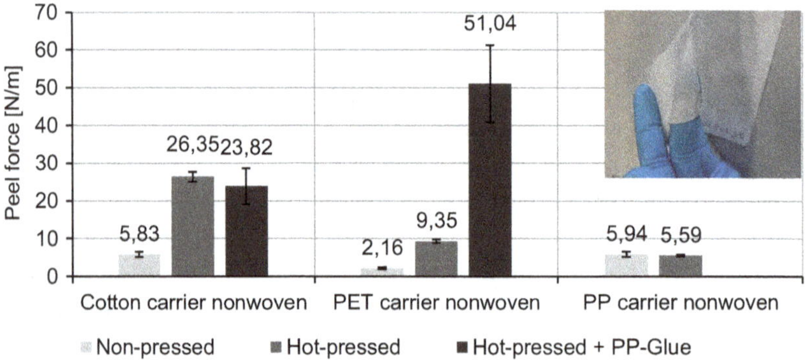

Figure 10.9: Comparison of the peel force (N/m) with different carrier nonwovens [28].

10.6 Applications of new types of fine dust filters

These new types of fine dust filters can be used in various areas. On the one hand, the filters can be integrated into industrial processes, thus reducing dust sources. Here, it is conceivable to convert the nonwoven into bag filters.

Furthermore, the filters can be used in the field of personal protective equipment. Currently, only full-face masks protect the wearer from breathing in fine dust and other substances, but these are uncomfortable. By integrating nanofiber nonwoven into a filtering facepiece, comfort could be combined with the protection of full-face masks. The goal should not be to obtain pure gas, as is the case in some industrial processes, but to enable achievement of the WHO limit values for each wearer. This could reduce the exposure of the population to PM.

In order to enable the widespread use of nanofiber nonwovens, the influence of the adhesive on the filtration line should be investigated and a filter structure consisting of several layers be developed. The scaling effects and further processing of the nonwovens must also be investigated on a pilot scale [28].

10.7 Nanofibers in water filtration

Like mentioned before, clean water is essential for life and threatened by different causes [2, 8]. Nanofiber technology helps ensuring access to clean water by filling a critical gap between conventional filter media and membranes. Before the development of nanofibers, only glass fibers and polytetrafluoroethylene membranes were used for submicron particle filtration. Now, nanofibers outperform these kinds of filters and membranes and are also an alternative to melt-blown nonwoven that are charged to filter fine particles [29].

The main properties of nanofibers, that make them very suitable for a wide range of water filtration applications, are the high specific surface area, low basis weight, the interconnected small pores and simple functionalization [30, 31].

For selective capture of molecules affinity membranes are used. These membranes can be made out of nanofibers for different applications. For example, heavy metals can be filtered from water by use of polyethylenimine nanofibers [30].

Microfiltration (MF) describes a pressure-driven process for particle sizes from 0.1 to 10 µm. It is important for water purification and wastewater treatment, for example, separation of algae and waterborne bacteria. Traditionally, porous membranes manufactured by phase inversion are used for MF. These have a mean pore size of 0.22 µm. Compared to conventional (porous) membranes, membranes made of nanofibers exhibit various advantages like an interconnected open-pore structure. For example, a nanofiber membrane based on Polyacrylonitrile (PAN) has a three times higher permeate flux, a lower pressure drop, 99.9999% retention of bacteria, 99.99% retention of bacteriophage compared with commercial MF membranes [30].

In the ultrafiltration process substances with a size from 1 to 100 nm are filtered from the liquid environment. Examples are viruses, emulsions, proteins and colloids. State of the art are thin-film composite membranes that are consisting of three fundamental layers:

- Ultrathin selective layer
- Middle porous support layer
- Nonwoven fabric layer

The middle layer can be replaced by nanofibers made by electrospinning. It is made of PAN, Polyvinylidene fluoride (PVDF), cross-linked Polyvinyl alcohol (PVA) or cellulose acetate (CA) and has a total thickness of 50 μm. The main advantages of these nanofiber layers are interconnected fibers and directed water channels by the combination of nanofibers and polymer matrix [30, 32].

In nanofiltration and reverse osmosis, particles smaller than 2 nm are being separated from liquids. These technologies are crucial for desalination and therefore to produce drinking water. Membranes for nanofiltration and reverse osmosis show the same layered structure as membranes for ultrafiltration. The main difference is that the first layer has a smaller pore size. Typically, phase inversion is used to produce CA membranes and interfacial polymerization is used for polyamide membranes. The deficit of CA membranes is their limitation of use regarding the pH-value. Cross-linked polyamides can be used to produce nanofiber membranes for nanofiltration and reverse osmosis [32].

Another important part of liquid purification is oil/water separation. At the moment, techniques like skimming, centrifugation, sedimentation and flotation are used. The deficits are high costs, low separation efficiency, complex equipment and secondary pollutants. In comparison to conventional membranes nanofiber membranes show better properties when the oil is present as a separate phase and when it is mixed with the water in an emulsion. For the first application, membranes are produced by spinning polystyrol onto a stainless steel mesh. With a PVDF-based or polysulfone membrane, it is also possible to retain the oil. For separation of emulsions a 3D aerogel based nanofiber membrane can be used [31].

10.8 Conclusion

PM has a huge impact on our society: 90% of the world's population is exposed to ambient and/or household air pollution, which accounts for an estimated seven million premature deaths worldwide every year [6]. Next to air, clean water is crucial for the health of people. Given the fact that 2.2 billion people had no access to clean drinking water in 2018, it is essential to enhance the quality of filtration processes for water [8]. Nonwovens are particularly suitable for separating PM, although the current disadvantage is a high pressure drop. This deficit can be eliminated by using nanofiber nonwovens, as they have a higher filtration capacity with a lower pressure drop. Therefore, no new materials need to be developed, but the production of nanofibers needs to be further developed in order to filter fine particles over a wide area.

10.9 Future developments

Global demand for clean air and water, stringent regulations and public pressure will continue to influence the filtration industry in the future. Not only particles like microplastic can have an impact on the public health but also drug residues, hormones and other chemicals are important. To remove these small particles, fine filter media have to be developed. Besides conventional filters, nontraditional systems like catalytic systems are relevant for this application [29, 31].

Growing areas of interest for filtration are healthcare, digital technology, sustainable materials and systems, zero emissions and high performance. The requirements for filter systems will rise in general, especially regarding reusability and longevity. To reach these goals, special materials, synthetic filter media and hierarchical structures are important. Special materials can have self-healing properties, and systems can be improved by use of biomimetic [29, 31, 33].

The usage of nanofibers will rise especially in healthcare and biomedical industry, e.g. for wound textiles, tissue engineering and drug delivery [29]. Furthermore, nanofibers can be used to produce superhydrophobic surfaces and can be functionalized to perform specific reactions (e.g., ion exchange) for use in filters to chemically remove particles, viruses, microorganisms [29, 34].

References

[1] Industrial Filtration Market worth $41.1 billion by 2025: 2020-11-17T16:35:44.000Z. https://www.marketsandmarkets.com/PressReleases/industrial-filtration.asp. Accessed 17 Nov 2020.920Z.

[2] Sparks, T.; Chase, G.: Filtration – Introduction, Physical Principles and Ratings. In: Filters and Filtration Handbook. Butterworth-Heinemann, Oxford, Waltham 2016; pp. 1–54: doi:10.1016/B978-0-08-099396-6.00001-0.

[3] EEA: Belastung der städtischen Bevölkerung in Deutschland durch Luftverschmutzung mit Schwebstaub* (Feinstaub und Grobstaub) von 2005 bis 2017; 02/2019.

[4] Bundesministerium für Umwelt, Naturschutz, nukleare Sicherheit und Verbraucherschutz (BMUV): www.bmuv.de. Nationales Luftreinhalteprogramm der Bundesrepublik Deutschland.

[5] Greenpeace (Peking): Entwicklung der durchschnittlichen Feinstaubwerte (PM2,5)* für Peking in den Jahren 2013 bis 2017, 03/2018.
[6] WHO: Ambient air pollution: A global assesment of exposure and burden disease, 2016.
[7] WHO: Hauptstädte mit der größten PM2,5-Feinstaubbelastung* weltweit im Jahr, 2019.
[8] United Nations: Goal 6 | Department of Economic and Social Affairs. 2021-01-27T12:57:08.000Z. https://sdgs.un.org/goals/goal6. Accessed 28 Jan 2021.835Z.
[9] Statista: Water consumption by countries | Statista: 2021-01-28T09:57:30.000Z. https://www.statista.com/statistics/263156/water-consumption-in-selected-countries/. Accessed 28 Jan 2021.592Z.
[10] Mao, N.: Nonwoven Fabric Filters. In: Advances in Technical Nonwovens. Woodhead Publishing, Duxford, Cambridge, Kidlington, 2016; pp. 273–310: doi:10.1016/B978-0-08-100575-0.00010-3.
[11] Sparks, T.; Chase, G.: Filter Media. In: Filters and Filtration Handbook. Butterworth-Heinemann, Oxford, Waltham, 2016; pp. 55–115: doi:10.1016/B978-0-08-099396-6.00002-2.
[12] Mullins, B. J.; King, A. J. C.; Mead-Hunter, R.; Heikamp, W.: Knitted Fibrous Filter Media. In: Fibrous Filter Media. Woodhead Publishing, Duxford, Cambridge, Kidlington, 2017; pp. 125–132: doi:10.1016/B978-0-08-100573-6.00004-6.
[13] HS-Luftfilterbau GmbH: Grundlagen Filtertechnik, 2021. https://www.luftfilterbau.de/de/filtertechnik/index.html
[14] Ultravation, I.: What is Pressure Drop? https://www.ultravation.com/news/2013/01/17/what-is-pressure-drop/, 2021.
[15] Mostinsky, I. L.: Filtration. In: A-to-Z Guide to Thermodynamics, Heat and Mass Transfer, and Fluids Engineering. Begellhouse, Redding, 2006. doi:10.1615/AtoZ.f.filtration.
[16] Zhao, X.; Wang, S.; Yin, X.; Yu, J.; Ding, B.: Slip-effect functional air filter for efficient purification of PM2.5. Scientific Reports, 2016; 6, 35472: doi:10.1038/srep35472.
[17] Brown, T. D.; Dalton, P. D.; Hutmacher, D. W.: Melt electrospinning today: An opportune time for an emerging polymer process. Progress in Polymer Science, 2016; 56, 116–166: doi:10.1016/j.progpolymsci.2016.01.001.
[18] Ko, F. K.; Wan, Y.: Introduction to Nanofiber Materials, 1st ed. Cambridge, UK: Cambridge University Press.
[19] Luo, C. J.; Stoyanov, S. D.; Stride, E.; Pelan, E.; Edirisinghe, M.: Electrospinning versus fibre production methods: From specifics to technological convergence. Chemical Society Reviews, 2012; 41, 4708–4735: doi:10.1039/c2cs35083a.
[20] Rupp, J.; Yonenaga, A.: Microfasern – Das neue Image der Chemiefaser. ITB International Textile Bulletin, 2000; 46, 12–24.
[21] Persano, L.; Camposeo, A.; Tekmen, C.; Pisignano, D.: Industrial upscaling of electrospinning and applications of polymer nanofibers: A review. Macromolecular Materials and Engineering, 2013; 298, 504–520: doi:10.1002/mame.201200290.
[22] Agarwal, S.: Electrospinning: A Practical Guide to Nanofibers. Berlin, Boston: De Gruyter, 2016.
[23] Wertz, J.; Schnieders, I.: NanoWeb: Advantages of a New and Advanced Nanofiber Coating Technology for Filtration Media Compared to the Electrospinning Process; 2008.
[24] Rakowska, P.D., Tiddia, M., Faruqui, N. et al. Antiviral surfaces and coatings and their mechanisms of action. Commun Mater 2, 53 (2021). https://doi.org/10.1038/s43246-021-00153-y
[25] Som, C.; Nowack, B.; Wick, P.; Krug, H.: Nanomaterialien in Textilien: Umwelt-, Gesundheits- und Sicherheits-Aspekte: Fokus: synthetische Nanopartikel. St. Gallen, 2010.
[26] direct Industry: http://www.directindustry.de/prod/mse-teknoloji-ltd-sti/product-174678-1778339.html
[27] TWE: Verfestigung von Vliesen. https://www.twe-group.com/entwicklung/verfestigung/
[28] Kruse, M.: Vliese in der Filtration – Zukunft oder Vergangenheit? Aachen, 11.07.2019.

[29] Davis, R.: High-Performance Materials, Nanofibers Emerge As Filtration Solutions | Textile World, 11/17/2020 13:35:47. https://www.textileworld.com/textile-world/features/2019/11/high-performance-materials-nanofibers-emerge-as-filtration-solutions/. Accessed 17 Nov 2020.

[30] Cheng, C.; Li, X.; Yu, X.; Wang, M.; Wang, X.: Electrospun Nanofibers for Water Treatment. In: Ding, B.; Yu, J.; Wang, X., editors: Electrospinning. San Diego: William Andrew, 2018; pp. 419–453: doi:10.1016/B978-0-323-51270-1.00014-5.

[31] Qin, X.; Subianto, S.: Electrospun Nanofibers for Filtration Applications. In: Electrospun Nanofibers. Woodhead Publishing, Duxford, Cambridge, Kidlington 2017; pp. 449–466: doi:10.1016/B978-0-08-100907-9.00017-9.

[32] Ma, H.; Chu, B.; Hsiao, B. S.: 15 – Functional Nanofibers for Water Purification. In: Wei, Q., editor: Functional Nanofibers and Their Applications. Oxford, Philadelphia: Woodhead Publishing, 2012; pp. 331–370: doi:10.1533/9780857095640.2.331.

[33] Unknown: Filter werden immer individueller: Interview mit Dr. Martin Lehmann. 2020-11-1716: 47. https://www.mann-hummel.com/de/das-unternehmen/magazine/automotive-news/automotive-news-ausgabe-012017/luftfilter-nach-mass/. Accessed 17 Nov 2020.

[34] Chase, G. G.; Swaminathan, S.; Raghavan, B.: Functional Nanofibers for Filtration Applications. In: Functional Nanofibers and Their Applications. Woodhead Publishing, Duxford, Cambridge, Kidlington 2012; pp. 121–152: doi:10.1533/9780857095640.2.121.

Annett Schmieder*, Christoph Müller

11 Technical textiles for machine elements

Keywords: five-category scheme, braided ropes, coating and particle technology, conveyors, cranes, discard criteria, high-performance fibers, HM-HT fiber ropes (high modulus – high tenacity), rope drives, ropes service life, running fiber ropes, signs of wear and aging, sheaves, steel wire ropes (SWR), textile mechanical components, traction, load bearing and safety elements, tribological stresses

11.1 Technical fiber braids in conveying engineering

11.1.1 Introduction

Steel wire ropes (SWR) have established as traction, load bearing and safety elements in conveyors, due to their high breaking strength and operational safety as well as out of many years of experience. However, SWR have disadvantages such as high dead weight, high bending stiffness and sensitivity to corrosion. In various areas of conveyor technology, SWR are currently reaching their technical limits due to increasing demands on rope drives. These technical limits can be seen in the example of the highest skyscraper in the world, the Burj Khalifa in Dubai at 828 m. The maximum height of the SWRs in the main elevators is 504 m [18, 22]. This limit is due to the high mass of the rope and the high safety factor to be observed. These cranes are used for the construction of high-rise buildings. The CC 8800–1 Twin is one of the largest cranes in the world and has a maximum tip height of 234 m and an approximately 13-km-long SWR with a net weight of approximately 44 t [11, 21].

Increasing requirements on the energy-efficient and resource-saving operation of technical systems demand new materials, technologies and processes. As a result, textile machine elements like high-modulus and high-tenacity (HM-HT) fiber ropes (high modulus – high tenacity) come into focus of considerations, see Figure 11.1. At present, HM-HT fiber ropes can be produced in such a way that they have the same or higher tensile strength than SWRs, but only have 20–25% of the weight of a SWR.

*Corresponding author: **Annett Schmieder,** Arbeitsgruppe Textile Maschinenelemente, Professur Förder- und Materialflusstechnik, Fakultät für Maschinenbau, Technische Universität Chemnitz, Reichenhainer Str. 70, 09126 Chemnitz, Germany, e-mail: annett.schmieder@mb.tu-chemnitz.de
Christoph Müller, Arbeitsgruppe Textile Maschinenelemente, Professur Förder- und Materialflusstechnik, Fakultät für Maschinenbau, Technische Universität Chemnitz, Reichenhainer Str. 70, 09126 Chemnitz, Germany, e-mail: christoph.mueller@mb.tu-chemnitz.de

https://doi.org/10.1515/9783110670776-011

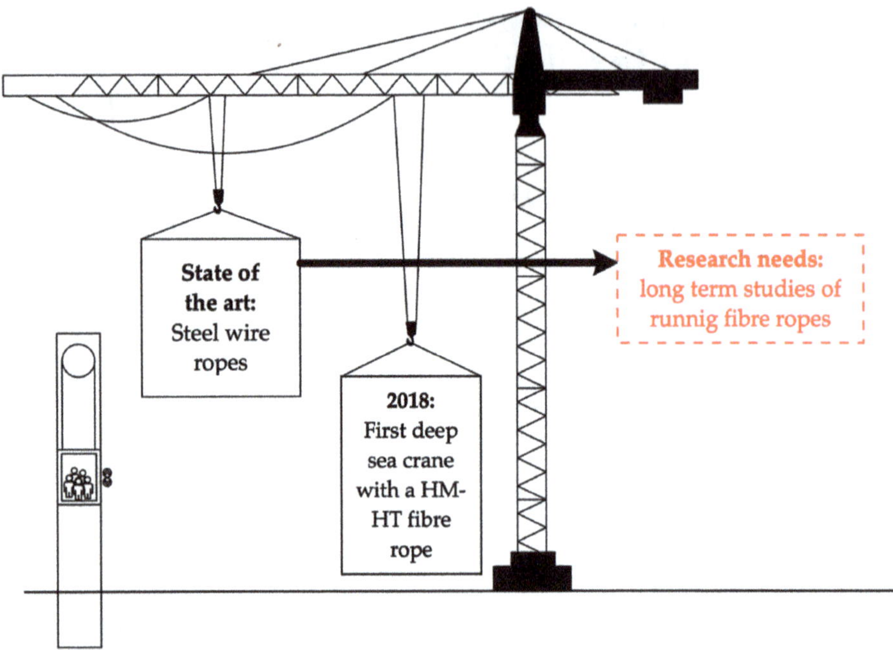

Figure 11.1: State of the art [21].

The advantages of using HM-HT fiber ropes compared to SWRs with a comparable breaking strength are:
- increase in the masses and lengths,
- reduction of the required drive power,
- reduction in bending stiffness,
- good chemical and corrosion resistance and
- good handling when replacing the suspension element [11, 21].

Despite their economic potential and availability, HM-HT fiber ropes have still not been widely used in safety-relevant applications for several years. And also with every innovation, new cases of damage are associated with insufficient experience and long-term studies.

The processing of small or very small particles, so-called particle technology, is nowadays an important component for setting certain properties for many modern materials in a wide variety of technology fields. This also includes the coating of running fiber ropes with dispersed systems that contain particles. In this way, several functions or properties can be combined in the fiber ropes. In general, this can be magnetism, electrical or thermal conductivity, luminescence, UV protection, flame retardancy, improvement of sliding properties, functional fillers and additives, functional surface coatings, recycling tasks, marker particles and so on [7, 26].

The following chapters provide an overview of the most important high-performance fiber materials that can be used for technical applications (see Section 1.1), the coating technologies of fiber ropes (see Section 1.4) and the main failure mechanisms of running fiber ropes (see Section 1.5). Finally, Section 1.6 shows an overview about fiber ropes in rope drives, the standards and guidelines, relevant patents and technology for discarding fiber ropes, different research studies and gives an outlook and research needs.

11.1.2 Terms and definitions of running HM-HT fiber ropes

The aim of recent research is to expand the level of knowledge with regard to the state of wear and aging. Also influencing factors on the lifetime is of interest. Most of all, there is still a lack of a valid discard criterion for running HM-HT fiber ropes (see Figure 11.2) as well as objective methods for lifetime detection. This chapter explained the most important terms and definitions of fiber ropes.

Figure 11.2: Braided fiber rope in running application [21].

In contrast to SWR, where only laid or twisted designs are possible, fiber ropes provide the possibility of braided and laid rope designs. As laid designs are often used with conventional natural and synthetic fiber types (e.g., mooring and towing lines), braided designs are predominantly used for the high-performance types so far.

Braided ropes in comparison to laid ropes are symmetrical with regard to their longitudinal axis which results in a twisting-free design under load. This is a central reason why braided HM-HT fiber ropes as so-called textile machine elements came into consideration for conveyors. Further reasons are no untwisting, a low tendency to kinking, outstanding load–mass relation, easy handling, high wear potential and a relatively equal stress distribution to the load-bearing strands [6, 14, 28]. Table 11.1 shows a few basic textile terms and definitions especially the difference between laid and braided ropes [4, 28].

Table 11.1: Textile specific terms and definitions [4, 14, 28].

Term	Definition
Fiber	"Unit of matter characterized by its flexibility, fineness and high ratio of length to maximum transverse dimension" [4].
Man-made fibers	**Man-made fibres** **Natural polymers** **Plant origin** • Elastodien (Rubber)* • Cellulose fibres: regenerated cellulose (cupro, viscose, modal, paper), cellulose ester (acetate, triacetate) • Alginate fibres **Animal origin** Regenerated protein fibres: plant protein (zein), animal protein (casein) **Inorganic raw materials** Glass fibres Metal fibres Carbon fibres **Synthetic polymers** **Polymerisat fibres** *Polyethylene**, Polypropylene**, Polychlorid*** **Polycondensate fibres** *Polyamide**, Polyester*** **Polyaddition fibres** *Polyurethane**, Elastane*** (*) Low part in rope industry (production of expander ropes). (**) Relevant man-made fibres for making ropes. "Fibers obtained by a manufacturing process, as distinct from materials which occur naturally in fibrous form" [4].
Synthetic fibers	"Fibers made from polymers which are chemical compounds made up of long chains of molecules" [4].
Strand	A strand is a textile structure, which is produced by twisting at least two rope yarns. In a further step, the strands are made into a rope by stranding or braiding [4, 21].
Rope	"Product obtained when three or more strands are twisted or braided or set in a parallel construction to provide a composite cordage article larger than 4 mm in diameter" [4].
Laid rope	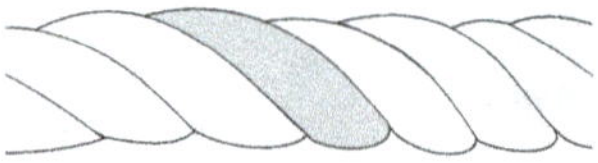 Laid fiber rope [21]. According to DIN EN ISO 1968 [4], stranding is a "process to shape a laid rope by twisting the strands together in such a way that a stable structure is created that does not open."
Braided Rope	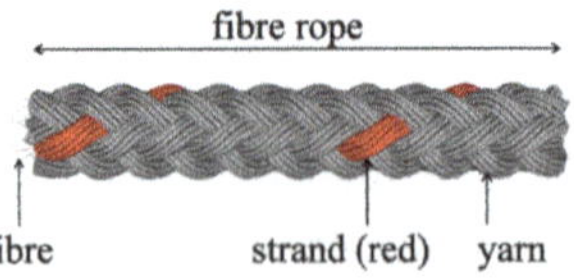 fibre rope fibre strand (red) yarn Braided fiber rope [21]. According to DIN EN ISO 1968 [4], braiding is a "method of crossing strands or braiding yarns in a helical manner." A braided rope has a stable structure and consists of at least three threads or strands, which are crossed diagonally [4, 14, 28]. Braided ropes are made by using a braiding machine.

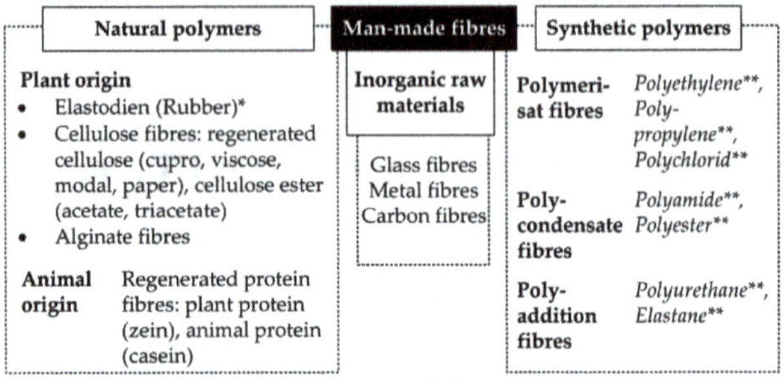

11.1.3 High-performance fibers

In the following explanations show various high-performance fibers and their special properties compared to conventional fibers. High-performance fibers perform significantly better compared to conventional fibers in specific properties, for example, strength, modulus of elasticity, temperature, flame resistance, chemical resistance and so on. Anisotropic thermoplastics with linear long-chain molecules and crystalline structures are called HM-HT polymer fibers. They characterize by a high degree of crystallization, a uniform orientation of the molecules and strong intermolecular bonds. This leads to very high axial stiffness and strength in HM-HT fibers [28]. The main types of high-performance or HM-HT fibers that mostly are used for conveyor applications are high-modulus polyethylene (HMPE), lyotropic liquid crystal polymers (LLCP) – aramid fibers and thermotropic liquid crystal polymers (TLCP).

Polyethylene, which has a high molecular weight (UHMW-PE: ultra-high-molecular-weight polyethylene), is the basic material for HMPE fibers. It is produced in a special form of the solution spinning process, the so-called gel spinning process [16, 23]. The molecular structure enables a tightly packed crystalline arrangement, when the fiber is exposed to stress in the chain direction. Due to the large coverage areas and very long molecular chains, high tensile strengths and modules are achieved in the fibers [5]. A gel-like mass is created. This improves the dimensional stability and stretchability of the fibers to be spun. In this way, an irreversible sliding of the molecular chains is possible due to the high degree of orientation of the chains, the weak intermolecular binding forces and the reduced tendency to become entangled. These irreversible changes in length (caused by molecular sliding processes even with low loads) can be reduced by incorporating co-polymers [28]. Highly modular PE as fibers is known under the trade names Dyneema® and Spectra®.

According to VDI 2500 [28], aramids belong to the LLCP. They are characterized by the binding of amide groups (CONH) to aromatic groups (long-chain polyamides). Fibers are divided into [5, 8, 16, 24, 28], meta-aramids (e.g., Nomex®, Teijinconex®), *para*-aramids (e.g., Kevlar®, Twaron®) and *para*-aromatic copolyamides (e.g., Technora®). The differences in between the types of aramid are based on the linking of the aromatic bridges in the macromolecules. The type of structural bond between the atoms or molecules can be in the meta- or *para*-position (1,3-position or 1,4-position). Fibers with para-position are used for ropes in conveyor technology. The amide groups are the connecting links, like it is in all polyamides (peptide bonds).

TLCP belong to the fully aromatic polyester and are made in the melt. The aromatic groups have strong intermolecular bonds, which give the fibers high tensile strength and modulus see Table 11.2 [28]. The melt cools down and the low-order state is as far as possible transferred to the solid state. The TLCP fibers from the Kuraray company are known under the trade name Vectran®.

Table 11.2: Properties of high-performance fibers [5, 8, 11, 16, 21, 23–25, 28].

Type of fiber	UHMW-PE fibers	LLCP (*para*-aramid copolymer)	TLCP (aromatic polyester copolymer)
Structural formula	$[CH_2 - CH_2]_n$	$[-NH-C_6H_4-NH-]_m[-NH-C_6H_4-O-C_6H_4-NH-]_n[-CO-C_6H_4-CO-]_{m+n}$	$[-O-C_6H_4-CO-]_m[-O-C_{10}H_6-CO-]_n$ 73 % HPS, 27 % HNS
Special features (chemical)	– Highly crystalline, highly stretched UHMW-PE – No functional groups, no side chains – Van der Waals forces as intermolecular forces – Crystallinity up to 95% – Orientation up to 98%	– Two-dimensional grid (layer structure, high anisotropy, tends to fibrillate) – H-bridges in the lattice, – Van der Waals forces between the layers (lower layer bonding forces and thus transverse forces than with *para*-aramids) – Up to 95% crystallinity	– Rod-shaped molecular shape – Extremely high tensile strength and a high modulus of elasticity parallel to the molecular axis
Density in g/cm^3	0.97–0.99	1.39	~1.4
Tensile strength in GPa	2.5–3.6	3.1–3.4	1.1–3.0
Modulus of elasticity in GPa	100–140	60–85	75–103

Elongation at break in %	3–4	4.6	3.3–3.7
Melting point, °C	144–152	–	~330
Decomposition temperature, °C	>300	500	–
Resistances + + + very good; + + good; + moderate; – bad; – – unstable	– Acid: + + +, – Alkalis: + + +, – Water: + + +, – UV radiation: + + to + (depending on the ambient temperature!)	– Organic acids: + +, – Concentrated inorganic acids: – –, – Concentrated alkalis: – –, – Water: good (also against steam/hydrolysis), – UV radiation: + to –	– Inorganic acids (conc. <90 %): + + to + + + – Inorganic bases (conc. <90 %): + + to + + + – Organic solvents: + + + – Water: + + +, also against steam/hydrolysis – UV radiation: +
Special properties	– Buoyant in the water	– Not buoyant in water	– Not buoyant in water
	– Increased strength and reduced elongation possible at all textile processing stages		

11.1.4 Coating and particle technology

Fiber ropes as so-called textile mechanical components mostly are functionalized by impregnations or coatings (see Figure 11.3). Functional improvements are the adjustment of tribological system by (i) impregnation/lubrification for reduced yarn/filament wear, (ii) impregnations/monolithic thermoplastic coatings for improved friction coefficients in the sheave/drum contact area and (iii) impregnations/coatings for UV protection and protection against abrasive particles. States of the art are coating systems on rope, strand or yarn level.

Figure 11.3: Incomplete wetted yarns in ropes (laid and braided design).

In rope coating plants there is often a parallel heat treatment implemented to improve/influence mechanical performance of the fibers. The advantages of this technology are (i) a broad variety of coating systems (particles and solvents) is applicable, (ii) relatively high process speeds (ranges from 5 to 25 m/min or even higher according to process parameters and rope diameter) and (iii) almost 100% of remaining coating amount in the rope until application (last subprocess in rope manufacturing, no abrasive loss of coating). The disadvantages of the respective rope coating technology are mostly described as an incomplete wetting, especially in the crossing over areas of the several yarns. Furthermore, the wetting in the inner cross section of the rope decreases with increasing rope diameter. Whereas in yarn coating plants a qualitative and complete wetting of the yarns is more or less guaranteed. In addition, (i) almost any coating system (solids content, solvents) is applicable, (ii) the coating is adjustable at yarn level, so single yarns in the rope design can be coated, (iii) yarn spreading is an option for optimal filament coating. The predominant shortcoming of the yarn coating technology is their lengthy process design and complex subprocess sequence (rewinding of the braiding bobbins). Typical systems for both, yarn and rope coating, are (i) polyurethane dispersions for abrasion-, cut- and UV protection, (ii) methacrylic esters-based dispersions for hydrophobic treatment, (iii) polyolefin waxes/oil-PTFE-dispersions for lubrification and (iv) polysiloxanes as processing aids.

Particulate materials and disperse systems play an important role in many industrial areas. Particle technology is also of crucial importance for setting the desired properties in ropes, especially in the high-performance field. Important process steps start with the preparation of primary and secondary raw materials up to the coating of fiber ropes with dispersions enriched with particles. In this way, fiber ropes can be developed and produced with specific property profiles. Finally, a product must be improved properties or functions after the particle technology without any negative impact on the rope properties or the use of the rope due to the particle dispersed systems and the resulting interparticle interactions. Particle technology and its materials are said to have great potential for the development of a new generation of fiber ropes with improved mechanical, thermal, tribological and other properties [7, 26].

11.1.5 General wear mechanisms in running ropes

11.1.5.1 Five-category scheme of factors influencing the rope service life

With every new innovation, there are also new cases of damage. In the case of high-performance fiber ropes and their various areas of application, there is still a lack of sufficient long-term studies. This chapter deals with the influencing factors that lead to damage or failure of a rope. Figure 11.4 shows the damage analysis of HM-HT fiber ropes in safety-relevant applications in a five-category scheme with the most important factors influencing the ropes service life. The categories are purpose/handling, rope parameters, machine parameters, environmental conditions and time. The illustration shows that some of the categories are directly related and influencing one another [11, 15, 17, 21, 28, 29].

11.1.5.2 Tribological stresses

In conveyor applications, fiber ropes are exposed to tribological stresses (running over sheaves or drums) in addition to other stresses, see Figure 11.5. This results in aging, wear, fatigue and/or overload signs. These different signs of damage in a running fiber rope are [11, 21]: fatigue of the fibers (e.g., fibrillation of the fibers, transverse cracks), deforming (e.g., twisting of the rope, rope jacket displacement), splitting off of the fibers, rupture of fibers, strands or entire rope and wear-induced changes in braiding length.

Tribological stresses are present particular within running ropes. However, the five-category scheme shows that all influencing variables must be considered in context [21]. The interaction between friction, wear, lubrication, forces, speeds and environmental conditions is called the tribological system (tribosystem). The tribological systems are made up of the system-structure and the system-function. These fulfill the

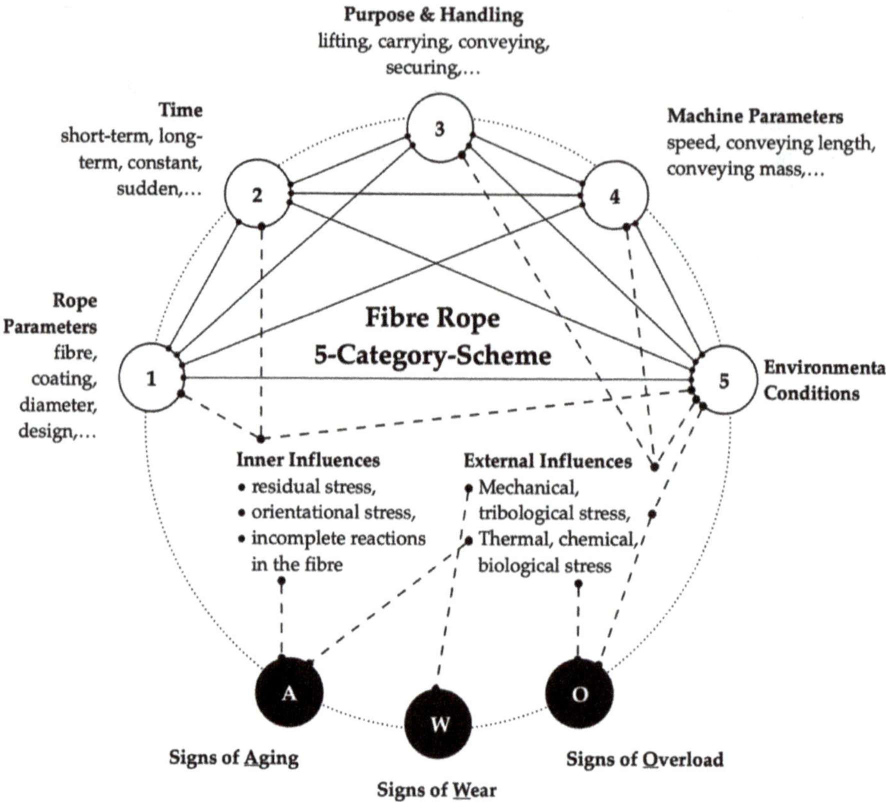

Figure 11.4: Five-category scheme of influences [21].

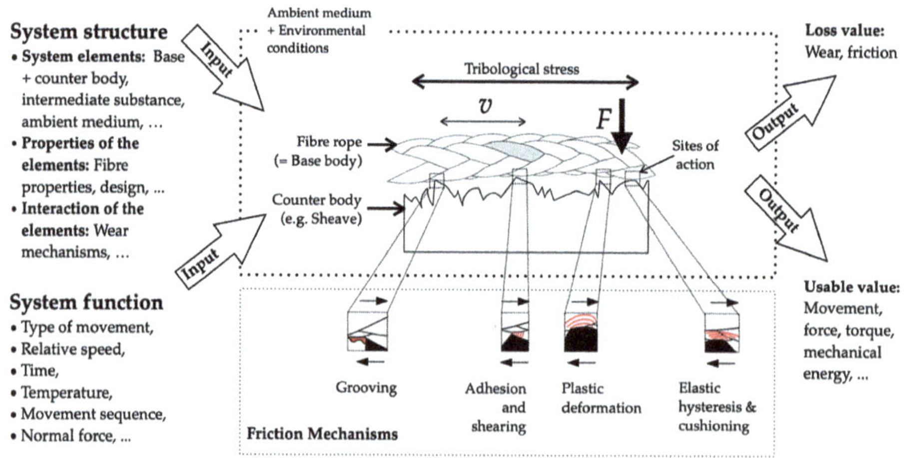

Figure 11.5: Tribological system of a running fiber rope [21].

energy-converting tasks that serve to transfer forces and moments [1, 2, 21]. The system-structure is made up of interactions and properties of the elements base body = fiber rope, counter body = depending on the consideration of the tribological system, the rope itself or the sheave, ambient medium = air, climate and the intermediate material, for example lubricant, . . . The system-function is determined by acting variables, such as speed $\boldsymbol{v}$ and force $\boldsymbol{F}$, environmental conditions, time, purpose and rope parameters. With every tribological system, useful and loss values arise [2]. Figure 11.5 is an example of the tribological system with its system structure and the system functions of a running rope [21].

The friction mechanisms (see Figure 11.5) are divided into the general wear mechanisms (see Table 11.3) adhesion, abrasion, surface disruption and tribochemical reaction. The force-related interactions (load and frictional forces) include abrasion and surface disruption. The adhesion and the tribochemical reaction are described by material interactions (interface reactions). Table 11.3 lists the wear mechanisms and their wear phenomena in running fiber ropes [21]. Table 11.4 shows typical signs of wear, aging and overloading of different high-performance fibers which have been processed in running ropes.

Table 11.3: Wear mechanism and signs of wear [1, 2, 9, 21].

Wear mechanism	Signs of wear
Adhesion – Formation or separation of adhesive bonds from contacting surfaces – Material interactions on the interfaces of the friction partners	– Indentations – Adherent particles (e.g., remains of the coating, fibril remains)
Abrasion – Friction between partners with different degrees of hard-ness – The harder friction partner (e.g., pulley) with its surface roughness hill damage the surface of the softer partner (e.g., fiber)	– Remains of fibrils and coating – Grooves and impressions in the fiber surface
Surface disruption – Tangential relative movements of the contact surfaces – Deformations can occur with energy losses – Fatigue of the surface due to recurring plastic or elastic effects	– Cracks on the parts
Tribochemical reaction – Chemical reactions, which can caused by friction on the contact surfaces	– Reaction products from fiber fibrils and coatings

Table 11.4: Specific signs of wear of UHMW-PE fibers and aramid fibers [14, 21, 27, 31].

Signs of wear of UHMW-PE fibers	Signs of wear of aramid fibers
Tensile load	
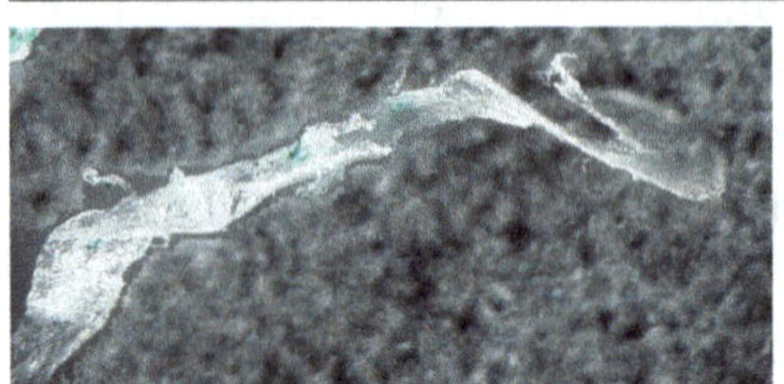 Reduction of fiber cross-section, viscoelastic deformation	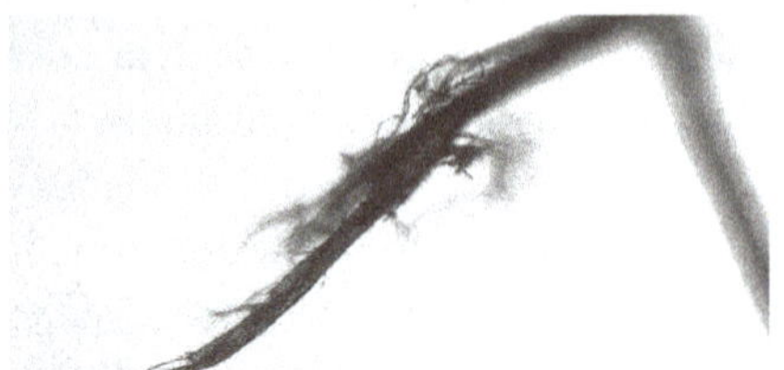 Fibrous fracture surface of the fiber, longitudinal structuring/splitting of the fiber
Compression	
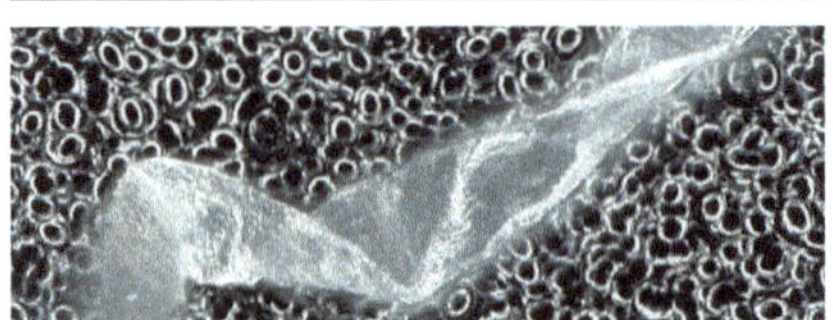 Transverse cracks, kink bands, viscoplastic deformations the fiber	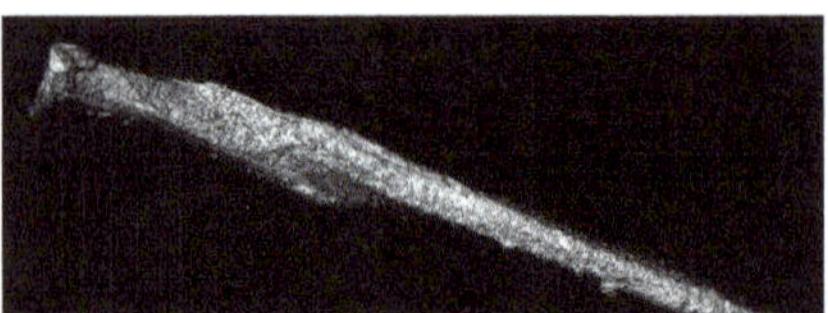 Inclined transverse cracks/breaks and kink bands of fiber, linear filament breaks
Alternating bending stress (CBOS test)	
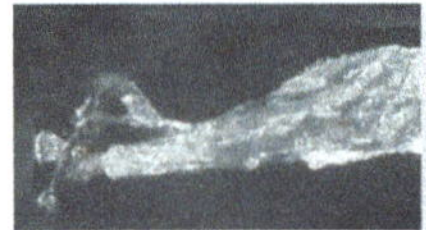 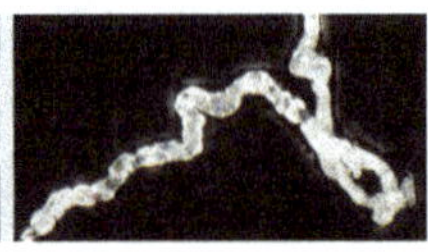 Fiber fracture with viscoplastic deformation and fibrillation, transverse cracks, kink bands	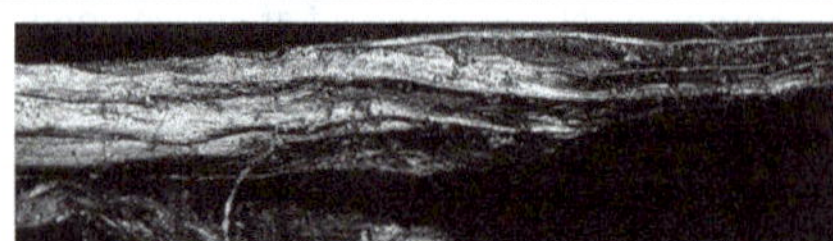 Transverse cracks, kink bands, linear fiber breakage
Fiber–fiber friction	
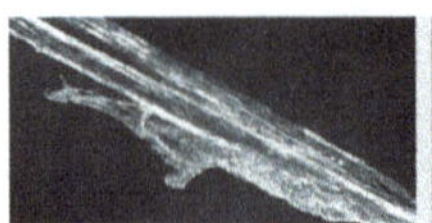 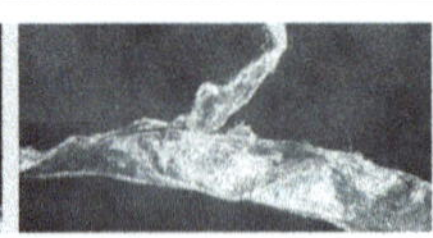 Viscoplastic deformation of the fiber break surface, abrasion of fiber coating and fibrillation	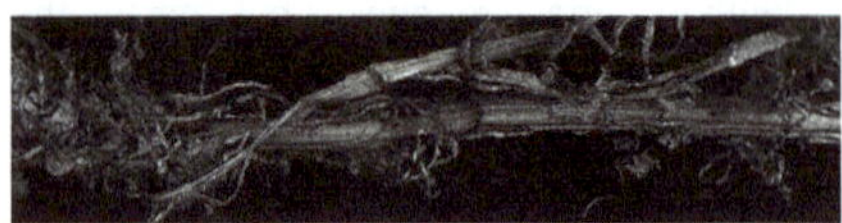 Fibrillation, fibrous/fibrillary structure of the fiber fracture surface

Table 11.4 (continued)

Signs of wear of UHMW-PE fibers	Signs of wear of aramid fibers
Environmental influences	
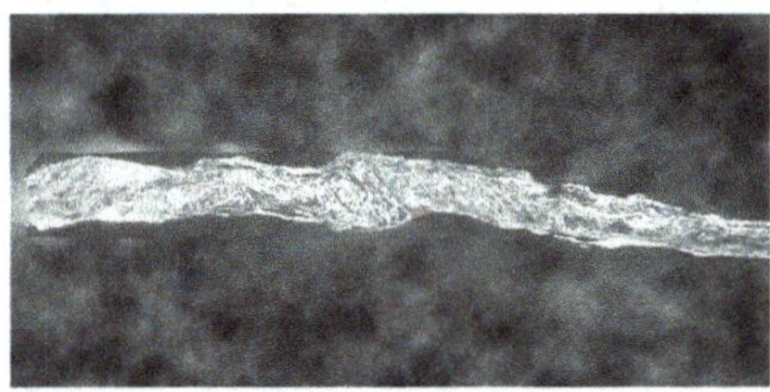	
E.g. thermal influence: kink bands, shrinkage, reducing fiber cross-section due to creep	E.g. chemical influence: fibrillation of aramid fibers salt crystals (alkaline environment)

11.1.6 Fiber ropes in rope drives

In the following explanations, the areas of application as well as the applicable standards and guidelines as well as an outlook for running fiber ropes are given, see Figure 11.6.

Figure 11.6: Example of a possible application (crane in New Zealand).

11.1.6.1 Brief summary

HM-HT fiber ropes can be used in a wider variety of rope driven conveyor systems. Due to their favorable breaking force/mass ratio, HM-HT fiber ropes are an attractive alternative to SWRs. Hence, the substitutional potential is limited to several disadvantages in fiber ropes, like (i) relatively high pricing induced by costly fibers und cost-intensive processing, (ii) temperature and degrading effects in plastics and (iii) safety issues caused by the mentioned above lack of valid discard criteria.

Further decisive advantages of HM-HT fibers compared to conventional tension member materials in conveying systems are (i) increase in lifting masses and highs/lengths, (ii) reduction of required drive power, (iii) reduction in bending stiffness, (iv) chemical and corrosion resistance and (v) good handling during replacement of the suspension element. Several rope manufacturers as well as conveyor systems providers all over the world pushed application-oriented research since the late 1980s.

Since then, a variety of application studies up to prototypical solutions was generated. A short list without any claim to completeness:
- mobile, tower, industrial and ship cranes,
- industrial hoisting equipment,
- power line and cableway construction,
- airborne wind energy systems (so-called power kites),
- elevator cables,
- hybrid mining cables (load-bearing fiber core) and
- winching systems (e.g., off-road or recovery vehicles).

11.1.6.2 Standards and guidelines

Standardization is the application of rules, guidelines or characteristics published by a recognized organization or body, primarily to increase quality and safety of products, processes and services, as well as their compatibility, repeatability and interchangeability. In general, there are three types of standards: de jure standards, de facto standards and voluntary standards. Next to standards, similar documents like guidelines, specifications and technical reports are covered as well. Content-related standards can be divided into general, product-related and application-oriented.

For fiber ropes, the international valid documents are prepared by the International Organization for Standardization (ISO) predominantly in Technical Committee ISO/TC 38, textiles except specific applications like the use of ropes in cranes or aerospace. In addition, there are several continentally and nationally responsible bodies, for example the European Committee for Standardization in Europe or ASTM International in the United States of America. Furthermore, several industry-driven associations like the Cordage Institute, the European Materials Handling

Federation or the Oil Companies International Marine Forum prepare and publish standards, specifications and guidelines as well. Table 11.5 provides a compilation of noteworthy international standards related to fiber ropes, yarn or fibers.

Table 11.5: Compilation of noteworthy international standards related to fiber ropes.

ISO 139	Textiles – standard atmospheres for conditioning and testing
ISO 1140	Fiber ropes – polyamide – 3-, 4-, 8- and 12 strand ropes
ISO 1141	Fiber ropes – polyester – 3-, 4-, 8- and 12 strand ropes
ISO 1181	Fiber ropes – manila and sisal – 3-, 4- and 8-strand ropes
ISO 1346	Fiber ropes – polypropylene splitfilm, monofilament and multifilament (PP2) and polypropylene high tenacity multifilament (PP3) – 3, 4, 8 and 12 strand ropes
ISO 1968	Fiber ropes and cordage – vocabulary
ISO 1969	Fiber ropes – polyethylene
ISO 2076	Textiles – man-made fibers – Generic names
ISO 2307	Fiber ropes – determination of certain physical and mechanical properties
ISO 5084	Textiles – determination of thickness of textiles and textile products
ISO 9554	Fiber ropes – general specification
ISO 10325	Fiber ropes – high-modulus polyethylene – 8-strand braided ropes, 12-strand braided ropes and covered ropes
ISO 10547 **ISO 10554**	Polyester fiber ropes – double braid construction Polyamide fiber ropes – double braid construction
ISO 10556	Fiber ropes of polyester/polyolefin dual fibers
ISO 10572	Mixed polyolefin fiber ropes
ISO/TS 14909	Fiber ropes for offshore station keeping – high-modulus polyethylene (HMPE)
ISO/TS 17920	Fiber ropes for offshore station keeping – aramid
ISO 18264	Textile slings – lifting slings for general purpose lifting operations made from fiber ropes – high modulus polyethylene (HMPE)
ISO 18692-1	Fiber ropes for offshore station keeping – general requirements
ISO 18692-2	Fiber ropes for offshore station keeping – polyester
ISO 18692-3	Fiber ropes for offshore station keeping – HMPE
ISO/TS 19336	Fiber ropes for offshore station keeping – polyarylate fiber
ISO/PRF TS 23624	Cranes – safe use of high-performance fiber ropes in crane applications

11.1.6.3 Relevant patents and technology for discarding fiber ropes

According to DIN EN 12927-6, discard criteria for ropes are defined by the "degree of deterioration at which the rope or the rope end fastening is declared unsuitable for further use" [3]. Knowledge and calculation of rope lifetime are important for the safe operation of rope drives. SWRs are well researched in this regard and have many years of experience. The lifetime of SWRs is assessed according to wear on the rope end fixings, loosening of the wires, fatigue wire breaks, corrosion, reduction in the metallic cross-section and abrasion on the ropes. In a row of current patents three branches of approaches for determination of discard criteria of fiber ropes are given: (i) derivation from operational data, (ii) integration of sensor elements in the rope and (iii) metrological recording of intrinsic properties of the rope and their change over time including the optically retrieved discard criteria. As an example for the latter, see Figures 11.7–11.10. The companies Teufelberger Fiber Rope GmbH (Austria) and Liebherr-Components Biberach GmbH (Germany) developed the product soLITE®. The rope has already been tested positive in various applications (as far as information is available here). The rope jacket shows visually the state of discard based on different degrees of damage. The rope manufacturer Samson offers a comparable visual system (see Figure 11.11) [20]. Several further property rights and technical approaches in the field of discard detection can be found in the respective literature (see Table 11.6).

Figure 11.7: soLITE® with 0% of the rope lifetime [21].

Figure 11.8: soLITE® with 20% of the rope lifetime [21].

Figure 11.9: soLITE® with 60% of the rope lifetime [21].

Figure 11.10: soLITE® with 95% of the rope lifetime [21].

Figure 11.11: Pocket guide from Samson [20].

Table 11.6: List of relevant property rights without any claim to completeness.

DE102013014265A1	Vorrichtung zur Erkennung der Ablegereife eines hochfesten Faserseils beim Einsatz an Hebezeugen
DE102013017110A1	Vorrichtung zur Erkennung der Ablegereife eines hochfesten Faserseils beim Einsatz an Hebezeugen
DE102015016382A1	Vorrichtung und Verfahren zur Bewitterung von Faserseilen sowie zur Prüfung und Beurteilung der Ablegereife
DE102018005926A1	Textiles Zug- und/oder Tragmittel und Verfahren zur Herstellung von textilen Zug- und/oder Tragmitteln

Table 11.6 (continued)

DE102018005927A1	Textiles Zug- und/oder Tragmittel und Verfahren zur Herstellung von textilen Zug- und/oder Tragmitteln
DE202011001846U1	Vorrichtung zur Erkennung der Ablegereife eines hochfesten Faserseils beim Einsatz an Hebezeugen
DE202016002171U1	Vorrichtung zur Überwachung von Betriebsdaten und/oder Bestimmung der Ablgereife eines Seils beim Einsatz an Hebezeugen
EP1930496B1	Erfindung zur Überwachung der Seillebensdauer eines Kunstfaserseil mit mindestens einer Litzenlage verseilter Garne (Litzen)
EP2740702B1	Vorrichtung zur Erkennung der Ablegereife eines hochfesten Faserseils für Hebezeuge
EP2848733B1	Synthetic rod for closure decoration and screening
EP 731209B1	Einrichtung zur Erkennung der Ablegereife bei Kunstfaserseilen
WO2020002615A1	Vorrichtung zum Einstellen der Ablegereifeerfassung hochfester Faserseile sowie Hebezeuge mit einer solchen Vorrichtung
WO2020002627A1	Verfahren zum Einstellen der Ablegereifeerfassung hochfester Faserseile sowie Faserseilsatz
WO2017067651A1	Vorrichtung zur Erkennung der Ablegereife eines hochfesten Faserseils für Hebezeuge
WO2017102821A1	Verfahren zur Bestimmung der Ablegereife eines Seiles aus textilem Fasermaterial
WO2015139842A1	Vorrichtung zur Bestimmung der Ablegereife eines Seils beim Einsatz an Hebezeugen
WO2010007112A1	Verfahren und Vorrichtung zur Ermittlung der Ablegereife eines Tragmittels eines Aufzugs

11.1.6.4 Research studies

Since the mid-2000s, research institutions have intensively researched on the determination of the discard status of high-performance fiber ropes. Special material and component properties as well as wear mechanisms were examined. Various research studies in extracts:

- development of laboratory testing technology,
- studies on fiber rope dimensioning,
- optimization,
- studies on visco-elastic component behavior in use and on the wear mechanisms.

Dissertation in the field of fiber rope research in extracts:
- see [10–13, 15, 17, 19, 21, 29],
- approaches to the metrological determination of the discard maturity – integration of optical waveguides in braids for spatially resolved strain measurement by means of Brillouin scattering and the ultrasonic testing of fiber ropes [12],
- complementary approach to damage analysis of discard detection on a laboratory scale – experiments on (i) electrical resistance measurements with copper strands, (ii) electrically conductive yarn and (iii) with bi-component fibers carried out in high performance fibers [21], methods of visual inspection in [21] (see [20, 30]) further developed into a color sensor measurement – results on a laboratory scale are promising, further development and practical tests are currently in progress.

11.1.6.5 Outlook and research needs

With growing innovation, new causes of damage go hand in hand and extensive research and testing becomes necessary. As mentioned above, yet there are no valid and objective discard criteria and measuring methods for running fiber ropes available in referring standards. Statements on the degree of damage to the rope are not well-defined. The lifetime of fiber ropes can only be precalculated for selected rope constructions based on tests [1]. Currently there is a lack of long-term studies, standards, test procedures as well as applicational knowledge and experience on:
- adequate test procedures and standards for determining the lifetime,
- tendency for application-relevant creep,
- alternating specific bending stress factors, for example counter-bending, bending length, groove type and so on,
- pulsating tensile stresses, when accelerating, decelerating and changing loads,
- transverse loading of the fibers, for example running of the rope in the groove of the traction sheave or in multilayer spooling.

References

[1] Brendel, H.; Winkler, H.; Hornung, E.; Leistner, D.; Neukirchner, J.; Schmidt, H.-J.; Winkler, L.: Wissensspeicher Tribotechnik: Schmierstoffe – Gleitpaarungen – Schmiereinrichtungen, 1. Aufl., Leipzig: VEB Fachbuchverlag, 1978.
[2] Czichos, H.; Habig, K.-H.; Celis, J.-P.: Tribologie-Handbuch: Tribometrie, Tribomaterialien, Tribotechnik. Wiesbaden: Springer Vieweg Verlag, 2015.

[3] DIN EN 12927-6: Sicherheitsanforderungen für Seilbahnen für den Personenverkehr – Seile: Teil 6: Ablegekriterien. Berlin: Beuth Verlag GmbH, 2005.
[4] DIN EN ISO 1968: Faserseile und Tauwerk – Begriffe, August 2005.
[5] Ehrenstein, G. W.: Polymer-Werkstoffe: Struktur – Eigenschaften – Anwendung, 2. völlig überarb. Aufl., Teilw. zugl.: Karlsruhe, Univ., Habil.-Schr 1978, München/Wien: Hanser Verlag, 1999.
[6] Firma Gleistein: Grundlagen des Seilhandwerkes, [online]. www.gleistein.com, Produktdatenblatt [14.08.2019].
[7] Fraunhofer isc: [online]. Fachkompetenz in der Partikelentwicklung für Forschung und Industrie. https://www.partikel.fraunhofer.de, Internetdokument [28.04.2021].
[8] Kanirope: Seil aus Aramid, [online]. www.kanirope.de [14.08.2019].
[9] Klucker, G.; Baumgartner, P.: Tribologische Untersuchungen zum Reibungs- und Verschleißverhalten von Modellfetten an einem Kugel-Scheibe-Tribometer, Diplomarbeit an der HAW Hamburg, 2012.
[10] Kretschmer, A.: Einflussfaktoren auf die Lebensdauer laufender Faserseile, Dissertation, Technische Universität Chemnitz, 2016.
[11] Heinze, T.: Zug- und biegewechselbeanspruchte Seilgeflechte aus hochfesten Faserseilen, Dissertation, TU Chemnitz, 2013.
[12] Helbig, M.: Grundlagenuntersuchungen zur zerstörungsfreien Prüfung von Seilen aus hochfesten Polymerfasern, Dissertation, Technische Universität Chemnitz, 2013.
[13] Mammitzsch, J.: Untersuchungen zum Einsatz von ultrahochmolekularen Polyethylenfasern in Seilen für die Fördertechnik, TU Chemnitz, 2015.
[14] McKenna, H. A.; Hearle, J. W.; O'Hear, N.: Handbook of Fibre Rope Technology, Cambridge England: Woodhead Publishing, 2004.
[15] Michael, M.: Beitrag zur Treibfähigkeit von hochfesten synthetischen Faserseilen, Dissertation, Technische Universität Chemnitz, 2011.
[16] Michaeli, W.; Wegener, M.:Einführung in die Technologie der Faserverbundwerkstoffe. München/Wien: Hanser Verlag, 1990.
[17] Novak, G.: Zur Abschätzung der Lebensdauer von laufenden hochmodularen Faserseilen, Dissertation an der Universität Stuttgart, 2017.
[18] OTIS: Burj Khalifa, www.otis.com, Dubai, Vereinigte Arabische Emirate, 2019.
[19] Putzke, E.: Charakteristik und Verhalten von synthetischen Faserstoffen in homogenen und heterogenen Wirkpaarungen, Dissertation, Technische Universität Chemnitz, 2017.
[20] Samson Rope Technologies: Rope users manual – Guide to rope selection, handling, inspection and retirement, 2019.
[21] Schmieder, A.: Schadensanalyse von hochfesten, laufenden Faserseilen, Dissertation, Technische Universität Chemnitz, 2020.
[22] Schneider, L.: Die zehn schnellsten Aufzüge der Welt, Fahrstuhlanlagen, 2017. www.ingen ieur.de.
[23] Schürmann, H.:Konstruieren mit Faser-Kunststoffverbunden. Berlin/Heidelberg/New York: Springer Verlag, 2007.
[24] Seilakademie: Das Seil Aufbau und Qualität, 2019, [online]. www.seilakademie.de [14.08.2019].
[25] Teijin: Technora: [online]. www.teijinaramid.com, Produktdatenblatt [14.08.2019].
[26] Teipel, U.: Produktgestaltung in der Partikeltechnologie. Chemie Ingenieur Technik, 2012; 84(3), 191. 2012 Wiley-VCH Verlag GmbH & Co. KGaA, Weinheim, Georg-Simon-Ohm Hochschule Nürnberg und Fraunhofer Institut für Chemische Technologie (ICT), Pfinztal.

[27] Bunsell, Anthony R.: Handbook of Properties of Textile and Technical Fibres, Second Edition. The Textile Institute Book Series. Pages 619–753. Woodhead Publishing is an imprint of Elsevier. Publisher: Mathew Deans. United Kingdom, January 2, 2018. ISBN: 978-0-08-1012727.

[28] VDI 2500: Fibre Ropes – Description – Selection – Dimensioning. Düsseldorf: Verein Deutscher Ingenieure e. V., 2020.

[29] Wehr, M.: Beitrag zur Untersuchung von hochfesten synthetischen Faserseilen unter hochdynamischer Beanspruchung, Dissertation an der Universität Stuttgart, 2017.

[30] WO2017067651A1: Vorrichtung zur Erkennung der Ablegereife eines hochfesten Faserseils für Hebezeuge, DPMA, 2016.

[31] Yang, H. H.: Kevlar Aramid Fiber. Chichester [u.a.]: Wiley, 1993.

Maroua Ben Abdelkader*, Nedra Azizi, Fabien Salaün, Mustapha Majdoub, Yves Chevalier

12 Cosmetotextiles

Keywords: cosmetic product, textile, functionalization, encapsulation, release, cosmetotextile

12.1 Introduction

A “cosmetotextile” is a cosmetic product attached to a garment fabric acting as a support [1]. The main purpose is the delivery of an active substance to skin over the full wearing time of the garment. Such “smart textiles” bring the consumer a supplementary cosmetic function to the textile. A typical product is the famous moisturizing and energizing tights launched quite early to the market by the Dim company [2]. Though a few cosmetotextile products have been commercialized before, it was felt as the first by many consumers. It has been followed by a wide series of active apparels from various companies. As typical example, tights functionalized by microcapsules containing caffeine slowly deliver this slimming substance to the legs (Figure 12.1). Since the first success stories of the 1990s, the market of cosmetotextiles is expanding fast. Its turnover grew from 100 M€ in 2010 (10% of the slimming market) to 500 M€ in 2013. It is especially driven by slimming products. The academic and industrial research is very active as revealed by the increasing number of scientific articles and patent applications [3].

The predominant technology for cosmetotextiles is functionalization of fabric with microcapsules containing active substances. The principle is shown in the sketch of Figure 12.2, together with a scanning electron microscopy (SEM) picture of such microcapsules bound to textile fibers.

Active substances are the same as for cosmetic products. A transfer from the fabric to the skin is required for the activity manifests. A delivery of active ingredient is

*Corresponding author: **Maroua Ben Abdelkader,** Laboratoire des Interfaces et Matériaux Avancés (LIMA), Faculté des Sciences, Université de Monastir, bd de l'Environnement, 5019 Monastir, Tunisia, e-mail: benabdelkadermaroua@gmail.com
Nedra Azizi, Mustapha Majdoub, Laboratoire des Interfaces et Matériaux Avancés (LIMA), Faculté des Sciences, Université de Monastir, bd de l'Environnement, 5019 Monastir, Tunisia
Fabien Salaün, ENSAIT, GEMTEX – Laboratoire de Génie et Matériaux Textiles, F-59000 Lille, France; Université de Lille nord de France, F-59000 Lille, France
Yves Chevalier, Université Claude Bernard Lyon 1, CNRS UMR 5007, Laboratoire d'Automatique, de Génie des Procédés et de Génie Pharmaceutique, 43 bd 11 Novembre, 69622 Villeurbanne, France

https://doi.org/10.1515/9783110670776-012

Figure 12.1: Advertisement for DIM slimming tights functionalized by caffeine microcapsules.

expected over the long time of wearing the apparel, so that sustained delivery is one characteristic. There are few cases where skin delivery is not intended. One example is fragrant textiles that release perfume molecules in the air surroundings. There is no need for skin absorption of the product; it should even be minimized.

Cosmetotextiles are derived from dermatological and transdermal patches that have been developed as medical devices since long times. Cosmetic and medical applications of textile finishes are mixed together under the common heading "medical applications" in many review papers because they are addressing upstream (academic) research results better than actual applications. As cosmetotextiles are intended to provide a cosmetic activity, the purpose is skin care instead of skin cure. With regards to the fundamental mechanisms of activity, there are many similarities between cosmetotextiles and pharmaceutical patches. As one main difference, the fabrics used in cosmetotextiles are garments that people use to wear, whereas the textile support for patches is a kind of a dressing (e.g., a bandage) that may be optimized with regards to its end-use and the activity of the drugs. Another characteristic is the focus on the active ingredient only, better than considering the full formulation of a cosmetic product including active ingredients and excipients that bring pleasant sensory properties. The active ingredient is delivered from the fabric as its primary function is not drug delivery. One consequence is a poor control over drug

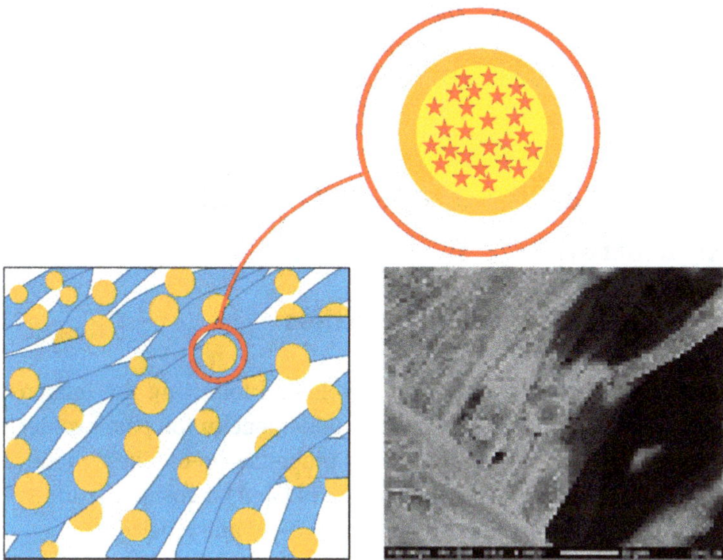

Figure 12.2: Principle of textile functionalization by binding microcapsules to the fibers surface (left), a zoom into the internal structure of a microcapsule with an internal core (yellow) containing the active substance (stars) surrounded by a polymer wall (orange), and SEM picture of such cosmetotextile (right).

delivery because of the variable uses of wears by consumers and the lack of well-defined contact time between the drug delivery system and the skin during wearing.

Another important difference stands in the regulations for cosmetic and medical devices. In the European Union, cosmetotextiles must comply with the requirements of the Cosmetic Regulation (EC) no. 1223/2009, as well as the several regulations governing textiles. The combination of these regulations is the essence of the definition given in 2006 by the Textile Industry and Clothing Standards Agency (BNITH) that mostly retained the terms of the European Cosmetics Directive 76/768/CEE of 1976 (the precursor of (EC) no. 1223/2009): "A cosmetotextile is a textile article that contains a substance or a preparation that is intended to be released sustainably on the different superficial parts of human body, especially the skin, and which claims one (or more) particular property(s) such as cleansing, perfume, change of appearance, protection, maintenance in good condition or correction of body odours."

With regards to the design of a cosmetotextile material, several requirements to be met are straightaway felt based on common sense:

- The drug delivery system should contact the skin so that drug can be transferred. Underwear with tight contact to the body are preferred; stockings are good candidates when legs are targeted.

- Slow release of drug is required for sustainability, even though this requirement is not based on the biological mechanism of action.
- Microcapsules bound to the fabric should resist washing.

12.2 Specific properties of cosmetotextiles and regulation

12.2.1 Specific properties of cosmetotextiles

Apart from the typical protective benefit of apparels and following the immense changes and innovations that the textile world has undergone in the past two decades, customers request novel cosmetic and wellness properties [4–6]. These products contain cosmetic preparations mainly for skin care uses [5]. As a cosmetotextile system, both performance and safety aspects should be specified and labeled.

12.2.1.1 Performance aspects

To satisfy the customers' demands, performance aspects of cosmetotextile products focus on three key points: cosmetic activities, durability and comfort.

12.2.1.1.1 Cosmetic activities/functionalities

Cosmetotextiles provide various activities that are claimed in the same way as a cosmetic product: perfume, slimming, moisturizing, refreshing, anti-aging, UV protection, softening, cleansing, changing appearance, protection, keeping good physical condition, correction of body odors, insect-repelling [5, 7] and so on.

Depending on their effect, ingredients functionalizing textiles are to be transferred in sufficient quantities into the wearer's skin for providing welfare, health or/and cosmetic benefits [7, 8]. Thus, the release of very small quantities of perfumes can cause noticeable effects, whereas skin care products (slimming and moisturizing) need to be transferred in larger quantities to the right location inside skin [9].

The cosmetotextile design also matters and may be used to maximize efficiency. The cosmetic ingredient and clothing must be designed in a complementary way so as to get optimum effects. For instance, the best slimming effects are obtained with cosmetotextiles implemented to tight jeans or elastic leggings with some degree of progressive nonmedical compression. This design guarantees close contact between the garment and the skin in all areas of the legs. A massage effect improves the transfer of the cosmetic ingredient from the textile to the skin [9].

12.2.1.1.2 Cosmetic durability

One specific property of cosmetotextiles is durability against washing, rubbing and abrasion according to the application field and the market demand [5]. As the intensity of the perceived effect progressively decreases during use, it is important to give customers the necessary information about the recommended procedures for optimum body care as well as the lifespan of the functional garment [10]. Washing, rubbing and abrasion fastness of cosmetotextiles are assessed according to standards as summarized in Table 12.1.

Table 12.1: Standards used for evaluation of washing, rubbing and abrasion fastness of cosmetotextiles.

Fastness	Standards	References
Washing	EN ISO 3175-2	[11]
	AATCC 61-2006	[12, 13]
	ISO 105-C10	[14–17]
Rubbing	EN ISO 12947-1	[18]
	EN ISO 12947-2	[11]
	BS 5690	[14]
Abrasion	ASTM D3884	[19]
	ASTM D4966	[12, 13]

The residual encapsulated cosmetic products can be evaluated by qualitative ways (survey, microscopic scanning [14–21]) and quantitative analyses [15–17, 21, 22], possibly supplemented by modeling [11].

12.2.1.1.3 Comfort

Customer comfort is considered as the main factor regarding the quality of clothing. Indeed, it is directly related to the feeling of consumer well-being through its physiological functions [5].

According to Bishop and Bartels, comfort involves multiple aspects that depend on many factors. These aspects are mainly based on both thermal comfort (thermophysiological wear comfort) and skin sensory comfort (mechanical sensations like sensitization and irritation) [23, 24].

12.2.1.2 Safety aspects

Due to the close contact of cosmetotextiles with skin, to avoid risks and/or danger, several safety aspects must be taken into account.

Apart from the biocompatible and biological aspects, the toxicity remains the most important concern. In the case of cosmetotextile articles, toxicity must be studied in two parts: individual toxicity of all the ingredients used in the functionalization formulation and that of the whole finished textile. Therefore, it is necessary to make sure that all ingredients and all chemicals and reagents used for the encapsulation/immobilization process, their by-products and their residues are safe to avoid skin damages like sensitization, irritation or epidermal cell toxicity [5, 25].

The identification of these potential negative problems is carried out by testing cosmetotextiles according to one of the following three standards (Table 12.2) [26].

Table 12.2: Standards used for toxicological evaluation of cosmetotextiles.

Standards	Details	Reference
EN ISO 10993	It is destined for medical devices and it contains two parts relevant for cosmetotextiles and intelligent dermotextiles: – Part 5: Cytotoxicity assessment by in vitro tests; – Part 10: Skin irritation and sensitization assessment by in vitro and in vivo tests.	[27]
The Organization for Economic Co-operation and Development (OECD) Guidelines for the Testing of Chemicals – section 406	It describes a skin sensitization test method inspired by Magnusson and Kligman's guinea pig maximization test (GPMT).	[28]
STANDARD 100 by OEKO-TEX®	It is a product label for textile products from all processing stages that have been tested for harmful substances.	[29]

12.2.2 Cosmetotextile regulations

Cosmetotextiles, as wearable body care, are a new generation of textile products appeared in recent decades especially in Europe. These products are a very attractive niche market for both textile and cosmetic companies. Since then, producers faced a lack of regulation concerning cosmetic textiles. It is noteworthy that until today, there is no specific regulation applicable to cosmetotextiles. The regulations on cosmetics and textiles must be satisfied. In fact, these products must comply both with textile products regulations and also with the cosmetic product directives taking into account the prohibited and restricted substances in the European Regulation [30]. In order to clarify all these aspects which must be taken into account when developing and launching cosmetotextiles on the market, the technical report CEN/TR 15917 was published in September 2009 [10, 31].

This guideline specifies general characteristics of such products and describes their recommended properties. General aspects, safety evaluation, claimed effects, care resistance and labeling are the five parts that have been set up. Four normative references (EN ISO 3175-1, EN ISO 3758, EN ISO 6330 and EN ISO 22716) are essential for the application of this document.

12.2.3 Labeling

According to European standard CEN/TR 15917 [31], product labels differ from marketing labels. Apart small items (hoses, stockings, etc.), the product label should be attached to the garment and not on the package as it is the case for marketing labels. Thus, the basic information that should be contained in a product label are essentially as follows [25] (Figure 12.3).

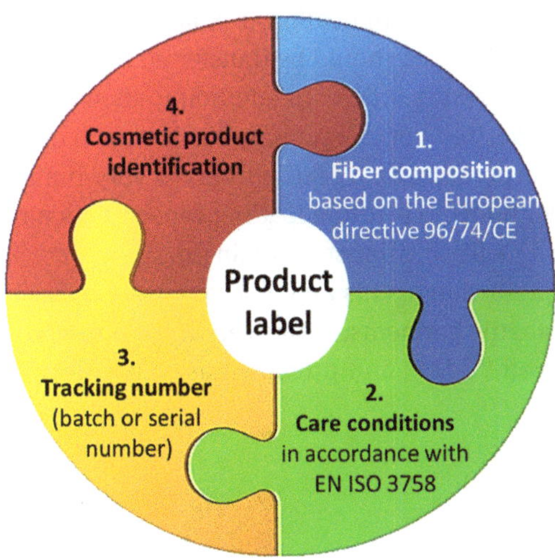

Figure 12.3: Basic information contained in a product label.

12.3 Processes of immobilization

Cosmetic textiles are garments containing cosmetic substance(s) or preparation(s). These ingredients are intended to be transferred to various parts of the epidermis for different functional properties as aesthetic effects, protection, comfort and skin care [32].

Cosmetotextiles materials can be prepared by using a variety of cosmetic-loading approaches such as incorporation into textile fibers, adsorption on fiber surface, binding to fiber surface, inclusion into cyclodextrins (CDs) grafted at fibers surface and encapsulation. The following sections discuss these methods in detail.

12.3.1 Incorporation into textile fibers

Cosmetic ingredients can be incorporated during various spinning processes (electrospinning, co-extrusion, dry or wet process) into chemical fibers like for example polyamide fibers (such as some nylons®), acrylic polymer fibers, polyolefin fibers and especially polyethylene or polypropylene fibers, polyester fibers, polyurethane fibers and aromatic polyimide-amide [9, 33, 34].

Depending on the functionality and the active material, the durability of the effect will be greater if the active material is incorporated into the polymer fiber matrix during the spinning process than by direct application to a fabric by topical treatment [7]. For such applications, no substance should be transferred to the skin, especially for an antimicrobial or biocidal textile article. In addition to their effects on textiles, these ingredients must not destroy microorganisms on human skin [5]. However, the process of cosmetic incorporation into textile fibers is very limited only to extremely robust ingredients with low to medium loading capacity. As a result, wash resistance is excellent with a negligible transfer to skin [9].

As examples of applications, UV protection agents and aloe vera were incorporated into the fiber matrix for cosmetotextile garments. Fibers serving as a reservoir for vitamins (C and E) have been named "V-Up" and promoted as "portable vitamins" [5, 9].

12.3.2 Adsorption on fiber surface

Immobilization of cosmetic ingredients can take place by adsorption on the fiber surface since textile fabrics are suitable for the application of different coatings. In fact, this technique is one of the simplest approaches. It involves spreading a polymer containing cosmetic ingredients substantially in the form of a thickened aqueous dispersion or a solution in a suitable solvent onto a fabric to form a continuous layer on one or both faces of the fabric [7]. A thin layer is deposited, so that the intrinsic properties of textiles, such as thickness, tear strength and flexibility, are retained [5]. Antimicrobial polymers [35–37] and active ingredients like silver, silver salts or other biocidal substances [38, 39] were coated on textile surfaces.

12.3.3 Binding to fiber surface

As for dyeing, cosmetic ingredients can be immobilized to the fiber surface by forming chemicals or physical reactions using padding or exhaustion treatment which is very common in the textile finishing industry [9].

- **Padding or continuous process:** this method is the simplest one to incorporate cosmetic ingredients on textile. The fabric is continuously immersed in an open finishing bath containing all chemicals such as solvent, cosmetic products, binders, softeners and wetting agents. After immersion, the fabric is squeezed by a pair of rollers in order to reach a defined and constant carrying rate. It then passes through an oven for drying and fixing ingredients on textiles (Figure 12.4) [7, 9, 26].

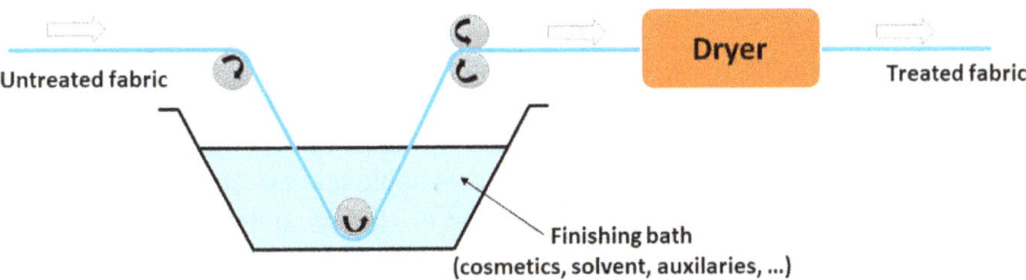

Figure 12.4: Padding process.

- **Exhaustion or discontinuous process:** the principle of this method is different from the previous one. Before their diffusion or adsorption, chemicals are transported to the fiber surface by the movements of the bath and/or the textile to be treated. For this immobilization process, two types of machine are available: circulating machines in which the fabric is stationary and the bath is moving, and circulating-goods machines in which both textile support and finishing bath are mobile [7]. This process is typical for knitted textiles and clothing, especially jeans. However, it is much more complicated because it requires very precise control in terms of temperature and pH, and greatly limits the choice of binders according to their exhaustion [9].

Cosmetic substances are attached to textile materials by covalent bonds, ionic interactions [40], van der Waals interactions or cross-linking. The last type is the most common. It involves an adhesive generally called a "binder," which is specifically adapted to the cosmetic and textile system and also has good skin compatibility. Usual binders are polyurethanes, polyacrylates, cross-linkable silicones and poly (ethylene-co-vinyl acetate. Those with low curing temperatures are preferred [9].

12.3.4 Inclusion into cyclodextrins grafted at fibers surface

A way for binding molecules to fibers relies on their complexation with CDs. CDs are cage-like molecules that can bind many organic molecules inside their internal cavity. Thus, CDs are attached to the surface of fibers, which allows the immobilization of molecules as their inclusion complex [41]. In one type of applications, active substances can be immobilized onto fabrics in the same way as for the direct adsorption presented in Section 12.3.2. CDs act as adsorption improvers; the binding mechanism remains adsorption to the fibers surface. In a second type of applications, CDs uptake smelly molecules from the skin surface to the fiber surface in order to prevent (reduce) undesirable body odors [42].

CDs are cyclic oligosaccharides having the shape of a hollow truncated cone [43]. Their internal cavity is rather hydrophobic, which allows binding hydrophobic molecules of the same size as the cavity by formation of an inclusion complex. The "hydrophobic cavity" actually is not very hydrophobic; its physicochemical properties as a "solvent" are close to that of methanol. As consequence, many organic molecules can bind CDs, including quite polar species such as amino acids. The driving force for binding is predominantly coming from hydrophobic interactions however. Three types of CDs are available with different number of glucose units and related cavity size (Figure 12.5). α-CD with six glucose units has the smallest cavity size that matches linear alkyl chains (*n*-alkanes, *n*-alkyl surfactants); β-CD with seven glucose units is the most widely used owing to its easy synthesis by a biotechnological process and low cost; γ-CD with eight glucose units has the largest cavity. Because of the low solubility of β-CD in water (18.5 $g{\cdot}L^{-1}$), its derivatives of higher solubility such as hydroxypropyl-CD (HP-CD) may be preferred. Derivatization does not change the properties of the cavity. Association constants in solution are well-documented, although subjected to a large scatter of experimental values [44].

CDs can easily be attached to cotton or wool fabrics with the monochlorotriazine coupling agent [45–47]. Several other chemical reactions are also possible [48]. Though there are commercial products claiming the use of CD bound to fabrics for their end-use as cosmetotextile (see Section 12.9), technical information is scarce. As consequence, mentions of such applications given in review papers mostly remain superficial [42, 49–51]. CD-based cosmetotextiles are mainly used for sustained delivery of fragrances and trapping mal-odors. Formation of an inclusion complex with CDs is not an encapsulation; guest molecules are in fast dynamic equilibrium with free molecules in the surrounding medium, skin moisture and sweat. Complexation by CDs reduces the chemical potential of the odorant molecules, thereby decreasing

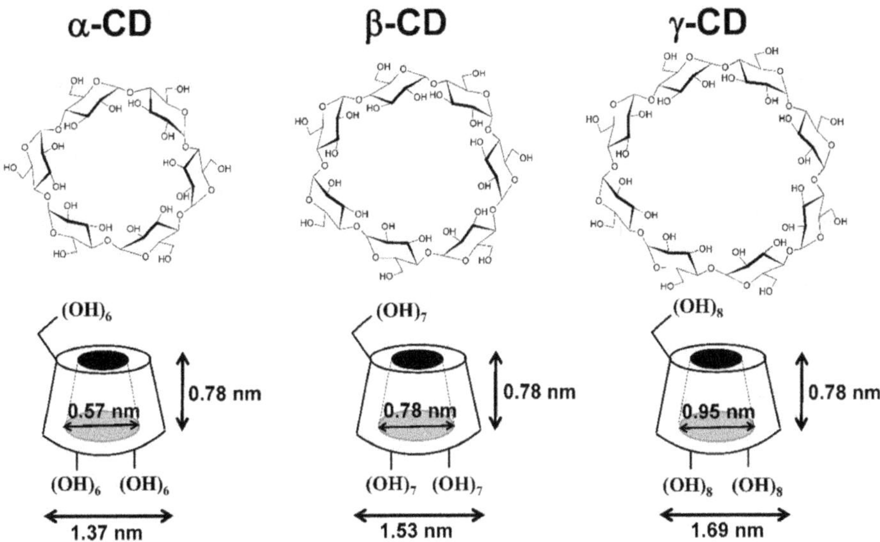

Figure 12.5: The three cyclodextrins with their typical dimensions: α-CD, β-CD and γ-CD.

their equilibrium partial pressure in the gas phase and consequently decreasing their rate of evaporation. Perception of odors is reduced and the effect lasts longer times. This is a definite benefit for perfumery. In case of preventing unpleasant odors coming from decomposition of sweat by the cutaneous microbiota, molecules trapped by the active textile are eliminated during cloth washing. This way to control body odors is more suitable than using a fabric with antimicrobial properties proposed as an alternative. Indeed, the useful functions of the cutaneous microbiota should be retained. Applications to bandages are at the borderline between cosmetics and medicine. An interesting example is a viscose bandage grafted with β-CD and loaded with the æscinanti inflammatory and vasoconstrictor active substance for the treatment of venous insufficiency and lymphedema in legs [52].

12.3.5 Encapsulation

A large number of cosmetic ingredients are unstable and sensitive to heat, oxidation, light, moisture, or being inherently volatile. That is why micropackaging or entrapment via microencapsulation is an alternative for providing satisfactory performance [6, 9].

Microencapsulation is a technique involving the production of microcapsules containing the cosmetic active ingredients inside a wall material. By this way, this technology enhances not only the protection of the cosmetic ingredients from hazardous environments and from interactions with other compounds in the system, but also allows them sustained release and/or stimuli-triggered release, long-lasting effect and it guarantees their durability [5–7]. For examples, fragrances may be retained on the garment for a considerably longer period (up to 40 washes) without completely losing the scent [16].

This versatile micropackaging technique has very broad applications with high loading capacity. The wash resistance of this kind of cosmetic immobilization is good with a fair to great skin transfer depending on the method of microcapsules fixing to textile [9]. The major advantage of this approach is that almost all substances can be encapsulated and protected against all degradation processes [5].

12.4 Microcapsules

Microencapsulation of solid, liquid or gaseous materials results in a solid powder form that is advantageous in many applications. Two different structures are possible namely the microsphere or matrix system where the core material is coated with the matrix polymer shell and the microcapsule or reservoir system where the active material is dissolved. Various configurations of microspheres and microcapsules are summarized in Figure 12.6.

In particular, immobilization of liquid compounds onto solid supports often requires their encapsulation as prerequisite to their deposition onto solid surface as an adhesive and durable coating [21]. This technique has been early applied to carbonless copy paper [54, 55] and it has currently found applications in many technical domains such as food, adhesives, cosmetics, pharmaceutics, phytosanitary products, medicine, liquid crystals labels and phase-change materials [56–61].

Encapsulation processes have been introduced in the textile industry for the implementation of specific properties to textile materials by coating them with various active substances [62]. Such application takes advantage of the solid powdered form of microencapsulated liquids for an easy deposition onto the textile fibers. Fabrics having long-lasting fragrance release properties are manufactured by coating fibers with microcapsules loaded with perfumes [62–71].

In recent years, textile materials have been found in applications in the cosmetics field. A new sector of cosmetic textiles is introduced and several commercial cosmetic textile products are currently available in the market. On contact with human body and skin, cosmetic textiles are designed to transfer an active substance for cosmetic purposes. Microencapsulation technology is an effective technique used to control the release properties of active ingredients that prolong the functionality of cosmetic textiles [72].

Various routes for the incorporation of cosmetic potential in textiles are available, of which microencapsulation and the use of CD as cage material are the most popular [15]. The content of moisture on skin surface is a key parameter in maintaining skin

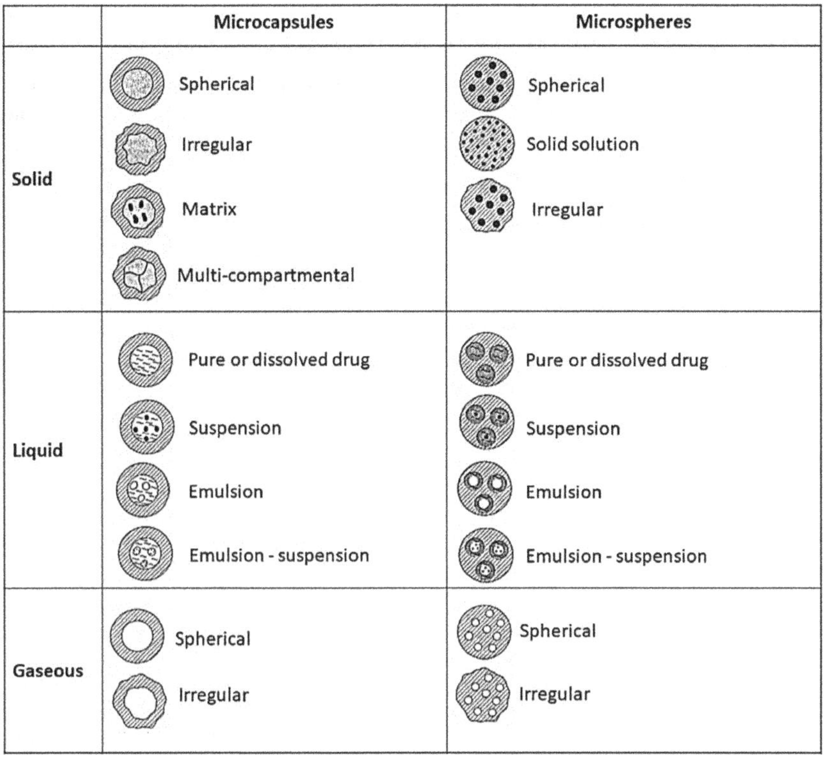

	Microcapsules	Microspheres
Solid	Spherical Irregular Matrix Multi-compartmental	Spherical Solid solution Irregular
Liquid	Pure or dissolved drug Suspension Emulsion Emulsion - suspension	Pure or dissolved drug Suspension Emulsion Emulsion - suspension
Gaseous	Spherical Irregular	Spherical Irregular

Figure 12.6: Various configuration of (a) microcapsules (b) microspheres [53].

elasticity and glowing potential. European standardization concerning the testing of multifunctional textiles felt the strong need to form test standards for cosmetotextiles. The European Union formed a working group, WG-25, to form test standards for cosmetotextiles [73]. The WG-25 has identified some areas where standardization is required immediately and formed five subgroups to work on different aspects of cosmetotextiles. Both subjective and objective evaluations of cosmetotextiles are possible to test various cosmetic effects such as chemical properties, toxicity/innocuousness, efficacy, perfume performance analysis and durability.

12.5 Encapsulation processes for cosmetotextiles

Microencapsulation processes are usually categorized into two groups: chemical processes and mechanical or physical processes. The first class of encapsulation involves polymerization during the process of preparing the microparticles. Examples of this class are generally known as interfacial polymerization or in-situ polymerization [74–

76]. The second type involves the precipitation of a polymeric solution wherein physical changes usually occur. Microencapsulation processes rely on several chemical methods such as in situ polymerization and coacervation [77]. Interfacial polymerization is a widely used method that allows the manufacture of microcapsules from an o/w emulsion by the formation of thick polymer walls around liquid droplets [78–86]. Encapsulation methods based on in-situ polymerization are open to health concerns coming either from residual monomers in case of their incomplete conversion.

A clearer indication as to which category an encapsulation method belongs is whether or not the capsules are produced in a tank or reactor containing a liquid, as in chemical processes, as opposed to mechanical or physical processes, which employ a gas phase as part of the encapsulation and rely chiefly on commercially available devices and equipment to generate microcapsules. The general classification of microencapsulation methods is described in Table 12.3.

Table 12.3: Classification of the microencapsulation methods [53].

Process name	Coating material	Suspended medium
Interfacial polymerization	Water soluble and insoluble monomers	Aqueous or organic solvent
Complex coacervation	Water soluble polyelectrolyte	Aqueous phase
Coacervation	Hydrophobic polymers	Organic solvent
Salting-out	Water soluble polymer	Aqueous phase
Solvent evaporation	Hydrophilic or hydrophobic polymers	Organic or aqueous phase
Hot melt	Hydrophilic polymers	Oil as continuous phase
Melt dispersion	Hydrophobic waxes	Aqueous solution
Melt dispersion	Hydrophobic polymers	Aqueous solution
Solvent extraction or diffusion	Hydrophilic or hydrophobic polymers	Organic solvent
Spray drying	Hydrophilic or hydrophobic polymers	Air, nitrogen

12.5.1 Interfacial and in situ polymerizations

In interfacial polymerization, the capsule shell is formed at or on the surface of a droplet or particle by polymerization of reactive monomers. A multifunctional monomer is dissolved in the liquid core material. A rapid polymerization reaction is then produced at the interface which finally generates the capsule shell [72].

New microcapsules based on renewable materials and containing perfume were designed for cosmetotextiles application. Such microcapsules contained neroline fragrance as core material and a bio-based polyurethane as wall material. The polymer wall was synthesized by interfacial polycondensation of isosorbide and methylene bis(phenyl isocyanate). Isosorbide-based polyurethane microcapsules were prepared by interfacial polycondensation method using the dibutyltindilaurate catalyst according to previous reports [21].

For in situ polymerization, capsule shell formation occurs because of the polymerization of monomers that is added to the encapsulation reactor, similar to interfacial polymerization. However, no reactive agents are added to the core material. polymerization of reagents located there produces a relatively low molar mass prepolymer.

12.5.2 Complex coacervation

This method takes advantage of the abilities of cationic and anionic water-soluble polymers to interact with water, forming a liquid, polymer-rich phase called complex coacervation. When the complex coacervate forms, it will be in equilibrium with a dilute solution called the supernatant. The supernatant acts as the continuous phase, whereas the complex coacervate acts as the dispersed phase. As the water-insoluble core materials are dispersed in the system, each droplet or particle of dispersed core material is spontaneously coated with a thin film of coacervate. The liquid film is then solidified to make the capsules harvestable (Figure 12.7).

12.5.3 Solvent evaporation method

Solvent evaporation/extraction is one of the widely investigated and employed methods for the microencapsulation of variety of bioactive materials [87]. The state of the art of microsphere preparation by solvent evaporation and extraction method had been reviewed by Freitas et al. and Arshady [88, 89]. The basic principle is the emulsification of oil-in-water. Initially polymer is dissolved in an organic solvent such as dichloromethane, chloroform or ethyl acetate and the active material is either dissolved or dispersed in this solution. It is called a dispersed phase. An aqueous or continuous phase constitutes the water containing emulsion stabilizers such as poly(vinyl alcohol) or poly(vinyl pyrrolidone). Then, the oil phase is emulsified into the aqueous phase under stirring, which is further continued for several hours in order to allow for the evaporation of the organic solvent. Finally, the particles are filtered, washed and dried to obtain a free-flowing powder (Figure 12.8).

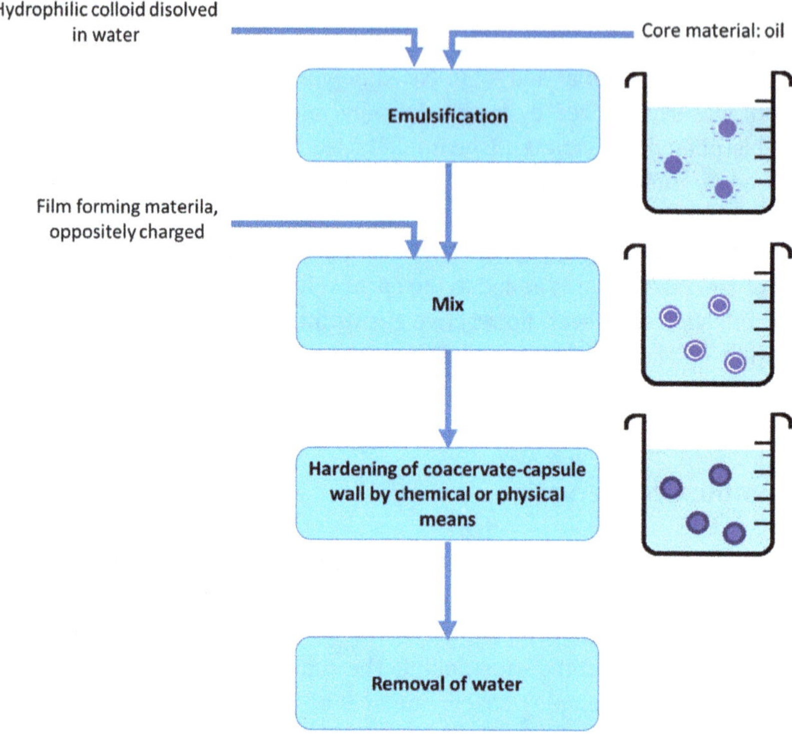

Figure 12.7: Schematic presentation of formation of microcapsules of o/w by complex coacervation.

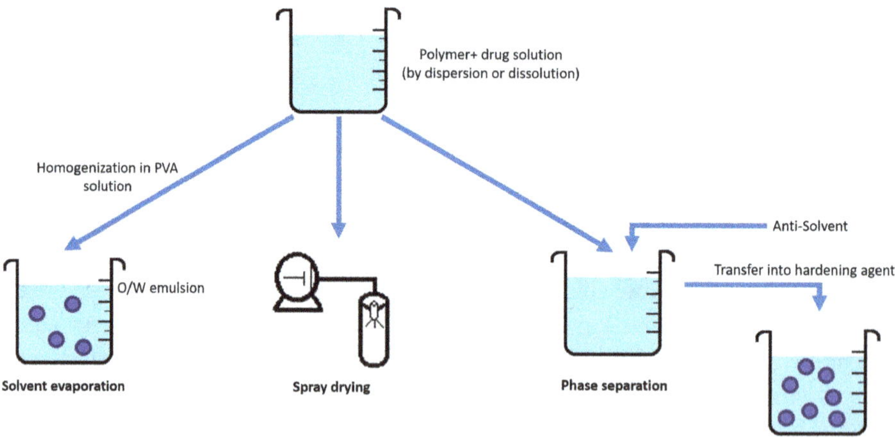

Figure 12.8: Schematic representation of various solvent-based methods for the microencapsulation.

12.5.4 Spray drying

Spray drying serves as a microencapsulation technique when an active material is dissolved or suspended in a melt or polymer solution and becomes trapped in the dried particles. In the widely used spray-drying process, the dry solid is formed by spraying an aqueous solution of the core material and the film-forming wall materials as fine droplets into hot air. The water then evaporates and the dry solid is usually separated by air separation.

12.5.5 Centrifugal extrusion

In centrifugal extrusion processes, liquids are encapsulated by using a rotating extrusion head with concentric nozzles. The fluid core material is pumped through a central tube while the liquefied wall material is pumped through a surrounding annular space.

A membrane of wall material is formed across a circular orifice at the end of the nozzle and the core material flows into the membrane, causing the extrusion of a rod of material. Droplets break away from the rod and hardening takes places on a passage through a heat exchanger. Solid capsules are removed by filtration or mechanical means and the immiscible carried fluid is reheated and recycled after passing through the files. Figure 12.9 shows a schematic diagram of a centrifugal two-fluid nozzle that was used to produce microcapsules.

12.5.6 Melt dispersion method

Melt dispersion method is an oil-in-water emulsification process, in which an oil phase constitutes molten polymer containing dissolved or dispersed drug and a continuous phase constitutes water containing emulsion stabilizer (Figure 12.10).

The early work concerning the melt dispersion method dates back to 1985 by Kagadis et al. [91]. They prepared ibuprofen-stearic acid wax microparticles by melt dispersion method. But the obtained microparticles were of large size. Benita et al. [92] encapsulated 5-fluorouracil using carnauba wax by the same method.

12.6 Binding of microcapsules

Microencapsulated textiles have been defined as those functional textiles obtained by surface functionalization of fabrics with microcapsules. A wide range of particles and actives substances can be employed in functional textile preparation. Therefore,

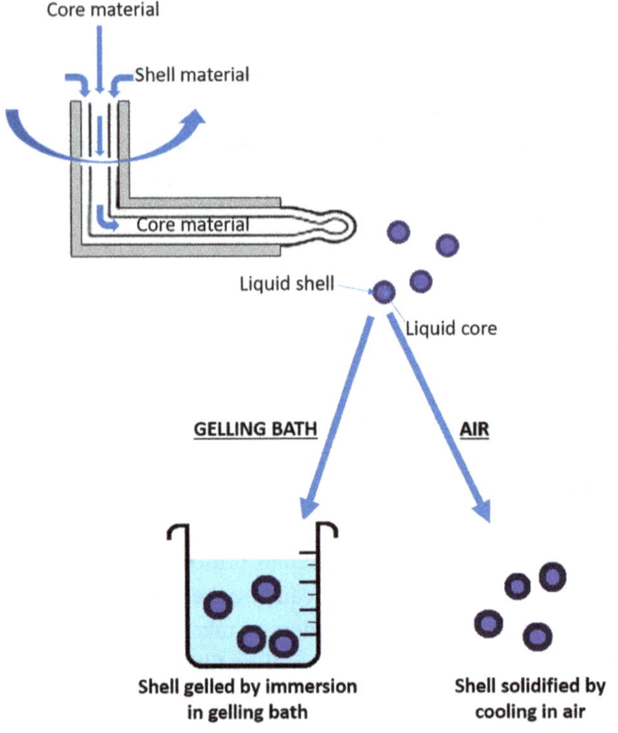

Figure 12.9: Schematic diagram of centrifugal two-fluid nozzle that was used to produce microcapsules [90].

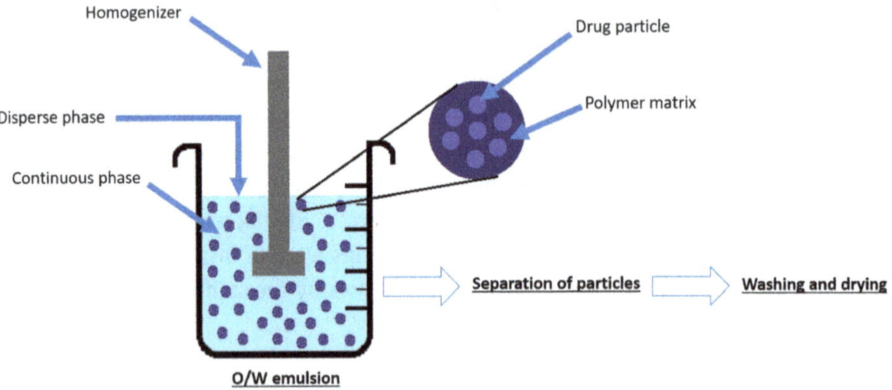

Figure 12.10: Schematic representation of principal steps involved in microparticles preparation by melt dispersion method.

the properties and application of such materials are various. The main driving force to integrate cosmetic microcapsules in textiles is that regardless of the improved efficacy of such formulations, they need to be topically applied to classical ointments and creams. Therefore, the treatment requires multiple applications, and its success is intrinsically related to the wearer's compliance. Functional textiles make the skin treatment wearable and require a single application overcoming the issue of poor wearer compliance.

The interactions between the microcapsules and fibers play a critical role in determining the properties of the functional textiles in terms of fastness and release behavior [93]. The use of polymeric binders and/or cross-linkers is a common strategy to promote the formation of physical and chemical interactions between the fabrics and the microcapsules [32]. In most of cases to ensure a permanent fixation, the fixation strategy depends on the fiber nature, thus cross-linking strategy is used for cotton, ionic interactions for polyamide, polyacrylonitrile and wool, electrostatic or van der Waals interactions for polyester, polyamide, and polyacrylonitrile, whereas covalent linking for cotton, polyamide and wool.

12.6.1 Covalent approaches

Chemical grafting is one the possible ways to avoid the hindrance of the active substance release while providing suitable washing durability [94]. This method can be divided into two main approaches, that is, (i) the use of a cross-linker or coupling agent, and (ii) the surface functionalized capsules with triazine moieties for example. In the first approach, cross-linking agents, such as glutaraldehyde, genipin [95], poly(carboxylic acid)s (tannic acid, citric acid, butanetetracarboxylic acid) [96–99] and urea derivative [94], are mainly employed. Among these, poly(carboxylic acid)s allows the reaction with the functional groups of the microcapsules shell and the hydroxyl groups of cellulose to form ester bonds [100]. The second approach is based on the surface functionalization of the capsules with chemical reactive groups, such as cyanuric chloride moieties, α-bromo-acrylic acid, adipic acid and dichloroquinoxaline with react with hydroxyl groups of the textile substrate via nucleophilic substitution reaction [101, 102]. "Click chemistry" may be also used to functionalize cellulosic fibers as described by Hamaide et al. [103].

12.6.2 Binder and film-forming properties

Most of the conventional methods require the use of a binder that ensures binding of the microcapsules to the fabric fibers. The binder supplements the lack of affinity between the particles and the substrate. Indeed, most of the available commercially capsules have nonreactive shells, therefore to improve their shelf-life and the treatment durability, the use of textile auxiliary compounds is required to bind the microparticles on the textile substrate. Most of them are based on the acrylic, silicone or polyurethane chemistry [67].

The choice of a binder is often dictated by its cost and chemical affinity with the capsules and the textile. Therefore, the permanence of the treatment and the microcapsule effectiveness is related to the lifetime of the coating on the fibers. Finishing treatments requiring a binder allow the formation of a continuous film, adhering to the substrate, and holding the textile microcapsules. The choice of the chemical nature of the binder can be based on two approaches. The first is a thermodynamic approach to wetting, allowing characterization of the interfaces between the textile and the microcapsules and between the binder and the microcapsules. It is based on measuring contact angles to determine dispersive and polar components and a fortiori to calculate the adhesion energy. The knowledge of the microcapsule wettability and free surface energy is necessary to establish the formulation for finishing treatments; it is based on measurements of the washing resistance of microcapsules once they are bonded to the textile with the binder. This type of information can help at selecting compounds if the interfacial interactions and compatibility of ingredients in the formulation are known.

Also, to minimize the loss of microcapsules during washing, a minimum amount of binder is required to adjust the binder/microcapsule ratio.

12.6.3 Electrostatic interactions

Most of the fixation processes of microcapsules induce the creation of covalent bonds with the fibers or a coating around the fibers, which entail a modification to the physical or chemical structure of the fiber, which can lead to degradation and/or breakage of the textile material. Ionic bonds between the microcapsules and the fibers may overcome these drawbacks, and allows a temporary fixation of the charged capsules onto the textile fabric. Fibers having a surface potential, such as polyamide in acidic medium, induce ionic bonds through electrostatic affinity with the textile during the bath exhaustion treatment. The binding due to electrostatic interaction between the negative charges of the fibers and the previously cationized microcapsules, allows keeping the physical properties of the textile intact.

The presence of cationic or anionic functional groups on the outer surface of the microcapsules may be induced during the microencapsulation process or via a post-treatment. Nevertheless, this strategy leads to poor washing durability, but allows the possibility for the wearer to reload the textile material with microencapsulated active ingredients.

Viladot et al. have proposed a way to avoid the loss of capsules during the washing cycle [104]. The active substance was encapsulated with a cationic polymer before the fixation operation by an exhaustion bath or a padding process.

Ripoll et al. have studied the functionalization of washing-resistant cosmetotextile based on electrostatic interactions is a new functionalization route for the preparation of active textiles, from polyamide fibers and amino-functionalized PMMA particles [105, 106]. They are related the adsorption process to the electrostatic interactions, which was favored by increasing the pH of the medium. Whereas, during desorption experiments, the presence of strong interactions between particles and fibers, limits to 20% the weight loss of particles after 10 washing cycles.

The use of a curing treatment after finishing one is also a commonly used strategy to increase the possible chemical reactions between the particles and the textile. Common curing treatments are thermos-curing, which exploits thermal activation, or photocuring which employs UV irradiation. In this context, particular care must be taken to avoid active substance degradation or microcapsules deterioration under the curing process conditions.

12.7 Laboratory and industrial processes of microcapsules deposition

The elaboration of microencapsulated textiles requires the application of microcapsules to the fabrics which is done by a finishing treatment [107]. The latter is an important unit operation of the textile supply chain whose scope is to impart the desired final properties to the material both from the aesthetic and technical point of view. It is generally performed at the end of the textile manufacturing process in the industrial case. Usually, the approach used to design functional textile is to employ some already known finishing technique using the microcapsules as a component of the finishing formulation. The rationale behind this approach is the easy scalability of the production by employing an industrially feasible process at the laboratory scale.

The selection of the appropriate textile finishing process is a key parameter to achieve the desired effectiveness and durability of the treatment containing the microcapsules. This requires consideration of the intrinsic characteristics of the textile fabric such as its chemical nature, fiber type and construction; the final performance requirements including the durability of the microcapsules to achieve the desired long-term effect, the availability of equipment or materials to functionalize the textile substrate, the cost-benefit balance, environmental considerations and legislation as well as compatibility with other formulations and/or finishing processes [108]. Most of the microcapsules, may be applied by conventional finishing techniques or during the rinse cycle of a washing machine [109, 110] on any kind of fabric (woven, nonwoven, knitted or garments) regardless of its nature (natural and synthetic).

The most common finishing techniques include bath exhaustion, padding, layer-by-layer finishing, chemical grafting, and the spraying method. These processes are often followed by drying and curing to effectively fix or link the microcapsules onto the textile surface [111]. In most cases, the microcapsules are dispersed into an aqueous solution so wet finishing processes are the most used ones.

– Impregnation
Impregnation is one of the more appropriate methods according due to the lack of affinity between the microcapsules and the textile substrate [62]. Microcapsules are used with a dispersant promoting their diffusion through the textile material, followed by the addition of a cross-linking agent to bind them to the fabric.

– Bath exhaustion
Bath exhaustion process is one of the first proposed finishing technologies. In this process, the fabrics or garments are introduced in a dyeing apparatus, and they are immersed in the finishing liquor whose temperature and pH are controlled to achieve the highest affinity between the fibers and the finishes. The microcapsules are dispersed in a solution with the binder, the auxiliaries and acids/bases for pH adjustment. The microcapsules fixation is controlled from the introduction of the solution at defined times, speeds, temperatures and pH adjustments to promote the diffusion of the particles on the textile surface. Bath exhaustion provides an excellent treatment uniformity; however due to the high liquor ratios required it also implies a significant microcapsules and water consumption [112]. During the absorption, the batch becomes clear, which allows following the process by turbidity measurements.

In this process, the bath ratios range from 1/30 to 1/60, autoclaves for cross coil or fabrics, jigger and tourniquet, or overflow and jet are the main apparatus used [113].

– Padding
Padding is one the best solutions for applying microcapsules to fabrics, since this process requires fewer chemical products and can be used at room temperature compare to bath exhaustion process [114]. Padding, as a continuous process, differs from the previous ones by the use of lower liquor amounts. This technology consists of quickly immersing the fabric in the finishing bath, containing a softener, and wetting agents, and then squeezing it between two rollers; the wet pick up is adjusted by controlling the pressure applied by the rollers. The finishing bath contains also a binding agent to ensure the adhesion of the finish. Furthermore, the affinity between fabric and microcapsules must be high to insure an effective and uniform padding finish [115].

– Spraying
Spray-based finishing consists of the nebulization of the finishing solution and direct application on the textiles. The spray method can be easily implemented in the process line where it is usually followed by a fixation treatment that aims to

enhance the finishing fastness [116]. It is generally a water saving approach [117]. Finishing by spray may be divided into two main processes, that is, (i) compressed air spraying and (ii) hydraulic spraying. The first one consists of applying the formulation to spray media using compressed air as a means of transporting them. Nevertheless, it presents the drawback to lead and irregular deposition of the finishing treatment onto the fabric. In the second one, the liquid is atomized under pressure, so that a minimum amount of the formulation is atomized to ensure a uniform deposition.

– Layer-by-layer deposition
Layer-by-layer deposition is a surface engineering technique that consists in dipping alternatively the textile in oppositely charged electrolyte solutions, targeting a precise number of monolayers on the fiber surface. Indeed such technique is quite versatile, given the large number of material combinations that could be deposited and is currently feasible at level of pilot scale plants [118].

12.8 Release mechanisms

The active ingredient has to be released from the microparticles bound to the fabric surface and reach its biological target for its activity can operate. Fast or sustained release may be willed. The site of activity depends on the type of active ingredient.

- Fragrances must be released in the environment close to the cosmetotextile and a long-lasting release is expected. Several other actives such as insect repellents act in the same way. Such cases only require the encapsulated active ingredient being released from the microparticles. The active ingredient should not penetrate the skin.
- Skin care ingredients must be transferred into the skin and reach their site of activity. The first step is the release of the active ingredient. It should be followed by a transfer into the skin by entering the *stratum corneum*; and finally, diffusion inside skin to the target should operate (Figure 12.11).

In both cases, release is required. It may occur by means of either passive diffusion out of the microcapsules or by mechanical breakage of the microcapsules. These two mechanisms are discussed in the following.

12.8.1 Passive diffusion

The active ingredient travels inside the polymer material, and it is released in the surrounding medium where diffusion takes place again. Diffusion in the surrounding medium depends on whether it is stirred or quiescent. It is most often quiescent.

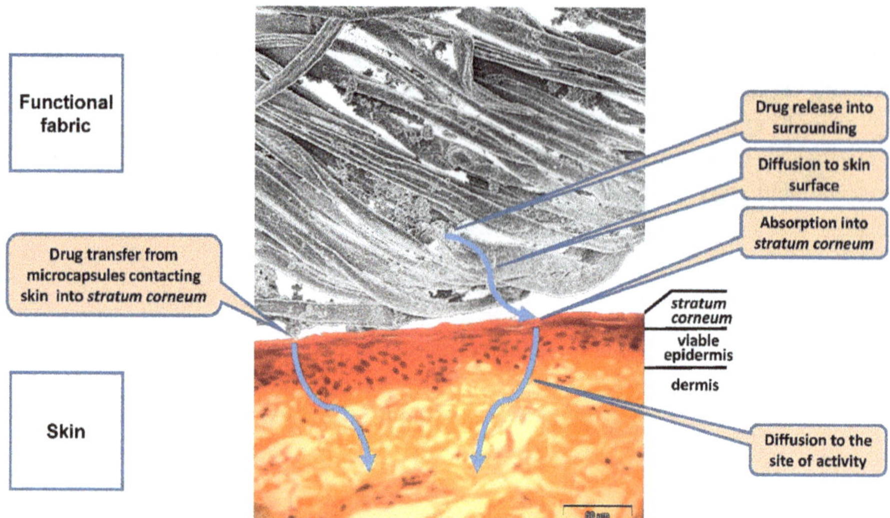

Figure 12.11: The complex paths of an active substance from the microcapsules of the cosmetotextile to the site of its biological activity inside the skin.

The structural design of microparticles (microcapsules or microspheres) diffusion inside a polymer material is the slowest rate-determining step. The diffusion rate depends on many parameters of the polymer:

- The chemical nature of the polymer: soft materials with high mobility allow fast diffusion. Crystalline polymers strongly hinder diffusion, so that diffusion takes place in the amorphous (glassy or rubbery) part of semicrystalline polymers.
- Low molar mass allows faster diffusion
- Tight cross-linking impedes diffusion
- The presence of small molecules (solvents) inside the polymer material accelerates diffusion. They can be ingredients of the surrounding medium that swell the polymer or plasticizers. The active ingredient can act as a plasticizer.
- The length of the diffusion path obviously matters. It is the size of microspheres or the thickness of microcapsules walls.

All these parameters may be adjusted in order to set the release rate from fast when immediate activity is intended to very slow in case of a long-lasting delivery.

In most cases, the active ingredient is partly encapsulated inside the polymer and partly adsorbed at the surface of microparticles. Adsorbed molecules are released immediately upon contacting an acceptor medium; the kinetics shows the burst release of this part followed by the slow release of the encapsulated part. Diffusion out of microparticles depends on the diffusion coefficient (diffusivity) of the active in the polymer medium. The kinetics can be calculated by using well-known equations available for diffusion in simple geometries [119, 120]. The choice of the equation describing the actual case with best closeness may be a difficult task because a balance should be found between the necessary simplicity of a model and the complexity of real situations. The equation for the release from a homogeneous spherical body loaded with active substance under sink condition (no limitation by the solubility in the acceptor medium) is given as an example [119, 120]:

$$\frac{M(t)}{M(\infty)} = 1 - \frac{6}{\pi^2} \sum_{n=1}^{\infty} \frac{1}{n^2} e^{-\frac{Dn^2\pi^2}{R^2} t}$$

where $M(t)$ is the released amount at time t, D is the diffusivity (m^2/s) and R is the radius of the microspheres.

12.8.2 Mechanical break up

Breaking the wall of microcapsules immediately releases the content. Microcapsules rupture is generally achieved by means of a mechanical stimulus as the apparel is manipulated. Release reaches completion within a short time. As common sense teaches, the mechanical strength of the polymer walls (tensile strength and wall thickness) is the key factor. A breakage after long times relies on a fatigue mechanism of the hard material upon long and repeated mechanical stresses. Rupture takes place at structural defects that are weak points in the walls. Control over the stress and strain at rupture of the walls requires a good mastery of the walls thickness and homogeneity during the manufacturing process.

12.8.3 Transfer to the skin

Once the active substance has reached the external surface of the microparticles, it is transferred to the skin and diffused inside it to its targeted site. The liquid medium that receives the active substance is either sweat or the hydrolipidic film coating the skin surface. Transfer to them is controlled by the partition equilibrium at both side of the microparticle surface. The partition coefficient of a hydrophobic active substance is such that it is mostly retained in the hydrophobic polymer material, so that transfer to the skin through the aqueous intermediate medium is unfavorable. A direct transfer from the microparticles to the skin may occur when microparticles are contacting the skin surface. Indeed, the *stratum corneum* at the skin surface is hydrophobic. Effective contact (transient adhesion) to the skin surface is better achieved with tight apparels such as stockings. Upon such shunt of the intermediate medium, the relevant parameter is the partition coefficient of the active substance between the microparticles and the *stratum corneum*. The chemical nature of the polymer can be chosen such that the adhesion energy and the partition coefficient are favorable.

12.9 Examples of marketed products

By modifying the structure of the textile fabric and/or the nature of the fibers, new cosmetotextile effects can be obtained. Examples of commercial items are summarized in Table 12.4.

Table 12.4: Commercial cosmetotextiles with manufacturer's product name, main structural modification and product functionality [26].

Manufacturer	Brand name	Main structural modification	Product functionality	Reference
Nike (USA)	Dri-fit fabrics	Microfibers of nylon, polyester and Spandex	These items are made from fibers of fine diameters which play a decisive role in creating the perfect levels of surface tension and adhesion between molecules. This creates a strong capillary action and promotes the rapid movement of sweat along these fibers.	[26]
Hefel Textile GmbH (Austria) HEFEL	Lyosilk®	Tencel and silk fiber	This product is composed of microfine Tencel fibers and pure silk. Individual delicate yarns are twisted together to form a soft and open silk thread for use as a weft. By the incorporation of pure silk, these actively breathable downy fibers are brighter, smoother and even more refined.	[113, 121, 122]
Solidea (Italy) SOLIDEA	MicroMassage Magic	Fiber composition (80% polyamide; 18% elastane; 2% cotton)	This product offers elegant shaping, toning and smoothing of the skin. The three-dimensional knitted structure lightly massages the skin by working with the natural movements of the body by promoting the circulation of the skin and fatty tissues and by stimulating the drainage of fluids causing the orange peel effect on the skin.	[26, 113, 122]
Spanx (USA) SPANX	–	Elastane fibers	These are controlled compression underwear for body shaping.	[26]

In addition to the novelty in fiber and fabric, several textile companies are developing finishes that provide users with a cosmetic effect. A wide range of finishes has been launched by various cosmetotextile manufacturers by integration of cosmetic ingredients. Among the most widespread technologies, microencapsulation offers many possibilities for improving the textile properties [122]. Since the first Japanese applications in the late 1980s by the Kanebo company, new cosmetotextile articles are increasingly popular and are developing considerably thanks to the diversity of cosmetic substances which can be encapsulated [32, 123]. In France, first applications dated from the late 1990s with Dim moisturizing and slimming Pantyhose functionalized by grafted microcapsules [32] and microencapsulated perfumed scarfs launched by Hermes in 1995. A wide range of cosmetic textiles all over the world has been commercialized by various manufacturers now expanding as shown in the Table 12.5 below.

The list of cosmetotextile products is even longer and constantly evolving stimulated by the needs of customers by addressing a social passion for research in terms of health, beauty and comfort.

12.10 Conclusion

Cosmetotextiles are cosmetic devices intended at applying cosmetic active substances over the skin surface for long times. The technology is mainly derived from medical devices such as bandages and patches. However, the purpose is different as it is claimed as "cosmetic," which means that cosmetotextiles comply with cosmetic regulation specifying their activity aiming at "body care." A specific feature of most cosmetotextiles, compared to cosmetic creams directly spread on skin, is a sustained delivery of an active substance. The main technique relies on binding microcapsules to the fabric surface. Main challenges are (i) the design of microcapsules in the same way as for classical microencapsulation, (ii) binding the microcapsules to textile fibers over long durations and resisting the washing processes. Surface physical chemistry (wetting and adhesion) is ubiquitous in designing the materials and the manufacture processes. The latter are mostly direct implementations of textile finishing processes such as those for fabric dyeing. Several cosmetotextile articles are currently on the market. This is a developing area, in particular because cosmetics is driven by a fast and continuous innovation that increasingly relies on associating new application devices to the products.

Table 12.5: Commercial cosmetotextiles with manufacturer's product name, main cosmetic component and product functionality.

Manufacturer	Brand name	Main cosmetic component	Product functionality	References
Ajinomoto (Japan) with Mizuno Corp (Japan)	Amino Veil	Arginine amino acid	Tennis and golf clothes were functionalized. Dissolved cosmetic ingredient into a wearer's perspiration enhances the material ability to absorb moisture and keeps the skin pH level balanced which promotes its regeneration.	[26, 113, 122]
Yonex (Japan)	–	Xylitol	Tennis and badminton clothes, containing xylitol, absorbs heat when in contact with water and offers a cool feel during sweating.	[73, 113, 122]
Fujibo formerly called Fuji Spinning (Japan)	V-Up	Provitamin C (dehydro ascorbic acid) soluble in sebum	Blouses (men's and women's) and shirts contain provitamin C that converts into vitamin C in the presence of sebum.	[5, 73, 113, 122]
		Vitamin E (tocopherol) or provitamin E (tocopheryl acetate)	No information on this fiber is available. Indeed, it may contain vitamin E or provitamin E.	[5]
Outlast (USA)	–	Phase-change microcapsules (PCM)	Textile structures containing PCM-based microcapsules are used to improve comfort and temperature control for consumers in bedding, medical supplies, sportswear and protective clothing.	[124]

Pulcra (Germany)	Skintex® Supercool	Retinol or chitosan microcapsules containing active ingredients (distilled oils of plants, fruits and leaves)	Functionalized fabrics provide gentle cosmetic care (stimulation of collagen production and reduction of fat absorption in the stomach) with special effects of invigorating aromas.	[26, 116, 122, 125, 126]
	Cyclofresh®	Cyclodextrin and perfume	This fabric is capable of releasing perfume and/or trapping unpleasant smells.	[73, 126]
Clariant (Switzerland) with Lipotec (Spain)	Quiospheres® moist	Antarcticine and Xpermoist microcapsules	Quiospheres moist are microencapsulated treated fabrics able to control the skin moisture level.	[26, 73, 122, 126]
	Quiospheres® slim	Liporeductyl and Relistase microcapsules	Quiospheres slim are microcapsulated treated fabrics that control the cellulite level of human body parts.	
Hefel Textile GmbH (Austria)	SeaCell® Active	Algae and silver particles	SeaCell® Active is a Lyocell fiber associated with algae fibers and silver particles. The amino acids and minerals contained in algae are released by the slight humidity of the body, which has a balancing effect on the organism. The mineral salts contained in SeaCell® algae mineralize the skin again. Thanks to its silver content, the fiber also has antimicrobial and fungicidal effects while preserving freshness.	[113, 121, 122]

(continued)

Table 12.5 (continued)

Manufacturer	Brand name	Main cosmetic component	Product functionality	References
Lytess (France) with L'Oréal (France) Mixa (France) Garnier (France) Biotherm (France) or Nivea (Germany)	–	Microcapsules containing cosmetic ingredients (caffeine, shea butter, red algae, mango butter, Copaiba/Flavenger oil, etc.)	Underwears (legging, body, tights, cheeky, shorty, etc.) containing cosmetic microcapsules for various effects (firming moisturizing, makeup remover, slimming, etc.).	[26, 32]
Edevan (Belgium, Portugal, UK, US)	eSCENTial®	Fragrance microcapsules	Reactive microcapsules containing fragrances and skin care ingredients are integrated into textile fabric. Cosmetotextile effects are divided in three categories: region specific, classics, sleep promoting and relaxing.	[7]

Schoeller (Switzerland) schoeller Switzerland	iLoad®	Skin care iLoad emulsion	Textile fabric is a refillable textile that can be reloaded during washing with the iLoad emulsion to ensure lasting application. It releases the active substances for skin care (wellness and anti-aging). Such cosmetotextiles can be used wherever textiles come into contact with human skin (in bedding or in pillowcases, pyjamas, underwear or hospital night shirts).	[7]
	Energear™	Titanium-mineral matrix	Through its use as a finishing technology on the fabric or as a print, titanium-mineral matrix reflects back the positive energy to the body in the form of far infrared rays. This additional energy can improve the performance, capacity and well-being of the body.	
Invista (USA) INVISTA with International Flavors & Fragrances (IFF) (USA) IFF	–	Aloe vera and chitosan with other PCMs	Legging and intimate clothing for men, women and yoga lines. These items offer cosmetic and wellness benefits (freshness, hydration and massage), stretching and recovery functions are guaranteed thanks to the use of Lycra fiber.	[73, 113, 122]
Dogi International Fabrics (Spain) DOGI international fabrics	–	Aloe vera nanoparticles	These smart fabrics with aloe vera nanoparticles provide moisturizing, calming, anti-aging and antioxidant effects.	[73, 113]

(continued)

Table 12.5 (continued)

Manufacturer	Brand name	Main cosmetic component	Product functionality	References
Richa (Belgium)	–	Phase-change microcapsules (PCMs)	Women's motorcycle pants have a thin lining of phase change material that can be removed in warm weather and replaced at lower temperatures.	[113, 122]
Nurel (Spain)	NOVAREL®	Microcapsules containing active ingredients (sweet almond oil, caffeine, vitamin E, retinol, rose hip oil, oleic acid, aloe vera, active minerals)	Microcapsules are incorporated into the polyamide (nylon) yarn during the spinning process. Thus, they are fixed into and outside the nylon filaments which provide for the skin cosmetic benefits during the garment use.	[122]
Lenzing (Austria)	Tencel®C	Chitosan	Stockings are soaked in chitosan. This textile is not only antibacterial, but it is very effective for the relief of itching, the conservation of humidity, the regulation and regeneration of cells as well as the protection of the skin. Another application is for healing. This type of fiber is used in clothing worn in contact with the skin and in furnishing articles such as bed sheets.	[122, 126, 127]
Nilit (Israel)	NILIT® Breeze	Inorganic microcapsules	Nylon-6,6 fibers containing inorganic microparticles are produced according to a special texturing process ensuring a lower body temperature. This material keeps wearers feeling refreshed and comfortable when the temperatures rise.	[122]

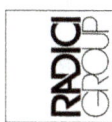

Rhodia (France) RHODIA SOLVAY GROUP	Emana®	Bioactive crystals from bio-ceramic	A bioactive yarn is created by a mixture of polyamide-6,6 and a polymer with bioactive crystals added from bio ceramic and integrated into the molecular structure of the fiber itself. This material increases skin elasticity, improves skin firmness and softness and even reduce the appearance of cellulite. Emana® helps also with enhancing sports performance, reducing muscle fatigue, accelerating recovery after exercise, and also the skin UV protection.	[7, 122]
Lanxess (Germany) LANXESS Energizing Chemistry	Bayscent® Aromatherapy	Microcapsules containing essential oils	Microcapsules are incorporated into fabrics which at the time of their break emit pleasant aromas, causing a great sense of well-being.	[126]
	Bayscent® Neutralizer	Cyclodextrin	The fabric is protected from unpleasant odors (in sleeping rooms, hotels or conference rooms, corridors, etc.) leaving a fresh and pleasant scent. Reactivation of this effect is possible by spray extraction.	
Radici (Italy) RADICI GROUP	Radyarn® and Starlight®	Nanostructured silver additive	Nanostructured silver additive is incorporated into PES yarn during the extrusion stage before spinning, which confers bacteriostatic properties inhibiting the proliferation of bacteria on the tissue. This material absorbs, reflects and emits far infrared showing thermophysiological comfort and improving microcirculation for sportswear applications.	[7]

References

[1] Nérée, L.: Cosmétotextiles, Cosmetic Textiles. In: Editorial of "Cosmetic Textiles Symposium". Organized by Interfilière Paris and La Fédération de la Maille & de la Lingerie, 2012.

[2] DIM, S. A.: Textile bioactif comportant dans ses fibres de la protéine de soie et des microcapsules de produit actif. patent FR2780073, 1998.

[3] Boh Podgornik, B.; Starešinič, M.: Chap 3 – Microencapsulation Technology and Applications in Added-value Functional Textiles. In: Tylkowski, B.; Giamberini, M.; Fernandez Prieto, S., editors: Microencapsulation. Berlin: De Gruyter, 2020; pp. 49–87.

[4] Cheng, S. Y.; Yuen, M. C. W.; Kan, C. W.; Cheuk, K. K. L.; Chui, C. H.; Lam, K. H.: Cosmetic textiles with biological benefits: Gelatin microcapsules containing vitamin C. International Journal of Molecular Medicine, 2009; 24(4), 411–419.

[5] Buschmann, H. J.; Dehabadi, V. A; Wiegand, C.: Medical, Cosmetic and Odour Resistant Finishes for Textiles. In: Paul, R., editor: Functional Finishes for Textiles. Woodhead Publishing Series in Textiles no 156, Cambridge: Woodhead Publishing, 2015; pp. 303–330.

[6] Petrusic, S.; Koncar, V.: Controlled Release of Active Aagents from microcapsules embedded in Textile Structures. In: Koncar, V., editor: Smart Textiles and Their Applications. Woodhead Publishing Series in Textiles no 178, Cambridge: Woodhead Publishing, 2016; pp. 89–114.

[7] Bonaldi, R. R.: Functional finishes for high-performance apparel. In: McLoughlin, J.; Sabir, T., editors: High-Performance Apparel. Cambridge: Woodhead Publishing, 2018; pp. 129–156.

[8] Paul, R.: Functional finishes for textiles: an overview. In: Paul, R., editor: Functional Finishes for Textiles. Woodhead Publishing Series in Textiles no 156, Cambridge: Woodhead Publishing, 2015; pp. 1–14.

[9] Mathis, R.; Mehling, A.: Textiles with Cosmetic Effects. In: Bartels, V. T., editor: Handbook of Medical Textiles. Woodhead Publishing Series in Textiles no 110, Cambridge: Woodhead Publishing, 2011; pp. 153–172.

[10] Almeida, L.: Environmental and Safety Issues regarding Functional Finishes. In: Paul, R., editor: Functional Finishes for Textiles. Woodhead Publishing Series in Textiles no 156, Cambridge: Woodhead Publishing, 2015; pp. 607–627.

[11] Teixeira, M. A.; Rodríguez, O.; Rodrigues, S.; Martins, I.; Rodrigues, A. E.: A case study of product engineering: Performance of microencapsulated perfumes on textile applications. AIChE Journal, 2012; 58(6), 1939–1950.

[12] Cheng, S. Y.; Yuen, C. W. M.; Kan, C. W.; Cheuk, K. K. L.; Tang, J. C. O.; Li, S. Y.: A comprehensive study of silicone-based cosmetic textile agent. Fibers and Polymers, 2009; 10(1), 132–140.

[13] Cheng, S. Y.; Yuen, C. W. M.; Kan, C. W.; Cheuk, K. K. L.; Tang, J. C. O.: Systematic characterization of cosmetic textiles. Textile Research Journal, 2010; 80(6), 524–536.

[14] Ben Abdelkader, M.; Azizi, N.; Chevalier, Y.; Majdoub, M.: Microencapsulated neroline with new epoxy resin shell based on isosorbide: Preparation, characterization and application to cosmetotextile. International Journal of Applied Research on Textile, 2013; 1(1), 59–61.

[15] Azizi, N.; Ben Abdelkader, M.; Chevalier, Y.; Majdoub, M.: New β-cyclodextrin-based microcapsules for textiles uses. Fibers and Polymers, 2019; 20(4), 683–689.

[16] Ben Abdelkader, M.; Azizi, N.; Baffoun, A.; Chevalier, Y.; Majdoub, M.: New microcapsules based on isosorbide for cosmetotextile: Preparation and characterization. Industrial Crops and Products, 2018; 123, 591–599.

[17] Ben Abdelkader, M.; Azizi, N.; Baffoun, A.; Chevalier, Y.; Majdoub, M.: Fragrant microcapsules based on β-cyclodextrin for cosmetotextile application. Journal of Renewable Materials, 2019; 7(12), 1347–1362.

[18] Sánchez-Navarro, M. M.; Pérez-Limiñana, M. Á.; Arán-Ais, F.; Orgilés-Barceló, C.: Scent properties by natural fragrance microencapsulation for footwear applications. Polymer International, 2015; 64(10), 1458–1464.
[19] Li, S.; Lewis, J. E.; Stewart, N. M.; Qian, L.; Boyter, H.: Effect of finishing methods on washing durability of microencapsulated aroma finishing. The Journal of the Textile Institute, 2008; 99(2), 177–183.
[20] Panisello, C.; Peña, B.; GilabertOriol, G.; Constantí, M.; Gumí, T.; Garcia-Valls, R.: Polysulfone/vanillin microcapsules for antibacterial and aromatic finishing of fabrics. Industrial & Engineering Chemistry Research, 2013; 52(29), 9995–10003.
[21] Azizi, N.; Chevalier, Y.; Majdoub, M.: Isosorbide-based microcapsules for cosmeto-textiles. Industrial Crops and Products, 2014; 52, 150–157.
[22] Azizi, N.; Ladhari, N.; Chevalier, Y.; Majdoub, M.: Study on perfume sustained-release cotton fabric with polystyrene-based microspheres. International Journal of Applied Research on Textile, 2014; 2(2), 42–51.
[23] Bishop, P.: Testing for fabric comfort. In: Hu, J., editor: Fabric Testing. Woodhead Publishing Series in Textiles no 76, Cambridge: Woodhead Publishing, 2008; pp. 228–254.
[24] Bartels, V. T.: Physiological comfort of biofunctional textiles. Current Problems in Dermatology, 2006; 33, 51–66.
[25] Decaens, J.; Vermeersch, O.: Specific testing for smart textiles. In: Dolez, P., Vermeersch, O., Izquierdo, V., editors: Advanced Characterization and Testing of Textiles. The Textile Institute Book Series, Cambridge: Woodhead Publishing, 2018; pp. 351–374.
[26] Barel, A. O.; Paye, M.; Maibach, H. I.: Handbook of Cosmetic Science and Technology, 4th ed. Boca Raton: CRC Press, 2014.
[27] Thangaraju, P.; Varthya, S.B; ISO 10993: Biological evaluation of medical devices. In: Timiri Shanmugam, P.S., Thangaraju, P., Palani, N., Sampath, T., editors: Medical Device Guidelines and Regulations Handbook. Cham, Switzerland: Springer, 2022; pp. 163–187.
[28] Kligman, A. M.; Basketter, D. A.: A critical commentary and updating of the guinea pig maximization test. Contact Dermatitis, 1995; 32(2), 129–134.
[29] Niinimäki, K.: Ecodesign and textiles. Research Journal of Textile and Apparel, 2006; 10(3), 67–75.
[30] Commission regulation (EU) 2019/831: Official Journal of the European Union, 2019; L137, 29–63. Available at https://eur-lex.europa.eu/legalcontent/EN/TXT/PDF/?uri=CELEX:32019R0831&from=EN
[31] CEN: CEN/TR 15917: Textiles – Cosmetotextiles, 2009. Available at www.cen.eu
[32] Salaün, F.: Microencapsulation technology for smart textile coatings. In: Hu, J., editor: Active Coatings for Smart Textiles. Woodhead Publishing Series in Textiles no 176, Cambridge: Woodhead Publishing, 2016; pp. 179–220.
[33] Abdullah, N. A.; Sekak, K. A.; Ahmad, M. R.; Effendi, T. B.: Characteristics of electrospun PVA-Aloe vera nanofibres produced via electrospinning. In: Proceedings of the International Colloquium in Textile Engineering, Fashion, Apparel and Design. Springer, Singapore, 2014; pp. 7–11.
[34] Jager-Lezer, N.: Cosmetic composition comprising fiber. Patent US7754196B2, 2010.
[35] Kenawy, E. R.; Worley, S. D.; Broughton, R.: The chemistry and applications of antimicrobial polymers: A state-of-the-art review. Biomacromolecules, 2007; 8(5), 1359–1384.
[36] Varesano, A.; Vineis, C.; Aluigi, A.; Rombaldoni, F.: Antimicrobial polymers for textile products. In: Méndez-Vilas, A., editor: Science Against Microbial Pathogens. Communicating Current Research and Technological Advances. Formatex Research Center, Spain, 2011; 3, pp. 99–110.

[37] Fuchs, A. D.; Tiller, J. C.: Contact-active antimicrobial coatings derived from aqueous suspensions. Angewandte Chemie International Edition, 2006; 45(40), 6759–6762.
[38] Mahltig, B.; Fiedler, D.; Böttcher, H.: Antimicrobial sol–gel coatings. Journal of Sol-Gel Science and Technology, 2004; 32(1), 219–222.
[39] Kalyon, B. D.; Olgun, U.: Antibacterial efficacy of triclosan-incorporated polymers. American Journal of Infection Control, 2001; 29(2), 124–125.
[40] Labay, C.; Canal, J. M.; Navarro, A.; Canal, C.: Corona plasma modification of polyamide 66 for the design of textile delivery systems for cosmetic therapy. Applied Surface Science, 2014; 316, 251–258.
[41] Leclercq, L.: Smart medical textiles based on cyclodextrins for curative or preventive patient care. In: Hu, J., editor: Active Coatings for Smart Textiles. Woodhead Publishing Series in Textiles no 176, Cambridge: Woodhead Publishing, 2016; pp. 391–427.
[42] Buschmann, H. J.; Knittel, D.; Schollmeyer, E.: New textile applications of cyclodextrins. Journal of Inclusion Phenomena and Macrocyclic Chemistry, 2001; 40(3), 169–172.
[43] Szejtli, J.: Introduction and general overview of cyclodextrin chemistry. Chemical Reviews, 1998; 98(5), 1743–1754.
[44] Rekharsky, M. V.; Inoue, Y.: Complexation thermodynamics of cyclodextrins. Chemical Reviews, 1998; 98(5), 1875–1918.
[45] Nostro, P. L.; Fratoni, L.; Ridi, F.; Baglioni, P.: Surface treatments on Tencel fabric: Grafting with β-cyclodextrin. Journal of Applied Polymer Science, 2003; 88(3), 706–715.
[46] Khanna, S.; Chakraborty, J. N.: Optimization of monochlorotriazine β-cyclodextrin grafting on cotton and assessment of release behavior of essential oils from functionalized fabric. Fashion and Textiles, 2017; 4(1), 1–18.
[47] Ibrahim, N. A.; Abdalla, W. A.; El-Zairy, E. M. R.; Khalil, H. M.: Utilization of monochlorotriazine β-cyclodextrin for enhancing printability and functionality of wool. Carbohydrate Polymers, 2013; 92(2), 1520–1529.
[48] Bhaskara-Amrit, U. R.; Agrawal, P. B.; Warmoeskerken, M. M. C. G.: Applications of β-cyclodextrins in textiles. AUTEX Research Journal, 2011; 11(4), 94–101.
[49] Buschmann, H. J.; Schollmeyer, E.: Applications of cyclodextrins in cosmetic products: A review. Journal of Cosmetic Science, 2002; 53(3), 185–192.
[50] Voncina, B.; Vivod, V.: Cyclodextrins in Textile Finishing. Valencia, Spain: Intech Open Science, Valencia, Spain, 2013; 3, 53–75.
[51] Singh, N.; Sahu, O.: Sustainable cyclodextrin in textile applications. In: Shahid-ul-Islam, Butola, B.S., editors: The Impact and Prospects of Green Chemistry for Textile Technology. The Textile Institute Book Series, Cambridge: Woodhead Publishing, 2019; pp. 83–105.
[52] Cravotto, G.; Beltramo, L.; Sapino, S.; Binello, A.; Carlotti, M. E.: A new cyclodextrin-grafted viscose loaded with aescin formulations for a cosmeto-textile approach to chronic venous insufficiency. Journal of Materials Science. Materials in Medicine, 2011; 22(10), 2387–2395.
[53] Mathiowitz, E.; Kreitz, M. R.; Brannon-Peppas, L.: Microencapsulation. In: Mathiowitz, E., editor: Encyclopedia of Controlled Drug Delivery. New York: John Wiley & Sons, 1999; pp. 493–546.
[54] Green, B. K.; Lowell, S.: Oil-containing microscopic capsules and method of making them. Patent US2800457A, 1957.
[55] White, M. A.: The chemistry behind carbonless copy paper. Journal of Chemical Education, 1998; 75(9), 1119–1120.
[56] Bouchemal, K.; Briançon, S.; Perrier, E.; Fessi, H.; Bonnet, I.; Zydowicz, N.: Synthesis and characterization of polyurethane and poly(ether urethane) nanocapsules using a new technique of interfacial polycondensation combined to spontaneous emulsification. International Journal of Pharmaceutics, 2004; 269(1), 89–100.

[57] Glenn, G. M.; Klamczynski, A. P.; Woods, D. F.; Chiou, B.; Orts, W. J.; Imam, S. H.: Encapsulation of plant oils in porous starch microspheres. Journal of Agricultural and Food Chemistry, 2010; 58(7), 4180–4184.
[58] Matsunami, Y.; Ichikawa, K.: Characterization of the structures of poly(urea–urethane) microcapsules. International Journal of Pharmaceutics, 2002; 242(1–2), 147–153.
[59] Suryanarayana, C.; Rao, K. C.; Kumar, D.: Preparation and characterization of microcapsules containing linseed oil and its use in self-healing coatings. Progress in Organic Coatings, 2008; 63(1), 72–78.
[60] Tseng, Y. H.; Fang, M. H.; Tsai, P. S.; Yang, Y. M.: Preparation of microencapsulated phase-change materials (MCPCMs) by means of interfacial polycondensation. Journal of Microencapsulation, 2005; 22(1), 37–46.
[61] Coffin, M. D.; McGinity, J. W.: Physical and chemical stability of aqueous colloidal Ddispersions. In: Whateley, T.L., editor: Microencapsulation of Drugs. Chur, Switzerland: Harwood Academic Publishers, 1992; pp. 197–214.
[62] Monllor, P.; Bonet, M. A.; Cases, F.: Characterization of the behaviour of flavour microcapsules in cotton fabrics. European Polymer Journal, 2007; 43(6), 2481–2490.
[63] Madene, A.; Jacquot, M.; Scher, J.; Desobry, S.: Flavour encapsulation and controlled release – A review. International Journal of Food Science &technology, 2006; 41(1), 1–21.
[64] Dieterich, D.: Aqueous emulsions, dispersions and solutions of polyurethanes; synthesis and properties. Progress in Organic Coatings, 1981; 9(3), 281–340.
[65] Nelson, G.: Microencapsulation in textile finishing. Review of Progress in Coloration and Related Topics, 2001; 31(1), 57–64.
[66] Pena, B.; Panisello, C.; Aresté, G.; Garcia-Valls, R.; Gumí, T.: Preparation and characterization of polysulfone microcapsules for perfume release. Chemical Engineering Journal, 2012; 179, 394–403.
[67] Rodrigues, S. N.; Martins, I. M.; Fernandes, I. P.; Gomes, P. B.; Mata, V. G.; Barreiro, M. F.; Rodrigues, A. E.: Scentfashion®: Microencapsulated perfumes for textile application. Chemical Engineering Journal, 2009; 149(1–3), 463–472.
[68] Specos, M. M.; Escobar, G.; Marino, P.; Puggia, C.; Defain Tesoriero, M. V.; Hermida, L.: Aroma finishing of cotton fabrics by means of microencapsulation techniques. Journal of Industrial Textiles, 2010; 40(1), 13–32.
[69] Specos, M. M.; García, J. J.; Tornesello, J.; Marino, P.; Vecchia, M. D.; Tesoriero, M. D.; Hermida, L. G.: Microencapsulated citronella oil for mosquito repellent finishing of cotton textiles. Transactions of the Royal Society of Tropical Medicine and Hygiene, 2010; 104(10), 653–658.
[70] Tzhayik, O.; Cavaco-Paulo, A.; Gedanken, A.: Fragrance release profile from sonochemically prepared protein microsphere containers. Ultrasonics Sonochemistry, 2012; 19(4), 858–863.
[71] Zhang, Y.; Rochefort, D.: Characterisation and applications of microcapsules obtained by interfacial polycondensation. Journal of Microencapsulation, 2012; 29(7), 636–649.
[72] Cheng, S. Y.; Yuen, C. W. M.; Kan, C. W.; Cheuk, K. K. L.: Development of cosmetic textiles using microencapsulation technology. Research Journal of Textile and Apparel, 2008; 12(4), 41–51.
[73] Singh, M. K.; Varun, V. K.; Behera, B. K.: Cosmetotextiles: State of art. Fibres & Textiles in Eastern Europe, 2011; 19(4), 27–33.
[74] Liang, M.; Davies, N. M.; Toth, I.: Increasing entrapment of peptides within poly(alkyl cyanoacrylate) nanoparticles prepared from water-in-oil microemulsions by copolymerization. International Journal of Pharmaceutics, 2008; 362(1–2), 141–146.

[75] Graf, A.; Jack, K. S.; Whittaker, A. K.; Hook, S. M.; Rades, T.: Protein delivery using nanoparticles based on microemulsions with different structure-types. European Journal of Pharmaceutical Sciences, 2008; 33(4–5), 434–444.

[76] Salaün, F.; Devaux, E.; Bourbigot, S.; Rumeau, P.: Preparation of multinuclear microparticles using a polymerization in emulsion process. Journal of Applied Polymer Science, 2008; 107(4), 2444–2452.

[77] Jyothi, N. V. N.; Prasanna, P. M.; Sakarkar, S. N.; Prabha, K. S.; Ramaiah, P. S.; Srawan, G. Y.: Microencapsulation techniques, factors influencing encapsulation efficiency. Journal of Microencapsulation, 2010; 27(3), 187–197.

[78] Arshady, R.: Suspension, emulsion, and dispersion polymerization: A methodological survey. Colloid and Polymer Science, 1992; 270(8), 717–732.

[79] Frere, Y.; Danicher, L.; Gramain, P.: Preparation of polyurethane microcapsules by interfacial polycondensation. European Polymer Journal, 1998; 34, 193–199.

[80] Hong, K.; Park, S.: Preparation of polyurethane microcapsules with different soft segments and their characteristics. Reactive & Functional Polymers, 1999; 42(3), 193–200.

[81] Jabbari, E.: Morphology and structure of microcapsules prepared by interfacial polycondensation of methylene bis(phenyl isocyanate) with hexamethylene diamine. Journal of Microencapsulation, 2001; 18(6), 801–809.

[82] Mirabedini, S. M.; Dutil, I.; Farnood, R. R.: Preparation and characterization of ethyl cellulose-based core–shell microcapsules containing plant oils. Colloids and Surfaces. A, Physicochemical and Engineering Aspects, 2012; 394, 74–84.

[83] Pearson, R. G.; Williams, E. L.: Interfacial polymerization of an isocyanate and a diol. Journal of Polymer Science: Polymer Chemistry Edition, 1985; 23(1), 9–18.

[84] Salaün, F.; Bedek, G.; Devaux, E.; Dupont, D.; Gengembre, L.: Microencapsulation of a cooling agent by interfacial polymerization: Influence of the parameters of encapsulation on poly(urethane–urea) microparticles characteristics. Journal of Membrane Science, 2011; 370(1–2), 23–33.

[85] Wagh, S. J.; Dhumal, S. S.; Suresh, A. K.: An experimental study of polyurea membrane formation by interfacial polycondensation. Journal of Membrane Science, 2009; 328(1–2), 246–256.

[86] Yan, N.; Ni, P.; Zhang, M.: Preparation and properties of polyurea microcapsules with non-ionic surfactant as emulsifier. Journal of Microencapsulation, 1993; 10(3), 375–383.

[87] O'Donnell, P. B.; McGinity, J. W.: Preparation of microspheres by the solvent evaporation technique. Advanced Drug Delivery Reviews, 1997; 28(1), 25–42.

[88] Arshady, R.: Microspheres and microcapsules, a survey of manufacturing techniques: Part III: Solvent evaporation. Polymer Engineering & Science, 1990; 30(15), 915–924.

[89] Freitas, S.; Merkle, H. P.; Gander, B.: Microencapsulation by solvent extraction/evaporation: Reviewing the state of the art of microsphere preparation process technology. Journal of Controlled Release, 2005; 102(2), 313–332.

[90] Benita, S., editor: Microencapsulation: Methods and Industrial Applications. Boca Raton: CRC Press, 2005.

[91] Kagadis, C. A.; Choulis, N. H.: Ibuprofen microcapsules from stearic acid. I: Effect of particle size. Pharmazie, 1985; 40(11), 807–808.

[92] Benita, S.; Zouai, O.; Benoit, J.-P.: 5-Fluorouracil: Carnauba wax microspheres for chemoembolization: An in vitro evaluation. Journal of Pharmaceutical Sciences, 1986; 75(9), 847–851.

[93] Salaün, F.; Devaux, E.; Bourbigot, S.; Rumeau, P.: Application of contact angle measurement to the manufacture of textiles containing microcapsules. Textile Research Journal, 2009; 79(13), 1202–1212.

[94] Liu, J.; Liu, C.; Liu, Y.; Chen, M.; Hu, Y.; Yang, Z.: Study on the grafting of chitosan–gelatin microcapsules onto cotton fabrics and its antibacterial effect. Colloids and Surfaces. B, Biointerfaces, 2013; 109, 103–108.
[95] Solé, I.; Vílchez, S.; Montanyà, N.; García-Celma, M. J.; Ferrándiz, M.; Esquena, J.: Polyamide fabric coated with a dihydroxyacetone-loaded chitosan hydrogel for a cosmeto-textile application. Journal of Industrial Textiles, 2020; 50(4), 526–542.
[96] Fiedler, J. O.; Carmona, Ó. G.; Carmona, C. G.; Lis, M. J.; Plath, A. M. S.; Samulewski, R. B.; Bezerra, F. M.: Application of Aloe vera microcapsules in cotton nonwovens to obtain biofunctional textiles. The Journal of the Textile Institute, 2019; 111(1), 68–74.
[97] Yang, Z.; Zeng, Z.; Xiao, Z.; Ji, H.: Preparation and controllable release of chitosan/vanillin microcapsules and their application to cotton fabric. Flavour and Fragrance Journal, 2014; 29(2), 114–120.
[98] Fan, F.; Zhang, W.; Wang, C.: Covalent bonding and photochromic properties of double-shell polyurethane-chitosan microcapsules crosslinked onto cotton fabric. Cellulose, 2015; 22(2), 1427–1438.
[99] Nada, A.; Al-Moghazy, M.; Soliman, A. A.; Rashwan, G. M.; Eldawy, T. H. A.; Hassan, A. A. E.; Sayed, G. H.: Pyrazole-based compounds in chitosan liposomal emulsion for antimicrobial cotton fabrics. International Journal of Biological Macromolecules, 2018; 107, 585–594.
[100] Julaeha, E.; Puspita, S.; Eddy, D. R.; Wahyudi, T.; Nurzaman, M.; Nugraha, J.; Herlina, T.; Al Anshori, J.: Microencapsulation of lime (Citrus aurantifolia) oil for antibacterial finishing of cotton fabric. RSC Advances, 2021; 11(3), 1743–1749.
[101] Cuevas, J. M.; Gonzalo, B.; Rodríguez, C.; Domínguez, A.; Galán, D.; Loscertales, I. G.: Grafting electrosprayed silica microspheres on cellulosic textile via cyanuric chloride reactive groups. Journal of Experimental Nanoscience, 2015; 10(11), 868–879.
[102] Frere, Y.; Danicher, L.; Merji, M.: Capsules à surface modifiée pour greffage sur des fibres. Patent FR2897617A1; Capsules with a modified surface for grafting onto fibres, 2007. WO2007096513A1 (2007), EP1994217A1 (2008), US8586143B2 (2013).
[103] Hamaide, T.; Fleury, E.; Bertholon, I.; Drockenmuller, E.; Tillement, O.; Roux, S.; Boulon, G.: Fibres de polysaccharides sur lesquelles sont greffées par "click chemistry" des nanoparticules. Patent FR2960244, 2010.
[104] Petit, J. L. V.; Gonzalez, R. D.; Botello, A. F.: Process of treatment of fibers and/or textile materials. Patent US9708757B2, 2017.
[105] Ripoll, L.; Bordes, C.; Marote, P.; Etheve, S.; Elaissari, A.; Fessi, H.: Polymer particle adsorption at textile/liquid interfaces: A simple approach for a new functionalization route. Polymer International, 2012; 61(7), 1127–1135.
[106] Ripoll, L.; Bordes, C.; Marote, P.; Etheve, S.; Elaissari, A.; Fessi, H.: Electrokinetic properties of bare or nanoparticle-functionalized textile fabrics. Colloids and Surfaces. A, Physicochemical and Engineering Aspects, 2012; 397, 24–32.
[107] Massella, D.; Giraud, S.; Guan, J.; Ferri, A.; Salaün, F.: Manufacture techniques of chitosan-based microcapsules to enhance functional properties of textiles. In: Crini, G., Lichtfouse, E., editors: Sustainable Agriculture Reviews 35. Cham, Switzerland: Springer International Publishing, 2019; pp. 303–336.
[108] Salaün, F.: Microencapsulation as an effective tool for the design of functional textiles. Advances in Textile Engineering, 2019; Open Access eBooks, pp. 1–26.
[109] Michael, D. W.: Microcapsules containing hydrophobic liquid core. Patent US5112688A, 1992.
[110] Aouad, Y. G.; Cabrera, T. E. B.; Boekley, L. J.; Brown, J. L.; Catalfamo, V.; DeNome, F. W.; Fossum, R. D.; Gizaw, Y.; Wahl, E. H.; Roselle, B. J.; Edwards, M. B.: Functionalized substrates comprising perfume microcapsules. Patent US7786027B2, 2008.

[111] Massella, D.; Argenziano, M.; Ferri, A.; Guan, J.; Giraud, S.; Cavalli, R.; Barresi, A. A.; S alaün, F.: Bio-functional textiles: Combining pharmaceutical nanocarriers with fibrous materials for innovative dermatological therapies. Pharmaceutics, 2019; 11(8), 403.
[112] Alonso, D.; Gimeno, M.; Sepúlveda-Sánchez, J. D.; Shirai, K.: Chitosan-based microcapsules containing grapefruit seed extract grafted onto cellulose fibers by a non-toxic procedure. Carbohydrate Research, 2010; 345(6), 854–859.
[113] Muñoz, V.; Gonzalez, J. S.; Martínez, M. A.; Alvarez, V. A.: Functional textiles for skin care active substance encapsulation. Journal of Textile Engineering & Fashion Technology, 2017; 2(6), 1–8.
[114] Bonet, M. Á.; Capablanca, L.; Monllor, P.; Díaz, P.; Montava, I.: Studying bath exhaustion as a method to apply microcapsules on fabrics. Journal of the Textile Institute, 2012; 103(6), 629–635.
[115] Souza, J. M.; Caldas, A. L.; Tohidi, S. D.; Molina, J.; Souto, A. P.; Fangueiro, R.; Zille, A.: Properties and controlled release of chitosan microencapsulated limonene oil. Revista Brasileira de Farmacognosia, 2014; 24(6), 691–698.
[116] Lam, P. L.; Li, L.; Yuen, C. W. M.; Gambari, R.; Wong, R. S. M.; Chui, C. H.; Lam, K. H.: Effects of multiple washing on cotton fabrics containing berberine microcapsules with anti-Staphylococcus aureus activity. Journal of Microencapsulation, 2013; 30(2), 143–150.
[117] De Falco, F.; Guarino, V.; Gentile, G.; Cocca, M.; Ambrogi, V.; Ambrosio, L.; Avella, M.: Design of functional textile coatings via non-conventional electrofluidodynamic processes. Journal of Colloid and Interface Science, 2019; 541, 367–375.
[118] Junthip, J.; Tabary, N.; Leclercq, L.; Martel, B.: Cationic β-cyclodextrin polymer applied to a dual cyclodextrin polyelectrolyte multilayer system. Carbohydrate Polymers, 2015; 126, 156–167.
[119] Crank, J.: The Mathematics of Diffusion, 2nd ed. Oxford:Oxford University Press, 1975.
[120] Siepmann, J.; Ainaoui, A.; Vergnaud, J. M.; Bodmeier, R.: Calculation of the dimensions of drug-polymer devices based on diffusion parameters. Journal of Pharmaceutical Sciences, 1998; 87(7), 827–832.
[121] Üreyen, M. E.: Spinning performance and antibacterial activity of SeaCell® active/cotton blended rotor yarns. Fibers and Polymers, 2009; 10(6), 768–775.
[122] Subramanian, K.; Govindan, N.: Integration of cosmetics with textiles: An emerging area of functional textiles – A review. Journal of Textile Engineering & Fashion Technology, 2018; 4 (4), 316–318.
[123] Carvalho, I. T.; Estevinho, B. N.; Santos, L.: Application of microencapsulated essential oils in cosmetic and personal healthcare products – A review. International Journal of Cosmetic Science, 2016; 38(2), 109–119.
[124] Onofrei, E.; Rocha, A. M.; Catarino, A.: Textiles integrating PCMs – A review. Buletinul Institutului Politehnic din Iasi, 2010; 60(2), 99–107.
[125] Mamta, K.; Saini, H. K.; Kaur, M.: Cosmetotextiles: A novel technique of developing wearable skin care. Asian Journal of Home Science, 2017; 12(1), 289–295.
[126] Persico, P.; Carfagna, C.: Cosmeto-textiles: State of the art and future perspectives. Advances in Science and Technology, 2013; 80, 39–46.
[127] Ben Abdelkader, M.: La Microencapsulation et le Textile : Progrès Scientifiques et Industriels. Chisinau, Moldova: Presses Académiques Francophones, 2017.

Michael Doser

13 Particles for medical textiles

Keywords: wound treatment/dressings, drug delivery/release, regenerative/medicine, biocompatibility, antimicrobial textiles

Textiles are very suitable materials for medical applications. Due to their mechanical properties, which can be adjusted in a wide range, they are not only used as sutures and wound dressing materials but also as life-time implants. Their low weight, their porosity and high flexibility are also strong advantages. Textile-based implants include vascular prostheses and patches, occluder, hernia meshes, dura replacements and many other applications but the main area, where fiber-based medical devices are used is wound care. Within the last 30 years, new applications for textiles came into the focus of research and development:

- In the late 80s the idea of cultivating and growing complete tissues and organs in a laboratory came up in the USA, called "tissue engineering." Resorbable fiber-based scaffolds, often nonwoven, were used as carriers for living cells. The high expectations of this concept could not be fulfilled at first. Today, the approach is a little bit different, focusing on the process of regeneration inside the body. For this purpose, fibers are often used as lead structures and various signal molecules are used to try to control the process in a targeted manner. This is called "regenerative medicine."
- Another new field for fiber based medical devices is telemedicine. It has been realized that fibers can be used as sensors, for example, to control vitality parameters, in some cases they can even be used as actors.

In all these cases, an actual trend for textile medical devices can be observed, which is well known from other textile applications like apparel: functionalization. More and more materials and devices are provided with additional functions. For medical devices, this includes the delivery of active agents to improve healing and regeneration processes and to avoid undesired developments like bacterial infections or tissue irritations with scar formation. Especially for these purposes, particles are of high interest. They can be loaded with drugs and other active substances or deliver active ions by themselves and they can be integrated into or attached to fibers or fabrics. This chapter describes several applications of particles in textile medical devices.

Michael Doser, Deutsche Institute für Textil- und Faserforschung Denkendorf (DITF), Körschtalstraße 26, 73770 Denkendorf, Germany, e-mail: michael.doser@ditf.de

https://doi.org/10.1515/9783110670776-013

13.1 Antimicrobial particles

By far, the most important application of particles in combination with textiles are antimicrobial effects. Recently, that has been extended to antiviral effects which came into focus following the Covid-19 pandemic. Particles with antimicrobial impact or often used in apparel fabrics, mainly to avoid odor formation as described in other chapters of this book. In the medical field, this is even more important: nosocomial infections (infections acquired at a stay or on the occasion of a treatment in a hospital) still have a high prevalence – the European Centre for Disease Prevention and Control estimates from 2016/2017 data that 4.5 million infections occur each year in European hospitals [1] and the US Center for Disease Control and Prevention reported about 1.7 million infections in 2018 from which 98,000 people died [2]. This means that up to 10% of all patients in hospitals may be affected.

The reasons for these high infection rates are manifold. Undetected infections are brought into the hospital by patients, visitors and staff, hygiene measures may not be sufficient and catheters are very common instruments penetrating the body and bringing germs into the patient. People are also often not aware, that the operation theatre is not a clean room notwithstanding that all equipment is sterilized and air is filtered. Thus, infectious microbes get into the wound where they often bind to implants and form long lasting biofilms. This is normally not a problem as long as the patient has no immune deficiencies.

The most commonly used material, beside synthetic antibiotics, to add antimicrobial properties to fibers is silver. This metal is known for very long times for its antimicrobial effects. Today it is mainly used in form of nanoparticles due to the large surface in a size range from 10 to 100 nm. They are very effective, for example, against formidable bacteria like *Staphylococcus aureus* [3]. But these particles are not easy to handle because they tend to agglomerate to larger microparticles thus reducing the surface area (Figure 13.1). On the other hand, this agglomeration is often estimated to be an advantage, because it is still unclear, if nanoparticles are toxic when taken up by living cells (see below: Toxicity). The biological activity of silver nanoparticles is not only influenced by their size and agglomeration, many other factors like particle morphology, agents used for their production, surrounding media and others are relevant. The particles are active in direct contact with bacteria and viruses, but when bound in or on the surface of materials like fibers they act through the release of active silver ions, mainly denaturating proteins in the cell and through the formation of reactive oxygen species (ROS) and radicals, causing increased oxidative stress and disruption of the microbial cell membrane. Thus, also viruses can be inactivated when silver ions bind, for example, to their spike proteins causing denaturation of these proteins and so prohibiting the binding of the virus to their target cells [4]. It is a major advantage of silver that bacteria normally do not become resistant to these ions. Very few cases of resistant germs had been reported (from silver containing wastewater) but because the resistance

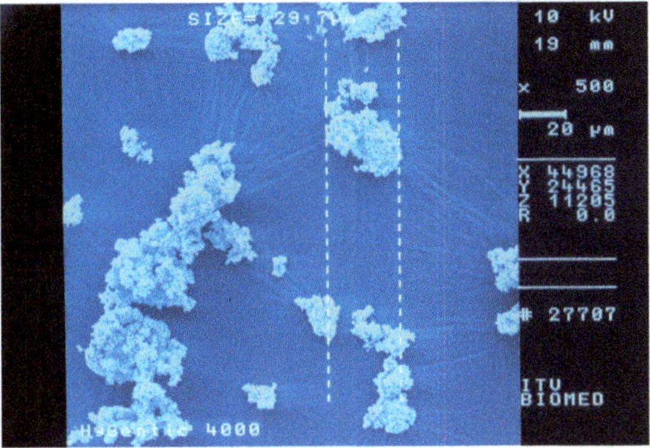

Figure 13.1: Agglomeration of nanosilver particles.

seems to be very energy consuming the bacteria also loose the resistance when silver ions are no longer present.

Normally silver particles are used for acute infections, for example, in a wound dressing. In a project financed by the German foundation for industrial research ("Development of a bioactive medical device to fight bacterial bladder infections," Stiftung Industrieforschung S. 751, 2008) the author and his team also tried the application for a typical chronic disease: many people suffer from chronic urinary tract infections, which mainly affect the bladder. To prevent these, a braided basket-shaped implant was developed (Figure 13.2), which could be placed in the bladder using a catheter. For the implant, absorbable polymers (co-polymer of caprolactone and trimethylene carbonate, polydioxanone and others) were compounded with silver nanoparticles. Loading levels of up to 20% silver were possible. Depending on the polymer, the antibacterial efficacy varied: in the case of polydioxanone, 6% silver was sufficient for a bactericidal effect because it absorbs a comparatively large amount of moisture; in the case of other polymers, higher concentrations were necessary. Unfortunately, in the following experiments it was detected that urine in the bladder contains a significant amount of proteins which bind most of the released silver ions. This reduced the antibacterial effect dramatically and is a general problem with silver and other metallic ions inside the body.

13.2 Radiation and particles in medical applications

Metallic particles are not only used for antimicrobial activities. Interesting applications are the usage in combination with X-rays.

It had been considered, if metallic particles can also be used for the protection against X-rays. Normally, aprons made of lead are used to protect parts of the body

Figure 13.2: Braided bladder implant made of resorbable fibers containing silver particles.

of the patient or the staff taking radiographies against X-rays. These aprons are very uncomfortable due to their weight, high density and their low flexibility. Therefor the idea had been followed to create a garment made of fibers filled with lead and tin particles [5]. The particles had been integrated in polypropylene monofilament yarns, which were then processed to woven garments. The authors found a satisfactory attenuation against X-rays which increased with lower filament diameters. Because lead is quite toxic, barium sulfate had been tested as alternative, mainly as a coating on the fiber surface. This material is well known as nontoxic, when it is used as contrast agent in medical applications. An interesting possibility to create lightweight fabrics seems to be the possibility to integrate $BaSO_4$ particles of 1–3 µm in diameter into viscose within the wet-spinning process. Knitted fabrics made of these fibers showed a protection against X-rays compared to 0.1 mm lead [6].

The standard application where $BaSO_4$ is already intensively used in medical devices is the use as a radiopaque material to allow visibility for materials which are normally invisible with X-rays, for example, polymers. The typical areas of application are dental and osseous tissues, where barium sulfate is included in bone cements, cages or plates for osteosynthesis. So, it was obvious to use $BaSO_4$ also for fiber-based implants. Most developments in recent years focused on radiopaque stents. Where it is already difficult to visualize metallic stent (due to their thin struts) when they should be placed in the right position in, for example, a blood vessel, this is impossible with polymeric stents. In a recent publication it could be demonstrated that $BaSO_4$ particles (40–60 µm) as a filler in coronary stents, made of polylactic acid (PLLA) not only allowed visualization but also reinforced the stent in a concentration up to 15% and thus increased the mechanical properties [7].

An interesting application of metallic nanoparticles is their use in cancer therapy. It has been demonstrated, that iron oxide (magnetite) nanoparticles can be used to destroy tumor tissue and allow a lower dose of radiation compared to a radiation therapy without particles [8]. Up to now, only the particles itself are used but it might be of interest to use this kind of nanoparticles also integrated in fibers for thermotherapies.

13.3 Drug delivery systems

A very important functionalization of medical devices is the delivery of therapeutic substances. It offers the possibility to apply a drug very specifically at the location, where it should be effective and the time of delivery as well as the concentration can be adjusted according the specific need of the therapy. Compared to a systemic application, for example, through the gut or the blood stream, the quantity of the drug can strongly be reduced and there will be significantly less undesired side effects. Another major advantage is also the possibility to maintain the concentration of the active substances constant within therapeutic limits over the necessary duration.

The simplest way to deliver a drug with a device would be to embed it into the matrix of the device if it is a polymeric material or to use this drug-loaded matrix as a coating of the device. This typically is done for instance with drug-eluting stents in coronary arteries [9], where cytostatic substances like sirolimus and paclitaxel are incorporated in a polymeric coating of the struts of metallic stents preventing restenosis. In the beginning nonbiodegradable polymers were used like poly(n-butyl methyl acrylate) (PBMA) but better results were achieved with biodegradable polymers. Today, drug-eluting stents mainly use PLLA or poly(lactic-co-glycolic acid) due to their better biocompatibility, complete degradation and mechanical properties. A long-time release of the drug over several months by diffusion has been demonstrated.

A major problem of embedding drugs in a polymeric matrix is the instability of many drugs against heat and/or organic solvents. Especially those drugs, which are of interest for local application like small, signaling proteins or nucleic acid-based drugs are very sensitive to such treatments. They should be kept preferentially in an aqueous environment and/or they are often stabilized with additional molecules. For stabilization drugs are, for example, embedded in calcium phosphate (CaP) *nanoparticles*, which might be coated with lipids, polyethyleneglycol or chitosan to improve their stability or their transfection rate [10]. Also, chitosan is used as reservoir for drugs with low stability [11]. It allows the production of *nanospheres* where drugs are embedded in the particle matrix as well as the production of *nanocapsules* with a drug containing solvent inside the particle. Chitosan may be used as a direct coating of filaments or spray dried in form of microparticles in a finishing process on a fabric. For the delivery in aqueous droplets, several other materials can also be used for microencapsulation like *liposomes* or *polymer-based microparticles and nanoparticles*. Other polymeric carriers of drugs are *dendrimers*. These are branched, star-shaped molecules. They often have functional groups at the outer branches, where active substances can be bound to covalently or by ionic or coordination interactions. Dendrimers can also function as micelles encapsulating the drugs [12].

Depending on their size many of the particles will be taken up into the cells by endocytosis which allows the direct delivery of the drug into the cytoplasm. Liposomes are well tolerated due to their natural composition similar to the cell membrane and polymer-based microspheres can be designed in many ways, for example, with a defined degradation time to determine the period of delivery of the drug. These particles can be used for the delivery of traditional drugs like antibiotics to fight against locally restricted infections [13]. But more interesting are new therapies with nucleic acid-based drugs. An actual example is the vaccines against the SARS-CoV-2 virus. This application is quite new, but nucleic acids are under investigation as therapeutics for more than 40 years for many other applications. For example, siRNA (small interfering RNA) is used to silence gene expression [14]. This allows reducing of inflammation which is a major problem in many regenerative processes because it results typically in the formation of dense and hard scar tissue instead of the soft original tissue.

In a European project the author and his team developed together with Elias Fattal from the University of Paris siRNA containing microparticles for the guided nerve regeneration. These particles were produced with a resorbable polymer, a block co-polymer of caprolactone, trimethylene carbonate and glycolid and prepared as microspheres by solvent emulsion and evaporation allowing the preparation of particles in a wide range of sizes. They were then integrated into a tubular solution-blown non-woven which is used as a nerve guide, bridging the two ends of a cut nerve. Figure 13.3 shows an example of such a particle-loaded non-woven.

Figure 13.3: Solution-blown non-woven with integrated resorbable microcapsules for drug delivery.

A major area of textile-based drug delivery is the transdermal delivery. Many applications are already existing on the market where textiles are involved: antimicrobial activity, pain reduction, hormone therapy, psoriasis and dermatitis treatment and so on [12]. In these cases, the delivery of active substances is not as easy as inside tissues where a direct contact between drug and target cell is possible. The skin is a very good barrier protecting us against environmental factors. The problem is that the outer layer of the skin, the stratum corneum, is mainly a non-aqueous lipid matrix, whereas the dermis below this layer is mainly aqueous and contains the typical target of the drugs, the cells and blood vessels. It has been demonstrated, that mainly dendrimeric particles are suitable carriers for the penetration through the skin layers, where the effectiveness is depending on size, surface charge, and hydrophobicity of the particles. Also, liposomes can promote the penetration of active substances through the outer skin layers [12].

Additionally, drug delivery is becoming more and more "intelligent" meaning that there isn't a constant delivery of substances from the first contact of the loaded textile with patients' tissue but a stimulus-driven delivery. The stimulus can be temperature or pH (as indicators for infection or irritation) but can also be light or mechanical forces (to induce delivery from outside the body) or biological substances like enzymes, glucose or signaling molecules which indicate a certain status of a disease [12]. As a result, molecules or particles may detach from the fibers they had been bound to or a spacer for binding to a surface is cleaved.

13.4 Regenerative medicine

A very promising medical field, where drug delivering particles play an important role is regenerative medicine. This discipline started in the 1980s in the USA under the term "tissue engineering" with the idea to grow tissues and organs from living human cells in a laboratory and to transfer these tissues into injured or destroyed areas of a patient or even to replace complete organs which were not functioning any more. The initial euphoria very quickly gave way to the understanding that laboratory-grown tissues are difficult to integrate into existing healthy tissue. Today, the approach is pursued as "regenerative medicine," trying to exploit the regenerative potential in humans to control regenerative processes in the body. For this, special scaffolds are used as guiding structures and textiles are particularly well suited for this application due to their similarity to natural structures: our connective tissue is for example a non-woven, made of collagen fibers. As the function of the scaffold is limited to the time of regeneration these materials are normally made of degradable polymers. To guide the regeneration process signaling molecules are applied, which cells are using to communicate among each other. Cells, cultivated outside the body are now rarely used. Instead, molecular signals attract cells or guide them into a certain direction or differentiate stem cells into the desired tissues.

The signaling molecules are mainly small proteins like cytokines and growth factors or small RNA molecules. All these are very sensitive to heat and solvents, when handled outside the body, and to digestion by enzymes inside the body. So, they are normally protected. For this, in synthetic systems often *liposomes* are used, but in recent years it had been discovered that cells themselves are also using membrane coated particles in their signaling pathways which are called *exosomes* [15]. Typically, exosomes are very small (40–160 nm) and may contain proteins in their membrane facilitating a covalent attachment of the particles to surfaces. Also, microparticles, made of resorbable polymers are used as mentioned before [16]. They allow a delayed release of molecules or even a controlled release when the capsule material is, for example, temperature- or pH-sensitive or contains components which can be degraded by specific enzymes. Even mesoporous silica nanoparticles are used as carrier systems for biological substances [17]. Their pore size can be adjusted to meet the demands for loading various molecules, they are degradable and as nanoparticles they can be taken up into cells and start their activity inside the cytoplasm.

In most cases, liposomes and other drug-loaded particles are just distributed into the porous structure of the (textile) scaffolds, where they nonspecifically adsorb to the surface of the filaments. Another possibility is the embedding of the particles into a matrix, typically a hydrogel like chitosan, alginate or carboxymethylcellulose. These drug-loaded scaffolds can then be used for the regeneration of nearly any injured tissue like skin (angiogenic growth factors, anti-inflammatory pharmaceuticals), bone (bone morphogenic proteins), peripheral and spinal cord nerve regeneration (nerve growth factors, antiphlogistic drugs) and many other tissues as well as for many kinds of diseases like cancer therapy.

13.5 Biocompatibility, toxicology

Safety is an essential requirement for all medical devices and a prerequisite for the approval of these products. Each country has its own regulations how it has to be demonstrated to guarantee a maximum of safety for their patients, for example, the MDR in Europe (Medical Device Regulations (EU) 2017/745, mainly Annex I) or 21 CFR 814 in the US (Code of Federal Regulations, Premarket Approval of Medical Devices). Beside special national requirements there are 2 common assessments to be done which are worldwide identical: a risk analysis based on the standard ISO 14971 (Application of risk management to medical devices) and the evaluation of the biocompatibility according the series of standards ISO 10993 (Biological evaluation of medical devices). Biocompatibility is much more than just the lack of harmful reactions: it also includes so called appropriate reaction of the body. This means, that any kind of material brought in contact with the patient's body will be recognized and cause a reaction of the body tissue and the immune system. Biocompatibility claims that it will control this reaction in the best way possible.

The strategy of evaluating biocompatibility of medical device has changed. In the past biological endpoints like cytotoxicity, skin irritation or implantation studies were the primary testing tools. The new version of ISO 10993-1 (Evaluation and testing within a risk management process), the guidance part of the standards, made a paradigm change in 2018 and focused on the chemical composition of the device in a first step. Only if not sufficient data are found in a risk analysis of the known and/or identified chemicals in the device additional biological tests have to be performed to assess the safety of the product. This procedure doesn't make it easier because the fabrication of the device often brings new chemicals into the materials or causes modification of chemicals (e.g., due to thermal degradation) or the biological reaction of a combinations of substances in a material is not known.

Within the Technical Committee developing the ISO 10993 standards it was also realized, that particles, mainly nanoparticles in medical device may play a crucial role, at least if they are not reliably fixed in the matrix of the device or on its surface but can be delivered into the patients' tissues. Due to their size they can easily be taken up by cells and their large surface might be very reactive. Therefore ISO/TR 10993-22 (Guidance on nanomaterials) was developed and published in 2017. It addresses the necessity to test these particles themselves instead of testing extracts which is normally done with medical devices. Moreover, it considers where nanomaterials are used in the other standards of the 10993 series, how nanomaterials are characterized, how they are released from devices and also factors which might influence the toxicokinetic of the particles itself of chemicals / ions released from the particles.

The mechanisms of action of silver nanoparticles are still not well understood, especially the effect of long-term exposure. The impact on the patient is dependent on the exposure route (dermal, oral, inhalation and intravenous) as well as on size, shape, dose and coatings of the particles [18]. Toxicity studies in vivo have demonstrated silver translocation, accumulation, and toxicity to various organs. Efficiency as well as toxicity of silver nanoparticles are dependent on the size of these particles. When nanosized silver colloidal solutions for application on textiles with particles sizes of 2–3 nm were compared with those of 30 nm diameter, the smaller particles were more effective against bacteria as expected due to their larger surface, but interestingly only the 30 nm particles showed a slight toxicity in an animal skin model [19]. Furthermore, it is known that silver can cause adverse effects like strong irritation and DNA damage through the formation of cytotoxic radicals and ROS [4].

In contrast the toxicity of $BaSO_4$ particles, even if they are applied as nanoparticles seems to be quite low, compared to NPs from other poorly soluble particles like TiO_2, but inflammation is possible also with these particles. Depending on the tissue where they have been delivered they will mostly be excreted within a few days but a relevant amount of Barium can finally be found in the bone [20].

13.6 Conclusion

Textiles are very suitable as medical devices due to their flexibility and similarity to body structures. These devices become more and more functionalized with diagnostic and therapeutic properties. For these applications, particles are very helpful components, for example, for visibility in radiography. By far the most important application is the delivery of biological active substances. This ranges from ions and radicals from silver and other metals with antibacterial and antiviral effects to the targeted release of growth factors that open up completely new healing perspectives in regenerative medicine. Particles therefore play a prominent role in research and development in medicine today, with great potential for future therapeutic options.

References

[1] ECDC: Healthcare associated infections. [Online], 2018. [Cited: 8 11, 2021.] https://www.ecdc.europa.eu/en/publications-data/infographic-healthcare-associated-infections-threat-patient-safety-europe

[2] Haque, M.; et al.: Health care-associated infections – An overview. Infection and Drug Resistance, 2018; 11, 2321–2333.

[3] Chen, J.; et al.: In situ synthesis and properties of Ag NPs/carboxymethyl-cellulose/starch-composite films for antibakterial application. Polymer Composites, 2020; 41, 838–847.

[4] Salleh, A.; et al.: The potential of silver nanoparticles for antiviral and antibacterial applications: A mechanism of action. Nanomaterials, 2020; 10, 1566.

[5] Mirzaei, M.; et al.: X-ray shielding behavior of garment woven with melt-spun polypropylene monofilament. Powder Technology, 2019; 345, 15–25.

[6] Qu, L.; et al.: Barium-sulfate/regenerated cellulose composite fiber with X-ray radiation resistance. Journal of Industrial Textiles, 2015; 45(3), 352–367.

[7] Ang, H. Y.; et al.: Radiopaque fully degradable nanocomposites for coronary stents. Scientific Reports, 2018; 8, 17409.

[8] Maier-Hauff, K.; et al.: Efficacy and safety of intratumoral thermotherapy using magnetic iron-oxide nanoparticles combined with external beam radiotherapy on patients with recurrent glioblastoma multiforme. Journal of Neuro-oncology, 2010; 103, 317–324.

[9] Rykowska, I.; Nowak, K.; Nowak, R.: Drug-eluting stents and balloons – Materials, structure designs, and coating techniques: A review. Molecules. 2020; 25, 4624.

[10] Xu, X.; et al.: Calcium phosphate nanoparticles-based systems for siRNA delivery. Regenerative Biomaterials, 2016; 3(3), 187–195.

[11] Atunes, J. C.; et al.: Bioactivity of chitosan-based particles loaded with plant-derived extracts for biomedical applications: Emphasis on antimicrobial fiber-based systems. Marine Drugs, 2021; 19, 359.

[12] Atanasova, D.; Staneva, D.; Grabchev, I.: Textile materials modified with stimuli-responsive drug carrier for skin topical and transdermal delivery. Materials. 2021; 14, 930.

[13] Pham, -D.-D.; Fattal, E.; Tsapis, N.: Pulmonary drug delivery systems for tuberculosis treatment. International Journal of Pharmaceutics 478. 2015; 478, 517–529.

[14] Fattal, E.; Fay, F.: Nanomedicine-based delivery strategies for nucleic acid gene inhibitors in inflammatory diseases. Advanced Drug Delivery Reviews. 2021; 175, 113809.

[15] Shafiei, M.; et al.: A comprehensive review on the applications of exosomes and liposomes in regenerative medicine and tissue engineering. Polymers, 2021; 13, 25–29.
[16] Sehgal, P. K.; Sripriya, R.; Senthilkumar, M.: Drug Delivery Dressings. In: Rajendran, S., editor: Advanced Textiles for Wound Care. s.l.: Woodhead Publishing Ltd., 2009; 223–253.
[17] Chen, L.; Zhou, X.; He, C.: Mesoporous silica nanoparticles for tissue-engineering applications. WIREs Nanomedicine and Nanobiotechnology. 2019; 11, e1573.
[18] Ferdous, Z.; Nemmar, A.: Health impact of silver nanoparticles: A review of the biodistribution and toxicity following various routes of exposure. International Journal of Molecular Sciences. 2020; 21, 2375.
[19] Lee, H. J.; Jeong, S. H.: Bacteriostasis and skin innoxiousness of nanosize silver colloids on textile fabrics. Textile Research Journal, 2005; 75(7), 551–556.
[20] Konduru, N.; et al.: Biokinetics and effects of barium sulfate nanoparticles. Part Fibre Toxicology, 2014; 11, 55–69.

Robert Tadej Boich*, Vadim Tenner

14 Application of nanotechnology in smart textiles

Keywords: smart textiles, nanoparticles, nanotechnology, electrically conductive fibers, electronics

This chapter describes the basic concepts of smart textiles, summarizes the main types of smart textiles and their applications and aims to provide useful information for practitioners in the development and production of smart textiles as well as the use of nanotechnology in combination with textiles.

According to the market research institute "IDTechEx," the market for smart wearables and thus the demand for the assembly of flexible electronic materials as well as textiles is estimated at over €32.5 billion in 2017 and will increase to €139 billion in 2026. Correspondingly, the sales volume is forecasted to increase and reach 5 billion by 2026. The term "smart wearables" include smart watches, fitness trackers, smart eyewear, smart clothing and more products. Textile-based smart wearables will represent 50% of total sales of all smart wearables. Accordingly, approximately 65 million textile smart wearables will be sold in Europe in 2025. In addition to wearables, home textiles, such as mattresses and curtains, are also important markets. The areas with the strongest projected annual growth rates for smart textiles are "Sports and Fitness" (+35% per year) and "Home and Lifestyle" (+68% per year).

The basic definition of smart materials is a concept in which the content of information science is integrated with the structure and function of materials. More specifically, smart textiles are defined as textiles made from smart textile fibers or refer to textiles made from yarn in certain technological processes (e.g., weaving) with added advanced functions or uses. According to many authors, smart textiles can sense different environmental conditions and especially intelligent textiles or e-textiles can not only sense environmental changes but also automatically respond to their environment or stimuli, such as thermal, chemical or mechanical changes, as well [1–3].

The European Committee for Standardization (CEN) defines Smart Textiles in the technical report (TR)16298:2011 more precisely as intelligent systems consisting of textile and non-textile components that actively interact with their environment, a user or an object. Smart textiles have a wide range of applications and will have a major impact on mobility, health, sports, civil engineering, entering completely new markets in both consumer and technical products.

*Corresponding author: Robert Tadej Boich, Institut für Textiltechnik of RWTH Aachen University, Otto-Blumenthal-Str. 1, 52074 Aachen, Germany, e-mail: robert.boich@ita.rwth-aachen.de
Vadim Tenner, Institut für Textiltechnik of RWTH Aachen University, Otto-Blumenthal-Str. 1, 52074 Aachen, Germany, e-mail: vadim.tenner@rwth-aachen.de

https://doi.org/10.1515/9783110670776-014

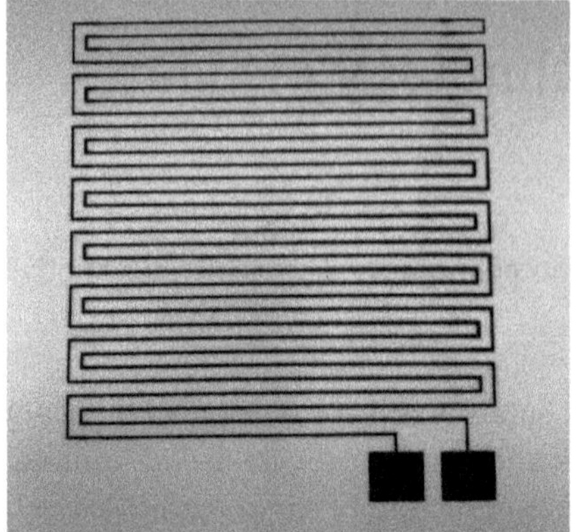

Figure 14.1: Printed conductive tracks on fabric with high elasticity.

There are various ways to obtain such a smart textile; for example, by integrating the conductive textile forms like woven, knitted, embroidered or coated fabrics, with electronic components, by printing electronic and conductive webs or by using nanotechnology. Compared to smart fibers, it is easier to achieve the purpose of "intelligence" by integrating non-textile materials into the textile materials themselves, such as properties of sensory elements, feedback elements and response elements and direct printing with conductive inks (Figure 14.1).

When combining electronics and textiles, there are different levels of integration of electronic components into textiles. The simplest version of integration is textile-adapted, the combination of two separate components, textiles and electronics, where the textile component is just a shell or a pocket for the electronics. A more advanced version is textile-integrated smart textiles, where individual functions are already fully realized in textiles made of electrically conductive fibers and with textile-technological techniques (e.g., conductors, heating loops and surfaces, resistors, capacitors and switches), but various interconnection methods are still needed to create the interface between the textile and the electronic component. Most intelligent textile materials are textile-based, where the electronic function is 100% performed by the textiles, such as conductor paths and sensors made of conductive yarns, generation of energy with piezoelectric fibers or polymer optical fibers for light transmission [4].

Recently, the electronic circuits are required to be printed onto a resin film for the sake of reducing the size and the weight and increasing the flexibility of the electronic parts (Figure 14.2). Also, nanosized metal particles with sizes ranging from a few nanometers to several dozen nanometers, accompanied by the current aggressive development of nanotechnology, are expected to aid the manufacturing of fine circuits (Table 14.1) [5].

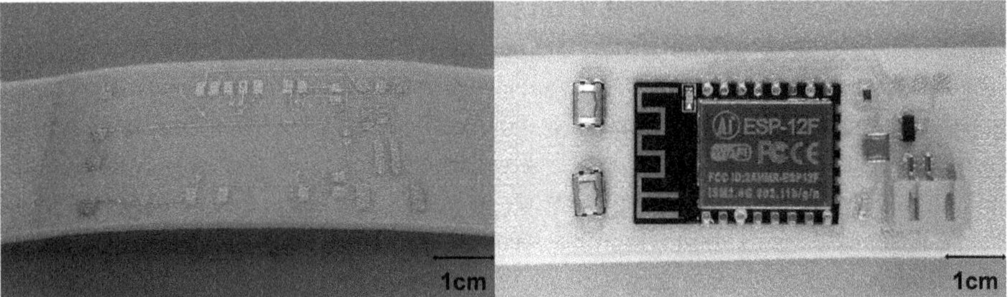

Figure 14.2: Printed electric circuit (left) and integrated electric components (right).

Table 14.1: Characteristics of nanoparticle samples used in the study [6].

Particle composition	Source	Round robin	Nominal size (nm)	Initial concentration (% solids)	Diluent
Gold	BBI	I and 2	30 ± 2 (TEM)	~0.01	H_2O
Carboxylated polystyrene	Invitrogen	I and 2	100 ± 11(TEM)	~0.1	H_2O
Aminated polystyrene	Polysciences	I and 2	100	~0.1	H_2O
Silica	Polysciences	I and 2	100	~0.1	H_2O
Polystyrene	Thermo Scientific	3	102 ± 3(TEM)	~1	H_2O
Gold	BBI	3	60 ± 3(TEM)	~0.01	H_2O
Gold	BBI	3	81 ± 4(TEM)	~0.01	H_2O
Polystyrene	Thermo Scientific	4	102 ± 3(TEM)	~1	H_2O, Ham's F10 Nutrient Mix, BSA
Polystyrene	Thermo Scientific	4	203 ± 5(TEM)	~1	H_2O
Gold	BBI	4	81 ± 4(TEM)	~0.01	H_2O

14.1 Electrically conductive fibers and yarns

The most important property for smart textiles is their electrical conductivity which can be achieved by using conductive materials. Besides using metals or conductive polymers such as polyaniline (PANI) or polypyrrole as intrinsically conductive materials in smart textiles, it is also possible to use other additionally processed yarns with extrinsic conductivity (addition of metal particles, carbon nanotubes and conductive polymers). Another possibility is the use of coated material as an electrical conductor, where the fibers can be coated with metals, carbon in the shape of fine spherical silver powder (average particle size 0.35 µm), nanoparticles (average particle size 20 nm) or conductive polymers, for example, with silver coating for silver-coated polyamide multifilament. Such yarns with all the "textile" properties can be processed very well in all textile processes, for example, knitting, weaving and embroidery, which is why metal-coated yarns and conducive polymers are usually used for industrial applications.

Conductive polymers are mostly used as antistatic film, as electromagnetic shielding, in electronic circuits, in shielding, in coated circuits in the electronics industry and in corrosion protection. Conductive polymers used for smart textile applications are as follows:

- **PANI**, also known as "metallic" plastic, can be used as a material for the manufacture of electronic devices such as sensors, radar-absorbing materials or corrosion protection. The material's internal structure makes it suitable for nanotechnology applications. However, PANI is difficult to process compared to other polymers.
- **Poly(ethylenedioxythiophene) (PEDOT)** has the chemical structure of being the most stable of all known conductive polymers and is used as a thin antistatic layer in photographic films made by the Bayer subsidiary Agfa-Gevaert N.V.
- Poly(3,4-ethylenedioxythiophene):poly(styrenesulfonate), also called PEDOT: PSS, is an electrically conductive polymer. It is produced by doping poly(3,4-ethylenedioxythiophene) with poly(styrenesulfonate) anions. The high electrical conductivity is achieved after doping with suitable solutions. PEDOT:PSS is also suitable as a coating material for textile substrates because of its corrosion resistance. Electrical and chemical transistors, electrodes for organic solar cells, organic light-emitting diodes (OLED or organic LED) and electrocardiographs can be realized in this way [7].

In various production processes, conductive yarns can be integrated into textile surfaces, and conductive webs, sensors and drives can be produced. These electrical components could be used to measure vital functions in the body, to stimulate muscles, to control functional components and as adaptive controls, screens or heating elements. The production technique could be knitting, weaving or the production of spacer fabrics, in which conductive yarns are integrated directly into the textile surface. Other textile techniques, such as embroidery or printing, allow the subsequent application of electronic elements to the textile surface.

14.2 Printed circuit boards on textiles

Various printing technologies on textile substrates with special ink have developed mainly due to the possibility of providing higher resolutions of conductive paths and thus the possibility of integration of much smaller surface mount devices. Here electrically conductive adhesive ink-form or paste is used with high density of nanosized (metallic) particles with a diameter of approx. 20 nm applicability to various printing methods.

The following two processes are particularly suitable for printing electronics on textiles:

- **Screen printing or stencil printing:** Functionalized inks in the form of a defined print image can be transferred to a substrate by the three printing processes such as rouleaux printing, stencil printing or rotary stencil printing. The screens or stencils used allow the ink to pass through only at defined points and create a high level of reproducibility. Depending on the functionalization, a subsequent heat treatment (e.g., conductive inks) is required to maintain high conductivity and fixation of the substrate. (Figure 14.3).
- **Inkjet printing:** This digital printing process eliminates the need for screens and is becoming increasingly popular in textile printing. A digital image is transferred to the substrate without contact in this process. Limitations arise from the printability of the substrate, which must be compatible with the ink used [4, 8, 9].

To achieve high conductivity, conductive tracks printed on textiles must be appropriately heat treated. The treatment parameters may vary depending on the ink used. However, ensuring resistance to washing and cracking under mechanical stress remains a major challenge in the functionalization of textiles.

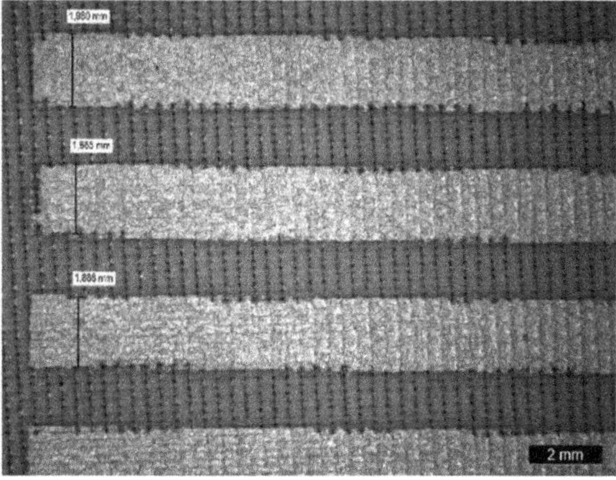

Figure 14.3: Screen-printed conductive tracks on textile.

The opportunities for the development of smart textiles are vast and it can be predicted that the development of smart textiles will become crucial for the economic growth of the textile and apparel industry. With the development achievements in the field of smart textiles, the textile industry is also keeping pace with the modern, fast-growing industries. Although the technological development of smart and intelligent textiles is extremely important for the industry, it is equally important that smart textiles used for medical, therapeutic and comfort purposes also contribute to a better quality of life.

14.3 Nanotechnology in textiles

Nano-functionalization can provide textiles with numerous properties. Fibers can be made stain-resistant, knit-free, antistatic and electrically conductive without degrading the physiological properties of clothing, for example. Increasing customer demand for durable and functional clothing represents the growth driver for this market. Further potential of functionalized clothing lies in its ability to respond to external stimuli through color, physiological or electrical signals [10].

The applications for electrical and photonic nanotechnologies are equally found in displays, drug release in medicine and sensor applications. Potential risk factors, on the other hand, are the release of nanomaterials during washing [10].

Nanotechnology is based on the effect of the drastic change in the properties of materials as soon as their particle size drops to a value below approx. 100 nm. This effect can be used to fundamentally change textile properties in a way that is not possible using traditional manufacturing processes [11].

For example, we can control or manipulate physical properties of material and obtain desired characteristics, for example, durability, stretchability, strength, water and fire retardancy, conductivity and antistatic protection, and with all these features they are excellent materials for advanced smart cloths.

Textiles can be equipped with nanoparticles in two ways. One method is to incorporate synthetic nanoparticles into the fiber or textile surfaces. The second method is to apply new coating technologies such as plasma polymerization, layer-by-layer or sol–gel. Furthermore, the electrospinning and split-spinning processes enable the production of nanofibers from conventional polymers (polyamide, polypropylene and polyester) [11].

For example, a paste containing silver oxide and silver nanoparticles or organic silver compounds and a paste containing silver nanoparticles and spherical silver powder with a diameter of a few microns have been suggested for this purpose [1–5]. Meanwhile, one method for manufacturing silver nanoparticles relies on the process of solid-state thermal decomposition [6, 7]. Silver nanoparticles obtained in this process are known to possess the following characteristics: (1) the protective layer of the metal silver core can be designed freely so that the silver nanoparticles are made suitable for any type of solvent, and (2) their mass manufacturability determines their low cost.

The resistivity of a conductive paste prepared with silver nanoparticles mixed with fine spherical silver powder was characterized, as well as the printing characteristics and the paste and film reliability with respect to its utilization as a low-resistivity curing paste suitable for implementing fine electrode layout. It was found that it is possible to develop a conductive paste which has a very low resistivity (10–6 Ωcm) at a curing temperature of 200 °C. It is also possible to form fine circuit lines 20 µm in width by utilizing brief screen printing. The paste obtained led to the improvement of stability with respect to viscosity and cured film resistivity. By combining silver nanoparticles prepared by solid-state thermal decomposition with the fine spherical silver powder prepared by chemical reduction, it is possible to manufacture in high volumes at low cost, which determines the superiority of the proposed method over the existing method for industrial application. Therefore, it will be expected that the paste can be extensively applied as an electronic layout technique in organic flexible circuit substrates, which must be cured at low temperatures.

The biggest advantage of nanofibers is that the surface reactivity and sensitivity to stimuli increase significantly already at fiber diameters of 10 nm. Nanofibers made of different polymers, such as polyacetylene, polypyrrole and PANI, or the use of functional components, such as graphene or carbon nanotubes, can be used for nanofunctionalization [12].

Also the size of a conductive fiber has an important effect on the response time to stimuli such as electricity, light, humidity and temperature. By reducing the diameter of the electrically conductive wire diameter, better rectifying properties as well as better electron transport in a nanowire can be achieved.

Nanoengineered textiles can lead to permanent positive enhance effects and do not lose their functions after mechanical stress such as stretching, washing and thus ensure a high durability of the fabrics.

14.4 Conclusion and outlook

Despite significant progress, smart textiles are still an ongoing research project and have not yet progressed beyond the prototype stage. Although textile-based electronics promise high user acceptance, there are several reasons that prevent them from gaining acceptance and becoming market-ready products, namely, the complexity of the products, the lack of common standards, the robustness and the life cycle of the technology. To overcome the known market barriers and bring textile, electrical and nanotechnology closer together, a coordinated and focused further development of materials, robustness and processes is needed, as well as collaboration, methodology and definition of good use cases and business models for the entire smart textiles system.

References

[1] Van Langenhove, C.; Hertleer, P.; Westbroek, J.: Priniotakis, 6 – Textile Sensors for Healthcare. In: Langenhove, L. V., editor: Woodhead Publishing Series in Textiles, Smart Textiles for Medicine and Healthcare. Woodhead Publishing, Cambridge, United Kingdom: University of Ghent Belgium 2007; pp. 106–122.

[2] van Langenhove, L.: 1 – Smart Textiles for Protection: An Overview. In: Chapman, R. A., editor: Woodhead Publishing Series in Textiles, Smart Textiles for Protection. Woodhead Publishing Cambridge, United Kingdom: University of Ghent Belgium, 2013; pp. 3–33.

[3] Koncar, V.: 1 – Introduction to Smart Textiles and Their Applications. In: Koncar, V., editor: Woodhead Publishing Series in Textiles, Smart Textiles and Their Applications. Woodhead Publishing Cambridge, United Kingdom: GEMTEX national research laboratory University of Lille/ENSAIT, 2016; pp. 1–8.

[4] Gehrke, I.; Tenner, V.; Lutz, V.; Schmelzeisen, D.; Gries, T.: Smart Textile Production: Overview of Materials, Sensor and Production Technologies for Industrial Smart Textiles. MDPI, Basel, Switzerland: Institute of Textile Technology RWTH Aachen University Germany, 2019.

[5] Yamamoto, Y; Ogihara,T: Electrical Properties of Conductive Paste with Silver nanoparticles and Its Application toFlexible Substrates. Key Engineering Materials Vols. 421–422 (2010) pp 297–300, Trans Tech Publications, Switzerland doi:10.4028/www.scientific.net/KEM.421–422. 297.

[6] Hole, P.; Sillence, K.; Hannell, C.; Maguire, C.; Roesslein, M.; Suárez, G.; Capracotta, S.; Magdolenova, Z.; Horev-Azaria, L.; Dybowska, A.; Cooke, L.; Haase, A.; Contal, S.; Manø, S.; Vennemann, A.; Sauvain, J.; Crosbie Staunton, K.; Anguissola, S.; Luch, A.; Wick, P.: Interlaboratory comparison of size measurements on nanoparticles using nanoparticle tracking analysis (NTA). Journal of Nanoparticle Research, 2013; 15. 10.1007/s11051-013-2101-8.

[7] Perrin, F.-X.; Oueiny, C.: Chapter 5 – Polyaniline-Based Thermoplastic Blends. In: Visakh, P. M.; Della Pina, C.; Falletta, E., editors, Polyaniline Blends, Composites, and Nanocomposites, Elsevier Amsterdam, Netherlands: TUSUR University, Russia, Universita degli Studi di Milano Italy, 2018; pp. 117–147.

[8] Van Langenhove, C.; Westbroek, H. P.; Priniotakis, J.: 6 – Textile Sensors for Healthcare. In: Langenhove, L. V., editor: Woodhead Publishing Series in Textiles, Smart Textiles for Medicine and Healthcare. Woodhead Publishing Cambridge, United Kingdom: University of Ghent Belgium, 2007; pp. 106–122.

[9] van Langenhove, L.: 1 – Smart Textiles for Protection: An Overview. In: Chapman, R. A., editor: Woodhead Publishing Series in Textiles, Smart Textiles for Protection. Woodhead Publishing Cambridge, United Kingdom: University of Ghent Belgium, 2013; pp. 3–33.

[10] Yetisen, A. K.; Hang, Q.; Manbachi, A.; Butt, H.; Dokmeci, M. R.; Hinestroza, J. P.; Skorobogatiy, M.; Khademhosseini, A.; Yun, S. H.: Nanotechnology in textiles. CS Nano, 2016; 10(3), 3042–3068.

[11] Mahmud, R.; Nabi, F.: Application of Nanotechnology in the field of textile. (Department of Textile Engineering, Bangladesh University of Business and Technology. IOSR Journal of Polymer and Textile Engineering (IOSR-JPTE), Jan.–Feb. 2017; 4(1), 01–06. e-ISSN: 2348–019X, p-ISSN: 2348–0181.

[12] Yilmay, N. D., editor: Smart Textile: Wearable Nanotechnology. Beverly, MA, USA: Wiley/ Scrivener Publishing, 2018.

Jeanette Ortega*, Thomas Gries

15 Electronic nanotechnologies in textiles

Keywords: CNT, carbon black, doping, particle, conductive, sensor fiber, ultrasound

The incorporation of nanomaterials into textiles is one approach to transform these standard fabrics into functional materials to be used in the field of smart textiles. The functional textiles are used to transmit data, provide power and act as sensors or actuators themselves. The application fields are not only limited to clothing for health monitoring but also can be extended to the monitoring of home textiles, such as carpets, or even for structures such as dykes, tunnels, wind turbines or pressure vessels. Fiber-based sensors and electronics have the advantage in these applications that the sensors and supporting structures, home textiles, geotextiles and fiber-reinforced composites, are fiber- or textile-based, meaning that the original functionality in the application is not altered. With non-fiber-based solutions, the sensor may have an additional interference on the functionality, altering the original structure.

Electrically conductive fibers and textiles are used for such sensing purposes. These fibers can be intrinsically conductive (the bulk material itself is conductive) or extrinsically conductive (material is added to the bulk to modify it) [1]. Materials which are considered intrinsically conductive are metal wires, carbon fibers and fibers made of selected conductive polymers [1]. In some textile applications such as wearable electronics, metal wires and carbon fibers lack the flexibility which would be necessary to be comfortable for the wearer. Whereas inherently conductive polymers retain this "textile" feel, they generally have lower conductivities than metal and carbon fibers and the production has not yet reached a mature industrial scale, keeping the costs relatively high in comparison to standard polymers [1].

Extrinsically conductive materials are modified by incorporating an intrinsically conductive material as a coating on the surface of the bulk material or as particles inside the bulk material. Metals, such as silver, are commonly used as a coating layer [2–4]. This is a relatively simple process and can be used on a variety of bulk materials, including synthetic and natural fibers. Unfortunately, these coatings are also subject to the wear and abrasion common for textiles. Through this wear and washing, the coating is removed and the electrical conductivity decreases [3]. The incorporation of electrically conductive particles to the bulk material allows

***Corresponding author: Jeanette Ortega,** Institut für Textiltechnik of RWTH Aachen University, Otto-Blumenthal-Straße 1, 52074 Aachen, Germany, e-mail: jeanette.ortega@ita.rwth-aachen.de
Thomas Gries, Institut für Textiltechnik of RWTH Aachen University, Otto-Blumenthal-Straße 1, 52074 Aachen, Germany

https://doi.org/10.1515/9783110670776-015

the conducting bodies to be inside the material, protecting them from abrasion. Contrary to a coating, this method of incorporating particles into the bulk material can only be used for synthetic and not for natural fibers.

Some of the most common additives used to impart electrical conductivity in synthetic fibers are carbon nanotubes (CNTs), graphene and carbon black (CB) [1]. As the names suggest, these are carbon-based materials which have the ability to conduct electrical charges. Although polymers are also carbon-based, they are considered electrical insulators due to the difference in the molecular structures. The carbon molecules in a polymer backbone generally form four bonds with adjacent molecules. These four bonds are the hybridized sp_3 bonds of the s- and p-orbitals (Figure 15.1, left) [1]. In conductive carbon materials, one of the p-orbitals is left open allowing it to be used as the charge carrier transporting electrons along the length of the material. This is the reason why CNTs, graphene and CB are conductive (Figure 15.1, right).

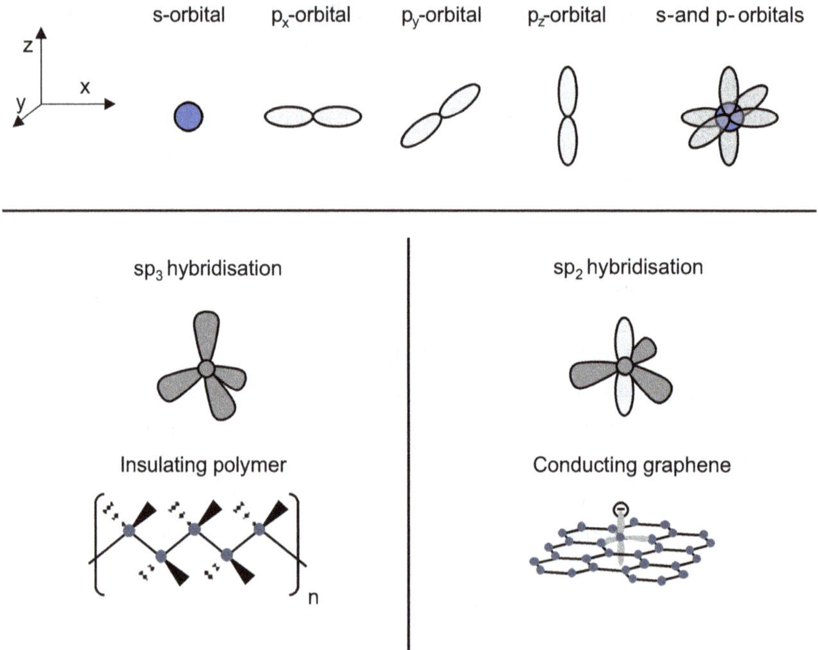

Figure 15.1: Molecular structure of insulating and conducting carbon-based materials.

It is necessary that a so-called percolation network of the particles is formed inside the fibers [5]. When coating is used, this continuous network is present, as long as there are no major cracks or voids in the coated layer. This can be easily examined and improved upon since the coating is the outermost layer of the fiber. For particles which are incorporated inside the fiber, the most common method to achieve this percolation network is to increase the amount of the conductive particles, statistically increasing the probability that the network will be formed (Figure 15.2).

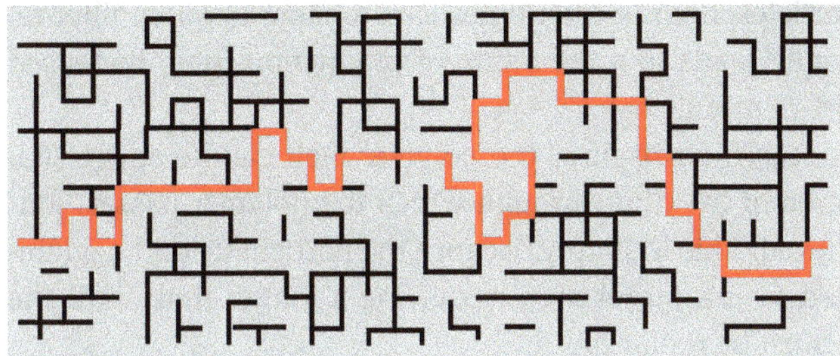

Figure 15.2: Theoretical percolation network (red) among individual particles (black).

Increasing the concentration of the particles brings challenges related to the dispersion of the particles in the polymer. Because these nanomaterials are so small, they have a large surface area in relation to their volume, the surface area to volume ratio. This high ratio is generally energetically unfavorable, resulting in the particles joining together to form agglomerates drastically reducing the surface area to volume ratio. This can be seen in Figure 15.3 for a simplified two-dimensional version. The agglomeration of the five individual particles decreases the boundary (surface area) to bulk (volume) ratio by 30%.

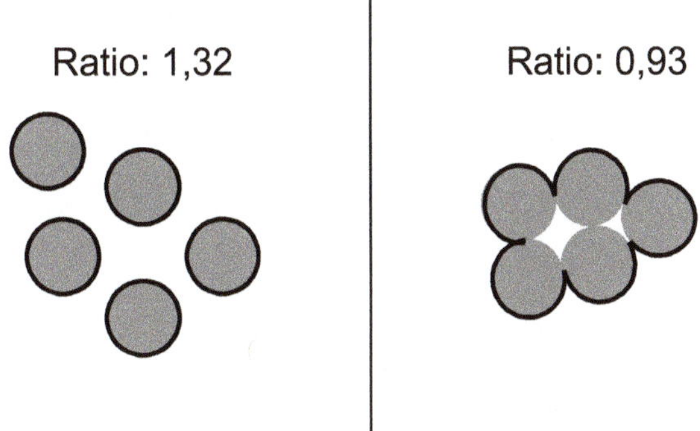

Figure 15.3: Comparison of the surface area to volume ratio for individual particles and agglomerates.

For many nanocomposite applications this would be considered a minor problem. For fiber extrusion, however, this brings major challenges to the production. Synthetic fibers are produced by a variety of spinning methods including melt spinning, solution spinning and electrospinning, of which melt spinning is the most prominent in the textile industry [6]. In melt spinning, filters are employed to remove impurities in the molten polymer and ensure the homogeneity of the extruded filament. When particles are introduced to this melt and they agglomerate as shown above, the usage time of the filter is reduced due to the larger dimension of the agglomerates in comparison to the individual particles. Shorter filter times lead to more frequent filter replacements which additionally leads to a reduction of production time, increased costs and eventually to a loss of profit.

To solve this challenge, the filter itself or the polymer melt can be addressed. By increasing the surface area of the filter through which the polymer passes, longer usage times can be achieved. This is simply because the particulate and agglomerate build up is distributed over a larger area, requiring a longer time until the filter is fully saturated and must be replaced.

Another approach to improve the production of particle filled fibers is to target the agglomerates themselves. This can be done with ultrasonication, in which ultrasound waves partially break up the agglomerates allowing for longer filter lifetimes and a better particle distribution in the polymer [7].

Besides increasing the concentration of the electrically conductive particles, the production process can be modified in such a way that the functionality of the already incorporated particles is effectively exploited. A very common step in the fiber production process is the stretching, or drawing, of the fiber. This increases the tensile strength of the fiber by aligning the polymer chains and increasing crystallinity. Additionally, the particles embedded in the polymer are aligned in the direction of the fiber axis. This alignment often leads to a breakdown of the needed percolation network (Figure 15.4). An electric charge can no longer be transferred along the length of the fiber and the electrical conductivity is drastically reduced [8].

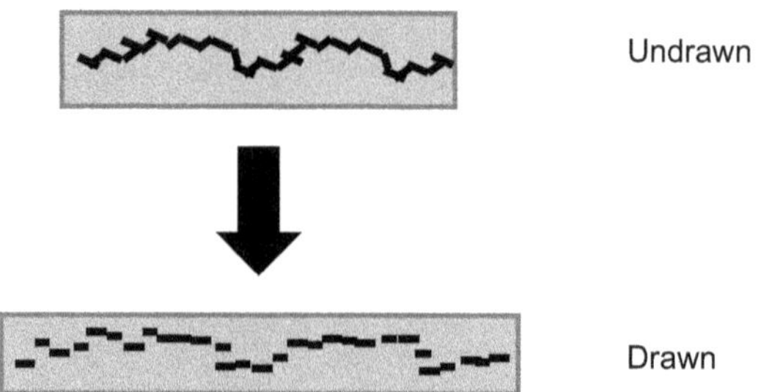

Figure 15.4: Breakdown of the percolation network after drawing.

In order to restore the electrical conductivity, the polymer with the added particles can be reheated, reintroducing the non-oriented, thermally stable configuration of the particles and polymer. When this nanocomposite polymer is spun as the core of a bicomponent fiber, the inner core will serve as the electrically conductive component and the outer layer, consisting only of polymer, provides the mechanical strength (Figure 15.5) [8].

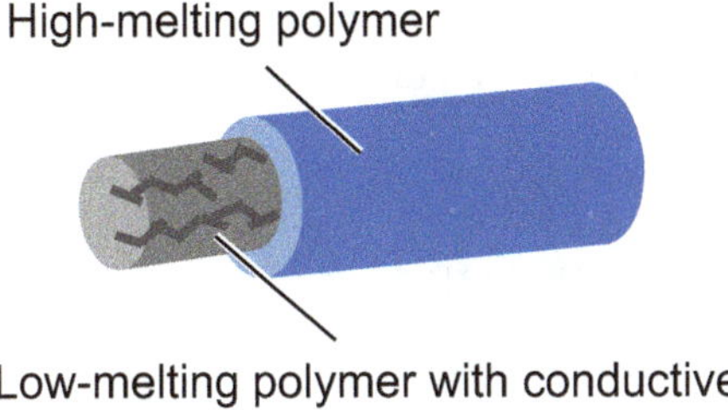

Figure 15.5: Schematic of a bicomponent fiber to improve the electrical conductivity.

Electrically conducting fibers are most well-known for the applications in health monitoring either for sports or medical applications, in which the electrical signals from muscles are detected and transmitted within the fibers. Most of the products on the market employ metal coated yarns, which often lack the washing and abrasion resistance needed for clothing applications. An example product for the measurement of muscle activity during sports or strenuous activity is shown in Figure 15.6.

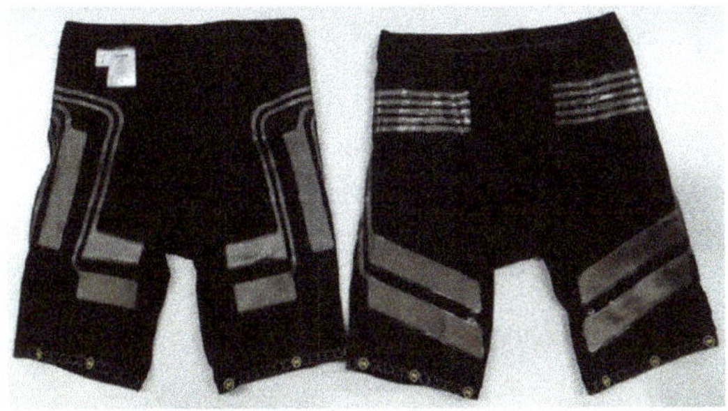

Figure 15.6: MShorts 3 from the Finnish company Myonctec Ltd [9, 10].

The future of electrically conducting fibers will continue to be rooted in the field of monitoring, but rather than human health, the focus will be on the monitoring of structural components. This application field can be further divided into geographic structures, such as dams, dykes, tunnels and streets or lightweight composite structures, such as wind turbines, aerospace components or hydrogen pressure vessels.

Man-made geographic structures are as old as humanity, with the construction of ancient dams and dykes in Syria, Spain and Rome. These structures are critical to the survival of civilizations as they are used to control the flow of water providing controlled irrigation and flood control [11]. Since the mid-twentieth century synthetic geotextiles have been implemented in dams and dykes to provide additional support to the structures. This lends the possibility to embed similar fiber-based sensors into the geotextiles to monitor the status of such structures. Examples of geotextiles for the construction of dams and dykes are seen in Figure 15.7.

Figure 15.7: Geotextiles from the German company Naue GmbH & Co. KG [12].

Fiber-reinforced composites, while already widely used in lightweight construction, are a very new development in comparison to geographic structures. In ancient times, natural fibers were mixed with soil and mud to construct stronger building materials. With the development of plastics in the twentieth century, the modern fiber-reinforced composites were developed. These composites consist of a fiber material, usually made of carbon, glass, aramid or basalt, along with the matrix material, usually epoxy, vinyl ester or, recently, thermoplastic materials [13]. Together these composites can be designed to be extremely lightweight, while also having the necessary mechanical strength and stiffness. Because of the lightweight strength, the aerospace industry is one of the largest application groups for fiber-reinforced composites [13].

These composites are often under extreme loading or environmental conditions, such as torsion or bending in wind turbines or airplane wings or high pressures and temperatures such as a hydrogen pressure vessel. This leads to the need for high safety factors in the design or construction and also extensive testing during use. Fiber-based sensors provide the opportunity to constantly monitor the condition of the component using nondestructive analysis and without the need for disassembly. Example composites structures, for which monitoring through fiber-based sensors is beneficial, are shown in Figure 15.8.

Figure 15.8: Fiber-reinforced composites. Left: wind turbine, right: hydrogen pressure vessel [14, 15].

The combination of data collection from various sensors along with the fusion and analysis of this data can not only provide information about the current state of a structure, but also predict the future status. This allows maintenance to be condition-based, rather than schedule-based, lowering costs due to unnecessary maintenance and allowing the altering of critical conditions in order to prevent catastrophes.

In order to fully develop the necessary fiber-based sensors for structural monitoring, many factors must be considered. Naturally the fibers must be able to be used in the sensing range needed. Due to the versatility of the synthetic fiber production, this can be tailored regarding the stiffness, elongation and electrical conductivity. Furthermore, the outer material of the fiber-based sensor must be tailored for the specific matrix, in the case of fiber-reinforced composites, but also for the application in geotextiles. It is pertinent that the fiber-based sensor adheres to the structure to be measured, in order to guarantee a correct sensor reading, but also important that outside influences, such as moisture, do not falsify the measurements taken. Lastly, the connection of the novel fiber-based sensors with transmitting devices needs to be addressed. It is crucial, that the information gathered is able to be sent to a central location to be combined and analyzed with other available sensors. This connection to a data transmitting device must also be mechanically stable in order to ensure a continuous reading of data. When the production, integration and connection of the sensor is achieved, structural health monitoring of geographic structures but also composites is possible.

15.1 Conclusion and outlook

Although the research of smart fibers has largely focused on clothing, the applications are much wider and offer possibilities for monitoring in technical applications as well. The development challenges must be addressed for each individual application, as the requirements can drastically differ. When considering the development for clothing of human health monitoring, washability and comfort will be two large hurdles to overcome, before products can be brought to market. In the case of smart geotextiles and structures, the long lifetime of the textiles as well as the diverse environments, ranging from dry to wet, hot to cold and acidic to basic, will require tailored solutions on the fiber level. Lastly, the integration of the sensor fibers into lightweight composite structures will require intense research, in order to ensure that the structural integrity of the final structure is not compromised through the sensor fiber application.

Through the incorporation of such sensor fibers into these technical applications, our society will continue to be more proactive, rather than reactive. These sensors, along with the software and data analysis will allow early predictions of accidents or failures, before major damage occurs. This will not only reduce the cost of damaged geostructures and the resulting cost to the surrounding area, but will also lead to a reduction in the necessary safety factors in aerospace structures, resulting in less needed material and therefore lower fuel costs. After the employment of sensor fibers in the mentioned applications, other fields will be identified, further increasing the demand and driving research activities.

References

[1] Steinmann, W.; Schwarz, A.; Jungbecker, N.; Greis, T.: Fibre-Table: Electrically Conductive Fibres. Aachen: Shaker Verlag, 2014.

[2] Simon, E. P.; Kallmayer, C.; Aschenbrenner, R.; Lang, K.-D.: Novel Approach for Integrating Electronics into Textiles at Room Temperature using a Force-Fit Interconnection. In: IEEE, Hrsg.: IMAPS-UK chapter (Veranstalter): EMPC-2011 18th European Microelectronics & Packaging Conference, Brighton, Vereinigtes Königreich, 12.–15. September 2011. Wien: IEEE Publishing, 2011.

[3] Uz Zaman, S.; Tao, X.; Cochrane, C.; Koncar, V.: Launderability of conductive polymer yarns used for connections of E-textile modules: Mechanical stresses. Fibers and Polymers, 2019; 20(11), 2355–2366. https://www.researchgate.net/deref/http%3A%2F%2Fdx.doi.org%2F10.1007%2Fs12221-019-9325-x

[4] Rotzler, S.; Kallmayer, C.; Dils, C.; von Krshiwoblozki, M.; Bauer, U.; Schneider-Ramelow, M.: Improving the washability of smart textiles: Influence of different washing conditions on textile integrated conductor tracks. The Journal of the Textile Institute, Latest Articles, 2020. https://doi.org/10.1080/00405000.2020.1729056

[5] Shante, V. K. S.; Kirkpartik, S.: An introduction to percolation theory. Advances in Physics, 1971; 20(85), 325–357. https://doi.org/10.1080/00018737100101261.

[6] Gries, T.; Veit, D.; Wulfhorst, B.: Textile Fertigungsverfahren: eine Einführung, Munich: Carl Hanser Verlag, 2015.

[7] Kammler, S.; Orth, M.; Bandelin, J.; Vad, T.; Gries, T.: Improvement of dispersion by sonication of Graphene-compounds in the melt spinning process. Scientific Federation Abode for Researchers: Global Conference on Carbon Nanotubes and Graphene Technologies, Mailand, Italien, 28.–29. März 2019.

[8] Lellinger, D.; Mroszczok, J.: Elektrisch leitfähige, textile Bikomponentenfasern auf Basis von Polymer-Nanoverbundwerkstoffen. AiF IGF Schlussbericht zu IGF-Vorhaben Nr. 18005N. Darmstadt: Forschungsgesellschaft Kunststoffe e.V., 2017.

[9] Myontec: MShorts 3. https://www.myontec.com/product-page/mshorts3

[10] Graepal, S.: $1,000 Shorts Coach Your Workout Gear Junkie, 20 October 2014. https://gearjunkie.com/apparel/myontec-mbody-smart-shorts

[11] Water Technology: The world's oldest dams still in use 20 October 2013. https://www.water-technology.net/features/feature-the-worlds-oldest-dams-still-in-use/

[12] Naue GmbH & Co. KG: Dyke Construction 20 October 2013. https://www.naue.com/applications/hydraulic-engineering/dyke-construction/

[13] Gao, F.: Advances in Polymer Nanocomposites Woodhead Publishing, 19 October 2012, ISBN: 9781845699406.

[14] The Manufacturer: From fishing to fiberglass: Hull embraces blade production 19 April 2018. https://www.themanufacturer.com/articles/fishing-fibreglass-hull-embraces-blade-production/

[15] Composites World: The markets: Pressure vessels 1 December 2015. https://www.compositesworld.com/articles/the-markets-pressure-vessels-2015

IV Emerging production technologies

Thomas Schneiders

16 Electrospinning of micro-/nanofibers

Keywords: electrospinning, microfibers, nanofibers, production technologies, extrusion

16.1 Introduction

Nanofibers and submicron fibers, as a textile structure in the form of a nonwoven fabric or yarn, show excellent properties for applications such as tissue engineering, drug delivery in medical products, filtration, catalysis and also technical textiles and protective clothing. These properties include a high specific surface area, interconnected pores and adjustable pore sizes on the nanoscale. Several technologies are used and developed to produce fibers with a diameter below one micrometer, the advantages and disadvantages of these technologies are shown in Table 16.1. Most processes have either a high investment and operating cost and are limited to few polymers or material combinations like phase separation and melt blowing or produce wide spread and nonuniform fiber diameters like flash spinning [1, 2].

Solution electrospinning (ES) is considered a viable method to produce nanoscale fibers due to its simplicity, cost efficiency and material compatibility. The attraction of a liquid by an electrostatic force was described the first time in 1600 by William Gilbert and applied for the production of nanofiber material the first time in 1887 described in a paper by Charles Vernon Boys. After singular research on the subject at the beginning of the twentieth century by Zeleny, Formhals or Taylor, electrospinning became more popular as a research topic after Reneker and his research group revived the topic in early 1990s. Since 1995 interest in electrospinning technology and the produced nanofibers increased significantly, with an exponential increase in publications about electrospinning every year. In the beginning research was driven by small-scale laboratory setups and the possibility to produce a wide range of synthetic and natural polymers. The focus was on the fiber production and its resulting morphology, explaining the influence of several process parameters on fiber diameter, uniformity and orientation. Research shifted toward more complex developments like multilayer und multimaterial fibers, drug delivery or three-dimensional structures. Looking at the past two decades, new production methods and alternative technologies to the nozzle-based electrospinning setup emerged, and scaling up production for industrial applications, and the use of this

Thomas Schneiders, Institut für Textiltechnik of RWTH Aachen University, Otto-Blumenthal-Straße 1, 52074 Aachen, e-mail: thomas.schneiders@ita.rwth-aachen.de

https://doi.org/10.1515/9783110670776-016

Table 16.1: Overview of technologies and their advantages and disadvantages for nanofiber production [1, 2].

Technology	Advantage	Disadvantage
Drawing	Long single fibers, cheap	Limited to viscoelastic material and orifice size of extrusion
Phase separation	Controlled pore size and structure	Long continuous fibers cannot be produced; limited to certain material pairings; high investment cost
Centrifugal spinning	High throughput, aligned fibers	Difficult continuous production
Melt blowing	High throughput, continuous stable production; well-developed process	High operating cost; limited to thermoplastic polymers; fibers >500 nm
Flash spinning	High throughput	Unprecise fiber diameter control
Solution electrospinning	High variety of polymers, high process control, tailoring of fiber properties (core–sheath, porous, multiple materials)	Use of solvents; low throughput (upscaling in development)

promising fiber production technology received more attention [3–6]. Current technologies can be differentiated into nozzle-based systems and free surface systems.

16.2 Basic principle

In comparison to conventional textile spinning the tensile force initiating the jet and fiber extrusion in ES is electrostatic. In an electrohydrodynamic process an ultrafine jet is ejected from a polymer solution under the influence of a strong electric field. A fiber is formed when the applied electrostatic force exceeds a critical threshold determined by solution viscosity, flow rate and geometry of extrusion orifice. Strong electric fields in the range of 100–500 kV/m are used. A basic needle-based setup for ES consists of a syringe needle connected to a high voltage generator, a syringe pump and a counter electrode functioning as a collector (Figure 16.1) [7].

Charges are separated in the extruded polymer solution droplet. Free charges interact with the applied electric field and are pulled to the counter electrode. With increasing electrostatic force, the Taylor cone is formed (Figure 16.2), named after the first person to describe this phenomenon, Geoffrey Ingram Taylor. If the electrostatic force exceeds surface tension a polymer jet is emitted along the electric field. Following the tip of the Taylor cone the jet initially follows a straight path before undergoing lateral acceleration resulting in a series of coils with the opening toward the collector.

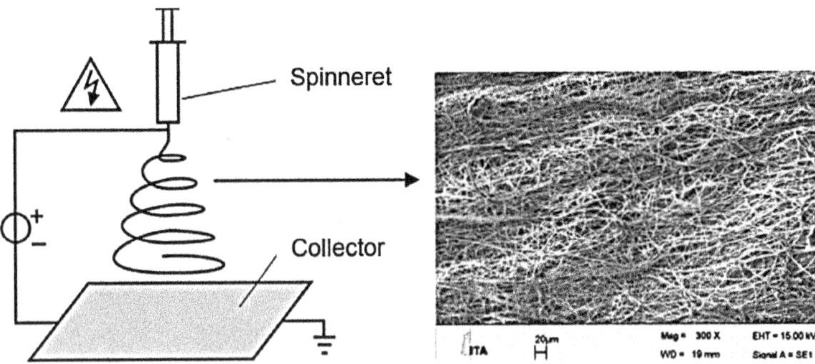

Figure 16.1: Electrospinning setup and generated nanofibers, based on [8].

This lateral acceleration is called bending (Earnshaw) instabilities. This acceleration leads to a decrease of jet diameter, thus an increase in surface and finally in the evaporation of the solvent. The polymer jet solidifies and a fiber is formed. The whipping from the instabilities stretches the fibers, resulting in fiber diameters in micro- and nanometer range. The fibers are gathered on the collector and due to the instabilities and coil movement assembled into a randomly oriented porous nonwoven mat [5, 8].

Several forces acting on the electrified jet are responsible for the formation of bending instabilities (Figure 16.2). Relevant forces affecting the liquid are surface tension, Coulomb forces, viscoelastic effects and electrical forces created by the imposed potential difference between the spinneret and the collector. The influence of gravitational and air drag forces is negligible. Surface tension steers the formation of a single or multiple jets to minimize surface area. Opposing to that electrostatic repulsion favors increased surface area and stretching into a thin jet, while the viscoelastic forces in the polymer solution counteracts the stretching of it. Manipulating this equilibrium of forces enables the development of a wide range of fiber morphologies, but demands an immense in-depth knowledge of the process and its variables [9, 10].

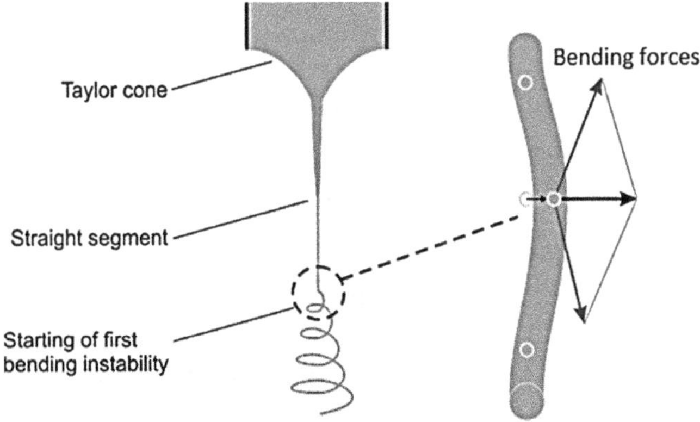

Figure 16.2: Formation of the Taylor cone and the bending instability, based on [11].

As in conventional spinning technologies the goal and measurement for a "good" process is continuous fiber formation and production of uniform fiber diameters. Further the produced fiber is supposed to have minimal bead defects or solvent residuals. Selection of materials as polymer, solvent and additives for solution ES has a significant influence on process quality, as in other fiber forming and extrusion technologies. Basic requirement is the use of a fiber-forming polymer with sufficient macromolecular overlap to build a coherent fiber after evaporation of the solvent and solidification of the jet. If the polymer does not possess mentioned properties, particles are formed under the influence of the electric field instead of a fiber. The resulting process is differentiated from electrospinning and is called electrospraying [12, 13].

Key processing parameters are applied voltage, flow rate, nozzle-collector distance, nozzle diameter or design of the collector. Additionally, ambient parameters such as temperature, humidity or air pressure have an influence on fiber morphology. An increase in polymer concentration or molecular weight increases viscosity and hence surface tension of a solution, while an increase in ambient temperature results in a decrease in solution viscosity. Finally, the electric field is a main factor in jet initiation and the origin of instabilities, thus in determining fiber morphology.

16.3 Technologies

Considering the complex jet initiation two different approaches can be differentiated (see Figure 16.3). On the one hand, there are *nozzle-based systems* and, on the other hand, *free surface systems*. In the nozzle-based technologies the surface for jet initiation and fiber forming is defined by an orifice which is used for polymer extrusion. Standard processes aim for one jet per orifice to have control of the produced fiber morphology, thus limiting the throughput of these technologies to the number of extrusion openings. Examples for such nozzles are single or multiple needles or extrusion plates with single or multiple holes. In free surface technologies, on the other hand, a large number of jets are formed on a free surface. The number of jets correlates with the size of the surface.

16.3.1 Basic nozzle-based setup

The basic setup for the nozzle-based spinning process gives the opportunity to adapt the process in many parts and thus influence fiber morphology and create the needed fiber properties. Figure 16.1 shows the fundamental setup. Only a needle or different orifice, a pump for polymer extrusion, a high voltage supply and a counter electrode for attracting the fibers are needed to start electrospinning, making it a cheap way to start with the technology. Such a setup can be found in a high

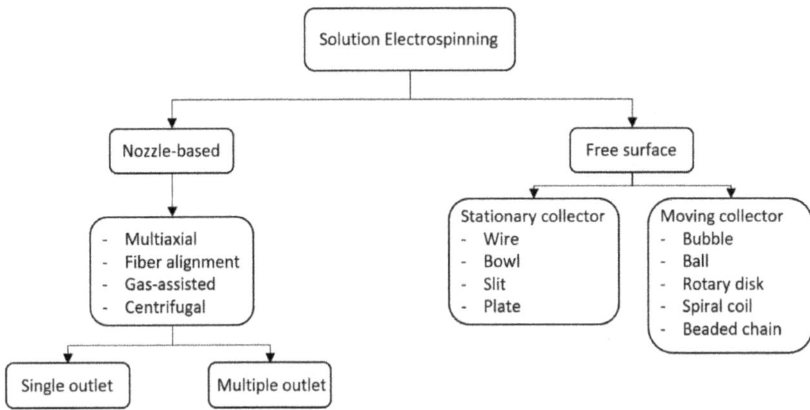

Figure 16.3: Overview of electrospinning technologies.

number of research facilities and laboratories for feasibility studies and researching basic principles and mechanisms. Adaptations to the different parts enable the production of more complex nanofibers, different materials at once and upscaling of the throughput [5, 8].

16.3.2 Multiple outlets

Throughput of a single needle is in the range of 0.1–10 mL of polymer solution per hour. That converts depending on the polymer concentration to a fiber production with roughly 0.1–1 g of polymer. Easiest approach to increase the production rate is to use multiple needles at the same time. The polymer solution is extruded through multiple needles connected to a single high voltage supply (Figure 16.4). Due to the high amount of extruded polymer solution higher voltage than for a single needle is required for a stable spinning process. Arrangement and distance of the needles can be very different, this can vary from linear arrangement to a round, symmetric or asymmetric positioning of the needles. Reason for that is the induced electrostatic field´s influence of each single needle on the adjacent needles and the form of the whole electrostatic field. Emitted jets repel each other due to same charge, which can lead to an increased variation in fiber morphology and quality. This led to the application of flat spinnerets, respectively, plate spinnerets with holes as outlets for the polymer extrusion, reducing the interaction of the single-needle electrostatic fields [14, 15].

Such collectors need to be designed to divert the polymer solution evenly and achieve a uniform flow rate over all needles, respectively, holes. Otherwise an uneven distribution will lead to a high variation in fiber diameters or beaded fibers. Another problem with multiple outlets is the possible clogging of single or multiple needles and the necessary cleaning step during the process, which again leads to

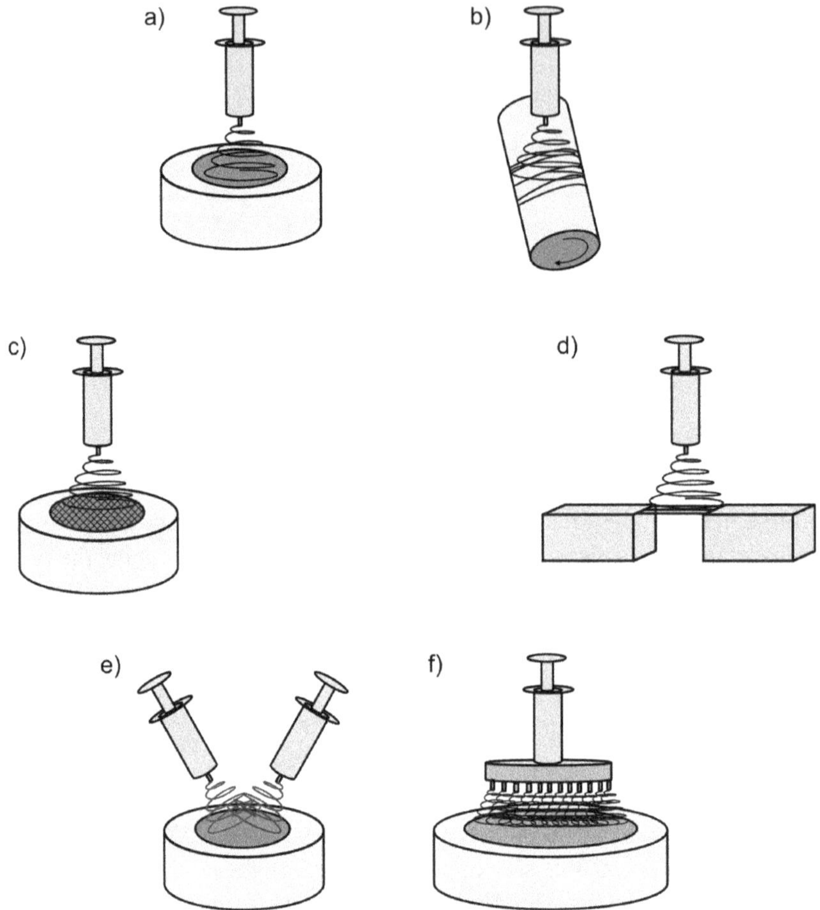

Figure 16.4: Nozzle-based electrospinning setups. (a) Basic setup, (b) rotating mandrel as collector for alignment, (c) spinning on a substrate, (d) gap spinning, (e) multimaterial spinning, (f) multinozzle spinning, based on [16].

an unstable process. Nevertheless, multiple needles are a promising way of upscaling throughput of an electrospinning process and shows the possibility to connect several multiple needle units in series or parallel for a continuous production [14, 15].

16.3.3 Multiaxial needles

Core–shell or core–sheath fibers extend the usable range of polymers and additives for electrospinning, as well as the applicability of electrospun nanofibers. Two or even three concentrically positioned needles are required in a coaxial or triaxial extrusion process (Figure 16.5). This is an extension of the basic needle setup for electrospinning

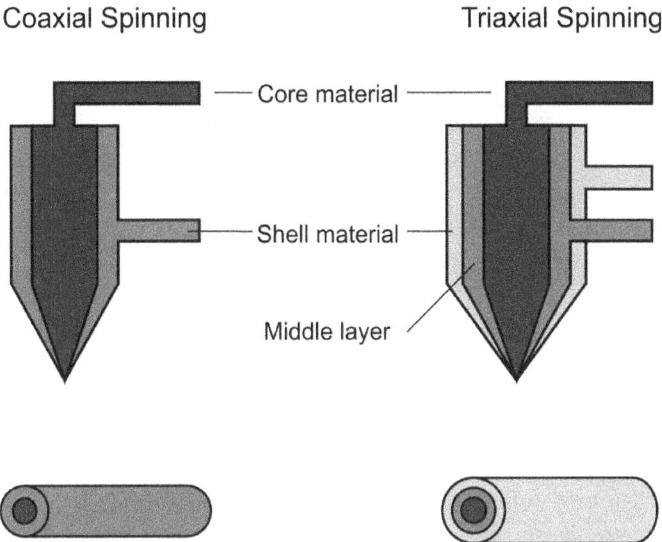

Figure 16.5: Coaxial (left) and triaxial (right) extrusion process for electrospinning. The result, core–sheath fibers are shown at the bottom, based on [16].

so that for each layer an independent syringe pump is needed. Hence, it is possible to process polymers, which are not fiber building on their own due to low molecular weight, for example. A spinnable solution is put as shell material and another less spinnable solution as core material. Torkamami et al. demonstrated with chitosan as core and polyethylene oxide as shell layer, that nanofibers can be produced from less spinnable solutions by multiaxial electrospinning [17]. Another application for the core–shell structure is the incorporation of drugs and biological agents, with the shell protecting a component in the core from an aggressive environment or enabling controlled drug release after a certain time [18, 19].

16.3.4 Fiber alignment

Mechanical characteristics and properties in textiles can be enhanced by orientating the fibers and aligning them in load direction. Orientation of nanofibers in electrospinning can be achieved in different ways.

16.3.4.1 Rotating mandrel

Besides enabling the production of uniform nanofiber sheets, a rotating mandrel is the most common tool for fiber alignment in electrospinning. (Figure 16.4) By rotating at high speeds, collected fibers are aligned along the circumference of the mandrel. Publications on this method report that there is a minimal rotational speed at which alignment occurs and a maximal rotational speed at which fibers break due to high stretching. Rotational speed between 500 and 4,500 rpm is used for fiber alignment [20]. A study by Kim et al. in 2004 showed the correlation of higher fiber alignment with higher rotational speeds and furthermore increased modulus and tensile stress in fiber direction at the example of poly(ethylene terephthalate) (PET) fibers [21].

16.3.4.2 Gap and magnetic electrospinning

Instead of the mechanical force exerted by a rotating mandrel collector, electrostatic or magnetically induced forces over a gap can be used for producing nanofiber mats with an increased fiber alignment. In gap electrospinning two conductive electrodes are positioned with an insulating gap between them. The induced electrostatic field is split between both electrodes and the resulting electrostatic force becomes direction dependent, pulling the charged fiber to both edges and aligning it alongside the gap. A similar setup is used in magnetic electrospinning. Two magnets are arranged at the collector region, so that parallel magnetic field lines are formed. The force induced to the fiber and responsible for alignment along the field lines on the collector is the Lorenz force instead of an electrostatic force. Both techniques can be used to produce highly orientated fiber mats with anisotropic characteristics, but are limited to the size of the small collector design and are complex to be used in an upscaled design [22–24].

16.3.4.3 Centrifugal electrospinning

Spinning methods for overcoming some limitations of the nozzle-based setup have been developed. In the basic electrospinning setup, the only significant force acting on the extruded polymer solution and the emitted jet is the electrostatic force. Centrifugal spinning aims at increasing the force acting on the droplet and jet to reduce necessary high voltage. By rotating the spinneret at thousands of rounds per minutes a centrifugal force is applied on the jet, reducing the voltage required to overcome surface tension and increasing fiber stretching during the process. Further benefit is the alignment of collected fibers, similar to when the collector itself rotates during the process, and less clogging of the orifices [25, 26]. Vass et al. reported an immense increase of productivity using a rotary spinneret, a throughput of 240 g/h was achieved [27].

16.3.5 Gas-assisted electrospinning

The use of a high speed gas stream has a similar goal as the centrifugal spinning, adding another force on the polymer solution besides the electrostatic force. This way polymer solutions with low conductivity, high viscosity or high surface tension can be processed and transformed into nanofibers. The gas stream is coupled to the spinneret and helps in overcoming the droplets surface tension and emitting a jet or multiple jets. This facilitates the scale up of nozzle-based systems as more jets are formed on a single nozzle and a higher flow rate can be achieved [28, 29].

16.3.6 Electrospraying

This technology can be used for nanoencapsulation of bioactives or drugs and production of nanoparticles. Nanoparticles as nanofibers have specific advantages compared to microparticles. Faster degradation and enhanced ability to reach their target are just two examples. The principle for electrospraying is very similar to electrospinning, a polymer solution is fed and charged under the influence of a high voltage. The difference is in the composition of the polymer solution, either polymer with a low-molecular weight (and thus a chain length too short to build fibers) are used or solutions with a low viscosity resulting in a sputtering of the emitted jet and dividing into separate particles [30, 31].

16.3.7 Needleless

As the nozzle-based technology and the use of multiple spinnerets has issues like spinneret clogging and jet repulsion, free surface electrospinning emerged on the way to scale up nanofiber production for an industrial application. A thin polymer solution film is formed on an open surface induced with a high voltage. Instead of defined orifices and jets, the jets are emitted randomly and in a higher number from the free surface than in the nozzle-based approach. First remarkable design of a free surface electrospinning setup was reported by Jirsak et al. in 2003 and later commercialized by the czech company Elmarco in their *Nanospider*™ technology. A mandrel rotates in a polymer solution bath and provides a constant thin solution layer on the mandrels surface (see Figure 16.6d). In this approach the collector is positioned above the emitter and the induced electrostatic field creates a high number of Taylor cones and jets on the free surface, spinning upside down. Hence, dripping polymer solution on the sample or product can be prevented [16, 32].

Throughput and production are increased due to the high amount of jets emitted compared to the nozzle-based approach. In a later development the moving emitter in form of a spinning mandrel is replaced by a stationary wire. A feeding system

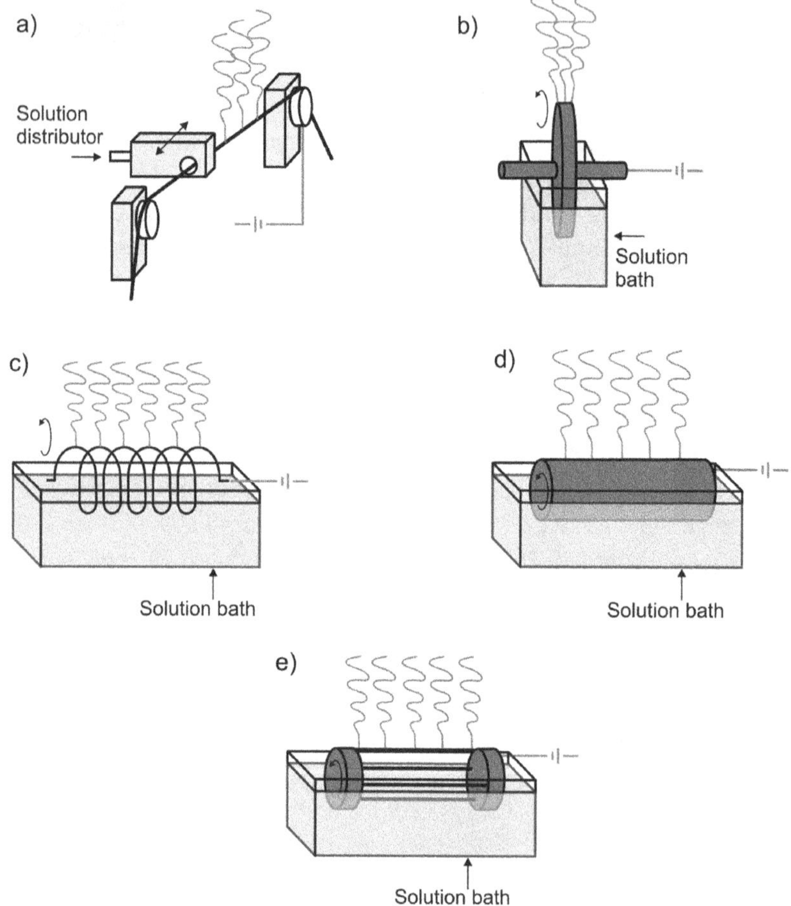

Figure 16.6: Free surface electrospinning setups: (a) static wire emitter, (b) disk emitter, (c) rotating coil emitter, (d) rotating mandrel emitter, (e) rotating wires emitter, based on [16].

covers the wire continuously in polymer solution, thus creating the free surface. Advantage of this compared to the mandrel setup is, that time dependency of the process and solution property change is reduced. Change of polymer solution over time and the resulting time dependency of the process are general disadvantages with free surface electrospinning. Typically for electrospinning of polymer solutions, organic solvents with a fast evaporation and volatile properties as acetone or dichloromethane are used. This results in changing polymer concentration and viscosity in the solution reservoir and film, thus making the process more difficult to control and to maintain product quality over time [16, 32].

Additional approaches are made to improve the electrostatic field of the emitter and thereby improving fiber quality and uniformity. For example spiral coils, mandrels with wires (Figure 16.6 c and e), beaded chains or bubbles are used [33–35].

16.3.8 Yarn production

In order to overcome the limitations of electrospun nonwovens, as inferior mechanical properties compared to other textile structures and difficult handling for further processing, great efforts have been made to modify the electrospinning process. Different kinds of modifications were successfully applied to enable the continuous production of nanofiber yarns instead of unoriented nonwovens. By applying twist cohesion, friction and lateral interaction between the fibers highly improve the structure's mechanical properties [36]. In the form of yarns, nanofibers can be further processed into more complex two- or three-dimensional structures by weaving, (warp) knitting or braiding. This way, load-efficient structures can be attained. The most important of these concepts are static and dynamic liquid support system, self-bundling, using pairs of oppositely charged nozzles along with a funnel-shaped collector or a rotating ring collector [37].

For nanofiber yarn fabrication a mesh has to be produced initially, the fibers aligned and then a twist inserted into the aligned fibers. An easy way to get a nanofiber yarn is to produce aligned nanofibers using a gap electrode and then twisting the fibers by hand or a roller, but this does not deliver continuous yarns or even a high throughput of material. The rotating funnel or ring-shaped collectors show promising results for a scale up of yarn production, subsequent inline stretching and heat setting of the yarns for improvement of the mechanical properties. Two oppositely charged nozzles aim at the opening of a rotating funnel. The fibers at-

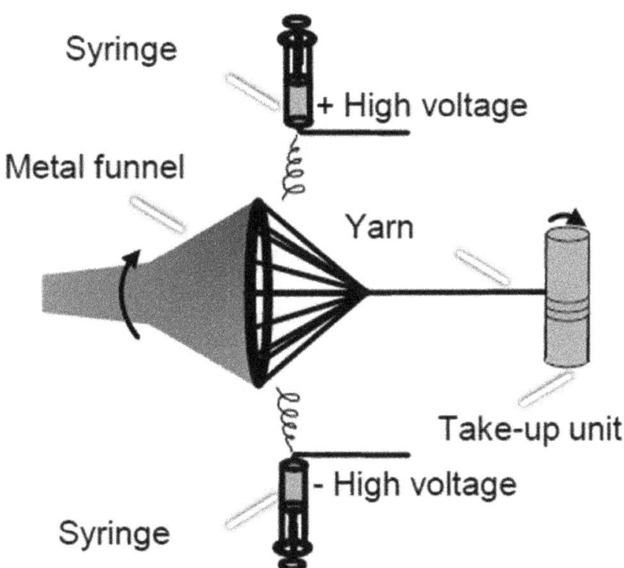

Figure 16.7: Electrospinning setup with two oppositely charged nozzles and a funnel collector, based on [38].

tract each other in this case and form a mesh at the funnel opening, which can be pulled to the take-up unit. Still adherent to the funnel ring, while rotating, twist is inserted into the yarn (figure 16.7) [38].

16.3.9 Alternating current spinning

The majority of research groups and publications uses direct current (DC) power supply for electrospinning. Amongst few other researchers Pokorny et al. showed, that alternating current (AC) can be used for nanofiber production as well. This approach may have some advantages over the DC electrospinning. In AC electrospinning multiple jets are emitted from each droplet and orifice, just as in the free surface technologies, thus enabling a higher fiber output. Due to the alternating charges the fibers attract each other and no counter electrode for collecting is needed. Instead a plume of fibers is produced close to the spinneret. In addition to the known factors of electrospinning like voltage, solution properties or climate conditions, waveform and frequency of the applied high voltage influence the electrospinning process with AC. As the fibers are produced midair instead of on a collector surface, this setup can be used for the direct manufacturing of nanofiber yarns by adding a winding and stretching unit [39, 40].

16.4 Manufacturers

Manufacturers in the field of electrospun nanofibers can be divided into equipment manufacturers and service providers for nanofiber production and development. In addition, there are companies that manufacture and sell their own products with nanofiber technology. A non-comprehensive list of manufacturers is given in Table 16.2.

Table 16.2: List of electrospinning machine manufacturers and service providers.

Company	Technology/type of machine	Services
Abalioglu Teknoloji (Turkey)	–	Product tailoring filter media, acoustic damping
Bioinicia (Spain)	Needle-based systems, lab scale up to high throughput industrial production, GMP conform	Contract development and manufacturing, process engineering
Contipro (Czech Republic)	Needle-based system, lab scale	–

Table 16.2 (continued)

Company	Technology/type of machine	Services
DiPole Materials (USA)	–	Design and manufacturing three-dimensional scaffolds and technical textiles
Electrospunra Nanofibers (India)	Needle-based system, lab scale and a small, portable electrospinning device	Consulting, trainings, scaffold production
Elmarco (Czech Republic)	Open surface system, lab scale to industrial high throughput scale, air conditioning units	–
Espin (USA)	Nozzle system, lab scale	Design, development, testing and launch of nanofiber products
FNM Co. (Iran)	Air-assisted needle-based system, lab scale up to industrial scale	–
IME Technologies (Netherlands)	Needle-based systems, research equipment and GMP conform industrial scale	Development of medical devices, drug delivery devices and tissue engineering
Inovenso (Turkey)	Needle-based systems, lab, pilot and industrial scale	Product development, contract manufacturing for scaffolds, analytical services
ITA GmbH (Germany)	–	Material & scaffold design, contract manufacturing and development, trainings
Nanofiber Solutions, Inc. (USA)	–	Design and manufacturing three-dimensional scaffolds
Nanoflux (Singapore)	Needle-based system, lab scale and mass production scale	–
Revolution Fibres (New Zealand)	–	Partnership & product development, products for industrial applications
SPUR (Czech Republic)	Needle-based system, lab scale	Trainings
The Electrospinning Company Ltd (United Kingdom)	–	Scaffold design, contract development & manufacturing

Table 16.2 (continued)

Company	Technology/type of machine	Services
Tong Li Tech (China)	Needle-based systems, lab scale up to pilot scale; portable electrospinning device	–
Yflow (Spain)	Coaxial needle-based systems, lab and industrial scale	Product development, customized machines and accessory, trainings

Meanwhile, there is a wide range of equipment manufacturers from the smallest laboratory devices and modular setups to production lines for industrial applications. The manufacturers are located all over the world with the majority in Europe and Asia. For example, there are the companies Tong Li Tech (China), FNM Co. (Iran), MECC CO. (Japan), Electrospunra Nanofibers (India) or Nanoflux (Singapore) from the Asian region, which offer comparatively inexpensive production machines. European manufacturers include the companies Contipro, Elmarco, SPUR (all Czech Republic), inovenso (Turkey), Bioinicia, Yflow (both Spain), IME Technologies (Netherlands) and Leonardo (Italy). Different sizes, production quantities and functional principles are offered, in the following exemplary products and companies.

16.4.1 Bioinicia (Spain)

Bioinicia offers a wide range of machines under their trade name *Fluidnatek®*. Basic benchtop laboratory machines or advanced research stations, as well as pilot-production plants for scaling up production are part of the available machines. The company is focused on a needle-based approach and targets upscaling using a modular multineedle system. In addition to electrospinning Bioinicia provides expertise in electrospraying for the encapsulation of functional ingredients and biological materials in particles. In February 2020 they are one of the first companies to be GMP approved for the manufacturing of pharmaceutical products. In cooperation with the company Dermtreat ApS the Rivelin® patch, a treatment for mucosal diseases, was developed and now produced on an industrial scale at Bioinicia´s facility [41, 42].

16.4.2 Elmarco (Czech Republic)

As mentioned earlier, Elmarco´s machines work on the free surface technology for electrospinning. Their product line *Nanospider*™ is available in R&D scale as well as industrial production scale. The largest, modular electrospinning unit NS 8S16000U is able to produce continuously with a width of 1.6 m and a maximum number of 32 spinning electrodes. Elmarco claims that this setup is capable of an annual throughput of 20,000,000 m^2 for polyamide 6 (PA6) as a coating with basis weight 0.03 g/m^2 and an uptime of 85%. Additionally, Elmarco offers suiting air conditioning units to control the climate conditions in the necessary range to produce nanofibers with a good quality [43].

16.4.3 Contipro (Czech Republic)

Contipro in contrast to Bioinica and Elmarco focuses on a benchtop system called 4SPIN® and a highly modular system. The device is set to work with a wide range of specially designed collectors for fiber alignment or complex structures and needle-based as well as needleless emitters [44].

16.4.4 IME Technologies (Netherlands)

The Dutch company was one of the first commercial electrospinning companies in Europe and is now focused on electrospinning platforms for medical development and production. They have a research line using a needle-based system which features a wide variety of collectors and emitters as rotating mandrels, rotating wires or coaxial needles and multiple needles. Newest roll out is the MediSpin® XL platform targeting a maximum of automation for a GMP conform production of medical products. On top of roll-to-roll continuous production it has online monitoring and process data capture to ensure reproducibility [45, 46].

In addition, as described above, there are companies that specialize in providing services as manufacturers of nanofibers and developers for applications and products. Some of the above mentioned equipment manufacturers also offer manufacturing and development in their portfolio. In this area there is a large number of suppliers from English-speaking countries such as Espin, DiPole Materials, Nanofiber Solutions, Inc. (all USA), Revolution Fibers (New Zealand) or The Electrospinning Company Ltd (United Kingdom), but also, for example, Abalioglu Teknoloji (Turkey) or ITA GmbH (Germany).

16.4.5 Nanofiber Solutions, Inc. and The Electrospinning Company Ltd

Nanofiber Solutions and The Electrospinning Company are good examples for companies, which specialize in the production of novel three-dimensional scaffolds for tissue engineering. The structure is used for the colonization of cells and tissue in medical technology. The fibers with diameters in the nanometer range and nonwoven structure mimic the structure of extracellular matrix and provide a suitable environment for the development and proliferation of various cells. Furthermore, they work with a broad range of polymers and equipment in biomaterial design, and are able to do extended research and development as well as translation into the clinic. The products can be used in cancer research, work with stem cells or the development of in vitro disease models, for example [47, 48].

16.4.6 Yflow (Spain)

The Spanish company Yflow offers its expertise for consulting in material and product development. The company benefits from the production of its own plants. Individual setups and solutions for production are developed and offered. In addition, the company focuses on the development of microcoatings, tubular membranes or continuous yarns made of nanofibers [49].

16.4.7 Revolution Fibres (New Zealand)

Revolution Fibres in New Zealand develops, produces and markets its own products. The product range covers filtration, composite reinforcement, skincare applications and acoustic insulation. They use industrial scale production lines as shown exemplary in Figure 16.8. Some of their cooperation and products are described in the following chapter [50].

Figure 16.8: Industrial-scale electrospinning unit at Revolution Fibres in New Zealand. Copyright Revolution Fibres 2020.

16.4.8 ITA GmbH (Germany)

German ITA GmbH is a service provider for electrospinning and does contract research and technology transfer mainly in medical fields, but lately expanded the applied topics to filtration, acoustics and machine development. Connected to the Institut für Textiltechnik of RWTH Aachen University the GmbH has access to state of the art research and developments, for example, to the latest yarn spinning results and developments. Services range from material development with drug or particle incorporation to product development for the medical market [51].

16.5 Applications

Nanofibers are already used in filtration, tissue engineering, composite reinforcement, skincare and acoustic insulation. Although most commercially available nanofiber products are available for air and water filtration, there are some other interesting and innovative products in other areas on the market.

16.5.1 Biomedical

In the past two decades a lot of biomedical applications using the electrospinning technology were researched on and tested. Two-dimensional and three-dimensional structures and scaffolds are used to boost repair and regeneration of various tissues like nerve, skin, heart or blood vessels. A good example is the ReDura™ patch by Medprin, which is used in neurosurgery and the repair of dural defects. A poly-L-lactic acid (PLLA) patch is used for mimicking extracellular matrix and degrading over time [52, 63].

ORTHOREBIRTH offers a commercial bone-void-filling material described as a cottony textile material. β-Tricalcium phosphate (β-TCP), a silicone-containing calcium carbonate and also PLLA are main ingredients for the product called ReBossis™. Textile morphology makes it easy to handle and the fiber and pore structure allow for new bone and capillary blood vessels to grow into the scaffold. Originally the product was made out of a conventional nonwoven until the electrospun nonwoven scaffold showed better regeneration results than the nonwoven [53].

Another innovative electrospun product is developed in Switzerland and the Netherlands by Xeltis, although not yet commercially available. But first results show a very promising technology for the natural restoration of cardiovascular function. A therapy called endogenous tissue restoration is applied to treat damaged heart valves or cardiovascular vessels. Instead of implanting a permanent device such as an artificial heart valve, stent or stentgraft, which can cause foreign body reactions, a degradable implant is put in place. In case of Xeltis and their RestoreX™ it is a fully polymer-based nanofiber device, that takes over the function of the heart valve or vessel. Over time the biomaterial degrades and natural restoration is enabled by the nanofiber scaffold. Goal is to have a restored anatomy in shape of the original implant by the time the implant has fully degraded. This would be a solution for a restored or repaired heart valve, that can grow with the patient [54, 55].

16.5.2 Furniture and acoustic

An interesting development was made by the company Revolution Fibres and IQ commercials. Revolution Fibres produces the Phonix™ membrane with layers of conventional fibers, an acoustic foam and a nanofiber layer (see Figure 16.9 left). The high specific surface area increases sound absorption while maintaining low profile of the membrane. This membrane is used in the Return Focus Pod™, an office and workspace solution to create single work stations with a high sound absorption and a light weight construction [56].

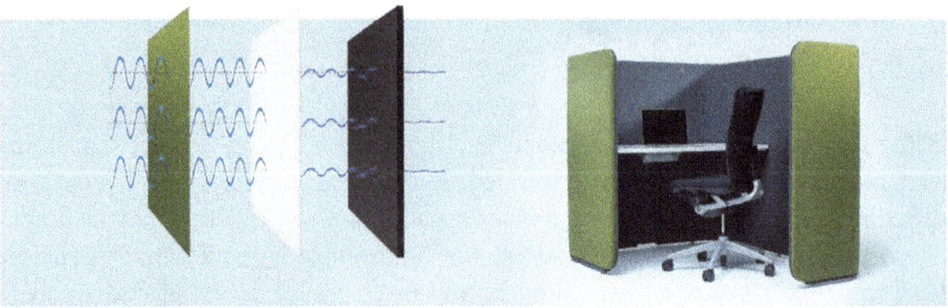

Figure 16.9: Example for furniture using nanofiber materials for functionalization. Return Focus Pod™ with a Phonix™ material layer. Copyright 2020 Revolution Fibres.

16.5.3 Filtration devices and media

So called particulate matter (PM) is a mixture of small particles of solid or liquid matter suspended in air. Particles with a diameter less than 2.5 µm (PM2.5) respectively 10 µm (PM10) are a threat for the human body. They can cause blockage of small blood vessels, congestive heart failure or pneumonia and lung cancer. Microfiber filters in conventional designs have two main problems, large pressure drop over the membrane and low filtration efficiency against small PM. Nanofiber media is either able to maintain the same filtration efficiency at a thinner layer thickness or increase filtration efficiency at a constant layer thickness [57, 58]. Furthermore, decreased fiber diameter has a positive effect on pressure drop over the membrane.

Air molecules are less influenced by impact and interception and pass the filtration device with less disturbance, resulting in a lower pressure drop. This consequently results in less energy needed to maintain stable flowrate over a filtration device or facilitates breathing through a filtration mask, for example [59, 60].

Different companies already promote face masks with a nanofiber core and varying classifications. Hong Kong-based company NASK offers a N95 face mask, FNM Co. has a wider range of FFP1 to FFP3 masks under the Name Respnano® available. Additionally, the products inofilter® and SETA™ are provided on rolls to use it in the production of filtering face pieces (Figure 16.10). The inofilter® material has a comparable structure to conventional meltblown microfiber filtration masks. The nanofiber layer is put into a sandwich structure with two outside layers of PET spunbond as protection for the ultrafine layer and prefiltration of bigger particles. Tebyetekerwa et al. published a guide in 2020 to incorporate such a nanofiber material in a community or do-it-yourself face mask and improve filtration efficiency against bacteria and viruses [60].

Figure 16.10: Filtration application with electrospun nanofibers. The SETA™ filter material can be applied for different air filtration applications as face filter masks or air cleaning for buildings. Copyright Revolution Fibres 2020.

Apart from filtration masks electrospun nanofibers are used in commercial products for air filtration of buildings or cars (ProTura™ by Clark Filter, exceed™ by exceed or MICROGRADE A-NF™ by MANN + HUMMEL) as well as for household water or pool filtration systems (Naked Filter by Naked Filter, Nanotrap™ by COWAY or NanoFiber™ by Astralpool) [67].

16.6 Conclusion

In the course of the last decade, electrospinning has been developed in many ways. Progress can be seen especially in the development of the process for new materials and geometries, the development of devices for increased throughput and product development out of the laboratories and into the commercialized market. A better understanding of the process and the acting forces and principles enables better control and adjustment of solution and process parameters. The range of materials that can be processed has been greatly expanded, so that interesting materials such as chitosan or silk fibroin can be spun, but also drugs and additives can be used, for example, for X-ray visibility. The use of nanofibers in biomedical applications can thus be expanded and new scaffold designs are possible. The spinning process has been continuously improved and adapted to achieve smaller fiber diameters, a specialized morphology or a greatly increased throughput. Methods such as centrifugal spinning or gas-assisted spinning stretch the resulting fibers in addition to the electrostatic force of electrospinning. Specially designed collector geometries or setups for yarn spinning allow to produce complex three-dimensional structures and scaffolds from the nanofibers, while multiaxial nozzles can be used to produce core–sheath fibers for biomedical applications. Multinozzle panels, free surface spinning or gas-assisted spinning increase the overall throughput possible in electrospinning, which was and still is a deficit of this technology for industrial scale production. Several machine manufacturers offer machines for such high throughput applications, promising the production on an industrial scale. Challenges that need to be addressed in the next years will be the precise and reproducible production with control of fiber and scaffold morphology while scaling up the process.

The number of commercialized products has also increased significantly in the area of product development. Products can be found in the area of biomedical applications in the form of scaffolds and stent applications, as well as furniture and fabrics, and to a large degree in filtration applications. Especially in the latter, products for filter masks, industrial air filters and water filtration systems are well established. It is expected that the market will grow in the future, both in the scope of the fields applied and in the number of products available and sold. Especially in the area of smart coatings and separators for batteries, an increase in commercial products and applications is predicted [61, 62].

Concluding solution electrospinning is a technology which gains more and more interest among researches but also for industrial partners and developers. Some companies have proven the applicability of nanofibers and electrospinning and opened the markets for future developments. Challenges are still to overcome as the upscaling of the process for a variety of applications and a wider range of materials, even though approaches and facilities for this already exist. Further a reliable online system for quality control of the manufactured nanofiber materials and an online method for monitoring fiber diameter and quality have yet to be introduced.

References

[1] Nune, S. K.; Rama, K. S.; Dirisala, V. R.; Chavali, M. Y.: Electrospinning of Collagen Nanofiber Scaffolds for Tissue Repair and Regeneration. In: Grumezescu, A. M.; Ficai, D., editors: Nanostructures for Novel Therapy. Amsterdam, Netherlands, Cambridge, MA: Elsevier, 2017.

[2] Kruse, M.: Anlagenentwicklung und -validierung zur Herstellung elektro-gesponnener Garne für den Einsatz im Tissue Engineering, 1. Auflage. Düren: Shaker, 2019.

[3] Haider, A.; Haider, S.; Kang, I.-K.: A comprehensive review summarizing the effect of electrospinning parameters and potential applications of nanofibers in biomedical and biotechnology. Arabian Journal of Chemistry, Article in Press, 2015; 11(8), 1–24.

[4] Mehetre, G.; Pande, V.; Kendre, K.: An overview of nanofibers as a platform for drug delivery. Inventi Rapid: Nano Drug Delivery Systems, 2015; 3, 1–6.

[5] Subbiah, T.; Bhat, G. S.; Tock, R. W.; Parameswaran, S.; Ramkumar, S. S.: Electrospinning of Nanofibers. Journal of Applied Polymer Science, 2005; 96, 557–569.

[6] S. De, V.; Daels, N.; Lambert, K.; Decostere, B.; Hens, Z.; S. Van, H.; K. De, C.: Filtration performance of electrospun polyamide nanofibres loaded with bactericides. Textile Research Journal, 2011; 82(1), 37–44.

[7] Xue, J.; Tong, W.; Dai, Y.; Xia, Y.: Electrospinning and electrospun nanofibers: Methods materials, and applications. Chemical Reviews, 2019; 119(8), 5298–5415.

[8] Laar, N.; Köppl, S.; Wintermantel, E.: Electrospinning. In: Wintermantel, E.; Ha, S.-W., editors: Medizintechnik – Life Science Engineering, 4th ed. Berlin; Heidelberg: Springer Verlag, 2008; pp. 381–401.

[9] Reneker, D. H.; Yarin, A. L.; Fong, H.; Koombhongse, S.: Bending instability of electrically charged liquid jets of polymer solutions in electrospinning. Journal of Applied Physics, 2000; 87, 4531–4547.

[10] Salas, C.: Solution Electrospinning of Nanofibers. In: Afshari, M., editor: Electrospun Nanofibers, 1st ed. Duxford, Cambridge, Kidlington: Elsevier Ltd., 2017; pp. 73–108.

[11] Yousefzadeh, M.: Modeling and Simulation of the Electrospinning Process. In: Afshari, M., editor: Electrospun Nanofibers, 1st ed. Duxford, Cambridge, Kidlington: Elsevier Ltd., 2017; pp. 277–301.

[12] Bhardwaj, N.; Kundu, S. C.: Electrospinning: A fascinating fiber fabrication technique. Biotechnology Advances, 2010; 28(3), 325–347.

[13] Nezarati, R. M.; Eifert, M. B.; Cosgriff-Hernandez, E.: Effects of humidity and solution viscosity on electrospun fiber morphology. Tissue Engineering. Part C, Methods, 2013; 19(10), 810–819.

[14] Khalf, A.; Madihally, S. V.: Recent advances in multiaxial electrospinning for drug delivery. European Journal of Pharmaceutics and Biopharmaceutics, 2017; 112, 1–17.

[15] Theron, S. A.; Yarin, A. L.; Zussman, E.; Kroll, E.: Multiple jets in electrospinning: Experiment and modeling. Polymer, 2005; 46(9), 2889–2899.

[16] Vass, P.; Szabó, E.; Domokos, A.; Hirsch, E.; Galata, D.; Farkas, B.; Démuth, B.; Andersen, S. K.; Vigh, T.; Verreck, G.; Marosi, G.; Nagy, Z. K.: Scale-up of electrospinning technology: Applications in the pharmaceutical industry Wiley interdisciplinary reviews. Nanomedicine and Nanobiotechnology, 2020; 12(4), e1611.

[17] Torkamani, A. E.; Syahariza, Z. A.; Norziah, M. H.; Wan, A. K. M.; Juliano, P.: Encapsulation of polyphenolic antioxidants obtained from Momordica charantia fruit within zein/gelatin shell core fibers via coaxial electrospinning Food. Bioscience, 2018; 21, 60–71.

[18] Zhang, Y. Z.; Wang, X.; Feng, Y.; Li, J.; Lim, C. T.; Ramakrishna, S.: Coaxial electrospinning of (fluorescein isothiocyanate-conjugated bovine serum albumin)-encapsulated poly(epsilon-caprolactone) nanofibers for sustained release. Biomacromolecules, 2006; 7(4), 1049–57.

[19] Jiang, H.; Hu, Y.; Zhao, P.; Li, Y.; Zhu, K.: Modulation of protein release from biodegradable core-shell structured fibers prepared by coaxial electrospinning. Journal of Biomedical Materials Research. Part B, Applied Biomaterials, 2006; 79(1), 50–57.
[20] Pan, H.; Luming, L.; Long, H.; Cui, X.: Continuous aligned polymer fibers produced by a modified electrospinning method. Polymer, 2006; 47(14), 4901–4904.
[21] Kim, K. W.; Lee, K. H.; Khil, M. S.; Ho, Y. S.; Kim, H. Y.: The effect of molecular weight and the linear velocity of drum surface on the properties of electrospun poly(ethylene terephthalate) nonwovens. Fibers and Polymers, 2004; 5(2), 122–127.
[22] Li, D.; Wang, Y.; Xia, Y.: Electrospinning of polymeric and ceramic nanofibers as uniaxially aligned arrays. Nano Letters, 2003; 3(8), 1167–1171.
[23] Li, D.; Wang, Y.; Xia, Y.: Electrospinning Nanofibers as uniaxially aligned arrays and layer-by-layer stacked films. Advanced Materials, 2004; 16(4), 361–366.
[24] Orr, S. B.; Chainani, A.; Hippensteel, K. J.; Kishan, A.; Gilchrist, C.; William, G. N.; Ruch, D. S.; Guilak, F.; Little, D.: Aligned multilayered electrospun scaffolds for rotator cuff tendon tissue engineering. Acta Biomaterialia, 2015; 24, 117–126.
[25] Huttunen, M.; Kellomäki, M.: A simple and high production rate manufacturing method for submicron polymer fibres. Journal of Tissue Engineering and Regenerative Medicine, 2011; 5(8), e239–e243.
[26] Weitz, R. T.; Harnau, L.; Rauschenbach, S.; Burghard, M.; Kern, K.: Polymer nanofibers via nozzle-free centrifugal spinning. Nano Letters, 2008; 8(4), 1187–1191.
[27] Vass, P.; Hirsch, E.; Kóczián, R.; Démuth, B.; Farkas, A.; Fehér, C.; Szabó, E.; Németh, Á.; Andersen, S. K.; Vigh, T.; Verreck, G.; Csontos, I.; Marosi, G.; Nagy, Z. K.: Scaled-up production and tableting of grindable electrospun fibers containing a protein-type drug. Pharmaceutics, 2019; 11(7), 329.
[28] Duan, G.; Greiner, A.: Air-blowing-assisted coaxial electrospinning toward high productivity of core/sheath and hollow fibers. Macromolecular Materials and Engineering, 2019; 304(5).
[29] Zhuang, X.; Shi, L.; Jia, K.; Cheng, B.: Kang, Weimin Solution blown nanofibrous membrane for microfiltration. Journal of Membrane Science, 2013; 429, 66–70.
[30] Gaskell, S. J.: Electrospray: Principles and practice. Journal of Mass Spectrometry, 1997; 32(7), 677–688.
[31] Zhang, S.; Campagne, C.; Salaün, F.: Influence of solvent selection in the electrospraying process of polycaprolactone. Applied Sciences, 2019; 9(3), 402.
[32] Radacsi, N.; Giapis, K. P.; Ovari, G.; Szabó-Révész, P.; Ambrus, R.: Electrospun nanofiber-based niflumic acid capsules with superior physicochemical properties. Journal of Pharmaceutical and Biomedical Analysis, 2019; 166, 371–378.
[33] Liu, Y.; He, J.-H.; Yu, J.-Y.: Bubble-electrospinning: A novel method for making nanofibers. Journal of Physics. Conference Series, 2008; 96(1), 12001.
[34] Liu, S.-L.; Huang, Y.-Y.; Zhang, H.-D.; Sun, B.; Zhang, J.-C.; Long, Y.-Z.: Needleless electrospinning for large scale production of ultrathin polymer fibres. Materials Research Innovations, 2014; 18(sup4), S4-833-S4-837.
[35] Miloh, T.; Spivak, B.; Yarin, A. L.: Needleless electrospinning: Electrically driven instability and multiple jetting from the free surface of a spherical liquid layer. Journal of Applied Physics, 2009; 106(11), 114910.
[36] Maleki, H.; Gharehaghaji, A. A.; Toliyat, T.; Dijkstra, P. J.: Drug release behavior of electrospun twisted yarns as implantable medical devices. Biofabrication, 2016; 8(3).
[37] Mooneghi, S. A.; Gharehaghaji, A. A.; Hosseini-Toudeshky, H.: Giti Torkaman Failure mechanism of polyamide 66 nanofiber yarns under fatigue and static tensile loading. Journal of Applied Polymer Science, 2015; 132.

[38] Ali, U.; Zhou, Y.; Wang, X.; Lin, T.: Direct electrospinning of highly twisted, continuous nanofiber yarns. The Journal of the Textile Institute, 2012; 103(1), 80–88.
[39] Pokorny, P.; Kostakova, E.; Sanetrnik, F.; Mikes, P.; Chvojka, J.; Kalous, T.; Bilek, M.; Pejchar, K.; Valtera, J.; Lukas Effective, D.: AC needleless and collectorless electrospinning for yarn production. Physical Chemistry Chemical Physics, 2014; 16(48), 26816–26822.
[40] Sun, B.; Long, Y. Z.; Zhang, H. D.; Li, M. M.; Duvail, J. L.; Jiang, X. Y.; Yin, H. L.: Advances in three-dimensional nanofibrous macrostructures via electrospinning. Progress in Polymer Science, 2014; 39(5), 862–890.
[41] N.N.: Fluidnatek® – Electrospinning & electrospraying equipment – Bioinicia, 2020. URL: https://bioinicia.com/electrospinning-electrospraying-lab-equipment/, Accessed on 16.07.2020.
[42] N.N.: Bioinicia is awarded GMP Certification for its industrial manufacturing plant for electrospun nanofibers – Bioinicia, 2020. URL: https://bioinicia.com/bioinicia-gmp-certification-for-electrospun-nanofibers/, Accessed on 16.07.2020.
[43] Elmarco, N. N.: Production profile Nanospider™ Production Line NS 8S1600U – Elmarco, 2020 URL: https://www.elmarco.com/getFile/case:show/id:448778/2019-10-04%2010:13:17.000000, Accessed on 16.07.2020.
[44] Contipro, N. N.: 4SPIN® From Spinnability to Productivity Portfolio – Contipro, 2020 URL: https://www.4spin.info/portfolio, Accessed on 16.07.2020.
[45] N.N.: Research Equipment – IME Medical Electrospinning, 2020. URL: https://www.ime-electrospinning.com/electrospinning-equipment/, Accessed on 16.07.2020.
[46] N.N.: MediSpin® XL – IME Medical Electrospinning, 2020. URL: https://www.ime-electrospinning.com/medispin-xl/, Accessed on 16.07.2020.
[47] N.N.: Scaffolds for Organ and Tissue Regeneration – Technology – Nanofiber Solutions, Inc., 2020. URL: https://nanofibersolutions.com/technology/, Accessed on 16.07.2020.
[48] N.N.: About the Electrospinning Company Design, development and manufacture of biomaterials for regenerative medical devices – The Electrospinning Company, 2020. URL: https://www.electrospinning.co.uk/about-us/company/, Accessed on 16.07.2020.
[49] N.N.: Coaxial Electrospraying & Electrospinning Machine & Equipment - Services –Yflow S.D., 2020. URL: http://www.yflow.com/services/, Accessed on 16.07.2020.
[50] N.N.: Extraordinary naofibre. Ingenious solutions. Product & Services Brochure – Revolution Fibres, 2020. URL: https://www.revolutionfibres.com/wp-content/uploads/2017/08/Revolution-Fibres-Products-Services_WEB.pdf, Accessed on 16.07.2020.
[51] N.N.: ITA GmbH is your partner for the transfer of research and development results – ITA GmbH, 2020. URL: https://ita-gmbh-aachen.com/en/, Accessed on 16.07.2020.
[52] Tao, X.; Shi, Z. D.; Xiong, L.; Ke, Y. Q.; Deng, K. X.; Liu, M.; Tian, Q.; Yuan, Y. Y.: In-vitro and clinical study on a novel synthetic absorbable dural substitute. Journal of Neurological Surgery. Part B, Skull Base, 2014; 75(S 02), a192.
[53] Uyar, T.; Kny, E.: Electrospun Materials for Tissue Engineering and Biomedical Applications: Research, Design and Commercialization. Elsevier Science, 2017.
[54] Brugmans, M. M. C. P.; Soekhradj-Soechit, R. S.; van Geemen, D.; Cox, M.; Bouten, C. V. C.; Baaijens, F. P. T.; Driessen-Mol, A: Superior tissue evolution in slow-degrading scaffolds for valvular. Tissue Engineering Tissue Engineering. Part A, 2016; 22(1–2), 123–132.
[55] Kort, B. D.; Lichauco, A.; Dekker, S.; Serrero, A.; Cox, M.; Smits, A.: Unraveling the mechanisms of endogenous tissue restoration after bioresorbable pulmonary valve implantation in sheep. Structural Heart, 2020; 4(sup1), 46.
[56] N.N.: Phonix Creating sound is something that comes naturally to us. Controlling sound is another story. – Revolution Fibres, 2020. URL: https://www.revolutionfibres.com/products/phonix/, Accessed on 16.07.2020.

[57] Koo, W. T.; Jang, J. S.; Qiao, S.; Hwang, W.; Jha, G.; Penner, R. M.; Kim, I. D.: Hierarchical metal-organic framework-assembled membrane filter for efficient. Removal of Particulate Matter ACS Applied Materials & Interfaces, 2018; 10(23), 19957–19963.
[58] Zhao, X.; Wang, S.; Yin, X.; Yu, J.; Ding, B.: Slip-effect functional air filter for efficient purification of PM 2.5. Scientific Reports, 2016; 6(1), 1–11.
[59] Akduman, C.; Akçakoca Kumbasar, E. P.: Nanofibers in face masks and respirators to provide better protection IOP Conference Series. Materials Science and Engineering, 2018; 460, 12013.
[60] Tebyetekerwa, M.; Zhen, X.; Yang, S.; Ramakrishna, S.: Electrospun nanofibers-based face masks. Advanced Fiber Materials, 2020; 2(3), 161–166.
[61] Persano, L.; Camposeo, A.; Tekmen, C.; Pisignano, D.: Industrial upscaling of electrospinning and applications of polymer nanofibers: A review. Macromolecular Materials and Engineering, 2013; 298(5), 504–520.
[62] Akampumuza, O.; Gao, H.; Zhang, H.; Wu, D.; Qin, X.: Raising nanofiber output: The progress, mechanisms, challenges, and reasons for the pursuit. Macromolecular Materials and Engineering, 2018; 303(1), 1700269.
[63] N.N.: ReDura™ Biomimetic-Synthetic-Absorbable Dural Substitute product brochure – Medprin, 2020. URL: https://pdf.medicalexpo.com/pdf/medprin-biotech/redura-brochure/76788-185418.html, Accessed on 16.07.2020.
[64] Mann+Hummel GmbH: Nano-fibre-coated air filter medium protects engines. Membrane Technology, 2012; 2012(12), 5–6.

Sofie Huysman

17 3D printing

Keywords: 3D-printing, additive manufacturing, polymers, extrusion, smart applications, adhesion, functional fillers

17.1 Introduction

Additive manufacturing (AM), also known as 3D printing, is an innovative technique of which the global importance is quickly growing. The 2019 report from the consulting firm Wohlers Associates– recognized as one of the preeminent 3D printer experts in the world – forecasts a $23.9 billion revenue growth in 3D printing industry by 2022, and $35.6 billion by 2024 [1].

3D printing is already performed with a variety of materials: metals, polymers, ceramics and concrete. Metal 3D printing is predominantly used for advanced applications in the aerospace industry, because traditional processes are more time consuming and costly. Ceramics are mainly used in 3D-printed scaffolds, while concrete is the main material employed in the AM of buildings. Polymers are considered as the most common materials in the 3D printing industry due to their diversity and ease of adoption to different 3D printing processes [2].

When making the link between 3D printing and textile applications, 3D printing with polymers seems to be the most obvious choice. To process polymers, various 3D printing techniques are available, which can be subdivided as follows: those for thermoplastic polymers, which are polymers that can be remelted and reshaped, and those for thermosets, which are polymers that remain in a solid state after being cured.

Typical techniques used for thermoplasts are material extrusion and powder fusion, while 3D printing techniques for thermosets are typically based on a photopolymerization process. The different techniques are presented in Table 17.1.

Sofie Huysman, Plastic Characterisation, Processing and Recycling, CENTEXBEL, Technologiepark 70, BE 9052 Zwijnaarde, e-mail: shu@centexbel.be

https://doi.org/10.1515/9783110670776-017

Table 17.1: Main polymer-based 3D printing techniques [3].

Thermoplasts		Thermosets
Material extrusion	Powder fusion	Photopolymerization
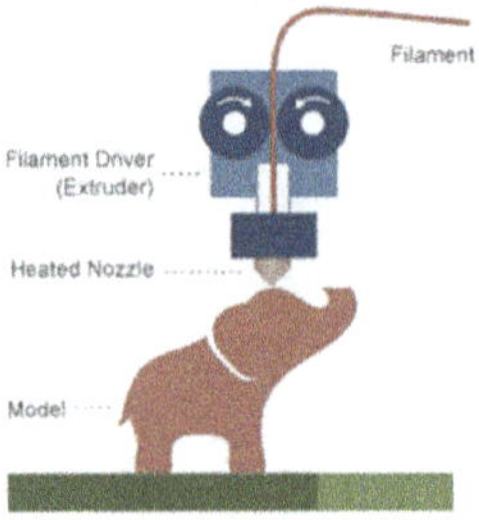	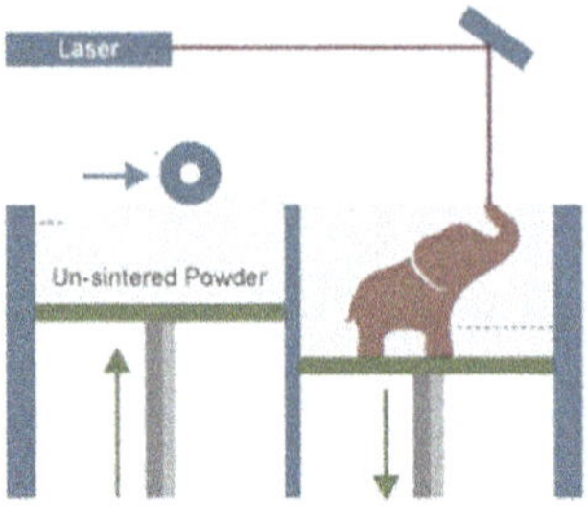	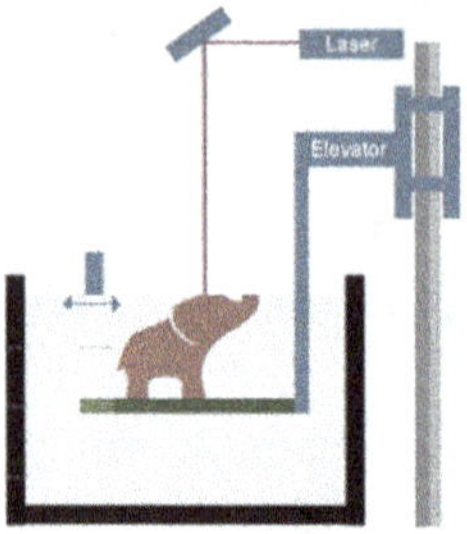
The polymer is selectively dispensed through a nozzle	Regions of a polymer powder bed are fused through a laser beam or a liquid bonding agent	A liquid photopolymer in a vat is selectively cured by light-activated polymerization [4]
– Fused filament fabrication (FFF) – 3D dispensing	– Selective Laser Sintering – Binder Jetting	– Stereolithography (SLA) – Digital light processing (DLP)

17.2 Overview of the main polymer-based techniques

17.2.1 Material extrusion

Fused filament fabrication or FFF is also known under the trademarked term "fused deposition modeling". FFF-based printers use a thermoplastic filament that is pushed through a heated extruder head (including one or more extrusion nozzles) using a drive wheel. The head can be driven both horizontally and vertically, creating one layer at a time before adjusting vertically to begin a new layer.

The used polymeric materials should be extrudable into filaments, not be too soft to avoid buckling between the drive wheels (typically hardness values above 80 Shore A), ductile enough to be spooled and strong enough to avoid shearing due to the pinching from the wheels. Yet, numerous thermoplastics-based materials are available as filaments for FFF. The most common polymers are acrylonitrile butadiene styrene (ABS) and polylactic acid (PLA). Other examples are polyamide (PA), polyurethane (TPU), polyethylene terephthalate and so forth. Many more materials and combinations are possible: the FFF technique offers large freedom in material choice [5].

This is one of the main reasons, next to its low costs and its simple and open source machine structure, that FFF is one of the most widespread AM methods. The following companies are just a few of the many producers of FFF printers: Ultimaker, RepRap, Makerbot Systems, Leapfrog, Prusa Printers and so on.

Some of the more recent 3D printers employ a screw-based extrusion system, making it possible use standard polymer pellets instead of filaments. The use of pellets paves the way to a much broader range of materials, since the mentioned criteria for a filament are not relevant anymore. If a polymer works to a certain extent with injection molding, then it will work with this technology.

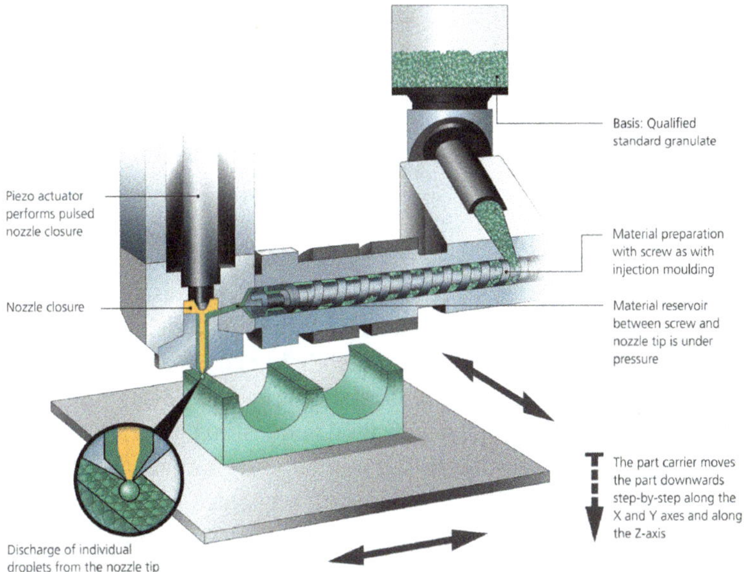
Basis: Qualified standard granulate
Piezo actuator performs pulsed nozzle closure
Material preparation with screw as with injection moulding
Nozzle closure
Material reservoir between screw and nozzle tip is under pressure
The part carrier moves the part downwards step-by-step along the X and Y axes and along the Z-axis
Discharge of individual droplets from the nozzle tip

Figure 17.1: Arburg Plastic Freeformer with extrusion system [6].

An example of a pellet 3D printer is the Arburg Freeformer [7], as shown in Figure 17.1: first, the polymer granulate is melted in a cylinder unit, after which it is transported by a screw toward a nozzle tip. By use of a discharge unit and variable pressure, the melt is deposited as tiny droplets in a layer-by-layer manner onto a platform. Another 3D printer starting directly from pellets instead is the Pollen AM. This printer is capable of printing with up to four different materials, and it is also capable of mixing two materials during the printing process [8]. The two mentioned pellet printers are meant to manufacture small objects. But there are also pellet 3D printers on the market for the manufacturing of large-format objects. Examples of such large printers are the Titan Robotics Atlas [9], Colossus [10] or the CFAM Prime [11].

17.2.2 Powder fusion

In the selective laser sintering (SLS) process, 3D parts are generated by solidifying successive layers of powder material on top of each other. Solidification is obtained by fusing or sintering selected areas of the successive powder layers using thermal energy supplied through a laser beam [12].

Polymers were the first and are still the most widely applied materials in SLS. An obvious condition for SLS polymer powders is that the particles tend to fuse or sinter when heat is applied. Further, the shape and size of the individual particles determines the behavior of the resulting powder to a great extent. Ideally, powders should have a high sphericity to facilitate a good flowability. The average particle size is recommended to be less than half of the build layer thickness, which is approximately 100–150 µm for SLS. The most optimal particle size distribution (PSD) however is still under discussed. Drummer et al. [13], for example, mention a PSD between 1 and 100 µm and a mean diameter larger than 10 µm. A review of different studies is given by Berreta et al. [14].

These specific criteria are the reason that only a few polymer materials are commercially available to date: polycarbonate (PC) and PA. Amorphous PC is able to produce parts with very good dimensional accuracy. However, because PC can only partially consolidate, the parts are only useful for applications that do not require strength and durability. Semi-crystalline PA on the other hand, can be sintered to fully dense parts with good mechanical properties [10]. Because of its relatively large temperature interval between the crystallization and melting point, premature crystallization (which leads to distortion of the printed part) and melting of particles in the direct neighborhood of the laser beam pattern (which leads to reduced resolution) can be avoided [15]. A wide variety of SLS printers are available on the market: from small-scale office printers (e.g., Sinterit Lisa, Vit Natural Robotics) to fully industrial printers (e.g., EOS P500, Promaker P1000) [16].

The Binder Jetting technique is similar to SLS, but instead of using thermal energy, sinterable powders are bound together with liquid binders. Example of 3D printers are the ExOne and the VoxelJet. Typical polymers are again PA and PC, and sometimes ABS [17].

17.2.3 Photopolymerization techniques

In stereolithography (SLA), a liquid photopolymer is selectively cured by UV light. A similar technique is digital light processing (DLP). It represents a sort of evolution of the SLA technique, using a projector screen instead of a laser. Examples of commercial SLA printers are iPro and ProJet from 3D systems, Form1 + from Formlabs or DWS from DigitalWax. Examples of DLP printers available on the market are Perfactory from EnvisionTec, Titan 1 from Kudo3D, and many more . . . [18]

The Polyjet technique from Stratasys is a special case of photopolymers-based printing: an inkjet head with several hundred nozzles is swept along the x-axis and ejects small droplets of photopolymer. After deposition of one layer, a UV-lamp flash cures the fresh layer and the process is repeated. The main advantage is that, with the use of multiple inkjet heads, it is possible to build multimaterial or multicolor structures, which is much more difficult in the case of SLA and DLP [4].

To summarize the main advantages and disadvantages of each technique, it can be said that both powder fusion and photopolymerization can reach a high level of complexity and accuracy, compared to material extrusion. This also implies that the equipment and materials are more expensive and complex, while material extrusion equipment is often cheaper and quite easy to use.

Another important advantage of material extrusion is the great variety of materials, while the other techniques are more limited in material choice, as they require special powders and photopolymers, which can also be more harmful for health and environment.

17.3 Linking 3D printing and textiles

The link between 3D printing and textiles can be very broad. The first thing that typically comes to mind is the *3D printing of garments*. Pioneering work in this field was done by Evenhuis and Kyttanen [19]. Their first printed dress, named the Draped Dress, is a complex model of interwoven links, resembling chainmail structures as used for armor in the Middle Ages [20]. To produce this structure, SLS technology was applied.

More recent examples include the Dutch designer Iris Van Herpen, known for her organic-inspired clothes, and the Israeli designer Danit Peleg, known for the dress Amy Purdy wore on the Paralympics Opening Ceremony in Rio in 2016 (see Figure 17.2). Danit Peleg's work is based on the FFF technique, using flexible filaments. Iris Van Herpen used Stratasys' Objet technology, working together with o.a. 3D printing innovator Materialise [21, 22].

These garments are still limited to the world of Haute Couture, o.a. because of their complex geometries, for which thousands of tiny elements have to be produced. Geometry is the main focus of these garments, not the development of new polymeric materials.

Another emerging trend is *3D printing onto textiles*. By directly 3D printing functional add-ons onto textiles, a new dimension can be given to customization of textile products. Last years, there is a clear trend in the European textile industry toward the production of products with higher added value to distinguish from large-scale standard production in low-wage countries. In sportswear & workwear, for example, there is an increasing demand for custom-fit products improving the

Figure 17.2: 3D-printed dresses designed by Danit Peleg (photographers: Daria Ratiner – Marina Ribas).

wearer's comfort. By 3D printing, protective parts or reinforcements that fit perfect to the body (e.g., elbow protection) can be incorporated into the textile. The sector also witnesses a high need of optimized products for medical and assistive applications with supporting or corrective parts. AM has the advantage of a very high freedom in design, making it suitable for unique structures.

The technique applied to print on textiles is almost exclusively material extrusion. The other techniques, based on the use of laser beams and/or liquid polymers, are less suitable to apply directly on fabric.

The main parameter to be taken into account when 3D printing directly on textiles is the adhesion between the printed polymer and the textile. This parameter is influenced by several factors, such as the type of polymer (e.g., TPU vs. thermoplastic styrenic elastomer), the material composition of the textile (e.g., polypropylene vs. PA), the textile structure (e.g., an open or dense knit) or the textile coating (e.g., untreated vs. rubber coated).

3D printing onto textiles is still in its development stage. Several research institutes are working on this topic and further exploring the potential: for example, Centexbel [23], the Research Institute for Textile and Clothing (FTB) of Hochschule Niederrhein [24, 25], RWTH Aachen [26], Fachhochschule Bielefeld [27, 28], . . .

It's a small leap of thought to go from 3D printing on textiles to smart textile applications. Smart textiles are able to sense and respond to changes in their environment, providing the wearer with increased functionality. Some main application fields are safety and protection (e.g., personal protective equipment (PPE)), medical support (e.g., monitoring and assistance for the elderly) and performance enhancing (e.g., fitness and sport). Basically, a smart textile product exists of four components: sensors, communication paths, a data processor and a power source. The challenge is to integrate these components into the textile. Connection between the electronic components, for example, is still challenging, because typical connection methods used in electronics such as soldering and bonding, cannot be used or are difficult to use on textiles [29].

The rapidly emerging 3D printing technology, with equipment becoming more and more affordable even for small companies, can be very promising in the area of smart textiles but is still in its infancy. Hereby some examples of recent developments:

Researchers from Tsinghua University in Beijing developed a 3D printer that deposits electronic flexible fibers onto textiles [30]. They used two different inks – a carbon nanotube (CNT) solution to build the conductive core of the fibers and silkworm silk for the insulating sheath. Injection syringes filled with the inks were connected to the coaxial nozzle, which was fixed on the 3D printer. This technique can bypass the conventional method of sewing electrical components, such as LED fibers, into fabrics. Another example is the research from Fachhochschule Bielefeld, which is focused on the use of 3D-printed connections to connect small electronic components on partly conductive textiles [26]. They made use of the FFF technology and conductive filaments. Swedish researchers from the university of Borås used this technology to print conductive features onto textile [31]. Also Centexbel and FTB are working together on preliminary research within this field, using filament and pellet-based 3D printing.

The availability and development of conductive materials for 3D printing is discussed in more detail in the next section.

17.4 Particle-enhanced materials in 3D printing

The link between 3D printing and textiles brings us to the emerging trend toward particle-enhanced polymer materials: there is a growing need for novel materials with enhanced properties to allow the manufacturing of functional products.

In material extrusion techniques, for example, particles are added for several reasons. Sometimes only for the visual aspect, for example, metal powders (steel, copper, bronze) for a metallic look, wood or cork powders for a natural look [32], but in other cases also to obtain special properties, such as improved mechanical strength or conductivity.

To improve the *mechanical strength* of 3D-printed parts, short fibers are typically used. Ideally, the fiber length is above the critical fiber length, which defines if the fiber is long enough to act as reinforcement in the polymer matrix. This value depends on the nature of the fiber and the matrix. Further, it has to be taken into account that mixing of fibers and polymers during extrusion leads to fiber breakage, reducing the final length of the fiber in the filament. Typically, the final fiber length varies between 0.1 and 1 mm.

In the review of Brenken et al., [33] different studies using discontinuous fibers are summarized, including a.o. chopped carbon fibers, glass fibers and silicon carbide. Further, with the trend toward a more sustainable society, development of natural fiber reinforced composites is gaining significant attention. Examples of short natural fibers used in FFF are jute, flax, hemp and wood [34].

In addition to short fibers, 3D printing with continuous fibers is being advanced as well. However, this requires special equipment, in which the fibers are coated in a curing agent, laid down into a thermoplastic matrix and extruded via a secondary nozzle. The best-known 3D printer is probably MarkForged, which is capable of printing continuous carbon fiber and Kevlar reinforced parts [35].

Perhaps more interesting for textile applications, in particular smart textiles, is conductivity. To introduce conductive features, the polymers are incorporated with conductive fillers. An important phenomenon related to conductivity within polymers is percolation: the sudden increase of conductivity when the concentration of filler exceeds a critical value. This critical value allows the filler particles contact one another and form a continuous path for electrons to travel through. Typical conductive fillers are carbon-based particles (e.g., carbon black, activated carbon, graphene, graphite, CNT), metallic powders (e.g., silver and copper,) and inherent conductive polymers (e.g., polyaniline). Most metal powders fall in the range of 0.5–100 μm [36]. Polyaniline has a typical particle size of 50–100 μm after synthesis, but after processing, for example, blending with other polymers, the size is often reduced to 10 μm or less [37].

Carbon-based fillers are most widely used to create conductive polymers. Carbon black is the cheapest material, but it has a quite high percolation threshold. Graphene requires lower concentrations but is more expensive and can experience problems during processing because of their clustering behavior. Carbon allotropes with a high aspect (length-to-diameter) ratio like CNTs are even more expensive; however, they benefit from very low percolation thresholds.

Commercially available conductive materials for extrusion-based 3D printing, more specifically FFF, are generally made from PLA filled with carbon black (Protopasta [38]), graphene (Black Magic 3D [39]) or undeclared carbon additives offered by some small distributors (e.g., FiloAlfa [40], 3R3D Technology Materials [41]). However, conductive polymers for 3D printing are still limited and require further development: there is a much broader range of conductive fillers yet to be explored.

For the SLS (powder fusion) technique, only a few filled systems are commercially available. Although in theory any polymer available in powder form can be processed by SLS, practice is still far from this situation. This is especially the case for filled material powders. Depending on the type of laser consolidation applied, the powder may not always be consolidated to full density or may even not be processable.

The main filled materials on the market are Alumide, an aluminum-filled PA12, or DuraForm HST, which is PA12 filled with glass fibers [42]. The addition of aluminum results among others in a metallic appearance, increased stiffness and increased thermal conductivity. The addition of glass fibers provides high stiffness and elevated temperature resistance.

The commercial availability of conductive materials is also very limited. An example of a material with conductive properties is carbon fiber-reinforced plastic (Carbon SLS) developed by Graphite AM [43], but little information is available on its composition. It is however the subject in several scientific studies. Athreya et al. [44], for example, analyzed carbon black as a filler. With a concentration of 4 vol%, an increase of the electrical conductivity five times as high as pure PA12 could be realized. Eshraghi et al. [45] looked into graphite nanoplatelets as a filler for PA12. With a concentration of 5 wt%, they observed an increase in conductivity by 3 orders of magnitude.

In photopolymerization techniques, particle-enhanced polymers are even less common. Fillers highly influence the refractive index of the photosensitive system, which is a crucial factor for absorbance and refraction behavior, and therefore curing depth [46].

Some examples from scientific studies: several fillers such as CNTs [47], graphene [48] and nanoclays [49] have been incorporated in photopolymer resin to enhance its mechanical properties, all resulting in increased strength and stiffness but reduced elongation. Conductive resins are usually obtained by adding (multiwall) CNTs to a thiolacrylate resin or a polystyrene matrix. Other fillers are silver nanoparticles, and more rarely, nanoparticles of metal oxides and metal alloys, such as barium titanate, palladium–copper or silver–copper, or the highly conductive polymer poly(3,4-ethylenedioxythiophene (PEDOT) (Table 17.2) [50].

Table 17.2: overview of fillers used in 3D printing.

Technique	Function	Examples
Material extrusion	Visual aspect	Metal powders Cork powders
	Mechanical properties	Short fibers (carbon, glass, natural fibers) Continuous fibers (carbon, Kevlar)
	Thermal and electrical conductivity	Carbon-based fillers, metallic powders Inherent conductive polymers
Powder fusion	Visual aspect Increased stiffness Thermal conductivity	Aluminum-filled PA12 (alumide)
	High stiffness Temperature resistance	PA12 with glass fibers (Duraform HST)
Photopolymerization	Mechanical properties	CNTs, graphene, nanoclays
	Electrical conductivity	Silver nanoparticles Nanoparticles of metal oxides/alloys

Summarized, we can conclude that at a commercial level, particle-enhanced print materials are most established for the material extrusion techniques. This is mainly related to the freedom in material choice coupled with this printing technology: it is possible to use a wide range of polymers, not being limited to the materials provided by the 3D printer manufacturer. For powder fusion and photopolymerization techniques, particle-enhanced materials are largely still at research level. Their availability on the market is low or even nonexisting.

However, also for material extrusion techniques, more research is ongoing (and required), in order to develop smart and functional materials, offering new possibilities toward electrical, thermal and mechanical properties. Since material extrusion techniques are also the most popular ones for use in combination with textile applications, this is an additional incentive for further progress and innovations.

References

[1] Wohlers, T. T.; Campbell, I.; Diegel, O.; Huff, R.; Kowen, J.: Wohlers Report 2019: 3D Printing and Additive Manufacturing State of the Industry, Wohlers Associates Inc, 2019.

[2] Ngo, T. D.; Kashani, A.; Imbalzano, G.; Nguyen, K. T. Q.; Hui, D.: Additive manufacturing: A review of materials, methods, applications and challenges. Composites Part B, 2018; 143, 172–196. DOI: 10.1016/j.compositesb.2018.02.012.

[3] https://www.printspace3d.com/3d-printing-processes/ [retrieved 6-01-2020].

[4] Ligon, S. C.; Liska, R.; Stampfl, J.; Gurr, M.; Mülhaupt, R.: Polymers for 3D printing and customized additive manufacturing. Chemical Reviews, 2017; 117(15), 10212–10290. DOI: 10.1021/acs.chemrev.7b00074.

[5] Gonzalez-Gutierrez, J.; Cano, S.; Schuschnigg, S.; Kukla, C.; Sapkota, J.; Holzer, C.: Additive manufacturing of metallic and ceramic components by the material extrusion of highly-filled polymers: A review and future perspectives. Materials, 2018; 11(5), 840. DOI: 10.3390/ma11050840.

[6] https://www.arburg.com/nl/nl/prestatiespectrum/additieve-productie/systeem-freeformer [retrieved 6-01-2020].

[7] https://www.arburg.com/products-and-services/additive-manufacturing/ [retrieved 06-01-2020].

[8] https://www.pollen.am/ [retrieved 06-01-2020].

[9] https://titan3drobotics.com/atlas/ [retrieved 06-01-2020].

[10] https://colossusprinters.com/printers/ [retrieved 06-01-2020].

[11] https://ceadgroup.com/solutions/gantry-based-solutions/cfam-prime/ [retrieved 06-01-2020].

[12] Kruth, J. P.; Wang, X.; Laoui, T.; Froyen, L.: Lasers and materials in selective laser sintering. Assembly Automation, 2003; 23(4), 357–371. DOI: 10.1108/01445150310698652.

[13] Drummer, D.; Medina-Hernández, M.; Drexler, M.; Wudy, K.: Polymer powder production for laser melting through immiscible blends. Procedia Engineering, 2015; 102, 1918–1925. https://doi.org/10.1016/j.proeng.2015.01.332

[14] Berretta, S.; Ghita, O.; Evans, K. E.: Morphology of polymeric powders in Laser Sintering (LS): From polyamide to new PEEK powders. European Polymer Journal, 2014; 59, 218–229. https://doi.org/10.1016/j.eurpolymj.2014.08.004
[15] Schmid, M.; Amado, A.; Wegener, K.: Polymer powders for selective laser sintering (SLS). AIP Conference Proceedings, 1664, 2015; https://doi.org/10.1063/1.4918516
[16] https://www.3dnatives.com/en/different-sls-3d-printers-220320184/ [retrieved 9-01-2020].
[17] https://www.lboro.ac.uk/research/amrg/about/the7categoriesofadditivemanufacturing/binderjetting/ [retrieved 9-01-2020].
[18] https://all3dp.com/2/stereolithography-3d-printing-simply-explained/ [retrieved 9-01-2020].
[19] Kyttanen, J.; Evenhuis, J.: Method and device for manufacturing fabric material. Patent WO2003082550 A22003, 2003.
[20] Lussenburg, K.; Van der Velden, N.; Doubrovski, Z.; Geraedts, J.; Karan, E.: Designing with 3D Printed Textiles: A case study of Material Driven Design. Paper presented at the International Conference on Additive Technologies, Vienna, 2014.
[21] Parraman, C.; Segovia, M. V. O.: 2.5D Printing: Bridging the Gap Between 2D and 3D Applications, 1st ed. Chichester, UK: John Wiley and Sons Ltd, 2018.
[22] https://www.materialise.com/en/cases/iris-van-herpen-debuts-wearable-3d-printed-pieces-at-paris-fashion-week [retrieved 10-01-2020].
[23] Deleersnyder, K.: Monofilament Development for 3D Print Technology and Implementation on Textiles. Presentation at the Global Fiber Congress, Dornbirn, 2016.
[24] Glogowsky, A.; Kletter, I.; Bretz, I.; Eloo, C.; Jelen, E.; Korger, M. 3D Printed Plastic/Textile Composites Modifications for Technical Applications. Presentation at the Global Fiber Congress, Dornbirn, 2018.
[25] Glogowsky, A.; Korger, M.; Meyer, J.; Steinem, C.; Sanduloff, S.; Raibe, M.; Ernst, M.: 3D Printing Potentials for Textile Applications – From Digital Model to Customized Production. Presentation at the ITMA Innovation Lab Speakers' Platform, 2019.
[26] Schmelzeisen, D.; Koch, H.; Pastore, C.; Gries, T.: 4D Textiles: Hybrid Textile Structures that Can Change Structural Form with Time by 3D Printing. In: Kyosev, Y.; Mahltig, B.; Schwarz-Pfeiffer, A., editors: Narrow and Smart Textiles. Cham, Switzerland: Springer International Publishing, 2018; pp. 189–201.
[27] Döpke, C.; Martens, Y.; Grimmelsmann, N.; Ehrmann, A.: 3D-Printing on Textile Fabrics. In: Mahltig, B., editor: Textiles: Advances in Research and Applications. New York: Nova Science Publishers, 2017.
[28] Grimmelsmann, N.; Lutz, M.; Korger, M.; Meissner, H.; Ehrmann, A.: Adhesion of 3D printed material on textile substrates. Rapid Prototyping Journal, 2018; 24(1), 166–170. DOI: 10.1108/RPJ-05-2016-0086.
[29] Grimmelsmann, N.; Martens, Y.; Schäl, P.; Meissner, H.; Ehrmann, A.: Mechanical and electrical contacting of electronic components on textiles by 3D-printing. Procedia Technology, 2016; 26, 66–71. DOI: 10.1016/j.protcy.2016.08.010.
[30] https://www.innovationintextiles.com/smart-textiles-nanotechnology/3d-printer-deposits-electronic-fibres-onto-fabrics/ [retrieved 13-01-2020].
[31] https://www.technicaltextile.net/news/bor-s-university-s-new-method-for-printing-on-textiles-252584.html [retrieved 13-01-2020].
[32] https://colorfabb.com/filaments/specials-filaments [retrieved 13-01-2020].
[33] Brenken, B.; Barocioa, E.; Favaloroa, A.; Kunc, V.; Pipes, B.: Fused filament fabrication of fiber-reinforced polymers: A review. Additive Manufacturing, 2018; 21, 1–16. DOI: 10.1016/j.addma.2018.01.002.

[34] Balla, K. V.; Kate, H. K.; Satyavolu, J.; Singh, P.; Tadimeti, J. G.: Additive manufacturing of natural fiber reinforced polymer composites: Processing and prospects. Composites Part B, 2019; 174, 106956. DOI: 10.1016/j.compositesb.2019.106956.
[35] https://markforged.com/learn/3d-printing-carbon-fiber-and-other-composites/ [retrieved 10-01-2020].
[36] https://www.horiba.com/uk/scientific/products/particle-characterization/applications/metal-powders/?tx_feloginhoriba_pi1%5Bforgot%5D=1
[37] https://cordis.europa.eu/project/id/BREU0469/fr [retrieved 31-08-2020].
[38] https://www.proto-pasta.com/pages/conductive-pla [retrieved 10-01-2020].
[39] https://www.blackmagic3d.com/ [retrieved 10-01-2020].
[40] https://www.filoalfa3d.com/en/filaments-175mm/171-conductive-pla-o-175-mm-8050327032354.html [retrieved 10-01-2020].
[41] https://www.3r3dtm.com/producto/conductive-pla [retrieved 10-01-2020].
[42] Lanzl, L.; Wudy, K.; Greiner, S.; Drummer, D.: Selective laser sintering of copper filled PA12: Characterization of powder properties and process behavior. Polymer Composites, 2019; 40(5), 1801–1809. DOI: 10.1002/pc.24940.
[43] https://www.graphite-am.co.uk/materials/materials/ [retrieved 13-01-2020].
[44] Athreya, S.; Kalaitzidou, K.; Das, S.: Processing and properties of carbon blackfilled electrically conductive nylon 12 nanocomposites produced by selective laser sintering. Materials Science and Engineering A, 2010; 527(10–11), 2637–2642. DOI: 10.1016/j.msea.2009.12.028.
[45] Eshraghi, S.; Karevan, M.; Kalaitzidou, K.; Das, S.: Processing and properties of electrically conductive nanocomposites based on polyamide-12 filled with exfoliated graphite nanoplatelets prepared by selective laser sintering. International Journal of Precision Engineering and Manufacturing, 2013; 14(11), 1947–1951. DOI: 10.1007/s12541-013-0264-y.
[46] Baumgartner, S.; Pfaffinger, M.; Busetti, B.; Stampfl, J.: Comparison of dynamic mask- and vector-based ceramic stereolithography. Ceramic Engineering and Science Proceedings, 2017; 38(3), 163–173. DOI: 10.1002/9781119474746.ch16.
[47] Eng, H.; Maleksaeedi, S.; Yu, S.; Choong, Y. Y. C.; Wiria, F. E.; Kheng, R. E.; Wei, J.; Su, P.; Tham, H. P.: Development of CNTs-filled photopolymer for projection stereolithography. Rapid Prototyping Journal, 2017; 23(1), 129–136. DOI: 10.1108/RPJ-10-2015-0148.
[48] Manapat, J. Z.; Mangadlao, J. D.; Tiu, B. D. B.; Tritchler, G. C.; Advincula, R. C.: High-strength stereolithographic 3D printed nanocomposites: Graphene oxide metastability. ACS Applied Materials & Interfaces, 2017; 9(11), 10085–10093. DOI: 10.1021/acsami.6b16174.
[49] Weng, Z.; Zhou, Y.; Lin, W.; Senthil, T.; Wu, L.: Structure-property relationship of nano-enhanced stereolithography resin for desktop SLA 3D printer. Composites Part A, 2016; 88, 234–242. DOI: 10.1016/j.compositesa.2016.05.035.
[50] Scordo, G.; Bertana, V.; Scaltrito, L.; Ferrero, S.; Cocuzza, M.; Marasso, S. L.; Romano, S.; Sesana, R.; Catania, F.; Pirri, C. F.: A novel highly electrically conductive composite resin for stereolithography. Materials Today Communications, 2019; 19, 12–17. DOI: 10.1016/j.mtcomm.2018.12.017.

V Toxicology/safety/ecological toxicity and environmental impact

Dana Kühnel*, Harald F. Krug, Andreas Mattern, Anita Jemec Kokalj

18 Human and environmental hazard of nanomaterials used in textiles

Keywords: nanoparticles, nanomaterials, fibers, hazard, release, exposure, uptake, human health, environmental health, safety, toxicity, ecotoxicity, fate, ion-releasing nanomaterials, non-soluble nanomaterials, carbon nanomaterials, carbon nanotubes, CNT, aspect ratio, WHO fiber dimension, life cycle, risk, risk assessment, adverse effects, oxidative stress, physicochemical properties, humans, daphnia, algae, fish

18.1 Introduction

As outlined in the previous chapters, various engineered nanomaterials and nanotechnological processes are used in textiles to achieve tailored functionalities. This concerns wearable fashion textiles as well as various technical textiles, such as geotextiles and agricultural textiles. Depending on the intended use of a textile product, the incorporation of the nanomaterials is realized in different ways, which influences a potential release in quantity as well as with regard to the product life phase when the release is most likely to occur (overview in Figure 18.1, [1]).

The safety of nanomaterials in textile applications is assessed according to various legal as well as regulatory frameworks. Details on risk assessment procedures that usually consider exposure and hazard (and provide guidance on risk management) are laid out in detail in Chapter 7. This chapter is focused on summarizing the state of knowledge regarding the potential hazards of nanomaterials most frequently used in textile applications. The information was retrieved from a publicly accessible

Acknowledgments: This work was partly funded by the BMBF, grant number FKZ 03XP0282 (project DaNa4.0), and the Slovenian Research agency and DAAD bilateral exchange agreement: BI-DE/20-21-008.

Note: Dana Kühnel and Harald F. Krug contributed equally to the manuscript.

***Corresponding author: Dana Kühnel,** Department of Bioanalytical Ecotoxicology, Helmholtz Centre for Environmental Research – UFZ, Permoserstrasse 15, D-04318 Leipzig, Germany, e-mail: dana kuenel@ufz.de
Andreas Mattern, Department of bioanalytical ecotoxicology, Helmholtz Centre for Environmental Research – UFZ, Permoserstrasse 15, D-04318 Leipzig, Germany, e-mail: andreas.mattern@ufz.de
Harald F. Krug, NanoCASE GmbH, St. Gallerstr. 58, 9032 Engelburg, Switzerland
Anita Jemec Kokalj, Department of Biology, Biotechnical Faculty, University of Ljubljana, Večna pot 111, 1000, Ljubljana, Slovenia

https://doi.org/10.1515/9783110670776-018

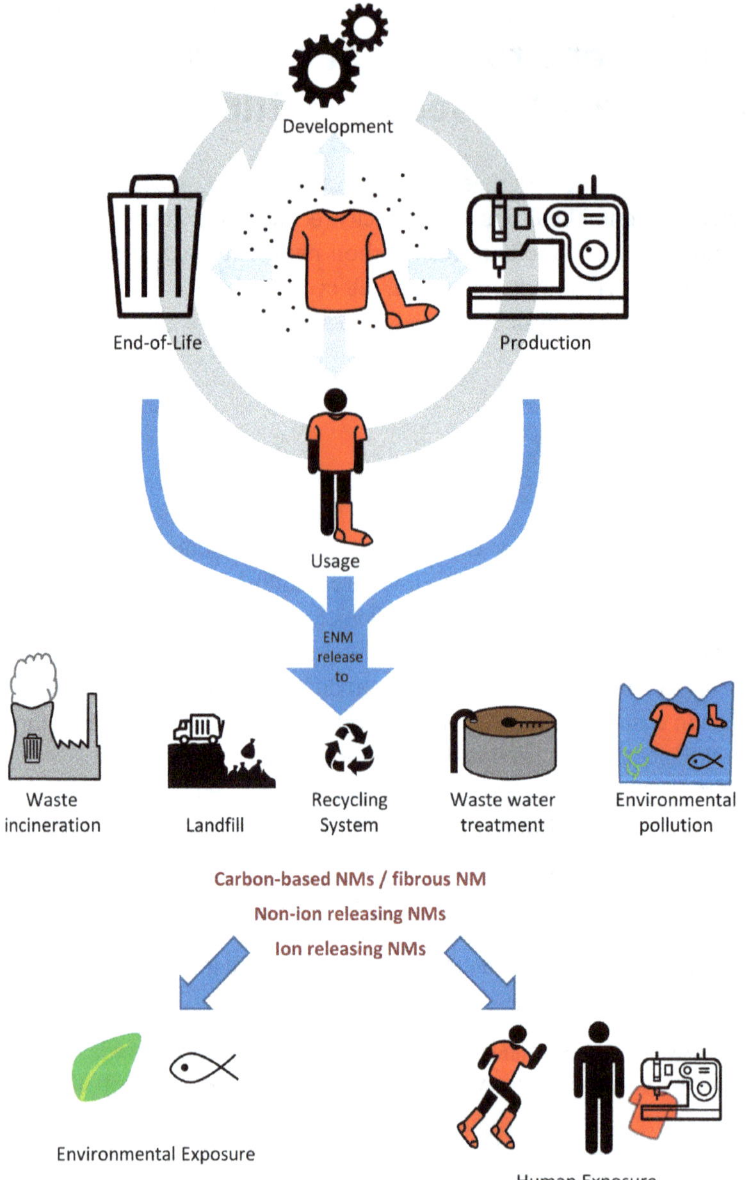

Figure 18.1: Setting the scene: Nano-enabled textiles may release nanomaterials during different stages of their life cycle, which leads to exposure of humans and the environment. This chapter deals with potential adverse effects arising from nanomaterials released from textiles and a rough assessment of potential hazards (NM - nanomaterial).

knowledge base, www.nanoobjects.info, which provides quality approved information on the hazard of various nanomaterials for human and environmental health. For the purpose of this chapter, information on the following nanomaterials was collected: silver (Ag), zinc oxide (ZnO), copper (Cu) and copper oxide; titanium dioxide (TiO_2), silicon dioxide (SiO_2), aluminum oxide (Al_2O_3), gold (Au), specific types of clay and carbon nanotubes (CNT), as well as carbon black (CB).

While the processing of nanomaterials in textiles is often designed to result in minimal release during use, release at other life cycle stages is more likely [2–5]. Upon release, the likelihood of exposure and uptake of textile-borne nanomaterials by humans or living organisms within the environment increases. For human beings, dermal contact to wearable textile products and inhalation of nanomaterial dust is the most probable. Aquatic organisms may be exposed via wastewater effluents that contain nanomaterials and/or ions released by abrasion during washing. Wastewater effluents may also contain nanomaterials released by textile industry. In soil, organisms are exposed to nanomaterials retained in wastewater sludge that is applied onto soil as fertilizer as well as through direct use of geotextiles and textiles used in agriculture as mulching [6]. For all these cases, it is plausible that after degradation or abrasion scenarios humans and the environment are exposed to the degradation products which may contain nanomaterials.

In general, ion-releasing materials are often more toxic than insoluble ones [7], nanofibers exert probably a higher potential for adverse effects than nanoparticles because of the slower clearance from human lungs upon inhalation [8], and surface properties may determine adverse effects [9, 10].

The most frequently used nanomaterials in textile applications and the respective purposes, as well as their categorization according to their physicochemical properties are shown in Table 18.1.

The exposure to textile-borne nanomaterials may impact human and environmental health. Two conditions need to be met: the respective nanomaterial (I) is able to trigger damage in organisms, meaning that it has a hazardous potential, and (II) is present in concentrations high enough to cause harm. The hazard potential of specific nanomaterials, for example, nanosilver [14], silicon dioxide [15] or TiO_2 [16], is assessed for decades now [17, 18]. For most of the nanomaterials applied in textile products and applications the hazard is already well characterized. It became clear that “nano” does not mean “toxic” per se and each of the materials must be considered separately for their possible effects on environment and health. Critical nanomaterial properties have been identified that translate into adverse effects in humans and the environment, as well as mechanisms typically leading to a biological damage. The relevant properties and as well as the mechanisms of action differ for humans and environmental organisms not only due to different exposure routes and conditions, but also due to different biology, behavior and habitat. This chapter aims to provide a summary on the potential hazard caused by the most frequently used nanomaterials in textile applications (see Table 18.1).

Table 18.1: Overview of nanomaterials probably used in textiles for various purposes (adopted from [11, 12*]).

Nanomaterial	Purpose of application
Nonsoluble materials: release mainly by material abrasion, and at the end-of-life phase	
Aluminum oxide (Al_2O_3)	Increased elasticity and resistance to fracture of fibers, heat protection
Gold (Au)	Electrically conductive fibers, fashion
Magnesium oxide (MgO)	Antimicrobial and antifungal effect, auxiliary agent, heat stability
Silicon dioxide (SiO_2)	Hydrophobic surface functionalization, increased wear and tear resistance, nanoencapsulation of molecules
Nanoclays (layered silicates)	UV protection, heat protection, flame retardancy, gas diffusion barrier, dye-carrier in synthetic textiles
Titanium dioxide (TiO_2)	UV protection, photocatalytic antibacterial and anti-dirt activity, anti-odor effect, hydrophilic surface functionalization
Carbon black (CB)	Black pigment, filler additive, electrical conductive fibers
Ion-releasing materials: release of ions and particles by abrasion, and at the end-of-life phase	
Silver (Ag)	Anti-odor effect, antibacterial textiles, fashion
Copper oxide (CuO)	Anti-microbial effect
Zinc oxide (ZnO)	UV protection, anti-microbial textiles, water repellent textiles, piezoelectric energy harvesting for smart textiles
Fibrous materials: release of fibers by abrasion, and at end-of-life phase**	
Carbon nanotubes (CNT)	Increased tensile strength of fibers, electrical conductive fibers, electrostatic dissipation, flame retardancy
Carbon fibers	Increased tensile strength of fibers

For humans, exposure via skin and lung is most likely. For the environment, the exposure situation is more complex. In general, exposure via the freshwater and terrestrial environment is more likely compared to marine environments.

*The original survey was conducted by C. Som et al. The purpose of the survey was to assess which nanomaterials will be applied in textiles in the near future [12, 13].

**There are also fibers made of inorganic nanomaterials, but to our knowledge those are not applied in textiles.

18.2 Hazard of textile-related nanomaterials for human health

18.2.1 General mechanisms of action

An overview of the known adverse effects of nanomaterials in textile applications is given in Table 18.2. The most relevant characteristics that drive a subsequent toxicity of nanomaterials are the release of toxic ions (e.g., silver ions – Ag^+), a critical crystal structure (e.g., anatase vs. rutile for titanium dioxide – TiO_2) and fiber-like morphologies (e.g., certain types of CNT). With regard to the health effects, different so-called end points have been considered, for example, acute toxicity, the ability of the nanomaterial to cross barriers in the body such as the gastrointestinal tract, or potentially genotoxic action. The nanomaterials show a wide variability in their characteristics, which translates into a quite different potential to initiate health effects (Table 18.2).

In recent years, there has been a tremendous effort to rank nanomaterials hazard according to their physicochemical properties [19–21]. Based on current knowledge on physicochemical properties, the nanomaterials relevant in textile applications can be classified according to their potential human hazard as demonstrated in Table 18.2.

18.2.2 Adverse effects of nanomaterials in textile applications on humans

18.2.2.1 Uptake and internalization, tissue barriers

A prerequisite for the toxic action of a nanomaterial is the uptake into the human body and its distribution inside the body. The human body has a number of barriers whose task is to selectively exchange substances with the environment, for example, oxygen and carbon dioxide at the air–blood barrier in the lung. Several studies on the potential of nanomaterials to cross these body barriers have been performed, as there is concern that the smallness and large surface area of the particles may foster transfer across barriers. The evidence for TiO_2, SiO_2 and CB to cross tissue barriers and damage these is valid for crossing the air–blood barrier and to some extent to cross the gastrointestinal tract barrier, but there is no evidence so far for a damage of this barrier tissue. In addition, many projects could demonstrate that there is no penetration of healthy skin by any of the nanomaterials [22]. The translocation of nanoparticles across the blood–brain barrier is very low [23]. For none of the listed nanomaterials there is evidence for a brain damaging effect. For CNT, the diversity of this material needs consideration. Long and rigid fibers exhibit a potential carcinogenic effect in the lungs similar to asbestos fibers due to the effect of frustrated phagocytosis [24].

Table 18.2: Assessment of different nanomaterials in relation to their human health effects based on published results with respect to different biological endpoints (adapted from [5]).

	Ag	ZnO	CuO	Al_2O_3[1]	Au	TiO_2	SiO_2 amorph.	Clay[3]	CNT	CB
Characteristic										
Ion release	+	+	+	+	–	–	–	–	–[4]	–
Critical crystal structure	–	–	–	–	–	±	–	–	–	–
Surface/catalytical activity	+	+	+	–	±	+	+	–	–	–
Critical morphology	–	–	–	–	–	–	–	±	+ (fiber)	–
Health effect										
Acute toxicity	–	+	+	–	––	––	––	––	±[4]	+
Chronic toxicity (long-term effects to be expected)	+	+	±	n.a.	±	±	–	––	+[4]	++
DNA damage	–	+	–	n.a.	–––	–	–	n.a.	–	+
Crossing and damaging tissue barriers	n.a.	n.a.	n.a.	–[1]	±	+	+	n.a.	–	+
Brain damage: damage of the central nervous system	n.a.	n.a.	n.a.	n.a.	n.a.	n.a.	n.a.	n.a.	n.a.	n.a.
Skin penetration and damage	––	––	––	n.a.	––	––	––	n.a.	–	–
Gastrointestinal tract damage	–	±	–	n.a.	–	–	–	–	–	–
Respiratory tract damage or inflammation	+	+	+	–	–	–	–	n.a.	+	+
Safety classification[2]	a	c	c	b	a	a	a	b	b	c

++ applies strongly to; + applies to; ± weak evidence available; – does weakly apply to; –– does not apply to; n.a. no data available (high uncertainty).

[1]AlOOH was explored in the lungs; [2]the original classification by Som et al. [5]: a – rather safe; b – uncertainty due to weak evidence; c – biological effects detectable; [3]specified as montmorillonite; [4]depends often on contaminants in the samples (especially transition metals such as iron, nickel and cobalt).

18.2.2.2 Non-soluble and inert nanoparticles – hazard

The non-soluble and inert nanoparticles belong to the group of granular biopersistent particles. These materials without known intrinsic toxicity may still induce health problems when inhaled. The biological response and the toxicokinetic behavior after inhalation are predominantly based on the surface (nanomaterials have a very high specific surface area) and volume (mass concentration). The most prominent effect of this material class is the induction of inflammatory processes in the lungs. In a situation of chronic exposure, for example, at the workplace, this may lead to inflammatory effects and subsequently to more severe health problems such as cardiovascular diseases or genotoxicity via oxidative processes. In that respect, the nanomaterials are well comparable to ambient ultrafine particles [25] and the main criterion for the induced adverse mechanisms is the biopersistence, meaning the nanoparticles deposit in the lungs and their uptake exceeds the clearance mechanisms [26].

18.2.2.3 Non-soluble and inert nanoparticles – linking hazard to release and exposure

Because adverse effects as described above are strongly dose-dependent and the overall dose for the onset is relatively high, the consumer will usually not be affected. However, this common response of the lungs to particle burden results in clear occupational exposure limits for the dust at workplaces, especially for the inhalable dust particles, so-called respirable fraction. This fraction consists of inhalable particles which may reach the alveolar space because of their smallness of less than 3 µm in diameter (for definitions and explanations, see [27]).

18.2.2.4 Soluble (ion-releasing) nanomaterials – hazard

The situation is different for metal particles or metal oxides which dissolve in an aqueous environment. It has been shown that silver as a particle or dissolved silver ions are released into artificial sweat in a substantial amount [28]. The same is true for metal oxides such as zinc or copper oxide. The common property of all these three materials is their antimicrobial activity, which makes them attractive to be incorporated into textiles but implies per se a biological effect. The biological activity is dependent on the amount and kinetics of metal ion release. Nevertheless, zinc and copper belong to the group of essential elements which means that several biological functions depend on the presence of these elements and humans rely on daily intake via the food. Zinc is needed for the regulation of transcription processes ("zinc-finger-proteins") but is involved in many other processes such as immune function, protein and DNA synthesis, and the activity of a multitude of enzymes.

Copper is essential for brain development and nerve cell communication, wound healing, some immune relevant processes and other biological mechanisms. Thus, both elements are not toxic at low concentrations but essential for human health. Overdosing, however, may lead to adverse effects [29, 30]. There is evidence that silver as well as the two metal oxides are released from finished textiles and can induce toxic effects in cells [3, 31–33].

18.2.2.5 Soluble (ion-releasing) nanomaterials – linking hazard to release and exposure

Nevertheless, for consumers the concentrations will not reach the toxic level simply by inhalation or skin exposure to silver, zinc oxide or copper oxide nanoparticles or their corresponding ions which are wear and tear particles during the use of their functionalized clothes. Again, workers in production plants are much more exposed to these materials and possibly the concentrations of dust particles in the air may reach critical concentrations.

18.2.2.6 Carbon (fibrous) materials including carbon nanotubes – hazard

Carbon fiber-reinforced polymers or CNT with specific properties are suspected to induce adverse effects comparable to those of asbestos [24, 34, 35]. Usually, foreign material which reaches the deep lung region will be removed by effective self-cleaning mechanisms, mainly by pleural macrophages that act against environmental pollutants in the lungs. The macrophages "eat" the foreign material (phagocytosis) and carry it out of the lung which is part of the lung clearance process. In the case of long and rigid fibers, such as asbestos fibers, the macrophages are not able to take up the fibers which lead to "frustrated phagocytosis" [24]. It is important to mention that this severe effect in humans is strongly dependent on the fiber length and rigidity of the material. The relevant parameters for this fiber-specific effect are the WHO fiber dimensions: >5 µm in length, the diameter is less than 3 µm, and the aspect ratio should be larger than 3:1 [36]. This has been demonstrated by comparing the effects of clearly defined silver nanowires, nickel nanowires and CNT [37].

18.2.2.7 Carbon (fibrous) materials including carbon nanotubes – linking hazard to release and exposure

Taken these studies into consideration a textile reinforced with carbon fibers or fibers made from other materials should never release fibers which exhibit the WHO-fiber dimensions. Although it was demonstrated that from CNT composites in consumer products, a significant release of CNTs is unlikely; thus, the exposure will be very low and to a non-pristine form of CNTs [38], the release of fiber-like material should be very low. On the other hand, during washing and mechanical stress abrasion and release may happen and CNTs are released in the environment, with subsequent exposure of organisms.

18.2.2.8 Where do humans get into contact with nanomaterials from textiles?

As already mentioned, workers and consumers have the highest likelihood to get into contact with nanomaterials released from textiles (Fig. 18.2).

During the last decade several workplace measurements have been conducted to describe the particle burden in the air at various workplaces [39–41]. Based on this, the occupational exposure limits have been discussed for a broad variety of nanomaterials relevant for textile manufacturing [42, 43]. These studies present data that very well confirm the above-mentioned difference between spherical particles and fibrous material (toxicity: fibers ≫ particles) and the higher toxic potency of ion-releasing materials compared to insoluble metal oxides. In any case the exposure via inhalation is the most relevant exposure pathway as the skin is a very effective barrier (Figure 18.2) [22].

For consumers, the situation is totally different to the workplace scenarios. Consumers will usually not be exposed to the pure nanomaterial but to abraded particles via dermal contact or inhalation. These wear and tear-related particles consist of a mixture of different materials such as polymers, natural fibers, finishing chemicals and integrated nanomaterials [44]. Accordingly, the particles released from fabrics are thus usually very different from the original nanomaterial embedded into the product and the total amount is very low over time, which leads to a low exposure of the consumer [45]. Accordingly, for consumers the ion-releasing nanomaterials can be considered the most relevant ones, and the exposure via the skin is assumed to be more relevant than inhalation (Figure 18.2). On the other side, the nanomaterials may be released during the washing of textiles into the environment leading to a general exposure of aquatic organisms possibly at a higher concentration level [46–48]. Effects toward environmental organisms will be further elaborated in the next section.

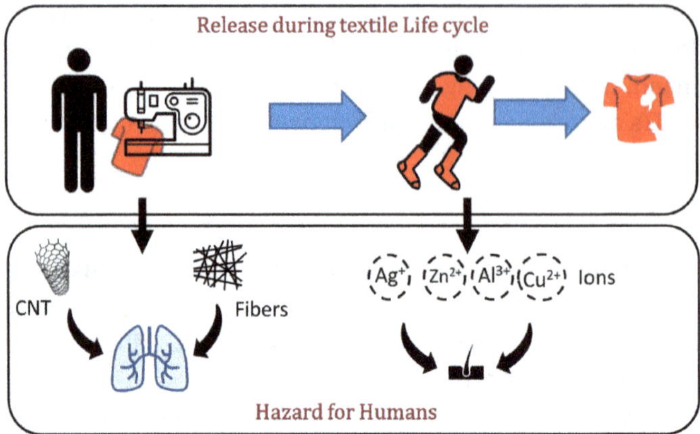

Figure 18.2: The most relevant exposure routes and mechanisms leading to hazardous effects of nanomaterials in humans. During the occupational exposure, inhalation of nanomaterials via the lung is most likely, whereas dermal exposure is most likely during the use phase of textiles. Here, toxic ions released by the nanomaterials are the main cause for toxicity.

18.3 Hazard of textile-related nanomaterials for environmental organisms

18.3.1 Categorization of engineered nanomaterials according to the mode of action and physicochemical properties

For most of the textile-related nanomaterials, there is a broad knowledge base regarding their effects on various terrestrial as well as aquatic environmental organisms [13, 18, 49–51]. From the numerous studies, several mechanisms of actions have been identified, as well as physicochemical properties of nanomaterials that are causative for adverse effects (see Table 18.3). Due to the diversity of organisms regarding their habitat, feeding strategies, size, behavior and other species-specific parameters the effects of nanomaterials may be quite different. Nevertheless, as for human health-related effects, a rough categorization of the textile relevant nanomaterials according to their properties and effects within the environment is possible [52, 53]. Interestingly, the resulting groups related to the two categories human health and environmental safety do not differ much. As demonstrated in a toxicity screening of seven different nanomaterials [49], the toxicity pattern for the nanomaterials was similar when comparing 14 different test species and cell lines. The toxicity decreased in the following order: Ag > ZnO > CuO > CNTs > Au > SiO_2 = TiO_2. According to existing test protocols, SiO_2 and TiO_2 were considered presumably safe (EC_{50} > 100 mg/L), but they were previously shown to have specific effects, such as body surface adsorption [54]. This latter end point is not regularly assessed in toxicity testing. Three nanomaterials proved toxic in all (Ag) or in the majority (ZnO and CuO) of assays. Hence, in summary, the toxicity of the nanomaterials engaged in the study was driven by two basic intrinsic properties of nanomaterials: solubility (Ag, CuO and ZnO) and aspect ratio (CNTs) [49], irrespective of whether human health or environmental health hazard was considered.

Accordingly, the main proposed mechanisms of toxic action of engineered nanomaterials for environmental organisms are:

- physical interaction of nanomaterials with organisms (e.g., [54]),
- radical production and oxidative damage,
- toxic effects of released ions from metal-based nanomaterials [49, 55, 56].

The following intrinsic properties of nanomaterials are hence important for categorization: release of ions, aspect ratio (morphology) and radical production [52]. For this reason, we have described the hazard of nanomaterials toward environmental organisms in the following groups: ion-releasing, non-ion-releasing (inert) and carbon-based nanomaterials.

Table 18.3: Assessment of different nanomaterials in relation to their potential to affect some of the most commonly measured end points in environmental organisms based on published results.

	Ag	ZnO	CuO	Al_2O_3	Au	TiO_2	SiO_2 amorph	Clay[1,3]	CNT[4]	CB
Characteristic										
Ion release	+	+	+	+	–	–	–	–	–	–
Surface/catalytical activity	+	+	+	–	+	+	+	–	–	–
Adverse effects										
Acute toxicity (growth inhibition, impaired mobility . . .)	+	+	+	–	+	–	–	+	–	–
Chronic/reproductive toxicity (long-term effects to be expected)	+	+	+	+	+	+	+	+	+	n.a.
Malformations (embryonic development)	+	+	+	–	–	–	–	–	+[5]/ –	n.a.
Reactive oxygen formation/oxidative stress, DNA damage, inflammation	+	+	+	+	+	+	+	+[3]/ –	+	–

Table 18.3 (continued)

	Ag	ZnO	CuO	Al_2O_3	Au	TiO_2	SiO_2 amorph	Clay[1,3]	CNT[4]	CB
Internalization, crossing body/tissue barriers	+[2]	+[2]	+[2]	–	+	–	–	–	+	+
Physical effects by attachment/ entrapment	+	+	+	n.a.	+	+	+	n.a.	+	+

– not commonly observed; + observed; n.a. no data available.
[1]Specified as montmorillonite; [2]mostly metal ions cross; [3]is dependent on the metal content of the clay, clays rich in Fe and Al are strong inductors of reactive oxygen species; [4]effects may be induced by remaining trace metals from synthesis (Co, Ni, Fe); [5]depends on fiber length, only observed for shorter fibers.

18.3.2 Hazard of ion-releasing nanomaterials for environmental organisms

The common denominator of these nanomaterials is the release of toxic metal ions, in this case Zn^{2+}, Ag^{+} and Cu^{2+}. These ions act as inhibitors of bacterial and fungal growth and are used for this purpose in textile applications to prevent hygienic and smell issues. At the same time, the unintentional release of ions to water during washing of textiles or contact with water is an important release path into the environment. All three nanomaterials are used in various textile applications mainly due to their antifungal, antiviral and antibacterial activities [57–59]. Colloidal silver, including formulations with silver nanoparticles, has been used commercially for almost 100 years, typically as a biocide [14, 58]. Copper compounds have been used for decades in paints for ship hulls to prevent the growth of algae, mussels and snails. Also, they have been proposed as nanopesticides [59]. Zinc compounds are often used as antimicrobial in personal medicine and cosmetics [60]. However, due to their antimicrobial properties they also affect the viability of beneficial bacterial communities such as for example in soil [61].

Besides the toxicity of Ag, ZnO and CuO via the released ions (Figure 18.3) [49, 55, 57] for all three additionally, particle-specific effects have been identified [62, 63]. This may be less relevant for textiles, as often the nanomaterials are tightly incorporated into the fabric, and release during use is less likely. A clear differentiation between the roles of dissolved ions and "nano-specific" effects in the observed toxic properties is often difficult to prove experimentally [57] and most probably the effect is a result of the interplay between different nanomaterial properties and ions.

Commonly reported modes of action at the cellular level for all three nanomaterials are oxidative stress [61, 64], destabilization of cell membrane [65], DNA damage and genotoxicity [63, 66]. For ZnO nanoparticles, specific toxic mechanisms could be attributed to their photocatalytic activity [56] leading to induction of oxidative damage [64, 67].

In general, Ag, ZnO and CuO nanoparticles rank as the first three most toxic nanomaterials in the environment (Ag>ZnO>CuO) [49, 55, 68]. The sensitivity of aquatic organisms to these metal-oxide nanomaterials differs widely, with crustaceans, algae and fish being among the most sensitive [49, 68]. Probably the most common test organisms in ecotoxicity testing are crustaceans such as water fleas. Many studies reported the effects of Ag [68], ZnO [69] and CuO nanoparticles [70] on water flea mobility. Since these organisms filter the water for food intake, nanoparticles enter directly into their bodies, but an efficient excretion from the gut was observed too [71]. Furthermore, some specific effects in water flea, like the ionoregulatory dysfunction, were reported in case of Ag nanoparticles [72]. For algae, a common consequence of nanomaterial exposure is a decrease in growth [70]. ZnO nanoparticles are absorbed onto algal cells resulting in shading effect [73]. In fish, cultured fish cells [74], embryos [75] or adult fish [76], accumulation of metals dissolved from nanoparticles was observed after Ag [77], ZnO [76] and CuO nanoparticle exposure [78]. Common effects for fish embryos are the attachment of nanomaterials on the chorion and delayed development, malformations of the skeleton and organs, as well as a slower heartbeat [75]. CuO nanoparticles also causes gill injury and acute lethality in zebra fish [79]. Silver nanoparticles accumulate in gills, intestines and liver and cause lesions in the liver, trigger stress response and damage in the gills thereby impairing oxygen uptake [77]. Excessively high ZnO concentrations caused developmental delays, damage of individual organs or the immune system of fish (liver, brain, gills) [76]. Additionally, zinc oxide nanoparticles accumulate in gills and digestive glands of mussels and snails and are toxic at high concentrations in these organisms [67].

In comparison to freshwater organisms, far less data regarding the hazard of ion-releasing nanomaterials is available for marine organisms [80] and terrestrial test species, but also here a number of adverse effects have been observed. As a direct input of textile-derived nanomaterials into marine environments is less likely we do not discuss the effects on marine organisms in detail. For terrestrial invertebrates, common test species include earthworms, nematodes, springtails, isopods and enchytraeids. Ion-releasing nanomaterials affect their reproduction and survival [81, 82] as well as avoidance of contaminated soil implying reduced habitat function [82]. Zinc oxide nanoparticles influence the nervous system and lower the survival rate of bees [83]. Clear decrease of plant growth was observed for all three nanomaterials [84, 85]. Particle uptake by the roots and the distribution in the plant has been demonstrated for some nanoparticles [86]. Significant differences in the sensitivity of individual plant species were observed.

The dissolution of alumina nanoparticles and consequent contribution toward toxicity remained largely unexplored owing to its presumed insoluble nature. However, Pakrashi et al. [87] showed that Al is released and induces the production of reactive oxygen species (ROS) and oxidative stress in algae [87, 88]. Furthermore, Al_2O_3 did not affect algal growth [89]. Also it was demonstrated to be genotoxic to the plant *Allium cepa* [88]. After acute exposure Al_2O_3 were not toxic to water flea, but significant effect on the reproduction was reported [90, 91]. There was no mortality of zebrafish after Al_2O_3 nanoparticles exposure [92]. Al_2O_3 nanoparticles decreased earthworm reproduction but did not affect mortality [93]. As well it reduced nematode growth and the number of their offspring [94]. Regarding the effects on plants, some studies showed effects of high concentrations of Al_2O_3 on their growth and seed germination [95] whereas others found no effect on their growth [96].

18.3.3 Hazard of non-ion-releasing nanomaterials: silicon dioxide, titanium dioxide, gold and clays

A number of nanomaterials do not release metal ions under physiological conditions. As for humans, non-ion-releasing nanomaterials were classified as less toxic to environmental organisms (e.g., induced effects at higher concentrations) in comparison to ion-releasing nanomaterials [49, 55]. Among non-soluble nanomaterials, most data have been generated for TiO_2, which probably relates to the fact that it is also the most commonly applied. For TiO_2 and SiO_2 it is generally assumed that their hazard to environmental organisms is low. In addition, as most of the nanomaterials are tightly incorporated into the fabric, the likelihood of being released into water is low. TiO_2 and SiO_2 nanomaterials are considered as photocatalytically active which means that they generate free radicals when exposed to light [97]. Into textiles, this introduces some self-cleaning properties, as the radicals will act against microbes. Therefore, phototoxicity was commonly reported for these nanomaterials and various test organisms [98, 99]. Furthermore, their effects on organisms are predominately driven by their specific physicochemical properties, such as aspect-ratio, physical entrapment, and photocatalytic activity [49, 55]. Induction of ROS, oxidative stress and oxidative DNA damage were commonly reported for TiO_2 and SiO_2 [100].

Physical interaction of nanomaterials with organisms was often reported (Figure 18.3). A clear example is entrapment of algae in agglomerates of TiO_2 [49], and SiO_2 [101]. Attachment of nanoparticles on algal surface results in so called shading-effect and consequently reduced growth, which has been evidenced for TiO_2 [102] and SiO_2 [101]. Physical interaction was also evidenced in case of water fleas, where the attachment on the body surface resulted in decreased moult and mobility, for example, for TiO_2 [54]. Accumulation in the intestine of water fleas was also evidenced and adverse effects on the morphology of gut epithelium was reported [92]. Mortality in water fleas was observed after chronic exposure to TiO_2 [103]. SiO_2 decreased the reproduction of water flea and caused damage to their intestine. The effect depended on the functionalization of SiO_2 with amino modified being more hazardous [104].

TiO_2 had no severe effects on adult fish [70] or their embryos and did not cross zebrafish embryo chorion [99]. SiO_2 did not induce severe effects on the development of zebra fish embryos [105], but some effects on behavior and cardiovascular system were found [106]. In some other fish exposed to SiO_2 blood composition was affected [107]. For other freshwater and saltwater organisms (mussels, snails) TiO_2 and SiO_2 nanoparticles [100], were often not acutely toxic.

Most of the tests with terrestrial organisms have been performed with earthworms, nematodes and plants. TiO_2 induced only sublethal effects on earthworms [108]. Nematode growth and the number of their offspring was reduced in case of TiO_2 and SiO_2 [77]. There are quite many studies that addressed the effects of TiO_2 nanoparticles on plants. Some studies showed the effects on their growth and seed germination. But, a number of studies showed no effect on their growth [109].

The toxicity of gold nanoparticles is low. Of the organisms studied, algae were the most sensitive to gold nanoparticles, most probably due to a shading effect [110]. As with other nanomaterials, also gold particles were observed to adhere to the carapace of water fleas, affecting both the swimming behavior and the moulting rate. In addition, gold particles are taken up by water fleas. They were detected in the intestine, but not in the surrounding tissue, and excreted after passage through the gut [111, 112]. Some studies showed no effects of AuNPs on the reproduction of water flea [113], while a recent study reported that AuNP caused mortality and reproduction impairment in this organism [114]. In embryonic and adult fish, neither malformations nor mortality was induced by gold nanoparticles despite evident particle uptake [115, 116]. Mussels also accumulated gold nanoparticles, mainly in their digestive glands. In addition, the particles triggered oxidative stress in the glands, but not in the gill and mantle tissue [110, 117]. AuNPs have the potential to induce oxidative stress in mussels [117]. A number of studies showed their genotoxic potential in vitro [118]. In bacteria, as well as cucumber and lettuce plants gold nanoparticles exerted no toxic effects [119, 120]. In soil containing gold nanoparticles, a slight reduction in the number of microorganisms was observed [120].

Nanoclays are used in textile applications or their antibacterial activity, in addition to their flame-retardant properties. Accordingly, they inhibit the growth of bacteria (e.g., *E. coli* [121, 122] and amoebae [123]). There is no indication of harmful effects in other organism groups so far [124].

18.3.4 Hazard of carbon-based NMs: carbon black, carbon fibers and CNTs

For the purpose of this chapter we combine data for single-walled and multi-walled CNTs, which differ in diameter and aspect ratio. Air-borne release of CNT from textiles (and in general) is considered to be very low, and production facilities employ effective air-filter technology. Hence hazard due to inhalation of CNTs, as likely for humans at the workplace, has not been studied in environmental organisms. Mechanical and indirect effects due to the fiber-like form are observed in many organisms (Figure 18.3), particularly in studies in which very high doses were tested. For low concentrations, as they are currently expected in the environment, no risk to environmental organisms is estimated [125].

In aquatic organisms, the fiber shape of CNTs can cause physical effects, for example, indirect toxic effects in bacteria by acting as needles which pierce the cell envelope. Such damage leads to a growth inhibition of bacteria [126, 127]. Green algae are indirectly affected by CNT, inhibiting algae growth by attaching to the cells and shading the light [128–130]. In addition, CNT induce oxidative stress in algae [129, 130]. For a couple of single-cell organisms uptake and excretion of CNTs has been observed without any effect on organisms' viability [131]. Likewise, CNTs are taken up and subsequently are secreted by water fleas [132, 133]. Effects on the gills of frog and fish have been observed. The CNTs "clog" the gills of larvae and adult animals, leading to irritation, accelerated respiration and signs of stress [134, 135]. Length-dependent toxicity toward zebra fish have been demonstrated [136].

As for the terrestrial environment, uptake of CNT from the soil by various crops such as wheat or rape was investigated. In general, only very small amounts of CNTs are taken up via the plant roots [137]. This led to the conclusion that a transfer and accumulation of CNT through the food chain via crop is considered unlikely. The growth of the plants was not affected, but there are indications of an increased stress of the plant [138].

Nano-scaled, pure carbon black exerts little toxicity toward organisms. As for CNTs, however, harmful effects are observed due to an attachment of carbon black particles to cells and bodies, as well as in combination with chemicals that are effectively bound by carbon black (so-called Trojan-horse effect) [130]. In fly, carbon black powder adhered quickly and firmly to the exterior of adult exposed animals, and removal by the natural grooming behavior was not possible. The particle-coating led to impaired mobility and killed the animals within hours by blocking the breathing holes [126]. Such an exposure scenario to large amounts of pure carbon black is, however, highly unlikely under real environmental conditions, as well as for carbon black stemming from textile products. Ground-dwelling amphipods also showed an increased mortality when exposed to very high concentrations of carbon black [127].

No uptake of nanoscale carbon black by brown algae was observed, but very high concentrations prevented the fertilization and development. However, germination and root growth were unaffected [131]. Mussels feed by filtering smallest particles from the water and also carbon black particles were present in the digestive tract of the animals [128, 129].

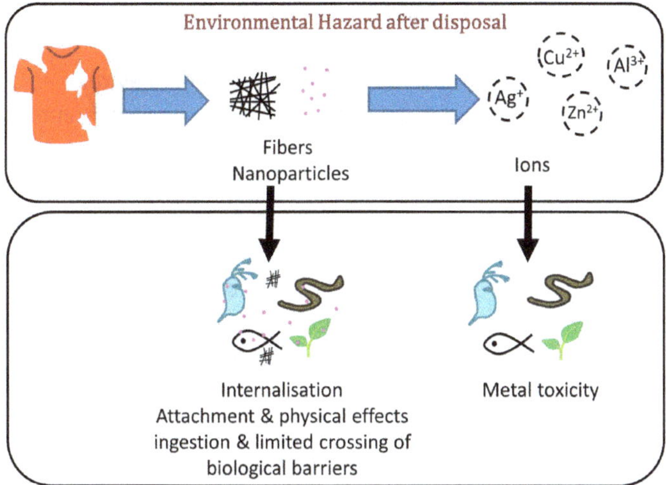

Figure 18.3: The most relevant mechanisms leading to hazardous effects of nanomaterials in environmental organisms. Indirect, physical effects are exerted due to attachment of nanomaterials, and ingestion and the crossing of biological barriers may occur. The release of toxic ions leads to direct toxic effects.

18.4 Conclusion

18.4.1 Hazard assessment

Nanomaterials are added to textiles for various purposes (e.g., [1]), and most of the materials used in textile applications have been assessed with regard to their hazard toward humans and the environment. By this assessment, textile-relevant nanomaterials can be classified with regard to their toxicity (low–high). Additionally, distinct mechanisms of action were identified, for example, the critical effect of CNT fibers for humans upon inhalation or the effects of dissolved toxic ions for Ag and Cu. Further, dose–response relationships were derived, informing on the critical doses for humans and environmental organisms, which provide the basis for the assessment of nanomaterial risks. For risk assessment, the hazard is set in relation to the amounts of the respective nanomaterial the organisms are exposed to (exposure assessment).

18.4.2 Release and exposure assessment

For human exposure, the amount of nanomaterials released from textile products will depend on the amount of nanomaterials released during the production process and the amount that gets into contact with the skin. For the environment, the nanomaterial amount released from the textile product during the product life cycle is decisive. Hence, robust estimates of release and the processes leading to release are needed (e.g., [139]). Next to release, also distribution and transformation under environmental conditions need consideration. This issue is specifically important for the environment (e.g., [140]). Transformation occurs by the influence of physical (e.g., adsorption), chemical (e.g., redox reactions) and biological (e.g., formation of bio-corona) processes and modifies particle surface or the whole particle [141]. As a consequence, the particle properties that determine transport and fate under environmental conditions, as well as subsequent toxicity, will change. Most toxicity studies so far were performed with as-produced nanomaterials, and hence the effects of transformation of the nanomaterials, which can be both passivating and activating are not considered [140, 142]. For example, conducting the experiments with Ag and CuO nanoparticles in the presence of natural water, reduced toxicity was observed presumably due to lower release of ions [143]. When CuO nanoparticles were coated with polymer a higher toxicity was observed presumably due to higher uptake of these nanoparticles [144]. As summarized in Table 18.4, the exposure of humans and the environment to nanomaterials released from textiles occurs during different phases of the life cycle of a textile product.

Table 18.4: Exposure potential for workers, consumers and the environment to nanomaterials from textile products (adapted from [44]).

Textile production sector	Worker	Consumer	Environment
Sports and outdoor clothing based on textile treatments	++	+	+
Sports and outdoor clothing where nanomaterials are incorporated into fibers	++	++	++
Medical textiles	++	++	+++
Geotextiles	++	+	+++

+ low, ++ middle, +++high.

18.4.3 Mixture toxicity

Further, organisms under realistic environmental conditions are never exposed solely to single substances, but rather to mixtures of several nanomaterials, in addition to chemicals (e.g., pollutants such as pesticides or residues from pharmaceuticals). In this context, we should not forget about the numerous chemicals that are used for textile production and finishing (e.g., dyes and plasticizers), and may hence be released together with nanomaterials. Many of the substances are also known to have hazardous properties. Also, as pointed our recently synthetic textiles release micro- or nanoplastic fibers [145]. From the interaction of engineered nanomaterials, nano(micro)plastics and chemicals, a variety of different mixtures with unknown human and environmental hazard effects may arise [146].

18.4.4 Risk of nanomaterials released from textile products

Taking all currently available information about release, exposure and hazard into account, the overall risk of nanomaterials can be evaluated for three groups of materials: (I) ion-releasing nanomaterials with evident metal human toxicity and ecotoxicity, (II) CNT fibers and other nanofibers for humans upon inhalation and (III) the other nanomaterials mentioned in this chapter. For all three groups, however, the amounts of nanomaterials incorporated into textiles are rather low. Next, release of nanomaterials is either low or only part of the contained nanomaterials (e.g., ions) is released. Upon transformation and transport, hence, the overall exposure for humans and environmental organisms is low.

(I) Due to only small amounts released, exposure for humans and the environment is low. Due to the evident ion toxicity, there is a low to medium risk, because hotspots such as effluents of wastewater treatment plants may require attention (see Figure 18.4).

(II) The release of CNTs from various textile applications is considered to be low. Due to the evident and known inhalation toxicity, and the high likelihood of workplace exposure, effective protection measures are in place that minimize the risk by reducing human exposure.

(III) The exposure to all other nanomaterials is low, and they pose a low hazard. Again, attention at hotspots of environmental release, for example, landfill may be required to prevent exposure for environmental organisms. For humans, occupational handling of nanomaterials poses the highest risk, because the exposure is likely to be highest at the workplace. At the same time, effective risk management measures are in place, like occupational exposure limits and personal protection equipment preventing exposure (i.e. masks or hoods to prevent inhalation). Hence, with the available information on hazards, this integrates into a low risk of adverse impacts of nanomaterials stemming from textile products (see Figure 18.4).

But of course the risk needs to be assessed nanomaterial-wise, especially for novel nanomaterials used in future textile applications. There are constantly novel products arising, for example, composite materials with halloycite nanotubes [147], and those new developments need attention with regard to hazard and risk assessment. Although the amount of different materials in use for textiles seem to be manageable from a toxicological point of view, the differences between the nanomaterial production processes or even different batches of the same material vary often so much that they may be handled as different materials. This and variations in the release from textiles, handling during processing and different purposes of the products contain an epistemic uncertainty in regard to the environment, health and safety risks which have not sufficiently been resolved. The highest uncertainty arises from the difficulty to reliably quantify nanomaterials released from textiles, also due to analytical detection limits. Further, the transformations that nanomaterials undergo prior to or after release need to be better understood, as these changes will also impact the amount and form of nanomaterial an organism is finally exposed to. In addition, uncertainties also remain for hazard assessment of nanomaterials due to the large number of results from different test procedures that lack standardization, although activities started to improve the situation, for example, first suggestions come from OECD providing guidance on nanomaterial testing [148] as well as from ISO (https://www.iso.org/committee/381983/x/catalogue/). Another issue is the missing analytics for the reliable detection of nanomaterials in exposed organisms and tissues and the evidence of a possible bioaccumulation. Efforts have been currently made to assess the applicability of current OECD guidelines for testing the bioaccumulation of nanomaterials in fish. The development of harmonized test methods for nanomaterials is also the aim of the EU project NanoHarmony (https://nanoharmony.eu/). On the other hand, another EU project named "Patrols" (https://www.patrols-h2020.eu/) aims to provide an innovative and effective set of alternative laboratory techniques and computational tools to predict potential human and environmental hazards resulting from nanomaterial

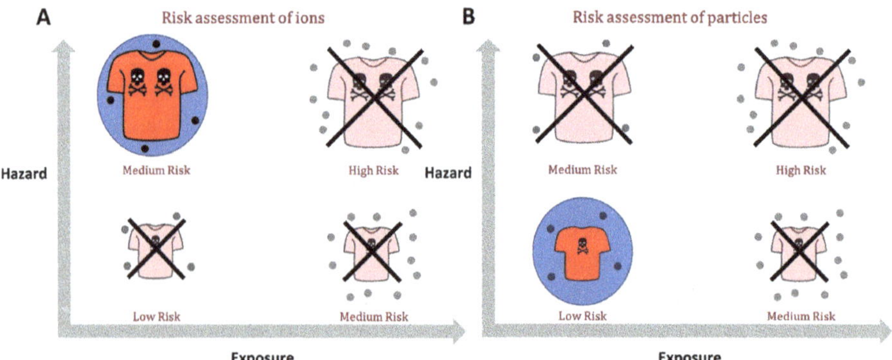

Figure 18.4: The assessment of risk is done by the weighting of exposure occurring for a substance and the hazard this substance can exert. As depicted, there are several options that translate into low, medium, or high risk. In the case of ion-releasing nanomaterials in textiles (A), the general risk is considered medium (highlighted panel), because the human and environmental exposure is low (indicated by the dots) while hazard may occur (indicated by only one skull on the T-shirt). In the case of the other relevant nanomaterials (B) both exposure and hazard are considered to be low, translating into a low risk (highlighted panel) (except for CNTs which possess a higher risk). However, one has to keep in mind that in individual cases, the risk may differ (e.g., high exposure at hotspots).

exposure more reliably. In general, the suitability of assays and related end points needs still careful consideration as most of the tests applied in hazard assessment were initially developed for chemicals with diverging properties compared to nanomaterials. Despite tremendous research efforts that had been done in the field of human and environmental toxicology during the last decade, environmental fate and exposure of environmental organisms are largely unexplored. Although nanomaterials have been the subject of biological research for more than 4 decades, there are still many unanswered questions that are important for a sound risk assessment for humans and the environment.

For more information on the toxicity of nanomaterials in textiles and safety of nanomaterials in general visit the publicly available knowledge base www.nanoobjects.info.

References

[1] Yetisen, A. K.; Qu, H.; Manbachi, A.; Butt, H.; Dokmeci, M. R.; Hinestroza, J. P., et al.: Nanotechnology in textiles. Acs Nano, 2016; 10(3), 3042–3068.

[2] Arts, J. H.; Irfan, M. A.; Keene, A. M.; Kreiling, R.; Lyon, D.; Maier, M., et al.: Case studies putting the decision-making framework for the grouping and testing of nanomaterials (DF4nanoGrouping) into practice. Regulatory Toxicology and Pharmacology, 2016; 76, 234–261.

[3] Mantecca, P.; Kasemets, K.; Deokar, A.; Perelshtein, I.; Gedanken, A.; Bahk, Y. K., et al.: Airborne nanoparticle release and toxicological risk from metal-oxide-coated textiles: Toward a multiscale safe-by-design approach. Environmental Science & Technology, 2017; 51(16), 9305–9317.

[4] Limpiteeprakan, P.; Babel, S.: Leaching potential of silver from nanosilver-treated textile products. Environmental Monitoring and Assessment, 2016; 188(3), 156.

[5] Som, C.; Wick, P.; Krug, H.; Nowack, B.: Environmental and health effects of nanomaterials in nanotextiles and facade coatings. Environment International, 2011; 37(6), 1131–1142.

[6] Müller, W. W.; Saathoff, F.: Geosynthetics in geoenvironmental engineering. Science and Technology of Advanced Materials, 2015; 034605,16(3).

[7] Zhang, H. Y.; Ji, Z. X.; Xia, T.; Meng, H.; Low-Kam, C.; Liu, R., et al.: Use of metal oxide nanoparticle band gap to develop a predictive paradigm for oxidative stress and acute pulmonary inflammation. Acs Nano, 2012; 6(5), 4349–4368.

[8] Braakhuis, H. M.; Park, M. V.; Gosens, I.; De Jong, W. H.; Cassee, F. R.: Physicochemical characteristics of nanomaterials that affect pulmonary inflammation. PartFibre Toxicol, 2014; 11, 18.

[9] Ali, A.; Ovais, M.; Cui, X.; Rui, Y.; Chen, C.: Safety assessment of nanomaterials for antimicrobial applications. Chemical Research in Toxicology, 2020; 33(5), 1082–1109.

[10] Varsou, D. D.; Afantitis, A.; Tsoumanis, A.; Papadiamantis, A.; Valsami-Jones, E.; Lynch, I., et al.: Zeta-potential read-across model utilizing nanodescriptors extracted via the nanoxtract image analysis tool available on the enalos nanoinformatics cloud platform. Small, 2020; 16(21), e1906588.

[11] Köhler, A. R.; Som, C.: Risk preventative innovation strategies for emerging technologies the cases of nano-textiles and smart textiles. Technovation, 2014; 34(8), 420–430.

[12] Som, C.; Halbeisen, M.: Einschätzungen Zu „nano“-textilien Und Entwicklungstrends – Umfrage, St. Gallen: Empa, 2008.
[13] Saleem, H.; Zaidi, S. J.: Sustainable use of nanomaterials in textiles and their environmental impact. Materials (Basel), 2020; 13(22):5134.
[14] Nowack, B.; Krug, H. F.; Height, M.: 120 years of nanosilver history: Implications for policy makers. Environmental Science & Technology, 2011; 45(4), 1177–1183.
[15] Krug, H. F.; Nau, K.: Reliability for Nanosafety Research – Considerations on the Basis of a Comprehensive Literature Review. ChemBioEng Reviews, 2017; 4(6), 331–338.
[16] Skocaj, M.; Filipic, M.; Petkovic, J.; Novak, S.: Titanium dioxide in our everyday life; is it safe? Radiology and Oncology, 2011; 45(4), 227–247.
[17] Krug, H. F.: Nanosafety research–are we on the right track? Angewandte Chemie International Edition in English, 2014; 53(46), 12304–12319.
[18] Skjolding, L. M.; Sorensen, S. N.; Hartmann, N. B.; Hjorth, R.; Hansen, S. F.; Baun, A.: Aquatic ecotoxicity testing of nanoparticles-the quest to disclose nanoparticle effects. Angewandte Chemie International Edition in English, 2016; 55(49), 15224–15239.
[19] Lynch, I.; Weiss, C.; Valsami-Jones, E.: A strategy for grouping of nanomaterials based on key physico-chemical descriptors as a basis for safer-by-design NMs. Nano Today, 2014; 9(3), 266–270.
[20] Arts, J. H. E.; Hadi, M.; Irfan, M.-A.; Keene, A. M.; Kreiling, R.; Lyon, D., et al.: A decision-making framework for the grouping and testing of nanomaterials (DF4nanoGrouping). Regulatory Toxicology and Pharmacology, 2015; 71(2,Supplement), S1–S27.
[21] Godwin, H.; Nameth, C.; Avery, D.; Bergeson, L. L.; Bernard, D.; Beryt, E., et al.: Nanomaterial categorization for assessing risk potential to facilitate regulatory decision-making. Acs Nano, 2015; 9(4), 3409–3417.
[22] Butz, T.; Reinert, T.; Pinheiro, T.; Moretto, P.; Pallon, J.; Kiss, A. Z., et al.: NANODERM – Quality of Skin as a Barrier to ultra-fine Particles, 2007.
[23] Oberdörster, G.; Sharp, Z.; Atudorei, V.; Elder, A.; Gelein, R.; Kreyling, W. G., et al.: Translocation of inhaled ultrafine particles to the brain. InhalToxicol, 2004; 16(6–7), 437–445.
[24] Boyles, M. S.; Young, L.; Brown, D. M.; MacCalman, L.; Cowie, H.; Moisala, A., et al.: Multi-walled carbon nanotube induced frustrated phagocytosis, cytotoxicity and pro-inflammatory conditions in macrophages are length dependent and greater than that of asbestos. Toxicology in Vitro, 2015; 29(7), 1513–1528.
[25] Stone, V.; Miller, M. R.; Clift, M. J. D.; Elder, A.; Mills, N. L.; Moller, P., et al.: Nanomaterials versus ambient ultrafine particles: an opportunity to exchange toxicology knowledge. Environmental Health Perspectives, 2017; 125(10), 106002.
[26] Laux, P.; Riebeling, C.; Booth, A. M.; Brain, J. D.; Brunner, J.; Cerrillo, C., et al.: Biokinetics of nanomaterials: The role of biopersistence. NanoImpact, 2017; 6, 69–80.
[27] WHO: Hazard prevention and control in the work environment: airborne dust. Geneva, 1999. Report No.: WHO/SDE/OEH/99.14 Contract No.: WHO/SDE/OEH/99.14.
[28] Wagener, S.; Dommershausen, N.; Jungnickel, H.; Laux, P.; Mitrano, D.; Nowack, B., et al.: Textile functionalization and its effects on the release of silver nanoparticles into artificial sweat. Environmental Science & Technology, 2016; 50(11), 5927–5934.
[29] Gaetke, L. M.; Chow-Johnson, H. S.; Chow, C. K.: Copper: Toxicological relevance and mechanisms. Archives of Toxicology, 2014; 88(11), 1929–1938.
[30] Plum, L. M.; Rink, L.; Haase, H.: The essential toxin: Impact of zinc on human health. International Journal of Environmental Research and Public Health, 2010; 7(4), 1342–1365.

[31] Han, J. W.; Jeong, J. K.; Gurunathan, S.; Choi, Y. J.; Das, J.; Kwon, D. N., et al.: Male- and female-derived somatic and germ cell-specific toxicity of silver nanoparticles in mouse. Nanotoxicology, 2016; 10(3), 361–373.
[32] Singh, G.; Beddow, J.; Mee, C.; Maryniak, L.; Joyce, E. M.; Mason, T. J.: Cytotoxicity study of textile fabrics impregnated with CuO nanoparticles in mammalian cells. International Journal of Toxicology, 2017; 36(6), 478–484.
[33] Verbic, A.; Gorjanc, M.; Simoncic, B.: Zinc oxide for functional textile coatings: Recent advances. Coatings, 2019; 9(9), 550.
[34] Chernova, T.; Murphy, F. A.; Galavotti, S.; Sun, X. M.; Powley, I. R.; Grosso, S., et al.: Long-fiber carbon nanotubes replicate asbestos-induced mesothelioma with disruption of the tumor suppressor gene Cdkn2a (Ink4a/Arf). Current Biology, 2017; 27(21), 3302–14 e6.
[35] Donaldson, K.; Murphy, F. A.; Duffin, R.; Poland, C. A.: Asbestos, carbon nanotubes and the pleural mesothelium: A review of the hypothesis regarding the role of long fibre retention in the parietal pleura, inflammation and mesothelioma. PartFibre Toxicol, 2010; 7, 5.
[36] Gebel, T.; Foth, H.; Damm, G.; Freyberger, A.; Kramer, P. J.; Lilienblum, W., et al.: Manufactured nanomaterials: Categorization and approaches to hazard assessment. Archives of Toxicology, 2014; 88(12), 2191–2211.
[37] Schinwald, A.; Murphy, F. A.; Prina-Mello, A.; Poland, C. A.; Byrne, F.; Movia, D., et al: The threshold length for fiber-induced acute pleural inflammation: Shedding light on the early events in asbestos-induced mesothelioma. Toxicological Sciences, 2012; 128(2), 461–470.
[38] Nowack, B.; David, R. M.; Fissan, H.; Morris, H.; Shatkin, J. A.; Stintz, M., et al.: Potential release scenarios for carbon nanotubes used in composites. EnvironInt, 2013; 59, 1–11.
[39] Plitzko, S.: Workplace exposure to engineered nanoparticles. Inhalation Toxicology, 2009; 21 (Suppl 1), (S1):25–9.
[40] Brouwer, D.: Exposure to manufactured nanoparticles in different workplaces. Toxicology, 2010; 269(2–3), 120–127.
[41] Kuhlbusch, T. A. J.; Asbach, C.; Fissan, H.; Gohler, D.; Stintz, M.: Nanoparticle exposure at nanotechnology workplaces: A review. Particle and Fibre Toxicology, 2011; 8, 22.
[42] Mihalache, R.; Verbeek, J.; Graczyk, H.; Murashov, V.; van Broekhuizen, P.: Occupational exposure limits for manufactured nanomaterials, a systematic review. Nanotoxicology, 2017; 11(1), 7–19.
[43] Pietroiusti, A.; Magrini, A.: Engineered nanoparticles at the workplace: Current knowledge about workers' risk. Occupational Medicine (London), 2014; 64(5), 319–330.
[44] Nowack, B.; Brouwer, C.; Geertsma, R. E.; Heugens, E. H.; Ross, B. L.; Toufektsian, M. C., et al.: Analysis of the occupational, consumer and environmental exposure to engineered nanomaterials used in 10 technology sectors. Nanotoxicology, 2013; 7(6), 1152–1156.
[45] Mackevica, A.; Foss Hansen, S.: Release of nanomaterials from solid nanocomposites and consumer exposure assessment – a forward-looking review. Nanotoxicology, 2016; 10(6), 641–653.
[46] Mitrano, D. M.; Lombi, E.; Dasilva, Y. A.; Nowack, B.: Unraveling the complexity in the aging of nanoenhanced textiles: A comprehensive sequential study on the effects of sunlight and washing on silver nanoparticles. Environmental Science & Technology, 2016; 50(11), 5790–5799.
[47] Windler, L.; Lorenz, C.; von Goetz, N.; Hungerbuhler, K.; Amberg, M.; Heuberger, M., et al.: Release of titanium dioxide from textiles during washing. Environmental Science & Technology, 2012; 46(15), 8181–8188.
[48] Benn, T. M.; Westerhoff, P.: Nanoparticle silver released into water from commercially available sock fabrics. Environmental Science & Technology, 2008; 42(11), 4133–4139.

[49] Bondarenko, O. M.; Heinlaan, M.; Sihtmäe, M.; Ivask, A.; Kurvet, I.; Joonas, E., et al.: Multilaboratory evaluation of 15 bioassays for (eco)toxicity screening and hazard ranking of engineered nanomaterials: FP7 project NANOVALID. Nanotoxicology, 2016; 10(9), 1229–1242.

[50] Nowack, B.; Ranville, J. F.; Diamond, S.; Gallego-Urrea, J. A.; Metcalfe, C.; Rose, J., et al.: Potential scenarios for nanomaterial release and subsequent alteration in the environment. Environmental Toxicology and Chemistry, 2012; 31(1), 50–59.

[51] Kwak, J. I.; An, Y.-J.: The current state of the art in research on engineered nanomaterials and terrestrial environments: Different-scale approaches. Environmental Research, 2016; 151, 368–382.

[52] Hund-Rinke, K.; Schlich, K.; Kühnel, D.; Hellack, B.; Kaminski, H.; Nickel, C.: Grouping concept for metal and metal oxide nanomaterials with regard to their ecotoxicological effects on algae, daphnids and fish embryos. NanoImpact, 2018; 9, 52–60.

[53] Kühnel, D.; Nickel, C.; Hellack, B.; van der Zalm, E.; Kussatz, C.; Herrchen, M., et al.: Closing gaps for environmental risk screening of engineered nanomaterials. NanoImpact, 2019; 15, 100173.

[54] Dabrunz, A.; Duester, L.; Prasse, C.; Seitz, F.; Rosenfeldt, R.; Schilde, C., et al.: Biological surface coating and molting inhibition as mechanisms of TiO2 nanoparticle toxicity in Daphnia magna. PloS One, 2011; 6(5), e20112.

[55] Juganson, K.; Ivask, A.; Blinova, I.; Mortimer, M.; Kahru, A.: NanoE-Tox: New and in-depth database concerning ecotoxicity of nanomaterials. Beilstein Journal of Nanotechnology, 2015; 6, 1788–1804.

[56] Ma, H.; Wallis, L. K.; Diamond, S.; Li, S.; Canas-Carrell, J.; Parra, A.: Impact of solar UV radiation on toxicity of ZnO nanoparticles through photocatalytic reactive oxygen species (ROS) generation and photo-induced dissolution. Environmental Pollution (Barking, Essex: 1987), 2014; 193, 165–172.

[57] Ivask, A.; Juganson, K.; Bondarenko, O.; Mortimer, M.; Aruoja, V.; Kasemets, K., et al.: Mechanisms of toxic action of Ag, ZnO and CuO nanoparticles to selected ecotoxicological test organisms and mammalian cells in vitro: A comparative review. Nanotoxicology, 2014; 8 (sup1), 57–71.

[58] Ivask, A.; ElBadawy, A.; Kaweeteerawat, C.; Boren, D.; Fischer, H.; Ji, Z., et al.: Toxicity mechanisms in escherichia coli vary for silver nanoparticles and differ from ionic silver. Acs Nano, 2014; 8(1), 374–386.

[59] Kanhed, P.; Birla, S.; Gaikwad, S.; Gade, A.; Seabra, A. B.; Rubilar, O., et al.: In vitro antifungal efficacy of copper nanoparticles against selected crop pathogenic fungi. Materials Letters, 2014; 115, 13–17.

[60] Khan, A. M.; Larson, C. P.; Faruque, A. S. G.; Saha, U. R.; Hoque, A. B. M. M.; Alam, N. U., et al.: Introduction of routine zinc therapy for children with diarrhoea: Evaluation of safety. Journal of Health, Population, and Nutrition, 2007; 25(2), 127–133.

[61] Dimkpa, C.; Calder, A.; Britt, D.; Mclean, J.; Anderson, A.: Responses of a soil bacterium, Pseudomonas chlororaphis O6 to commercial metal oxide nanoparticles compared with responses to metal ions. Environmental Pollution (Barking, Essex: 1987), 2011; 159, 1749–1756.

[62] Mortimer, M.; Kasemets, K.; Vodovnik, M.; Marinšek-Logar, R.; Kahru, A.: Exposure to CuO nanoparticles changes the fatty acid composition of protozoa Tetrahymena thermophila. Environmental Science & Technology, 2011; 45(15), 6617–6624.

[63] Gomes, S. I. L.; Soares, A. M. V. M.; Scott-Fordsmand, J. J.; Amorim, M. J. B.: Mechanisms of response to silver nanoparticles on Enchytraeus albidus (Oligochaeta): Survival, reproduction and gene expression profile. Journal of Hazardous Materials, 2013; 254–255, 336–344.

[64] Kaya, H.; Aydın, F.; Gürkan, M.; Yılmaz, S.; Ates, M.; Demir, V., et al.: Effects of zinc oxide nanoparticles on bioaccumulation and oxidative stress in different organs of tilapia (Oreochromis niloticus). Environmental Toxicology and Pharmacology, 2015; 40(3), 936–947.
[65] Lee, W.-M.; An, Y.-J.: Effects of zinc oxide and titanium dioxide nanoparticles on green algae under visible, UVA, and UVB irradiations: No evidence of enhanced algal toxicity under UV pre-irradiation. Chemosphere, 2013; 91(4), 536–544.
[66] Atha, D. H.; Wang, H.; Petersen, E. J.; Cleveland, D.; Holbrook, R. D.; Jaruga, P., et al.: Copper oxide nanoparticle mediated DNA damage in terrestrial plant models. Environmental Science & Technology, 2012; 46(3), 1819–1827.
[67] Trevisan, R.; Delapedra, G.; Mello, D. F.; Arl, M.; Schmidt, É. C.; Meder, F., et al.: Gills are an initial target of zinc oxide nanoparticles in oysters Crassostrea gigas, leading to mitochondrial disruption and oxidative stress. Aquatic Toxicology (Amsterdam, Netherlands), 2014; 153, 27–38.
[68] Bondarenko, O.; Juganson, K.; Ivask, A.; Kasemets, K.; Mortimer, M.; Kahru, A.: Toxicity of Ag, CuO and ZnO nanoparticles to selected environmentally relevant test organisms and mammalian cells in vitro: A critical review. Archives of Toxicology, 2013; 87(7), 1181–1200.
[69] Adam, N.; Vergauwen, L.; Blust, R.; Knapen, D.: Gene transcription patterns and energy reserves in Daphnia magna show no nanoparticle specific toxicity when exposed to ZnO and CuO nanoparticles. Environmental Research, 2015; 138, 82–92.
[70] Griffitt, R. J.; Luo, J.; Gao, J.; Bonzongo, J.-C.; Barber, D. S.: Effects of particle composition and species on toxicity of metallic nanomaterials in aquatic organisms. Environmental Toxicology and Chemistry, 2008; 27(9), 1972–1978.
[71] Skjolding, L. M.; Winther-Nielsen, M.; Baun, A.: Trophic transfer of differently functionalized zinc oxide nanoparticles from crustaceans (Daphnia magna) to zebrafish (Danio rerio). Aquatic Toxicology, 2014; 157, 101–108.
[72] Zhao, C. M.; Wang, W. X.: Importance of surface coatings and soluble silver in silver nanoparticles toxicity to Daphnia magna. Nanotoxicology, 2012; 6(4), 361–370.
[73] Suman, T. Y.; Radhika Rajasree, S. R.; Kirubagaran, R.: Evaluation of zinc oxide nanoparticles toxicity on marine algae chlorella vulgaris through flow cytometric, cytotoxicity and oxidative stress analysis. Ecotoxicology and Environmental Safety, 2015; 113, 23–30.
[74] Wise, J. P.; Goodale, B. C.; Wise, S. S.; Craig, G. A.; Pongan, A. F.; Walter, R. B., et al.: Silver nanospheres are cytotoxic and genotoxic to fish cells. Aquatic Toxicology (Amsterdam, Netherlands), 2010; 97(1), 34–41.
[75] Wu, Y.; Zhou, Q.; Li, H.; Liu, W.; Wang, T.; Jiang, G.: Effects of silver nanoparticles on the development and histopathology biomarkers of Japanese medaka (Oryzias latipes) using the partial-life test. Aquatic Toxicology, 2010; 100(2), 160–167.
[76] Ates, M.; Arslan, Z.; Demir, V.; Daniels, J.; Farah, I. O.: Accumulation and toxicity of CuO and ZnO nanoparticles through waterborne and dietary exposure of goldfish (Carassius auratus). Environmental Toxicology, 2015; 30(1), 119–128.
[77] Wu, Y.; Zhou, Q.: Silver nanoparticles cause oxidative damage and histological changes in medaka (Oryzias latipes) after 14 days of exposure. Environmental Toxicology and Chemistry, 2013; 32(1), 165–173.
[78] Zhao, J.; Wang, Z.; Liu, X.; Xie, X.; Zhang, K.; Xing, B.: Distribution of CuO nanoparticles in juvenile carp (Cyprinus carpio) and their potential toxicity. Journal of Hazardous Materials, 2011; 197, 304–310.
[79] Griffitt, R. J.; Hyndman, K.; Denslow, N. D.; Barber, D. S.: Comparison of molecular and histological changes in zebrafish gills exposed to metallic nanoparticles. Toxicological Sciences: An Official Journal of the Society of Toxicology, 2009; 107(2), 404–415.

[80] Canesi, L.; Corsi, I.: Effects of nanomaterials on marine invertebrates. The Science of the Total Environment, 2016; 565, 933–940.
[81] Amorim, M. J. B.; Scott-Fordsmand, J. J.: Toxicity of copper nanoparticles and CuCl2 salt to Enchytraeus albidus worms: Survival, reproduction and avoidance responses. Environmental Pollution, 2012; 164, 164–168.
[82] Shoults-Wilson, W. A.; Reinsch, B. C.; Tsyusko, O. V.; Bertsch, P. M.; Lowry, G. V.; Unrine, J. M.: Effect of silver nanoparticle surface coating on bioaccumulation and reproductive toxicity in earthworms (Eisenia fetida). Nanotoxicology, 2011; 5(3), 432–444.
[83] Milivojević, T.; Glavan, G.; Božič, J.; Sepčić, K.; Mesarič, T.; Drobne, D.: Neurotoxic potential of ingested ZnO nanomaterials on bees. Chemosphere, 2015; 120, 547–554.
[84] Yin, L.; Colman, B. P.; McGill, B. M.; Wright, J. P.; Bernhardt, E. S.: Effects of silver nanoparticle exposure on germination and early growth of eleven wetland plants. PLOS ONE, 2012; 7(10), e47674.
[85] Shi, J.; Abid, A. D.; Kennedy, I. M.; Hristova, K. R.; Silk, W. K.: To duckweeds (Landoltia punctata), nanoparticulate copper oxide is more inhibitory than the soluble copper in the bulk solution. Environmental Pollution, 2011; 159(5), 1277–1282.
[86] Wang, Z.; Xie, X.; Zhao, J.; Liu, X.; Feng, W.; White, J. C., et al.: Xylem- and phloem-based transport of CuO nanoparticles in maize (Zea mays L.). Environmental Science & Technology, 2012; 46(8), 4434–4441.
[87] Pakrashi, S.; Dalai, S.; Prathna, T. C.; Trivedi, S.; Myneni, R.; Raichur, A. M., et al.: Cytotoxicity of aluminium oxide nanoparticles towards fresh water algal isolate at low exposure concentrations. Aquatic Toxicology, 2013; 132, 34–45.
[88] De, A.; Chakrabarti, M.; Ghosh, I.; Mukherjee, A.: Evaluation of genotoxicity and oxidative stress of aluminium oxide nanoparticles and its bulk form in Allium cepa. The Nucleus, 2016; 59(3), 219–225.
[89] Velzeboer, I.; Hendriks, A. J.; Ragas, A. M. J.; Meent, Dvd.: Nanomaterials in the environment aquatic ecotoxicity tests of some nanomaterials. Environmental Toxicology and Chemistry, 2008; 27(9), 1942–1947.
[90] Zhu, X.; Zhu, L.; Chen, Y.; Tian, S.: Acute toxicities of six manufactured nanomaterial suspensions to Daphnia magna. Journal of Nanoparticle Research, 2009; 11, 67–75.
[91] Nogueira, D. J.; Vaz, V. P.; Neto, O. S.; da Silva, M. L. N.; Simioni, C.; Ouriques, L. C., et al.: Crystalline phase-dependent toxicity of aluminum oxide nanoparticles toward Daphnia magna and ecological risk assessment. Environmental Research, 2020; 182:108987.
[92] Zhu, X.; Zhu, L.; Duan, Z.; Qi, R.; Li, Y.; Lang, Y.: Comparative toxicity of several metal oxide nanoparticle aqueous suspensions to Zebrafish (Danio rerio) early developmental stage. Journal of Environmental Science and Health. Part A, Toxic/Hazardous Substances & Environmental Engineering, 2008; 43(3), 278–284.
[93] Coleman, J. G.; Johnson, D. R.; Stanley, J. K.; Bednar, A. J.; Weiss, C. A.; Boyd, R. E., et al.: Assessing the fate and effects of nano aluminum oxide in the terrestrial earthworm, Eisenia fetida. Environmental Toxicology and Chemistry, 2010; 29(7), 1575–1580.
[94] Wang, H.; Wick, R. L.; Xing, B.: Toxicity of nanoparticulate and bulk ZnO, Al2O3 and TiO2 to the nematode Caenorhabditis elegans. Environmental Pollution (Barking, Essex: 1987), 2009; 157(4), 1171–1177.
[95] Yang, L.; Watts, D. J.: Particle surface characteristics may play an important role in phytotoxicity of alumina nanoparticles. Toxicology Letters, 2005; 158(2), 122–132.
[96] Klančnik, K.; Drobne, D.; Valant, J.; Dolenc Koce, J.: Use of a modified Allium test with nanoTiO2. Ecotoxicology and Environmental Safety, 2011; 74(1), 85–92.

[97] Friehs, E.; AlSalka, Y.; Jonczyk, R.; Lavrentieva, A.; Jochums, A.; Walter, J.-G., et al.: Toxicity, phototoxicity and biocidal activity of nanoparticles employed in photocatalysis. Journal of Photochemistry and Photobiology C: Photochemistry Reviews, 2016; 29, 1–28.

[98] Hund-Rinke, K.; Simon, M.: Ecotoxic effect of photocatalytic active nanoparticles (TiO2) on algae and daphnids. Environmental Science and Pollution Research International, 2006, 13, 225–232.

[99] Bar-Ilan, O.; Louis, K. M.; Yang, S. P.; Pedersen, J. A.; Hamers, R. J.; Peterson, R. E., et al.: Titanium dioxide nanoparticles produce phototoxicity in the developing zebrafish. Nanotoxicology, 2012; 6(6), 670–679.

[100] Canesi, L.; Ciacci, C.; Vallotto, D.; Gallo, G.; Marcomini, A.; Pojana, G.: In vitro effects of suspensions of selected nanoparticles (C60 fullerene, TiO2, SiO2) on Mytilus hemocytes. Aquatic Toxicology, 2010; 96(2), 151–158.

[101] Hoecke, K. V.; Schamphelaere, K. A. C. D.; Meeren, P.; Lcucas, S.; Janssen, C. R.: Ecotoxicity of silica nanoparticles to the green alga pseudokirchneriella subcapitata: Importance of surface area. Environmental Toxicology and Chemistry, 2008; 27(9), 1948–1957.

[102] Hartmann, N. B.; Von der Kammer, F.; Hofmann, T.; Baalousha, M.; Ottofuelling, S.; Baun, A.: Algal testing of titanium dioxide nanoparticles–testing considerations, inhibitory effects and modification of cadmium bioavailability. Toxicology, 2010; 269(2–3), 190–197.

[103] Wiench, K.; Wohlleben, W.; Hisgen, V.; Radke, K.; Salinas, E.; Zok, S., et al.: Acute and chronic effects of nano- and non-nano-scale TiO2 and ZnO particles on mobility and reproduction of the freshwater invertebrate Daphnia magna. Chemosphere, 2009; 76(10), 1356–1365.

[104] Puerari, R. C.; Ferrari, E.; Oscar, B. V.; Simioni, C.; Ouriques, L. C.; Vicentini, D. S., et al.: Acute and chronic toxicity of amine-functionalized SiO2 nanostructures toward Daphnia magna. Ecotoxicology and Environmental Safety, 2021; 212, 111979.

[105] Pham, D.-H.; De Roo, B.; Nguyen, X.-B.; Vervaele, M.; Kecskés, A.; Ny, A., et al.: Use of zebrafish larvae as a multi-endpoint platform to characterize the toxicity profile of silica nanoparticles. Scientific Reports, 2016; 6, 37145.

[106] Duan, J.; Yu, Y.; Li, Y.; Yu, Y.; Sun, Z.: Cardiovascular toxicity evaluation of silica nanoparticles in endothelial cells and zebrafish model. Biomaterials, 2013; 34(23), 5853–5862.

[107] Krishna Priya, K.; Ramesh, M.; Saravanan, M.; Ponpandian, N.: Ecological risk assessment of silicon dioxide nanoparticles in a freshwater fish Labeo rohita: Hematology, ionoregulation and gill Na +/K + ATPase activity. Ecotoxicology and Environmental Safety, 2015; 120, 295–302.

[108] Hu, C. W.; Li, M.; Cui, Y. B.; Li, D. S.; Chen, J.; Yang, L. Y.: Toxicological effects of TiO2 and ZnO nanoparticles in soil on earthworm Eisenia fetida. Soil Biology & Biochemistry, 2010; 42 (4), 586–591.

[109] Seeger, E.; Baun, A.; Kaestner, M.; Trapp, S.: Insignificant acute toxicity of TiO 2 nanoparticles to willow trees. Journal of Soils and Sediments, 2009; 9, 46–53.

[110] Renault, S.; Baudrimont, M.; Mesmer-Dudons, N.; Gonzalez, P.; Mornet, S.; Brisson, A.: Impacts of gold nanoparticle exposure on two freshwater species: A phytoplanktonic alga (Scenedesmus subspicatus) and a benthic bivalve (Corbicula fluminea). Gold Bulletin, 2008; 41(2), 116–126.

[111] Lovern, S. B.; Owen, H. A.; Klaper, R.: Electron microscopy of gold nanoparticle intake in the gut of Daphnia magna. Nanotoxicology, 2008; 2(1), 43–48.

[112] Li, T.; Albee, B.; Alemayehu, M.; Diaz, R.; Ingham, L.; Kamal, S., et al.: Comparative toxicity study of Ag, Au, and Ag–Au bimetallic nanoparticles on Daphnia magna. Analytical and Bioanalytical Chemistry, 2010; 398(2), 689–700.

[113] Botha, T. L.; Boodhia, K.; Wepener, V.: Adsorption, uptake and distribution of gold nanoparticles in Daphnia magna following long term exposure. Aquatic Toxicology, 2016; 170, 104–111.

[114] Pacheco, A.; Martins, A.; Guilhermino, L.: Toxicological interactions induced by chronic exposure to gold nanoparticles and microplastics mixtures in Daphnia magna. The Science of the Total Environment, 2018; 628–629, 474–483.

[115] Bar-Ilan, O.; Albrecht, R. M.; Fako, V. E.; Furgeson, D. Y.: Toxicity assessments of multisized gold and silver nanoparticles in zebrafish embryos. Small, 2009; 5(16), 1897–1910.

[116] Browning, L. M.; Lee, K. J.; Huang, T.; Nallathamby, P. D.; Lowman, J. E.; Xu, X.-H. N.: Random walk of single gold nanoparticles in zebrafish embryos leading to stochastic toxic effects on embryonic developments. Nanoscale, 2009; 1(1), 138–152.

[117] Tedesco, S.; Doyle, H.; Blasco, J.; Redmond, G.; Sheehan, D.: Oxidative stress and toxicity of gold nanoparticles in Mytilus edulis. Aquatic Toxicology, 2010; 100(2), 178–186.

[118] Vales, G.; Suhonen, S.; Siivola, K. M.; Savolainen, K. M.; Catalán, J.; Norppa, H.: Genotoxicity and cytotoxicity of gold nanoparticles in vitro: Role of surface functionalization and particle size. Nanomaterials, 2020; 10(2), 271.

[119] Barrena, R.; Casals, E.; Colón, J.; Font, X.; Sánchez, A.; Puntes, V.: Evaluation of the ecotoxicity of model nanoparticles. Chemosphere, 2009; 75(7), 850–857.

[120] Shah, V.; Belozerova, I.: Influence of metal nanoparticles on the soil microbial community and germination of lettuce seeds. Water, Air, and Soil Pollution, 2009; 197(1), 143–148.

[121] Rawat, K.; Agarwal, S.; Tyagi, A.; Verma, A. K.; Bohidar, H. B.: Aspect ratio dependent cytotoxicity and antimicrobial properties of nanoclay. Applied Biochemistry and Biotechnology, 2014; 174(3), 936–944.

[122] Morrison, K. D.; Misra, R.; Williams, L. B.: Unearthing the antibacterial mechanism of medicinal clay: A geochemical approach to combating antibiotic resistance. Scientific Reports, 2016; 6(1), 19043.

[123] Toledano-Magaña, Y.; Flores-Santos, L.; Montes de Oca, G.; González-Montiel, A.; Laclette, J.-P.; Carrero, J.-C.: Effect of Clinoptilolite and Sepiolite Nanoclays on Human and Parasitic Highly Phagocytic Cells [Research Article], 2015 [updated 2015/05/19].

[124] Kansara, K.; Kumar, A.; Karakoti, A. S.: Combination of humic acid and clay reduce the ecotoxic effect of TiO2 NPs: A combined physico-chemical and genetic study using zebrafish embryo. Science of the Total Environment, 2020; 698, 134133.

[125] Jackson, P.; Jacobsen, N. R.; Baun, A.; Birkedal, R.; Kühnel, D.; Jensen, K. A., et al.: Bioaccumulation and ecotoxicity of carbon nanotubes. Chemistry Central Journal, 2013; 7(1), 154.

[126] Liu, S.; Wei, L.; Hao, L.; Fang, N.; Chang, M. W.; Xu, R., et al.: Sharper and faster "nano darts" kill more bacteria: A study of antibacterial activity of individually dispersed pristine single-walled carbon nanotube. Acs Nano, 2009; 3(12), 3891–3902.

[127] Chung, H.; Son, Y.; Yoon, T. K.; Kim, S.; Kim, W.: The effect of multi-walled carbon nanotubes on soil microbial activity. Ecotoxicology and Environmental Safety, 2011; 74(4), 569–575.

[128] Schwab, F.; Bucheli, T. D.; Lukhele, L. P.; Magrez, A.; Nowack, B.; Sigg, L., et al.: Are carbon nanotube effects on green algae caused by shading and agglomeration? Environmental Science & Technology, 2011; 45(14), 6136–6144.

[129] Wei, L.; Thakkar, M.; Chen, Y.; Ntim, S. A.; Mitra, S.; Zhang, X.: Cytotoxicity effects of water dispersible oxidized multiwalled carbon nanotubes on marine alga, Dunaliella tertiolecta. Aquatic Toxicology, 2010; 100(2), 194–201.

[130] Long, Z.; Ji, J.; Yang, K.; Lin, D.; Systematic, W. F.: Quantitative investigation of the mechanism of carbon nanotubes' toxicity toward algae. Environmental Science & Technology, 2012; 46(15), 8458–8466.

[131] Chan, T. S. Y.; Nasser, F.; St-Denis, C. H.; Mandal, H. S.; Ghafari, P.; Hadjout-Rabi, N., et al.: Carbon nanotube compared with carbon black: Effects on bacterial survival against grazing by ciliates and antimicrobial treatments. Nanotoxicology, 2012; 7(3), 251–258.
[132] Petersen, E. J.; Pinto, R. A.; Mai, D. J.; Landrum, P. F.; Weber, W. J.: Influence of polyethyleneimine graftings of multi-walled carbon nanotubes on their accumulation and elimination by and toxicity to Daphnia magna. Environmental Science & Technology, 2011; 45 (3), 1133–1138.
[133] Kennedy, A. J.; Gunter, J. C.; Chappell, M. A.; Goss, J. D.; Hull, M. S.; Kirgan, R. A., et al.: Influence of nanotube preparation in Aquatic Bioassays. Environmental Toxicology and Chemistry, 2009; 28(9), 1930–1938.
[134] Mouchet, F.; Landois, P.; Sarremejean, E.; Bernard, G.; Puech, P.; Pinelli, E., et al.: Characterisation and in vivo ecotoxicity evaluation of double-wall carbon nanotubes in larvae of the amphibian Xenopus laevis. Aquatic Toxicology, 2008; 87(2), 127–137.
[135] Smith, C. J.; Shaw, B. J.; Handy, R. D.: Toxicity of single walled carbon nanotubes to rainbow trout, (Oncorhynchus mykiss): Respiratory toxicity, organ pathologies, and other physiological effects. Aquatic Toxicology, 2007; 82(2), 94–109.
[136] Cheng, J. P.; Cheng, S. H.: Influence of carbon nanotube length on toxicity to zebrafish embryos. International Journal of Nanomedicine, 2012; 7, 3731–3739.
[137] Larue, C.; Pinault, M.; Czarny, B.; Georgin, D.; Jaillard, D.; Bendiab, N., et al.: Quantitative evaluation of multi-walled carbon nanotube uptake in wheat and rapeseed. Journal of Hazardous Materials, 2012; 227–228, 155–163.
[138] Khodakovskaya, M. V.; Silva, Kd.; Nedosekin, D. A.; Dervishi, E.; Biris, A. S.; Shashkov, E. V., et al.: Complex genetic, photothermal, and photoacoustic analysis of nanoparticle-plant interactions. Proceedings of the National Academy of Sciences, 2011; 108(3), 1028–1033.
[139] Gagnon, V.; Button, M.; Boparai, H. K.; Nearing, M.; O'Carroll, D. M.; Weber, K. P.: Influence of realistic wearing on the morphology and release of silver nanomaterials from textiles. Environmental Science: Nano, 2019; 6(2), 411–424.
[140] Mohan, S.; Princz, J.; Ormeci, B.; DeRosa, M. C.: Morphological transformation of silver nanoparticles from commercial products: Modeling from product incorporation, weathering through use scenarios, and leaching into wastewater. Nanomaterials (Basel, Switzerland), 2019; 9(9), 1258.
[141] Abbas, Q.; Yousaf, B.; Amina, A. M. U.; Munir, M. A. M.; El-Naggar, A., et al.: Transformation pathways and fate of engineered nanoparticles (ENPs) in distinct interactive environmental compartments: A review. Environment International, 2020; 138, 105646.
[142] Gorka, D. E.; Lin, N. J.; Pettibone, J. M.; Gorham, J. M.: Chemical and physical transformations of silver nanomaterial containing textiles after modeled human exposure. NanoImpact, 2019; 14, 100160.
[143] Heinlaan, M.; Muna, M.; Knobel, M.; Kistler, D.; Odzak, N.; Kuhnel, D., et al.: Natural water as the test medium for Ag and CuO nanoparticle hazard evaluation: An interlaboratory case study. Environmental Pollution, 2016; 216, 689–699.
[144] Saison, C.; Perreault, F.; Daigle, J. C.; Fortin, C.; Claverie, J.; Morin, M., et al.: Effect of core-shell copper oxide nanoparticles on cell culture morphology and photosynthesis (photosystem II energy distribution) in the green alga, Chlamydomonas reinhardtii. Aquatic Toxicology, 2010; 96(2), 109–114.
[145] Cesa, F. S.; Turra, A.; Baruque-Ramos, J.: Synthetic fibers as microplastics in the marine environment: A review from textile perspective with a focus on domestic washings. Science of the Total Environment, 2017; 598, 1116–1129.

[146] Naasz, S.; Altenburger, R.; Kuhnel, D.: Environmental mixtures of nanomaterials and chemicals: The Trojan-horse phenomenon and its relevance for ecotoxicity. Science of the Total Environment, 2018; 635, 1170–1181.
[147] Path, N. V.; Netravali, A. N.: Direct assembly of silica nanospheres on halloysite nanotubes for "green" ultrahydrophobic cotton fabrics. Advanced Sustainable Systems, 2019; 3(8), 1900009.
[148] OECD: Guidance Document on Aquatic and Sediment Toxicological Testing of Nanomaterials. Nr 317. 2020(ENV/JM/MONO(2020)8): OECD, Paris.

Ellen Bendt*, Maike Rabe, Sabrina Kolbe, Susanne Küppers, Stefan Brandt, Jens Meyer, Malin Obermann, Karin Ratovo

19 Micro/nanoplastics

Keywords: textile-based microplastic, fibrous microplastic household washing studies, fiber shedding, PES fleece fabric, improved materials, performance profiles

19.1 Microplastics

Micro-sized particles of plastics, called “microplastics,” have turned out to be an environmental problem in all water bodies. There is currently no standard definition of microplastics. The most common definition is that plastic particles with a diameter <5 mm are microplastics [1], but differentiations between large microplastics (5–1 mm), microplastics (1 mm to 1 µm) and nanoplastics (<1 µm) can also be found in the literature [2].

A distinction is made between primary and secondary microplastic particles. The primary microplastic particles are usually present as so-called microbeads, for example, in cosmetics, whereas secondary microplastic particles result from the comminution of larger plastic parts such as PET bottles [1], or the abrasion of synthetic textiles which can result in atmospheric microplastics as well.

The abrasion of synthetic textiles is often referred to as microfibers, which is actually a completely unrelated technical term for a certain type of fibers. Correctly, the residue should be referred to as fibrous microplastic must be considered separately and not be equated with each other.

Microfibers are fibers whose fineness is less than 1 dtex [3, 4] (1 g fiber material is needed to obtain 10 km fibers). Generally, the fineness of microfibers is between 0.3 and 1.0 dtex, the fiber diameter between 3 and 10 µm depending on the density. They are usually made of synthetic polymers such as polyester (PES), polyacrylate and polyamide, but regenerated natural polymers such as cellulose can also be processed into microfibers (viscose, modal or Lyocell) [5]. The fineness of natural fibers is lower than that of man-made fibers and only the finest natural fibers come close to the size of the microfiber range. Microfibers themselves as well as any other synthetic fibers can become fibrous microplastic particles when broken down mechanically or by decomposition.

***Corresponding author: Ellen Bendt,** Fachbereich Textil- und Bekleidungstechnik, Hochschule Niederrhein, Webschulstr. 31, 41065 Mönchengladbach, Germany,
e-mail: ellen.bendt@hs-niederrhein.de
Maike Rabe, Sabrina Kolbe, Susanne Küppers, Stefan Brandt, Jens Meyer, Malin Obermann, Karin Ratovo, Fachbereich Textil- und Bekleidungstechnik, Hochschule Niederrhein, Webschulstr. 31, 41065 Mönchengladbach, Germany, e-mail: ellen.bendt@hs-niederrhein.de

https://doi.org/10.1515/9783110670776-019

For a distinction between fibrous microplastic and microfibers, a clear definition is needed, even more as the proportion of microfibers among all synthetic fibers is continuously increasing. Within the BMBF research project "TextileMission" (13NKE010B, https://textilemission.bsi-sport.de), the following attempt of a definition of fibrous **microplastics** is made:

- Microfibers are not microplastics, in general, but are defined as fibers with a linear density between 1 and 0.3 dtex
- Fibrous microplastics are fibre fragments and fibers with different diameters, a fiber length below 5 mm and an aspect ratio of at least 3:1

When microplastics are fragmented due to physical or chemical stress, nanoplastics can be obtained.

19.2 Nanoplastics

As with microplastics, there is no uniform definition of textile-based nanoplastics. Various definitions are proposed in the literature: Gigault et al. defined nanoplastics as particles unintentionally produced through physical and mechanical breakdown, showing colloidal behavior with a size range of 1–1,000 nm [6, 7]. Other working groups define nanoplastics as particles with a size smaller than 100 nm in one dimension [8, 9]. The EU adopted a definition of a nanomaterial in 2011 (Recommendation on the definition of a nanomaterial (2011/696/EU)). These nanomaterials are defined as follows: "The International Organisation for Standardisation defines the term 'nanomaterial' as 'material with any external dimensions in the nanoscale or having internal structure or surface structure in the nanoscale'. The term 'nanoscale' is defined as size range from approximately 1 nm to 100 nm" [10]. In general, the produced nanoparticles like latex particles are not suitable for toxicological investigations because they show another behavior as degraded nanoplastics. But it is very difficult to detect and identify such nanoparticles in the environment [9].

19.3 Toxicology of micro/nanoplastics

Microplastics and nanoplastics can cause toxicological problems if they enter any environment. The best studied route is the aquatic one. Various organisms such as *Daphnia*, mussels, zooplankton and algae can actively ingest nanoplastic particles or adsorb them on their surfaces [11]. Feeding experiments with fish have shown that toxicological responses typically arise from smaller particles, but with particles >100 µm in size show no significant effect in some studies [12]. Based on the concentration of ingested particles, both acute toxicity and chronic effects have been observed.

A major problem regarding toxicity is the property of adsorption of organic pollutants or chemical substances on the surface of the particles which could potentially enhance the uptake in an organism and enhance toxicity [12]. Possible effects can be the hindering of the photosynthetic activity of green algae or the inhibition of growth due to the uptake of sharp polyethylene microplastics by various species [13].

Microplastic and nanoplastic accumulation in mammalian and human tissues would probably have negative but unclear long-term consequences [12]. The main uptake route in humans is eating and drinking of contaminated food and/or beverages [11] but also dermal absorption and inhalation are possible routes, and it was found in human stool samples in relatively high concentrations [12]. In human beings, exposure to nanoplastics may be described as (1) intentional with the use of personal care products or biomedical applications, (2) unintentional through intentional plastic use like plastic water bottles, (3) unintentional exposure by inhalation as part of air pollution or digestion through food production [7].

The uptake routes are generally not only dependent on size and surface chemistry but also cell-type specific [11]. The adverse effects of nanoplastics are cytotoxicity, inflammation and the production of reactive oxygen species (ROS) [11] but there is currently a lack of data on toxicity to humans in vivo [12]. There are some experiments with cell cultures where cytotoxic effects as well as ROS production and proinflammatory responses are reported [12].

19.4 Quantifying the problem

It should be noted that the textile value chain has numerous interfaces with the hydrosphere and thus also many entry routes into it. Even if various studies have shown that the abrasion of car tires, cosmetics or even paints and lacquers are a major cause of the environmental microplastic problem [1], the textile industry has to find solutions.

There are no exactly reliable research results on how many microplastic particles from the textile sector are released into the environment, especially by washing or wearing textiles. One possible entry route for fibers into the environment is household laundering. Most of the fibrous microplastic from washing machine effluents is filtered out in sewage treatment plants (where available), but according to several studies, between 5–16% cannot be retained and directly enter water bodies. The main part of the fibrous microplastic that is retained ends up in sewage sludge. However, this can become another source when used as a fertilizer. From there, fibrous microplastic can further spread, e.g. via wind or groundwater.

In general, it is important to note that due to different test procedures and methods the results of similar research projects on household laundry in the past are not really comparable.

Also in Germany, the exact quantities on microplastic emissions are not known: The German Federal Environment Agency reports that 80–400 tons of textile-based microplastic particles are released into the environment per year [14].

To get further and more detailed information, the Federal Ministry of Education and Research in Germany (BMBF) started the initiative "Plastics in the Environment" with a network of 20 different joint research projects, all focusing on microplastic emissions. The aim of the subproject called "TextileMission" was to quantify fibrous microplastic emissions from household laundry, and to determine their pathways and fate in the marine muss weg, geht um Umwelt und Gewässer allgemein environment.

To study the phenomen in depth a textile material with high potential of fibrous microplastic emissions and a great market importance was selected.

PES fleece with its open and fluffy surface – very popular for sportswear, outdoor clothing and fashion items – seems to cause several problems. Therefore, it is important to know the general construction and treatment of fleece fabrics. A fleece fabric is a multilayer plush knit. It contains at least two yarns, one on the inside and another one on both outsides, as shown in Figure 19.1. In order to reach volume, hand feel and performance, a mechanical treatment like brushing and/or shearing of the fabric is needed. Those strong mechanical treatments destroy the surface and could therefore lead to a higher release of microplastics (Figure 19.2). For this reason, the main focus of the TextileMission project was on PES fleece garments.

Figure 19.1: Visible multilayer structure of a fleece fabric with black (inside) and turquoise (surface) yarns after brushing.

After material selection, several washing studies were performed. Common front-loading washing machines were used to simulate the average German household washing procedures: washing load of 3.4 kg (±0.1), washing temperature of 40 °C, 900 rpm, liquid washing agent, washing liquor of 48 L and washing cycle duration of 1 h and 59 min.

Figure 19.2: Fleece fabric after (left) and before brushing (pictures from Niederrhein University, 2020).

The subsequent microplastic quantification was done by filtration of the total washing liquor. For this purpose, a five-stage filter cascade, designed by the Technische Universität Dresden (TUD), was installed (Figure 19.3).

(A) (B)

Figure 19.3: Washing procedures (A) filtration (B) washing process (pictures from Carlos Albuquerque, 2018).

The used set of steel filters had the following pore sizes: 1.5 mm, 0.5 mm, 0.15 mm, 50 µm and 5 µm.

Figure 19.4: Fibrous microplastic residues on finest filter with 5 µm, after the second washing cycle of 1.4 kg (±0.1), black PES fleece garment (pictures from Niederrhein University, 2018).

The obtained fibrous microplastic particles (Figure 19.4) were analysed via optical- and scanning electron microscopy, µ-FTIR and TED-GC/MS (the latter by Bundesanstalt

für Materialforschung und –prüfung, BAM). The results verified the particles as 99 % PET and gave information on typical size ranges.
A typical progression of fibrous microplastic release after 10 washing and drying cycles is depicted in Figure 19.5.

The highest discharge was detected during the first two washing cycles. Starting with the third wash, the values for fiber discharge decreased slightly until a plateau phase with a low fiber release was reached. In total, a fibrous microplastic release of 50–600 mg could be collected over 10 washing cycles. When drying in the dryer, the course of fibrous microplastic emissions during the cycles followed a similar trend to that of the washing cycles, but in each case, the overall fibrous microplastic discharge in the drying process was significantly higher compared to the washing process. If the laundry was not dried in the dryer but air-dried, the amount of fibrous microplastic discharged in the washing process increased. However, the discharge output was lower than in the respective washing and drying process combined.

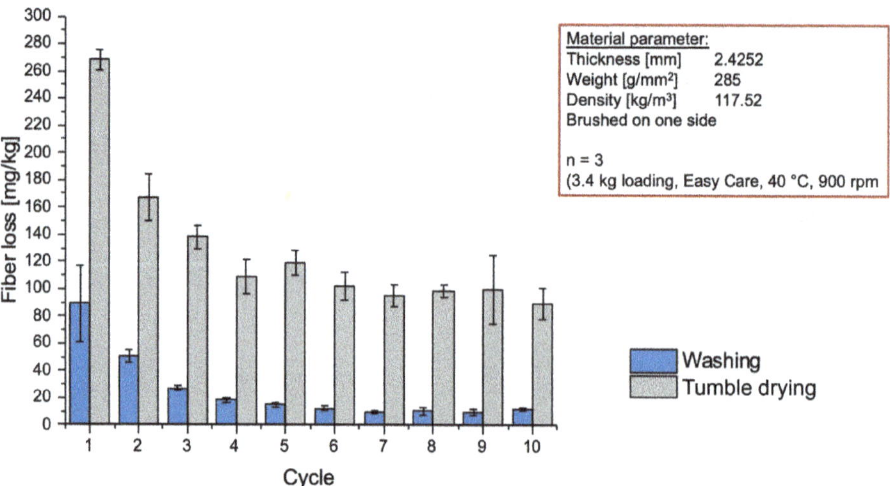

Figure 19.5: Progression of fibrous microplastic release after 10 washing and tumble drying cycles after washing and drying of 3.4 kg (±0.1) fleece jackets (picture from Niederrhein University, 2018).

Washing trials with a lower load of 1.4 kg (±0.1), compared to 3.4 kg (±0.1), showed the same behavior in fibrous microplastic emission, but the overall discharge became higher due to the higher mechanical stress of the textiles caused by the higher drop height in the spinning washing drum.

Figure 19.6 shows the washing results using different fleece jackets compared to a smooth plain PES garment without mechanical finishing. The total discharge of 10 washing and 10 drying cycles shows that the smooth garment has a similar fibrous microplastic output as a fleece material with low fiber discharge.

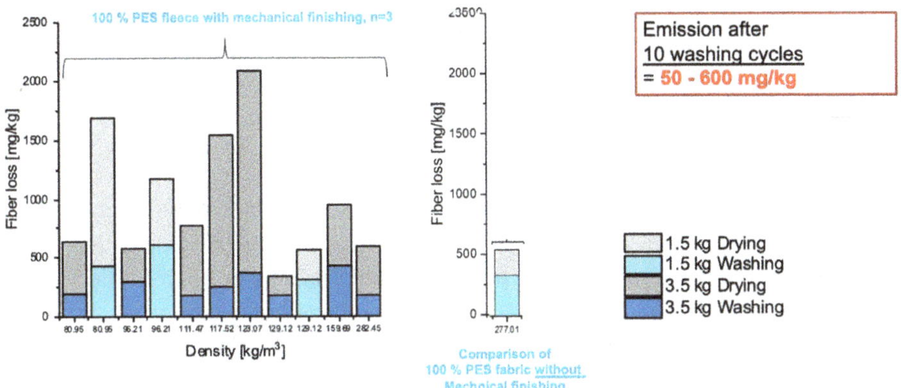

Figure 19.6: Fiber release after 10 washing and tumble drying cycles of different fleece jackets with a machine load of 3.4 kg (±0.1), and 1.4 kg (±0.1), in comparison to a 100% PES material without mechanical finishing (1.4 kg) (pictures from Niederrhein University, 2019).

An increased amount of microplastic release in the first washing cycle was detected for all tested materials. The mechanical brushing and shearing which is needed to get the voluminous surface of fleece could lead to an increased shedding potential. As an interesting fact shown in Figure 19.6, even plain-knitted fabric that did not undergo any brushing and shearing showed a high amount of microplastic release in the first wash. From this it can be concluded that, in addition to the mechanical finishing treatment of the material, the processes of cutting, making and trimming (CMT) of a garment also plays an important role in influencing fibrous microplastic discharge.

19.5 Solutions to reduce environmental impact

There are several concepts/solutions to reduce the textile-based impact of microplastics in the environment.

1. *Wastewater treatment plants*

In countries where sewage treatment plants are used for water treatment, fibrous microplastic entry into the aquatic system is significantly reduced. The European Union postulates an average retention rate of 53–84% [15], in Germany of 84–95% [15, 16] and in the canton of Zurich in Switzerland of 93% [17]. However, in Switzerland, it is postulated that fibers and fibrous microplastic (retention rate of 76%) are much more difficult to retain than particles (97%) or spherules (87%) [17]. A further problem is the distribution of sewage sludge with the microplastic particles contained therein for landscaping and agriculture [18]. In Germany, it is allowed to use 28% of the sewage sludge in this form [18]. The regulation on the reorganization of sewage sludge utilization of

27 September 2017 regulates that large communal sewage treatment plants which treat the wastewater of more than 100,000 or 50,000 inhabitants may only utilize sewage sludge for agricultural or landscaping purposes until 2029 or 2032, respectively [19]. Another solution could be the introduction of an additional fourth filter stage.

2. *Use of filters in washing machines*
Another way to reduce the release of microplastics into the environment could be to use filters which are directly connected to the washing machine outflow [20]. According to studies by the Research Institute for Textile and Clothing of Hochschule Niederrhein (FTB), most fibrous microplastic can be collected using filters with a pore size of 5 µm (see Figure 19.4). However, due to the high load of dirt, hair, natural fibres, microplastic particles, etc. in washing machine effluent, filters with a very large surface area are needed, which would then also need to be replaced regularly, to prevent clogging. Furthermore, the pumps used in current washing machines are not designed to generate the high pressures needed for filtration processes, which makes retrofitting difficult.

3. *Reducing fibre discharge of textiles*
One solution to reduce fiber shedding could be the development of new or improved materials with lower shedding potential. In the "TextileMission" project at Hochschule Niederrhein, the approaches within the development are twofold:
1. Improvement of the common/classic PES fleece
2. Combining PES yarn in the inside with cellulosic yarn on the outsides in order to lower the PES shedding

For this purpose, an industrial-scale circular knitting machine for plush knitting (Mayer Cie. MPU 1.6, gg24) was used to successively vary different parameters in the knitting process. It was found that a variation of yarn type and yarn tension, stitch size and fabric take down impacts the shedding of the material.

In general, fleece garments made out of synthetic fibres are widely used for outdoor garments, sportswear as well as fashion items due to their special properties like high insulation, light weight and rapid drying. There is a wide range of performance characteristics that different types of fleece materials fulfill. Fleece fabrics can be both highly insulating with high air permeability and medium insulating with very low air permeability.

It should be noted that a change in the material and production parameters always has a direct impact on the properties and performance profile of the fabric (Figure 19.7). In particular, the yarns used have an effect not only on the shedding potential but also on the touch and physiology of the fabric.

Due to this wide range of properties, the development of alternative materials cannot aim at one product that is suitable for all purposes.

In addition to the different parameters of the knitting process, the following steps of coloration, chemical and mechanical finishing of the fabric like dyeing, shearing

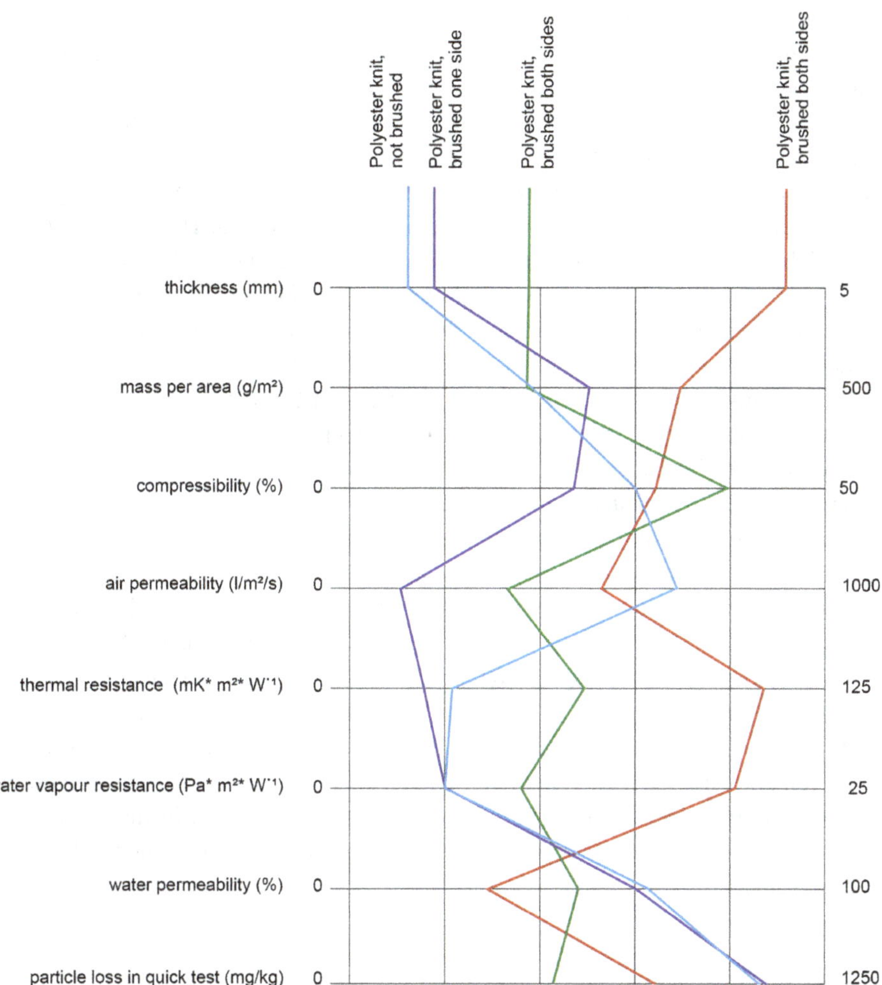

Figure 19.7: Performance profiles of different knitted fabrics made of PES in relation with fiber shedding picture Niederrhein University 2020.

and brushing, as well as cutting, making and trimming (CMT) of the garment also impact the microplastics shedding immensely. Therefore, different fabric treatments and cutting technologies, both mechanical and in combination with thermal and laser cutting, as well as different assembling techniques such as ultrasonic welding were also analyzed.

4. *Use of biodegradable polymers*

There are more and more polymers and -products that are marketed as "bio". However, this claim is misleading in some cases. For example, there are polymers that are labeled "biobased." These are polymers that are produced at least in part from biomass -

but this does not necessarily mean that they are also biodegradable. On the other hand, there are fossil-based plastics that are biodegradable, i.e. can be decomposed by microorganisms into water and carbon dioxide. So it is possible that, on the one hand, biobased materials are not biodegradable because of their chemical structure but, on the other hand, materials from fossil sources are very well biodegradable. Furthermore, environmental conditions also play an important role, which is why even supposedly biodegradable materials could not be decomposed for decades. The most commonly used biobased and simultaneous biodegradable polymer in the textile industry is polylactic acid (PLA) [4]. PLA is a linear aliphatic polyester [3], and its properties such as density or moisture regain are similar to those of the most common polyester, polyethylene terephthalate (PET). Due to all similarities, PLA has the potential to replace PET in textile industry. However, PLA behaves similar to PET in the environment, and biological degradation usually only takes place at temperatures above 60 °C which makes it only a limited alternative fiber solution for alternative sustainable materials [20].
In summary, it can be stated that the release of microplastics by textiles varies greatly and can be detected along the entire process chain from yarn to fabric production, the various finishes and further garment processing. The production line must be seen as a whole, and a holistic approach must be found to successfully reduce the microplastics shedding of fleece and other fabrics or garments made of synthetic fibers. The amount of discharge can be influenced by various measures, but this also changes the appearance and performance of the material, which is particularly important for activewear and sportswear. In addition to household laundry, the various wet treatments during textile production, such as dyeing and rinsing processes, are obviously another problematic interface with the hydrosphere. In the future, this will be the next starting point for a sustainable reduction of emissions of fibrous microplastics.

References

[1] Boucher, J.; Friot, D.: Primary Microplastics in the Oceans – A Global Evaluation of Sources. International Union for Conservation of Nature and Natural Resources, 2017 Gland, Switzerland.

[2] ISO/TR 21960: Technical Report, Plastics – Environmental aspects – State of knowledge and methodologies, ISO, 2020/02.

[3] Koslowski, H.-J.: Chemiefaser – Lexikon, 12., erweiterte Auflage. Frankfurt am Main: Deutscher Fachverlag, 2009. ISBN 978-3-87150-876-9.

[4] Bobeth, W. (Hrsg.): Textile Faserstoffe. Beschaffenheit Und Eigenschaften. Berlin / Heidelberg / New York: Springer-Verlag, 1993. ISBN 3-540-55697-4.

[5] https://www.ivc-ev.de/live/index.php?page_id=12. Industrievereinigung Chemiefaser e.V., Stand 18.10.2018.

[6] Gigault, J.; ter Halle, A.; Baudrimont, M.; Pascal, P.-Y.; Gauffre, F.; Phi, T.-L.; El Hadri, H.; Grassl, B.; Reynaud, S.: Current opinion: What is nanoplastic?. Environmental Pollution, 2018; xxx, 235 1–5.
[7] Stapelton, P.: Toxicological considerations of nano-sized plastics. AIMS Environmental Science, 2019; 6(5), 367–378.
[8] Schwaferts, C.; Niessner, R.; Elsner, M.; Ivleva, N. P.: Methods for the analysis of submicrometer- and nanoplastic particles in the environment. Trend in Analytical Chemistry, 2019; 112, 52–65.
[9] Ekvall, M. T.; Lundquist, M.; Kelpsiene, E.; Sileikis, E.; Gunnarsson, S. B.; Cedervall, T.: Nanoplastics formed during the mechanical breakdown of daily-use polystyrene products. Nanoscale Advances, 2019; 1, 1055–1061.
[10] Recommendation 2011/696/EU: Official Journal of the European Union, 20.11.2011, Commission Recommendation of 18 October 2011 on the definition of nanomaterial (Text with EEA relevance) (2011/696/EU).
[11] Lehner, R.; Weder, C.; Petri-Fink, A.; Rothen-Rutishauser, B.: Emergence of nanoplastic in the environment and possible impact on human health. Environmental Science & Technology, 2019; 53(4), 1748–1765.
[12] Qian Ying Yong, C.; Valiyaveetill, S.; Tang, B. L.: Toxicity of microplastics and nanoplastics in mammalian systems. International Journal of Environmental Research and Public Health, 2020; 17, 1509.
[13] Jungblut, S.; Liebich, V.; Bode-Dalby, M.: YOUMARES 9 – The Oceans: Our Research, Our Future. Proceedings of the 2018 conference for YOUng MArine RESearcher in Oldenburg, Germany Springer Open, 2020. ISBN 978-3-030-20388-7, 2020.
[14] Miklos, D.; Obermaier, N.; Jekel, M.: Mikroplastik: Entwicklung eines Umweltbewertungskonzepts_Erste Überlegungen zur Relevanz von synthetischen Polymeren in der Umwelt, Masterarbeit, Umweltbundesamt, 2015.
[15] Simon, H.; Chris, S.; Olly, J.; Molly, H.; Peter, K.; Ayesha, B.; George, C.: Investigation options for reducing releases in the aquatic environment of microplastics emitted by (but not intentionally added in) products-final report: Eunomia Research & Consulting Ltd.; ICF, Feb. 2018. [cited 2019 Jan 23]. Available from: URL: https://www.eunomia.co.uk/reports-tools/investigating-options-for-reducing-releases-in-the-aquatic-environment-of-microplastics-emitted-by-products/.standard for a quick test.
[16] Jürgen, B.; Ralf, B.; Leandra, H.: Kunststoffe in der umwelt: Mikro- und Makroplastik: Ursachen, Mengen, Umweltschicksale, Wirkungen, Lösungsansätze, Empfhelungen: Fraunhofer-Institut für Umwelt-, Sicherheits- und Energietechnick UMSICHT, June 21. 2018. Available from: URL: https://www.umsicht.fraunhofer.de/content/dam/umsicht/de/dokumente/publikationen/2018/kunststoffe-id-umwelt-konsortialstudie-mikroplastik.pdf
[17] Cabernard, L.; Durisch-Kaiser, E.; Vogel, J.-C.; Rensch, D.; Niederhauser, P.; Gewässerschutz, A.: Mikroplastik in Abwasser U. Gewässern. Aqua & Gas 2016 [cited 2019 Mar 20]; (7/8). Available from: URL: http://www.bildungsdirektion.ch/internet/baudirektion/awel/de/wasser/gewaesserschutz/abwasserreinigung/tagungen_und_referate/_jcr_content/contentPar/downloadlist/downloaditems/1251_1540474652419.spooler.download.1469019564271.pdf/Mikroplastik+in+Abwasser+u+Gew%C3%A4ssern.pdf
[18] BMU Statistisches Bundesamt, 2018: Entsorgung und Verwertung von Klärschlamm [cited 2019 Mar 21]. Available from: URL: https://www.bmu.de/themen/wasser-abfall-boden/abfallwirtschaft/statistiken/klaerschlamm/entsorgung-und-verwertung-von-klaerschlamm-infografik/.

[19] Faktenpapier „Mikroplastik beim Waschen und Pflegen von Textilien sowie beim Reinigen von Oberflächen im Haushalt“ (TEIL 2), Eintrag von Mikroplastik aus Textilien. Forum Waschen, 2020, Frankfurt am Main.
[20] Ministry of Ecological and Solidarity Transition (France): Anti-waste law for a circular economy, January 2020.
[21] Zhang, Y. et. al., Microplastics from textile origin - emission and reduction, Green Chemistry, 2021, 23, 5247–5271
[22] Abschlusspublikation “TextileMission”, Bundesverband der Deutschen Sportartikel-Industrie e.V., June 2021: https://textilemission.bsi-sport.de/fileadmin/assets/Abschlussdokument-2021/TextileMission_Abschlussdokument_Textiles_Mikroplastik_reduzieren.pdf

VI Governance/legislation/EU legal framework-REACH

Frauke Averbeck*, Angelina Gadermann

20 Nanomaterials and REACH

Keywords: REACH regulation, amendment, substances in nanoform, regulation (EU) 2018/1881, REACH annex revision

Nanomaterials have never been excluded from the Regulation (EC) 1907/2006 (REACH), nevertheless, REACH has been amended to explicitly address their specific properties as well as their hazard and risk assessment.

20.1 The REACH regulation

The REACH (**R**egistration, **E**valuation and **A**uthorisation of **Ch**emicals) Regulation (EC) no. 1907/2006 lays down the requirements for the registration of chemicals and for their evaluation as well as, if necessary, for regulatory risk management.

Chemicals manufactured in or imported into the EU above 1 tpa (ton per annum) have to be registered before they are placed on the market. For this purpose, manufacturers and/or importers are required to submit a dossier containing all relevant information on the substance and demonstrating its safe use to the European Chemicals Agency (ECHA). The information requirements to be fulfilled for this purpose increase with increasing tonnage and are listed in specific Annexes to the REACH Regulation: While Annex VI prescribes how a substance is identified and described under REACH, Annexes VII to X lay down the information needed to clarify the physicochemical and the (eco)toxicological properties of the substance. When this information is not available, testing has to be performed or, in case of animal studies, a testing proposal needs to be submitted. The hazard information together with information on exposure of the substance provides the basis for the chemical safety assessment (CSA) according to Annex I, which has to be performed for substances placed on the market above 10 tpa.

REACH also provides authorities with regulatory instruments to ban or limit the use of a substance under certain circumstances: A substance can become subject to authorization or restriction. These provisions as well as some other provisions of the REACH regulation (such as passing information down in the supply chain) apply regardless of the tonnage (i.e., also below 1 tpa).

***Corresponding author: Frauke Averbeck,** Bundesanstalt für Arbeitsschutz und Arbeitsmedizin – BAuA, Bundesstelle für Chemikalien, Chemikalienbewertung und Risikomanagement, Friedrich-Henkel-Weg 1 – 25, 44149 Dortmund, Germany, e-mail: averbeck.frauke@baua.bund.de
Angelina Gadermann, Bundesanstalt für Arbeitsschutz und Arbeitsmedizin – BAuA, Bundesstelle für Chemikalien, Chemikalienbewertung und Risikomanagement, Friedrich-Henkel-Weg 1 – 25, 44149 Dortmund, Germany

https://doi.org/10.1515/9783110670776-020

20.2 Textiles in REACH

Even though the focus of REACH is mainly on chemical substances and their safe use, there are also links to textiles: Firstly, chemical substances are obviously used in the production processes of textiles, for example, as solvents, as processing aids or for dyeing of textiles. Secondly, REACH also includes provisions for substances in articles, and hence for substances in textiles as these fall under the legal definition of an article under REACH:

> *Article = an object which during production is given a special shape, surface or design which determines its function to a greater degree than does its chemical composition.*

Only under certain specific circumstances, chemical substances in articles are subject to registration obligations, whereas the main provision on articles within REACH is focusing on the duty to communicate information on and notification of hazardous substances in articles as well as on the restriction of substances in articles (see figure 20.1).

In practice, this means that a supplier of an article needs to inform its recipient in case a so-called substance of very high concern[1] (SVHC), included in the candidate list, is present in the article in a concentration above 0.1% w/w. A consumer can also request this information from the supplier and retailer and has to be informed within 45 days. In addition, a notification to ECHA becomes necessary for producers or importers of articles if the 0.1% w/w threshold as well as a quantity totaling over 1 ton per SVHC per year is reached. Next to this, a number of entries in Annex XVII of the REACH Regulation, in which banned or restricted substances are listed, are dealing with articles, in general, or textiles, in particular.

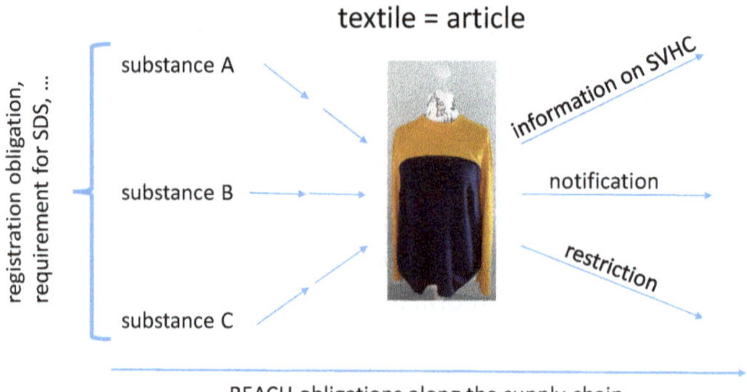

Figure 20.1: Obligations related to substances in textiles according to REACH.

1 Defined in Article 57 of the REACH Regulation: Substances with CMR, PBT, vPvB properties or other properties of equivalent level of concern.

20.3 Nanomaterials and REACH

There is (and always has been) general agreement in the EU that nanomaterials are substances within the meaning of REACH. This means that the obligations outlined above have in principle always also applied to nanomaterials. However, there was neither a need to indicate if a substance is placed on the market as nanomaterial nor was there an explicit obligation to properly characterize and address nanomaterials separately in the hazard and risk assessment of a substance.

With the amendment of the REACH Regulation introduced by Regulation (EC) No. 2018/1881[2] which applies since 1 January 2020, a binding and harmonized regulation of nanomaterials has been put in place. In the negotiations about possible changes to the REACH Regulation, the right balance had to be found between the benefit for human health and the environment from the introduction of new information requirements and the associated costs for industry. This was especially considered when new studies are to be performed for already registered nanomaterials. Therefore, the introduced changes mainly focus on the definition of a nanomaterial in the sense of REACH as well as on the adaptation of the existing information requirements with regard to nanomaterials. Effectively, only a few new information requirements specifically addressing nanomaterials have been added. In 2020, also the Regulation (EU) 2020/878[3] entered into force in which inter alia nanospecific requirements for REACH Annex II describing safety data sheet (SDS) provisions are specified.

20.4 Introduced amendments in detail

The nanospecific amendments only concern the Annexes of the REACH Regulation, in particular Annexes I, II, III, VI and VII–XII. Amendments to Annexes III, XI and XII are only minor and will not be further addressed here. For a better understanding table 20.1 provides an overview of the different REACH Annexes and their content.

Table 20.1: Overview of REACH Annexes and their content.

Annex of REACH	Content
I	General provisions for assessing substances and preparing chemical safety reports
II	Guide to the compilation of safety data sheets (SDS)

2 https://eur-lex.europa.eu/legal-content/EN/TXT/PDF/?uri=CELEX:32018R1881&from=EN.
3 https://eur-lex.europa.eu/legal-content/EN/TXT/PDF/?uri=CELEX:32020R0878&from=DE.

Table 20.1 (continued)

Annex of REACH	Content
III	Criteria for substances registered in quantities between 1 and 10 tons
IV*	Exemptions from the obligation to register in accordance with Article 2(7)(a)
V*	Exemptions from the obligation to register in accordance with Article 2(7)(b)
VI	Information requirements referred to in Article 10 (general information needed for registration and evaluation)
VII	Standard information requirements for substances manufactured or imported in quantities of 1 ton or more
VIII	Standard information requirements for substances manufactured or imported in quantities of 10 tons or more
IX	Standard information requirements for substances manufactured or imported in quantities of 100 tons or more
X	Standard information requirements for substances manufactured or imported in quantities of 1,000 tons or more
XI	General rules for adaptation of the standard testing regime set out in Annexes VII to X
XII	General provisions for downstream users to assess substances and prepare chemical safety reports
XIII*	Criteria for the identification of persistent, bioaccumulative and toxic substances, and very persistent and very bioaccumulative substances
XIV*	List of substances subject to authorization
XV*	Dossiers
XVI*	Socioeconomic analysis
XVII*	Restriction on the manufacture, placing on the market and use of certain substances, mixtures and articles

*No nanospecific amendments introduced.

Please note that in the following, the Annexes are quoted in order of importance of the amendment and not in numerical order.

20.5 Annex VI

The most important change is the inclusion of a definition into Annex VI. In the negotiations, how to best include nanomaterials in REACH, it was decided that the nanoproperty should be regarded as a characterizer. Therefore, and in order to be coherent with existing definitions under REACH, like the definition of a substance, nanomaterials are considered to be a specific form of a substance. Hence, based on the Commission recommendation for a definition of the term "nanomaterial" of October 2011[4] Annex VI now includes a definition for a "nanoform of a substance":

> A nanoform is a form of a natural or manufactured substance containing particles, in an unbound state or as an aggregate or as an agglomerate and where, for 50% or more of the particles in the number size distribution, one or more external dimensions is in the size range 1 nm-100 nm, including also by derogation fullerenes, graphene flakes and single wall carbon nanotubes with one or more external dimensions below 1 nm.
>
> For this purpose, "particle" means a minute piece of matter with defined physical boundaries; "agglomerate" means a collection of weakly bound particles or aggregates where the resulting external surface area is similar to the sum of the surface areas of the individual components and "aggregate" means a particle comprising of strongly bound or fused particles.

Next to this definition, several characterization parameters (size, shape, surface area and functionalization) were included in Annex VI. These parameters need to be provided by each individual registrant for each nanoform or set of nanoforms. In order to avoid that each individual nanoform has to be fully characterized (and hence tested), the possibility to form "sets of similar nanoforms" was introduced by the legislator. The condition for clustering similar nanoforms into one set is that the complete hazard, exposure and risk assessment can be performed jointly for all nanoforms covered by the set.

A set can be built either by a single registrant manufacturing different nanoforms or by different registrants of a joint submission. The latter option of clustering similar nanoforms of different registrants into sets is exemplified schematically in Figure 20.2. For each individual set, a scientific justification why the different nanoforms can be clustered together needs to be provided.

The amended Annex VI does not only provide clear legal requirements for characterization of nanomaterials to registrants but also enables ECHA to request this information in evaluation processes.

20.6 Annexes VII to X

Annexes VII to X of REACH contain the tonnage-based information requirements that need to be submitted with the registration of a substance. The Annexes contain two-column tables: while column 1 lists the information requirements, the information in column 2 provides further specifications or possibilities for adaptations and/

4 https://eur-lex.europa.eu/legal-content/EN/TXT/PDF/?uri=CELEX:32011H0696&from=EN.

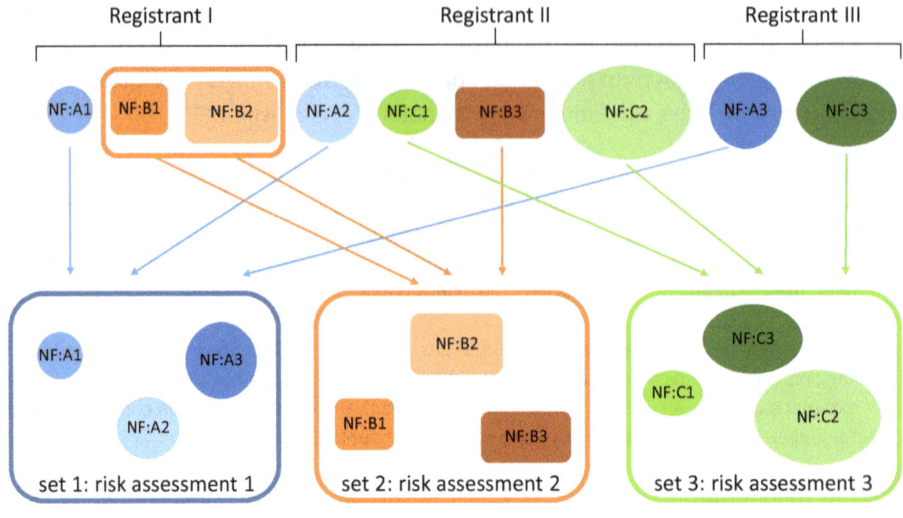

Figure 20.2: Nanoforms manufactured by a single registrant, covered in a joint registration.

or waiving of the requirement (Table 20.2). Most of the nanospecific changes to Annexes VII to X concern column 2 information, meaning that it is specified how nanospecific aspects should be taken into account when considering the need for respective testing. For example, some studies for bulk substances can be waived according to column 2 if the substance is poorly soluble in water. For nanoforms of substances, the amended column 2 now specifies that waiving based on poor water solubility alone is not justified.

Table 20.2: Example of specifying information requirements for nanoforms laid down in Annex VII for water solubility.

Column 1 **Standard information required**	**Column 2** **Specific rules for adaptation from column 1**
7.7. Water solubility *For nanoforms, in addition, the testing of the dissolution rate in water as well as in relevant biological and environmental media shall be considered.*	7.7. The study does not need to be conducted if: – the substance is hydrolytically unstable at pH 4, 7 and 9 (half-life less than 12 h), or – the substance is readily oxidizable in water. If the substance appears “insoluble” in water, a limit test up to the detection limit of the analytical method shall be performed. *For nanoforms, the potential confounding effect of dispersion shall be assessed when conducting the study.*

Only two new nanospecific endpoints have been introduced into the Annexes: an information requirement for dustiness testing in Annex VII and for further nanospecific physical–chemical testing in Annex IX without further specification which information this could be.

Furthermore, some additions were introduced to existing endpoints, for example, the obligation to also test the dissolution rate in different media in addition to water solubility testing according to Annex VII (see also table 20.2).

20.7 Annex I

Annex I of REACH lays down the process and requirements for the CSA. Regarding nanomaterials, amendments to Annex I mainly specify how nanoforms of substances have to be taken into account in this process and that the CSA has to be performed separately for each individual nanoform or set of nanoforms, which is covered by a registration.

20.8 Annex II

Annex II of REACH lays down the requirements for SDS. SDSs provide the user of a substance or a mixture with information on the substance or mixture and instructions for the safe use. SDSs are not required for articles.

The amendments for nanomaterials introduced into Annex II mainly concern sections 3 and 9 of the SDS. Here information on the characterization parameters described earlier for Annex VI as well as on the physicochemical properties, where relevant (e.g., when determining the solubility), needs to be included.

20.9 Big achievement – further challenges

While the amendment of the REACH Regulation with regard to nanoforms of substances can be considered as a huge achievement, further challenges are still ahead.

For many of the studies in the Annexes VII to X, which are now already obligatory for nanoforms of substances, the test protocols still have to be adapted to nanomaterials. This work is currently ongoing under the framework of the OECD and its Working Party on Manufactured Nanomaterials.

Registrants also need to update their registration dossiers with nanospecific information. To support in particular small and medium-sized enterprises in this obligation, guidance needs to be developed by regulators on how the new information requirements can be fulfilled. First guidance documents on how to register and identify/characterize nanoforms already exist.[5] More guidance dealing with the specific toxicological and ecotoxicological information to be generated and submitted was developed or updated in the last years.

Furthermore, a regulatory gap still remains for forms of substances with a particle size above 100 nm up to 10 µm which are biopersistent and have fibrous morphology. It is well known that such forms of substances pose a risk to human health as they can reach the alveolar region of the lung and cause inflammations. These forms are, however, not covered by the REACH definition of a nanoform. Hence, only the standard information requirements for bulk substances apply which are not sufficient for the identification of the hazard of these forms of substances.

5 https://echa.europa.eu/documents/10162/13655/how_to_register_nano_de.pdf/fd174fd9-d099-b8cc-a8d9-890ec662c6c6 and https://echa.europa.eu/documents/10162/22308542/howto_prepare_reg_dossiers_nano_en.pdf/.

Abdelqader Sumrein

21 Nanoregisters in Europe

Keywords: registry, inventory, REACH, ECHA, EUON, transparency

21.1 Introduction

The development of nanomaterials and nanotechnology has been recognized as holding the potential to address a variety of problems that our societies are currently facing. Nevertheless, in the European Union (EU), it was recognized the need to consider the potential risks for human health and the environment from nanomaterials and nanotechnology.[1] The need for potential regulatory action on nanomaterials was realized, and already in 2005 it was foreseen that the REACH regulation,[2] under development at that stage, "may cover some aspects on nanoparticles produced in very high quantities."[3] While the REACH regulation, in its initial wording, did not contain specific provisions for the nanomaterials, the European Commission considered that nanomaterials, being chemical substances, are nevertheless covered by the provisions of the REACH regulation, and that data on the hazards of nanomaterials must be included as part of the REACH registration dossier for substances covering nanomaterials.[4] The European Parliament, in a Communication to the Commission, considered it necessary to address nanomaterials explicitly within the scope of the REACH regulation, and called on the Commission to compile an "inventory" of nanomaterials on the EU market, and to make this inventory publicly available.[5] The European Commission also clarified what is considered as a nanomaterial for regulatory purposes by issuing an EC recommendation for the definition of a nanomaterial.[6] The European Commission updated this definition in June 2022 (Source:[7] This activity showed a recognition in the European public debate on the need for transparency on the identify of nanomaterials on the EU market, as well as for information on their safety.

The annexes of the REACH regulation were eventually amended in December 2018 to include explicit information requirements for nanoforms of substances, to be submitted as part of a REACH registration dossier.[8] In the interim, several European countries explored the creation of national inventories of nanomaterials, in response to the perceived lack of information on nanomaterials at the European level. For example, the lack of specific requirements in the REACH regulation was cited as one of the reasons for the creation of a nanoregister in France.[9] Eventually, three countries, Denmark,

Corresponding author: Abdelqader Sumrein, Directorate of Submissions and Interaction, European Chemicals Agency, P.O. Box 400, FI-00121 Helsinki, Finland, e-mail: abdelqader.sumrein@echa.europa.eu

https://doi.org/10.1515/9783110670776-021

Belgium and France, proceeded to create stand-alone inventories for nanomaterials in their respective countries. In addition, Sweden and Norway amended their existing product registers to allow for the collection of information on products containing nanomaterials. However, as these are not stand-alone nanoregisters, and due to the limited information available on the data collected to date on nanomaterials by these product registers, they are not discussed further within this chapter.

In addition to the earlier national nanomaterial registers, the EU Cosmetics Regulation requires that cosmetic products containing nanomaterials must be notified to the European Commission via the Cosmetics Products Notification Portal (CPNP), before being placed on the market, and the European Commission is responsible for publishing a catalogue of all nanomaterials used in cosmetic products on the EU market based on notifications submitted by companies. Finally, the European Commission tasked the European Chemicals Agency (ECHA) in December 2016 with the creation and management of a "European Union Observatory for Nanomaterials" (EUON), with the aim of increasing the transparency of information on the safety and markets of nanomaterials at the EU.

Together, these initiatives, whether termed "registries," "inventories," "catalogue" or "observatory," collect and provide information on nanomaterials on the EU market. There are nevertheless important differences between the different initiatives in terms of the scope, actors, and requirements. These different initiatives, including the key distinctions, are described in more detail further. Finally, the amendments to the REACH regulation requiring submission of information on nanomaterials as part of the requirements can be seen as an EU-wide register of nanomaterials, albeit one that has a much broader scope than the above initiatives. As the REACH requirements for nanomaterials are addressed in a separate chapter in this book, it will not be further discussed in this chapter.

21.2 The French nanoregister (R-Nano)

Following a series of public debates on nanomaterials and nanotechnology, France became the first European member state to establish a requirement for the notification of nanomaterials on the French market. The French inventory was established in 2012 based on a ministerial decree, which entered into force in January 2013.[10] The first deadline for notifications was 30 June 2013, with annual updates required after that. This relatively early start, coupled with the limited number of exemptions compared to other inventories, means that of all the different national inventories available, the French inventory has the longest history of operation, and also provides the most information publicly of the existing national inventories.

The inventory requires companies that produce, distribute or import nanomaterials, as well as private or public laboratories, to submit information on nanomaterials on the French market, with a notification threshold of 100 g of the nanomaterial. In the French inventory, a nanomaterial is defined based on the EC recommendation for the definition of a nanomaterial, with the only deviation that the French decree is limited to manufactured nanomaterials, and thus excludes from its scope naturally occurring or incidental nanomaterials. The French inventory does not have any exemptions for specific sectors, unlike other notification schemes.

The French notification system requires companies to submit information on a variety of elements. These can be divided into:

1. Identity of the declaring party (including company, sites and roles in the supply chain)
2. Information on the identity of the nanomaterials, including its chemical name, chemical formula, CAS and EC numbers, size of the particles, number size distribution of the particles, information on aggregation and agglomeration state, shape, a qualitative description of surface treatment, as well as information on the purity and whether the substance is used in a mixture
3. Information to be provided if available: including whether the substance has been registered under the REACH regulation, the presence of impurities, the crystalline state of the material, specific surface area and surface charge
4. Quantity of the substance at nanoscale either produced, distributed or imported during the previous year
5. Information on uses of the substance at nanoscale
6. Identity of professional users

The decree requires companies to update their notifications annually using an online IT tool, found at https://www.r-nano.fr/. The notification system itself is managed by the French Agency for Food, Environmental and Occupational Health & Safety. The agency creates annual reports (in French) providing public information from the notifications. Importantly, the decree allows for submitters to request for confidentiality for several aspects of the information submitted, including all information under point 3 above. This means that in practice, only a limited portion of the submitted information is available for public analysis or use.

The number of submissions in 2013, the first year of the inventory, was 3,419 submissions. In subsequent years, the number of submissions has ranged from a low of 10,122 submissions in 2019, to a high of 14,583 submissions in 2015. These submissions correspond to a total of 390 different nanomaterials. However, an examination of the list of substances in the latest report shows that a limited number of the identity of the chemical is either ambiguous, or does not refer to a chemical.

The majority (63% in 2019) of the substances declared on the French inventory are below the 1 ton/annum threshold that would require registration under the REACH regulation. However, it should be noted that the tonnages reported in the French inventory reflect only the tonnage of the nanomaterial, whereas the REACH regulation takes into account the total tonnage of the substance imported or produced on the EU market, including both the non-nanoforms and the nanoforms of the substance, in calculation of the tonnage threshold for registration. Therefore, the declarations in the French registry alone would not be suitable for producing an accurate picture of the EU market for nanomaterials. The highest volume nanomaterials in the inventory are carbon black, silicon dioxide, calcium carbonate and titanium dioxide. The details of the identity, tonnages and uses of the individual substances can be found in the Annex of the annual report.[11]

A limited number of nanomaterials declared to the French registry are used in the textile industry, based on the notifications provided. A list of the nanomaterials declared to be used in the textiles industry is included in the following table:

Nanomaterial
Aluminum oxide
Silicon dioxide
Carbon black
Silicic acid, magnesium salt
3,4,5,6-Tetrachloro-*N*-[2-(4,5,6,7-tetrachloro-2,3-dihydro-1,3-dioxo-1H-inden-2-yl)-8-quinolyl] phthalimide
N,N′-Phenylene-1,4-bis[4-[(2,5-dichlorophenyl)azo]-3-hydroxynaphthalene-2-carboxamide]
Pigment Violet 23
Ethene, homopolymer, oxidized

21.3 Belgian nanoregister

The Belgian register was established based on the Royal Decree of 27 May 2014 concerning the placing of substances produced in nanoparticular state on the market, and is managed by the Federal Public Service of Health, Food Chain Safety and Environment (FPS).[12] The decree covers “substances produced in a nanoparticular state,” and the definition of nanomaterial is nearly identical to the EU recommendation for the definition of a nanomaterial; however, it excludes natural, nonchemically modified substances, and substances where the nanomaterial particles are considered a by-product of human activity. The requirement to submit notifications for substances entered into force on 1 January 2016. For articles and for mixtures, the deadline was initially 1 January 2017, but was later amended to 1 January 2018. The decree also contains specific requirements for the notification of articles and complex objects containing nanomaterials, however, those provisions are set to enter into force “at a later date . . . following an evaluation for the articles." So far, no date has been set for entry into force of the requirements for articles and complex mixtures.

The Belgian register requires notification of substances and mixtures containing nanomaterials, as well as articles and complex objects containing one or more nanomaterials. The requirements and conditions differ slightly for substances and mixtures compared with articles and complex objects. However, in all cases, the decree exempts the following products from registration:

1) Biocidal products
2) Medicinal products
3) Foodstuffs and objects intended to come into contact with foodstuffs
4) Feed
5) Veterinary medicinal products
6) Processing aids
7) Pigments, when placed on the market in a mixture, an article or a complex object

The decree requires companies that place nanomaterials on the Belgian market either as substances or mixtures, or articles in a quantity of 100 g or more annually to submit a notification for the nanoform. The requirement to register applies only where the company in question produces the substance or mixture themselves, or places the substance or mixture on the market exclusively for professional uses. In other words, distributors of substances and mixtures who place nanomaterials on the market for consumer users (or consumer and professional users) are exempt from registering. The requirement to submit a notification also applies to articles and complex objects that contain nanomaterials.

The decree requires companies to submit the following information on substances and mixtures containing nanomaterials:

1) Administrative information identifying the registrant
2) Mandatory information on the identity of the substance: including information on the chemical name, formula, CAS and EC numbers, average and median particle size, the number-based particle size distribution, average aggregate size, shape and description of the surface treatment
3) Information to be provided if available: the REACH registration number, type and quantity of impurities (where applicable), crystallographic phase, average specific surface area and zeta potential
4) Quantity of the substance produced per year
5) Information on uses of the substance: including all foreseeable uses as well as trade name
6) Identity of professional users (if known)
7) Uses of the mixture (only for mixtures)

Companies are required to update the information annually between 1 January and 31 March, electronically, via an online portal managed by the FPS. The FPS has published annual reports in French and Dutch for the years 2016–2018. It should be noted that as with the French register, a significant portion of the substance information submitted is treated as confidential. Consequently, only information on the chemical name, as well as information on the quantities (for some notifications) as well as uses (in the form of NACE codes) are publicly available.

The latest publicly available report indicates a total of 1,322 registrations were submitted by 1 April 2019 for the year 2018. Based on this information, 27,590 tons of nanomaterials were imported, 61,029 tons were manufactured and 4,000 tons were distributed on the Belgian market.[13]

The Belgian inventory includes nanomaterials used in the textiles industry, as evident from the latest report published, as the report indicates that the declarations have reported nanomaterials using Sector of Use 5 (Manufacture of textiles, leather and fur), Product Category 34 (Textile dyes and impregnating products) and Article Category 5 (Fabrics, textiles and apparel). However, the public registry information does not list which specific nanomaterials included this use information.

21.4 Danish nanoregister

Denmark issued a statutory order in 2014 establishing a nanoproduct register, and that entered into force in June 2014,[14] with the first registration deadline being August 2015, with annual updates required thereafter. Similarly to the French and Belgian registers, the Danish register follows the EC recommendation for the definition of a nanomaterial. The order requires companies that manufacture or import mixtures or products intended for sale for private use (i.e., consumer products) to submit notifications to the register. This requirement applies to products only where the nanomaterial itself is expected to be released during normal use, or where the nanomaterial itself is not released, but substances in soluble form (e.g., ions) that are classified as carcinogenic, mutagenic or toxic to reproduction or as environmentally hazardous substances can be released from the nanomaterial. However, the order itself exempts a variety of mixtures and products from the requirement to register. Substances exempt are:

1) Food and food contact materials
2) Feeds
3) Medicines
4) Medical equipment
5) Cosmetic products
6) Pesticides
7) Waste
8) Mixtures and articles where the nanomaterial is covered by Annexes IV and V of Regulation no. 1907/2006/EC of the European Parliament and of the Council (REACH)

9) Mixtures and articles where the nanomaterial is not deliberately made in nanosize
10) Goods where the nanomaterial is included in a solid matrix, unless wear, washing, breakage and similar normal use of the product results in the release of free nanomaterial
11) Products where the nanomaterial is used as an ink directly on the product or on labels on the product, including newspapers, magazines, magazines, packaging that is not dyed through and the like
12) Textiles where the nanomaterial is used as printing ink or for dyeing through textiles
13) Paints, wood preservatives, adhesives and fillers containing nano-sized pigment, where the pigment is added solely for the purpose of coloring the mixture
14) Rubber articles or rubber parts of articles containing carbon black

The Danish nanoregister requires the submission of the following information as part of the register:
1) Administrative information on the registrant (e.g.name, address, type and size of company)
2) Information on the product (product name, whether the product is manufactured or imported into Denmark, volume of the product on the market, whether the product/mixture is also used professionally)
3) Information on the nanomaterial within the product/mixture (including chemical name, CAS and EC numbers, chemical formula)

In addition, companies may submit additional optional information as part of the registration. This optional information includes:
1) Category of the product (including product category, process category, environmental release category and article category)
2) Amount of the nanomaterial in the product/mixture
3) Physicochemical characterization of the nanomaterial (including particle size and size distribution, aggregation, agglomeration, form of the nanomaterial, specific surface area, crystallinity, surface chemistry and surface charge)

Compared to the French and Belgian registers, only a limited amount of information is publicly available regarding the data collected as part of the notifications, and accessibility to the information is restricted by the order to the Danish Environmental Protection Agency and the Danish Working Environment Authority. However, a report commissioned by the Danish Ministry for the environment in 2017 shows that during the first 3 years of operation, very few nanomaterials were registered, with only 10 nanomaterials registered in the first year (2015) followed by 9 nanomaterials in the second year (2016) and 6 nanomaterials in the third year (2017).[15] No information is available on the identity or volumes of the nanomaterials covered by the notifications.

21.5 The Cosmetic Products Notification Portal (CPNP)

The EU Regulation EC 1223/2009 on cosmetics includes specific rules on the use of nanomaterials within cosmetic products. However, unlike the registries discussed earlier, the regulation does not use the EC recommendation for the definition of a nanomaterial. Instead, the regulation defines nanomaterials as “an insoluble or biopersistent and intentionally manufactured material with one or more external dimensions, or an internal structure, on the scale from 1 to 100 nm.”[16]

While Article 13 of the regulation requires that all cosmetic products be notified through the CPNP before being placed on the EU market, Article 16 of the regulation includes specific requirements for nanomaterials, in addition to the notifications required under Article 13. The regulation requires that cosmetic products containing nanomaterials must be notified 6 months prior to the placement on the market, except where they have been notified by the same person before 11 January 2013. This does not apply to colorants, preservatives and UV filters, including those using nanomaterials, which require prior authorization by the European Commission before their use in cosmetic products. The European Commission may also request the Scientific Committee for Consumer Safety to perform a risk assessment of the nanomaterial if it has concerns regarding its safety.

The following minimum information must be submitted as part of the notification to the Commission:

1) Chemical identity of the nanomaterial (IUPAC name and other descriptors in the regulation)
2) Physicochemical properties of the nanomaterial, including its particle size
3) An estimate of the quantity of the nanomaterial contained in the cosmetic product to be placed on the market per year
4) The toxicological profile of the nanomaterial
5) The safety data of the nanomaterial
6) Information on reasonably foreseeable exposure conditions

The European Commission is required to publish a catalogue of cosmetic products using nanomaterials, including those used as colorants, preservatives and UV filters, and to regularly update this catalogue based on the information submitted through the CPNP. The Commission published the first draft of the inventory in December 2016. The information published is limited to an identification of the substance, information on the product category, as well as the foreseen exposure route. The Commission updated the catalogue in December 2018. This latest update includes 27 nanomaterials in total.

21.6 The European Union Observatory for Nanomaterials (EUON)

The EUON was established by the European Commission in 2016 and is managed by the ECHA as a Delegated Task from the European Commission to the Agency. The Observatory was established following a study by the European Commission to assess the impact of possible legislation to increase transparency on nanomaterials on the EU market.[I] The study considered a variety of options to enhance the level of transparency on nanomaterials on the EU market, taking into account lessons learned from previous schemes. The study examined several different options, including establishing recommendations for national inventories, creation of an EU wide registry (either based on substances or applications) or creation of an observatory with the observatory ending as the preferred model.

The observatory approach differs significantly from national inventories in two key areas: first, it aims to cover the entire EU market, and second, it does not establish a legal requirement for companies to notify their nanomaterials to the EUON. Rather, the EUON itself collects existing information in the public domain on nanomaterials, and complements these with studies aimed at addressing specific knowledge gaps.

Although there is no specific requirement established by the EUON for companies to submit information on nanomaterials, the EUON itself does collect and publicize information on nanomaterials on the EU market. Since June 2019, the EUON has offered a search portal that combines public information from the French and Belgian nanomaterial inventories, as well as the notifications of nanomaterials in cosmetics submitted to the CPNP. In addition, the portal combines this information with data from REACH registration for nanomaterials.

The EUON's search provides information on the identity of the substance, where it is notified/registered (France, Belgium, CPNP and/or REACH registration). In addition, the information is linked to the public version of the REACH registration dossier for the substance, where this is available, and therefore provides information on physicochemical properties, environmental fate and (eco)toxicology of the substance and its nanoforms. To date, the EUON database covers 333 substances, including 80 substances with REACH registration dossiers, 257 substances from the French inventory, 154 substances from the Belgian inventory and 24 substances notified to the CPNP. The EUON is updated with information from REACH registrations whenever new ones are submitted, and information from the French and Belgian inventories as

I Study to assess the impact of possible legislation to increase transparency on nanomaterials on the market, European Commission Directorate-General for Internal Market, Industry, Entrepreneurship and SMEs, https://op.europa.eu/en/publication-detail/-/publication/d42fe639-b080-11e6-aab7-01aa75ed71a1.

well as the CPNP notifications are updated regularly whenever those sources publicize new information.

It should be noted that some substances appearing in the underlying national nanomaterial inventories are not listed on the EUON. Before publishing the information, the EUON matches the substances reported in the data sources with ca. 300,000 substances found in ECHA's chemicals database. Where no match is found, the original data source is not reported on the EUON page. The reason for the absence of a match is generally because the name in the original source does not correspond to a uniquely identifiable chemical.

The EUON, in addition to providing information on nanomaterials within the EU market from existing sources (REACH registrations, or national inventories), also performs studies aimed at addressing data gaps on nanomaterials in the EU. The studies themselves cover a variety of topics, including studies on the markets for nanomaterials in the EU, as well as hazards and risks of nanomaterials. Such studies occasionally create inventories or lists of nanomaterials in specific sectors. An example of this is a study performed on nanomaterials used as pigments in the EU. As part of this work, the study compiled an inventory of nanomaterials used as pigments, which is available on the EUON website.[17]

21.7 Conclusions

Different initiatives exist in EU/EEA member states as well as within the EU to collect information on nanomaterials. These initiatives differ in their scope as well as the information collected. Due to the confidentiality rules surrounding some of these registries, only a limited portion of the information collected is available to the public; however, information from the French and Belgian registries combined with information from notifications to the Commission's CPNP and registrations under the REACH regulation show that there are over 300 nanomaterials on the EU market today in different volumes.[18]

A recent study examining the public's perception of nanomaterials, relying on respondents in five EU countries, showed that when presented with information that a particular product contains nanomaterials, consumers in these countries react with caution toward the product.[19] However, the study also indicated that the public's risk perception of nanomaterials decreases as their level of awareness of nanomaterials increases. While it is not possible to evaluate the impact of individual nanoregisters, the results of this study do suggest that the increased availability of information on nanomaterials from the combined work of these registers may foster the public's acceptance of products utilizing nanomaterials/nanotechnology in the EU.

Notes

1 Communication from the Commission – Towards a European strategy for nanotechnology, COM (2004) 338 final, https://eur-lex.europa.eu/LexUriServ/LexUriServ.do?uri=COM:2004:0338:FIN:EN:PDF

2 Regulation (EC) no. 1907/2006 of the European Parliament and of the Council of 18 December 2006 concerning the Registration, Evaluation, Authorisation and Restriction of Chemicals (REACH), establishing a European Chemicals Agency, amending Directive 1999/45/EC and repealing Council Regulation (EEC) no. 793/93 and Commission Regulation (EC) no. 1488/94 as well as Council Directive 76/769/EEC and Commission Directives 91/155/EEC, 93/67/EEC, 93/105/EC and 2000/21/EC: https://eur-lex.europa.eu/legal-content/EN/TXT/?uri=CELEX:32006R1907

3 Communication from the Commission to the Council, the European Parliament and the economic and social committee “Nanosciences and nanotechnologies: An action plan for Europe 2005–2009” COM(2005) 243 final, https://ec.europa.eu/research/industrial_technologies/pdf/policy/nano_action_plan2005_en.pdf

4 Communication from the Commission to the European Parliament, the Council and the European Economic and Social Committee, Regulatory Aspects of Nanomaterials, COM(2008) 366 final, https://ec.europa.eu/research/industrial_technologies/pdf/policy/comm_2008_0366_en.pdf

5 European Parliament resolution of 24 April 2009 on regulatory aspects of nanomaterials (2008/2208(INI)) https://www.europarl.europa.eu/sides/getDoc.do?type=TA&reference=P6-TA-2009-0328&language=EN

6 COMMISSION RECOMMENDATION of 18 October 2011 on the definition of nanomaterial (2011/696/EU) https://eur-lex.europa.eu/legal-content/EN/TXT/PDF/?uri=CELEX:32011H0696&from=EN

7 COMMISSION RECOMMENDATION of 10 June 2022 on the definition of nanomaterial COMMISSION RECOMMENDATION of 10 June 2022 on the definition of nanomaterial https://eur-lex.europa.eu/legal-content/EN/TXT/PDF/?uri=CELEX:32022H0614(01)&from=EN)

8 Commission Regulation (EU) 2018/1881 of 3 December 2018 amending Regulation (EC) No 1907/2006, https://eur-lex.europa.eu/legal-content/EN/TXT/?uri=uriserv:OJ.L_.2018.308.01.0001.01.ENG&toc=OJ:L:2018:308:TOC

9 See FAQ Question 1 on the French nano register: https://www.r-nano.fr/?locale=en#help

10 Ministerial Order of 6 August 2012 on the content and the conditions for the presentation of the annual declaration on substances at nanoscale, in application of articles R. 523–12 and R. 523–13 of the Environment code, file://echa/data/users/u09045/Roaming%20Profile/Downloads/arrete_final_ENG.pdf

11 2019 report on the declarations of substances imported, manufactured or distributed in France in 2018, https://www.ecologie.gouv.fr/sites/default/files/Rapport%20R-nano%202019.pdf

12 27 May 2014 – Royal Decree concerning the placing on the market of substances produced in nanoparticular state; https://www.health.belgium.be/sites/default/files/uploads/fields/fpshealth_theme_file/10._en_kb_20140527_unofficial_translation.pdf

13 Annual report for the year 2018, https://www.health.belgium.be/fr/rapport-annuel-nanoregistry-2018

14 Bekendtgørelse (644/2014) Executive Order on the register of mixtures and products containing nanomaterials as well as manufacturers and importers’ reporting obligation to the register, https://www.retsinformation.dk/eli/lta/2014/644

15 Christensen, F.M., 2017. Comparative Analysis of Nano-Registers in EU/EEA Member States. COWI A/S, Denmark. https://mst.dk/media/145582/final_nabotjek_nanoregistre.pdf

16 Regulation (EC) No 1223/2009 OF The European parliament and of the Council of 30 November 2009 on cosmetic products https://eur-lex.europa.eu/LexUriServ/LexUriServ.do?uri=OJ:L:2009:342:0059:0209:en:PDF

17 https://euon.echa.europa.eu/nano-pigments-inventory
18 As of 25 March 2021, a total of 327 nanomaterials can be found on the EU market, as seen in the EUON's search for nanomaterials portal: https://euon.echa.europa.eu/search-for-nanomaterials
19 UNDERSTANDING PUBLIC PERCEPTION OF NANOMATERIALS AND THEIR SAFETY IN THE EU, Miloš Tengler, Jana Hamanová, Tomáš Novotný, Simona Popelková, Jan Holomek, Viktorie Kováčová, Filip Oliva, Jana Trávníčková, Dušan Bašić, Kryštof Dibusz,, ECHA-20 R-13-EN, DOI: 10.2823/82474, https://euon.echa.europa.eu/documents/23168237/24095696/nano_perception_study_en.pdf/f7d43c54-8a78-061b-49b0-de619454e137

Luis Almeida*, Delfina Ramos

22 Occupational health aspects (EU)

Keywords: occupational Health and Safety, EU legislation, standards, nanomaterial, textiles

Occupational Health and Safety aspects are considered as very important for the European Commission. In this chapter, an overview of European legislation in this area is presented, with special emphasis on the "Framework Directive." The role of the European Agency for Safety and Health at Work (EU-OSHA) is presented, including relevant activities and publications. The European definition of nanomaterial emphasizes the concerns on health and safety and is an important basis for all the legal requirements related to nanomaterials. This chapter also includes a section on guidance to nanomaterials at workplace, as well as standardization on nanotechnology, especially at European level and the use of the control banding approach for occupational risk management applied to engineered nanomaterials, including a case study in a textile finishing company.

22.1 Introduction: history

Occupational Safety and Health (OSH) has been since the beginning an integral part of the European project. Created in 1951, the European Coal and Steel Community (ECSC), the foundation stone of what would later become the European Economic Community and then the European Union (EU), brought together Europe's coal and steel industries in an effort to mend a fractured Europe and pursue a collaborative future. In doing so, two of the most dangerous working environments at the time suddenly took center stage. This triggered one of the key objectives of the ECSC: to ensure "the equalisation and improvement of living conditions of workers" in the aforementioned industries.

It was not until 1985, however, that the tripartite approach to OSH gained momentum in Europe. Thanks to the initiative of Jacques Delors, the President of the European Commission at that time, the concept of "social dialogue" was given a constitutional mandate in the EU, paving the way for the publication in 1989 of the Framework Directive (89/391/EEC). This significant milestone for OSH not only established minimum

***Corresponding author: Luis Almeida,** Department of Textile Engineering and Centre for Textile Science and Technology, University of Minho, Guimarães, Portugal, e-mail: lalmeida@det.uminho.pt
Delfina Ramos, Centre for Research and Development in Mechanical Engineering (CIDEM), School of Engineering of Porto (ISEP), Polytechnic of Porto - School of Engineering (ISEP), Centre for Research and Development in Mechanical Engineering (CIDEM), School of Engineering, University of Minho, Guimaraes, Portugal

https://doi.org/10.1515/9783110670776-022

safety and health requirements in the European Union, introducing common principles and minimum standards that applied across all European Union. It also placed the revolutionary notion of risk assessment at its core.

In the wake of the Framework Directive, the European Commission launched a Europe-wide campaign aimed at putting OSH and the detrimental effects of unsafe working environments well and truly in the spotlight: it declared 1992 the European Year of Safety, Hygiene and Health at Work. This was in response to shocking figures from Europe's workplaces – 4 million workplace accidents, 8 thousand of which were fatal, were reported annually in the early 1990s, costing Europe 20 million Euro. The European Year aimed to raise awareness of OSH risks and relevant legislation and highlight the work being done to improve safety standards.

The 1992 European Year and the increasing volume of legislation on safety and health at work laid the foundation for the creation of an agency dedicated to OSH. The decision to set up the Agency was made at a Council Summit in October 1993, and Regulation (EC) no. 2062/1994 establishing the European Agency for Safety and Health at Work (EU-OSHA) was subsequently adopted in 1994 with a light amendment entering into force in 2019.

22.2 Directive 89/391 – OSH "Framework Directive"

Article 153 of the Treaty on the Functioning of the European Union gives the EU the authority to adopt directives in the field of safety and health at work. The Framework Directive (89/391), with its wide scope of application, and further directives focusing on specific aspects of safety and health at work are the fundamentals of European security and health legislation.

The "Framework Directive" of OSH on the introduction of measures to encourage improvements in the safety and health of workers at work has been published on 12 June 1989 (89/391/EEC). It is the most important legal European document in the area of OSH, which is still valid, with minor updates. The document, in all European languages, can be consulted at:

https://eur-lex.europa.eu/legal-content/EN/ALL/?uri=CELEX%3A31989L0391

In the website it is also possible to check the present consolidated version, which includes all the updates and amendments.

The Framework Directive includes basic obligations for employers and workers. Nevertheless, the workers' obligations shall not affect the principle of the responsibility of the employer.

It is the employer's obligation to ensure the safety and health of workers in every aspect related to work and he may not impose financial costs to the workers to achieve this aim. Alike, where an employer enlists competent external services or persons, this shall not discharge him from his responsibilities in this area.

The general principles of prevention listed in the directive are as follows:
- avoiding risks,
- evaluating the risks,
- combating the risks at source,
- adapting the work to the individual,
- adapting to technical progress,
- replacing the dangerous by the non- or the less dangerous,
- developing a coherent overall prevention policy,
- prioritizing collective protective measures (over individual protective measures),
- giving appropriate instructions to the workers.

Here are some highlights of the Directive, highlighted by EU-OSHA:

The term "working environment" was set in accordance with International Labour Organization (ILO) Convention No. 155 and defines a modern approach taking into account technical safety as well as general prevention of ill-health.

The Directive aims to establish an equal level of safety and health for the benefit of all workers. The only exclusions are domestic workers and certain public and military services.

The Directive obliges employers to take appropriate preventive measures to make work safer and healthier.

The Directive introduces as a key element the principle of risk assessment and defines its main elements (e.g., hazard identification, worker participation, introduction of adequate measures with the priority of eliminating risk at source, documentation and periodical reassessment of workplace hazards).

The obligation to put in place prevention measures implicitly stresses the importance of new forms of safety and health management as part of general management processes. The implementation of ISO 45001:2018 (Occupational health and safety management systems – Requirements with guidance for use) is an effective means for the organizations to properly address the legal requirements.

The Framework Directive 89/391 had to be transposed into national laws by the end of 1992. The repercussions of the transposition on national legal systems varied across Member States. In some Member States, the Framework Directive had considerable legal consequences due to inadequate national legislation while in others no major adjustments were necessary.

All the countries that became members of the EU after 1992 were also obliged to transpose this directive. But it is important to emphasize that although the EU directives must be transposed to national legislation, which must comply with the minimum standards, Member States may adopt stricter rules to protect workers. As a result, national safety and health legislation varies across Europe, so it is important, especially for companies who have business in more than one Member State, to be aware of the requirements of national legislations.

22.3 Other relevant EU directives

Apart from the Framework Directive, there are 23 other directives covering specific risks connected with safety and health in the workplace. Among those, the following EU sirectives are particularly relevant for occupational safety and health in relation to exposure to particles:

Directive 2009/104/EC (16 September 2009) – **use of work equipment** – concerns the minimum safety and health requirements for the use of work equipment by workers at work.

Directive 99/92/EC (16 December 1999) – **risks from explosive atmospheres** – presents the minimum requirements for improving the safety and health protection of workers potentially at risk from explosive atmospheres.

Directive 89/656/EEC (30 November 1989) – **use of personal protective equipment** – minimum health and safety requirements for the use by workers of personal protective equipment at the workplace. This Directive has now been replaced by the Regulation (EU) 2016/425 on personal protective equipment.[1]

Directive 89/654/EEC (30 November 1989) – **workplace requirements** – concerns the minimum safety and health requirements for the workplace.

Directive 98/24/EC (7 April 1998) – **risks related to chemical agents at work**. This Directive lays down minimum requirements for the protection of workers from risks to their safety and health arising, or likely to arise, from the effects of chemical agents that are present at the workplace or as a result of any work activity involving chemical agents. This Directive provides for the drawing up of indicative and binding occupational exposure limit values as well as biological limit values at Community level.

Directive 2000/39/EC (8 June 2000) – **indicative occupational exposure limit values** – establishes a first list of indicative occupational exposure limit values, in relation with Directive 98/24/EC mentioned above. The list has been updated already four times by Directives 2006/15/EC, 2009/161/EU, 2017/164/EU and 2019/1831. The limit values are established in relation to a reference period of 8 h (considering a time-weighted average) and referred to as long-term exposure limit values, and, for certain chemical agents, these relate to shorter reference periods, in general 15 min (also time-weighted averages), referred to as short-term exposure limit values, to take account of the effects arising from short-term exposure.

1 Note that unlike European Directives, an EU Regulation, which has to be approved by the European Commission and by the European Parliament, becomes law in all EU countries without any need of transposing into national legislations.

Directive 2000/54/EC (18 September 2000) – **biological agents at work**. This Directive lays down minimum requirements for the health and safety of workers exposed to biological agents at work. Note that these agents can for instance be carried out by particles in suspension in the air.

In 2004 the European Commission issued a Communication (COM [2004] 62) on the practical implementation of the provisions of several directives, including 89/391 EEC (framework directive), 89/654 EEC (workplaces), 89/655 EEC (work equipment), 89/656 EEC (personal protective equipment), 90/269 EEC (manual handling of loads) and 90/270 EEC (display screen equipment). This Communication stated that there was evidence of the positive influence of EU legislation on national standards for occupational safety and health made up of both national implementing legislation and practical application in enterprises and public sector institutions.

In general, this report concluded that EU legislation had contributed to promote a culture of prevention throughout the European Union as well as to rationalize and simplify national legislative systems. At the same time, however, the report highlighted various flaws in the application of the legislation that were holding back achievement of its full potential. It also noted cases where infringement proceedings had been opened. In fact, most of the Member States failed to meet the commitments taken as regards the delays for transposition. Also, in most of the countries there were deficiencies in the transposition of EU Directives and especially the Framework Directive. Several infringement processes have been opened against Member States.

The European Commission has also carried out an evaluation of the practical implementation of the EU occupational safety and health (OSH) directives in EU Member States [1]. This extensive report of a work carried out for two years concludes that the Framework Directive has achieved its stated objective of introducing measures to encourage improvements in safety and health at work. It is overall implemented and complied with, remains relevant and has led to positive workplace impacts as well as safety and health impacts and has contributed to leveling the playing field by setting common requirements for occupational safety and health in the EU.

This report presents several suggestions to improve the Framework Directive and related Directives. In the Framework Directive it should be stated clearly what is meant by risk assessment and by risk prevention measures. This definition should not be repeated in the specific Directives. One possible recommendation is to assemble the provisions of the Framework Directive and the 23 specific Directives into one completely new Directive with annexes. The strength is the ease of updating annexes when needed compared with that of updating specific Directives. However, the weakness is a heavy administrative burden arising from its implementation in national legislations.

Labor inspectorates perform a major role in promoting and checking the observance of occupational safety and health (OSH) legislation at national level. They also have a major role in terms of awareness campaigns, especially in terms helping micro and small enterprises to comply with legislation. EU gives some funding for labor inspectorates to complement the budgetary constraints at national levels.

The Communication “Safer and Healthier Work for All – Modernization of the EU Occupational Safety and Health Legislation and Policy,” adopted in January 2017, also puts an important emphasis on the role of labor inspectorates, identifying and deterring undeclared work, verifying compliance with labor law and ensuring the observance of OSH legislation [2].

Although the responsibility for enforcing EU occupational health and safety legislation rests with each individual Member State, the European Commission seeks reassurance that the legislation is being applied and enforced effectively and efficiently. The role of SLIC (Senior Labour Inspectors Committee) is essential to encourage consistent and effective enforcement of community-based legislation with respect to occupational health and safety.

22.4 Activities of European Agency for Safety and Health at Work (EU-OSHA), including specific information related to the textile sector

The motto of EU-OSHA is: “We work to make European workplaces safer, healthier and more productive – for the benefit of businesses, employees and governments. We promote a culture of risk prevention to improve working conditions in Europe.” EU-OSHA, established in 1994 and located in Bilbao, Spain, is an important support of the European Commission to promote occupational safety and health.

The main activities include the promotion of healthy workplaces campaigns. These campaigns are promoted since 2000 (formerly under the title “European Weeks for Safety and Health at Work”) and cover different topics. They are carried out in conjunction with national focal points, which exist in all EU and EFTA members, as well as in candidate countries.

EU-OSHA provides very relevant information related to occupational safety and health, namely through the website https://osha.europa.eu/. All the European legislation related to OSH can be easily be consulted, including not only EU Directives but also European Guidelines (non-binding documents which aim to facilitate the implementation of European directives) and European Standards. The relevant national legislation can also be consulted. Note that although the EU directives must be transposed to national legislation, member States may adopt stricter rules to protect workers, but their legislation must comply with the minimum standards. As a result, as mentioned above, national safety and health legislation varies across Europe.

EU-OSHA manages the OSHwiki – a collaborative online encyclopedia of accurate and reliable information on OSH. For instance, in OSHwiki can be found relevant information on nanomaterials, including risk assessment and management, health and safety hazards of nanomaterials, occupational exposure (sources, entry routes, occupational exposure monitoring and health surveillance), as well as control measures and occupational exposure limits (although there are still no regulatory occupational exposure limit values specific for nanoparticles in EU legislation, some information on recommendations at national level are presented).

In the website of EU-OSHA, some relevant publications related to the textile sector can be found, including:

- Occupational safety and health in the textiles sector [3]. This "E-facts sheet" includes a description of the main hazards and risks in the textiles sector, a check list of hazards in textiles as well as some considerations about managing the safety and health of women in the textiles sector.
- Workplace exposure to nanoparticles [4]. This extensive report includes considerations on occupational exposure to nanoparticles, health effects of nanomaterials, safety hazards and handling of nanomaterials, regulatory background, policies, research, standardization and collaboration programs. This report has several references on nanoparticles used in the textile industry.
- Implementation of OSH Management System in the Teofilów textile works [5]. This case study concerns the Polish company ZTK Teofilów S.A., employing 450 people and involved in the production and sale of top-quality knitted fabrics. Two years after the implementation of the OSH management system the number of occupational accidents had fallen to one-third of its previous level.
- Use of nanomaterials in textile finishing [6]. This report presents a case study in the German company Schmitz-Werke, which decided to use a finishing agent for textiles containing nanomaterial in order to improve the quality of its awnings to repel water, dirt and oils by creating a self-cleaning surface. The company has assessed potential risks from the use of the finishing agent and monitored the efficiency of the existing workers' protection measures at the workplaces in the concerned work areas (mixing station of the finishing agents, foulard and stentering frame).

22.5 European definition of nanomaterial

In 2011 the European Commission has published the following recommendation concerning the definition of nanomaterial – 2011/696/EU (Commission Recommendation of 18 October 2011, see https://eur-lex.europa.eu/eli/reco/2011/696/oj):

> **Nanomaterial** means a natural, incidental or manufactured material containing particles, in an unbound state or as an aggregate or as an agglomerate and where, for 50% or more of the

> particles in the number size distribution, one or more external dimensions is in the size range 1–100 nm.

This definition is the same as universally adopted (e.g., in ISO standards) but includes the following additions:

> In specific cases and where warranted by concerns for the environment, **health, safety** or competitiveness the number size distribution threshold of 50% may be replaced by a threshold between 1 and 50%.
>
> (. . .) fullerenes, graphene flakes and single wall carbon nanotubes with one or more external dimensions below 1 nm should be considered as nanomaterials.

This definition is very important as it provides a general basis for regulatory instruments across many areas. "Member States, the Union agencies and economic operators are invited to use the following definition of the term 'nanomaterial' in the adoption and implementation of legislation and policy and research programmes concerning products of nanotechnologies." So, all the legal requirements should take into account this definition, including the exceptions presented above. The special concern with health and safety is very clear.

It is written in the recommendation that "By December 2014, the definition (. . .) will be reviewed in the light of experience and of scientific and technological developments. The review should particularly focus on whether the number size distribution threshold of 50% should be increased or decreased." But in fact, up to now (August 2020) this revision has not yet been made, although a lot of technical work has been done by the European Joint Research Centre (JRC) to support the European Commission concerning the possible revision.

JRC has published relevant reports related to this definition. Here are the most important:

Toward a review of the EC Recommendation for a definition of the term "nanomaterial." This is a series of three documents that could be a basis for the revision of the definition of nanomaterial:

Part 1: Compilation of information concerning the experience with the definition [7]. This first report compiles information concerning the experience with the definition regarding scientific-technical issues that should be considered when reviewing the current EC definition of nanomaterial, published in 2011.

Part 2: Assessment of collected information concerning the experience with the definition [8]. This report provides the JRC assessment of feedback on the experiences of stakeholders with the EC nanomaterial definition.

Part 3: Scientific-technical evaluation of options to clarify the definition and to facilitate its implementation [9]. This report provides the JRC's scientific-technical evaluation of options to clarify the EC Recommendation on a definition of nanomaterial. The evaluation shows that the scope of the definition regarding the origin of nanomaterials should remain unchanged, addressing natural, incidental as well as manufactured nanomaterials. Moreover, because of the regulatory purpose of the definition, there is little evidence to support deviating from size as the sole defining property of a nanoparticle or from the range of 1–100 nm as definition of the nanoscale. Besides the need for clarification of some terms used in the definition additional implementation guidance would be useful. The role of the volume-specific surface area deserves clarification and a method to prove that a material is not a nanomaterial would be helpful. A strategy on how to avoid unintended inclusion of materials and the list of explicitly included materials deserve also attention.

An overview of concepts and terms used in the European Commission's definition of nanomaterial [10]. This report supports the implementation of the European Commission's Recommendation on a definition of nanomaterial. It addresses its key concepts and terms and discusses them in a regulatory context.

Identification of nanomaterials through measurements [11]. This report addresses identification of nanomaterials according to the European Commission's Recommendation on the definition of nanomaterial by measurements and discusses options and points to consider when assessing whether a particulate material is a nanomaterial or not.

22.6 Nanomaterials at workplace

The European Commission has published in 2014 two useful guides on the protection of the health and safety of workers from the potential risks related to nanomaterials at work, one for employers and health and safety practitioners, the other for workers.

The **guidance for employers and health and safety practitioners** [12] has been prepared in relation with the Chemical Agents Directive 98/24/EC. This guide presents the following steps in the risk assessment and management process:

Step 1 – Identification of the manufactured nanomaterials (MNMs)

Step 2 – Hazard assessment: general risk considerations, categorizing level of concern – shape and solubility, dustiness and flammability

Step 3 – Exposure assessment

Step 4 – Categorization of risk (control banding)

Step 5 – Detailed risk assessment

Step 6 – Risk management includes general principles, hierarchy of controls and risk management measures. The measures are presented for the four risk levels of exposure to manufactured nanomaterials. This section also includes guidance on information to the employees, instruction and training, as well as health surveillance of the workers, in conjunction with the requirements of the Chemical Agents Directive.

Step 7 – Review.

This guide also includes useful annexes:
- Concerns over hazards and risks of nanomaterials
- Further guidance on the industrial use of nanomaterials
- Examples of applications of MNMs
- Legislation applicable to nanomaterials
- Challenges in monitoring exposure to nanomaterials

The **guidance for workers**, Working Safely with Manufactured Nanomaterials [13], provides readily understandable information in questions and answers format. The following questions are answered:
- What are nanomaterials and nano-enabled products?
- What is the basis for current concerns about manufactured nanomaterials?
- Do I need to treat all manufactured nanomaterials as special cases?
- How can I tell if I am using nanomaterials or nano-enabled products, and how may I be exposed?
- What actions are necessary to enable safe working with manufactured nanomaterials and nano-enabled products?

The penetration routes for nanoparticles in the human body include inhalation, ingestion and dermal penetration. Skin contact is particularly relevant in textiles [14], not only for consumers but also for workers in the textile, clothing and distribution sectors.

22.7 Standardization

The standard ISO 45001:2018 – Occupational Health and Safety Management Systems –presents very useful information on how to manage in the companies this topic. It is aligned with ISO 9001:2015 and other management systems.

Concerning nanotechnology, ISO has been developing a large set of standards, especially within ISO Technical Committee 229, created in 2005. Working group 3 deals specifically with Health, Safety and Environmental Aspects of Nanotechnologies.

At European level, the Technical Committee CEN TC 352 has been created in 2006. Under AFNOR/UNMZ (France/Czech) Secretariats, CEN/TC352 is engaged in standardization in the field of nanotechnologies. This includes the development of a set of standards addressing the following aspects of nanotechnologies:
- classification, terminology and nomenclature;
- metrology and instrumentation, including specifications for reference materials;
- test methodologies;
- modeling and simulation;
- science-based health, safety and environmental practices;
- nanotechnology products and processes.

Under CEN/TC 352 coordination, several CEN and ISO Technical Committees are involved in the execution of Mandate M/461 from the European Commission: Several European standards have already been published under this Mandate and several other are in preparation. Many of them have a relation with occupational health and safety.

At present (August 2020), 21 European standards have been published (15 out of these are EN/ISO documents, in conjunction with ISO/TC229) and 16 are under preparation. The updated list can be consulted at https://standards.cen.eu/.

Related to ISO/TC229, up to now (August 2020) a total of 81 standard documents have been published and 34 are under development, covering namely health and safety aspects. The following documents are especially relevant:

- ISO/TR 12885:2018 – Nanotechnologies – Health and safety practices in occupational settings
- ISO/TS 12901–1:2012 – Nanotechnologies – Occupational risk management applied to engineered nanomaterials – Part 1: Principles and approaches
- ISO/TS 12901–2:2014 – Nanotechnologies – Occupational risk management applied to engineered nanomaterials – Part 2: Use of the control banding approach
- ISO/TR 13121:2011 – Nanotechnologies – Nanomaterial risk evaluation
- ISO/TR 13329:2012 – Nanomaterials – Preparation of material safety data sheet (MSDS).

In 2006 OECD (Organization for Economic Co-operation and Development) has established the Working Party on Manufactured Nanomaterials (WPMN) as a subsidiary body of the OECD Chemicals Committee. This program concentrates on human health and environmental safety implications of manufactured nanomaterials. Since then OECD has published almost 100 documents under the series of Safety of Manufactured Nanomaterials, some of which can be related to standards. The full list of all the freely downloadable documents can be consulted at: http://www.oecd.org/env/ehs/nanosafety/publications-series-safety-manufactured-nanomaterials.htm

It is important to emphasize the so-called Malta Initiative (which arose during the Maltese EU Council Presidency in 2017), involving 18 European countries, several Directorates-General of the European Commission, the European Chemicals Agency (ECHA), authorities, research institutions, NGOs, universities and industry work together on a voluntary and self-organized basis. The aim of this initiative is to make legislation enforceable, in particular in the chemicals sector. For this purpose, it is necessary to ensure that the essential test, measurement and verification procedures are available. Currently, the work is focused on amending the OECD Test Guidelines in the area of nanomaterials to ensure that a nanomaterial-adapted REACH Regulation will become enforceable.

22.8 Use of the control banding approach in occupational risk management applied to engineered nanomaterials

The control banding approach for occupational risk management applied to engineered nanomaterials, according to ISO/TS 12901-2:2014, is a pragmatic approach useful for the control of workplace exposure to possibly hazardous agents with unknown or uncertain toxicological properties and for which quantitative exposure estimations are lacking.

The control banding process, according to ISO/TS 12901-2:2014, includes the following elements:

- Information gathering
- Assignment of the nano-objects to a hazard band (on the basis of a comprehensive evaluation of all available data on each material, taking into account parameters such as toxicity, in vivo biopersistence and factors influencing the ability of particles to reach the respiratory tract, their ability to deposit in various regions of the respiratory tract, their ability to elicit biological responses)
- Description of potential exposure characteristics (assigning an exposure scenario at a workplace to an exposure band, taking into account the physical form and amount of the nano-object, dust generation potential of processes and actual exposure measurement data)
- Definition of recommended work environments and handling practices (control banding)
- Evaluation of the control strategy or risk banding

Ramos et al. [15] have applied this methodology in a case study in a textile finishing company involving two chemical finishes containing nanomaterials: mosquito repellence and antibacterial finish. The risk analysis concerned mainly four workers involved either in the preparation of the finishing baths and on the conducting of the stenter frame. Following the application of control banding method, measures to mitigate risks have been envisaged: appropriate ventilation and use of adequate personal protective equipment. Hazards related to one of the chemicals are higher and also require the use of a closed booth and a smoke extractor.

There is now a specific REACH registration system for nanomaterials, which entered into force from January 2020 (Commission Regulation [EU] 2018/1881 of 3 December 2018), so it is recommended that the suppliers of chemicals which incorporate nanomaterials include more information on the hazards and measures for risk mitigation in the safety data sheets, based for instance on the recommendations presented in ISO/TR 13329. This information is essential for implementing the control banding approach.

More recently, the new Commission Regulation (EU) 2020/878 of 18 June 2020, amending Annex II to REACH, which gives the requirements for compiling safety data sheets used to provide safety information on hazardous chemical substances and mixtures in the EU. This regulation presents more detailed requirements to be included in the safety data sheets of chemicals that include nanoforms. The Regulation applies from 1 January 2021 with a transitional period until 31 December 2022.

22.9 Conclusions

This chapter presented an overview of the European legislation related to Occupational Health and Safety relevant for the textile industry and nanomaterials. It is suggested to follow the activities of the European Agency for Safety and Health at Work (EU-OSHA). The website of EU-OSHA is the best place to follow all the relevant EU legislation, with links to legislation at national level. The European Commission is replacing as far as possible all the EU Directives into EU Regulations, avoiding in this way the problems of transposition into national laws in the different countries.

The increasing concerns related to health and safety of nanomaterials are leading to the emergence of legislation that poses more and more restrictions. It is essential that all stakeholders keep aware of all the updated legal requirements, considering these not only as limitations but also as opportunities of improvement.

References

[1] European Commission, Employment, Social Affairs & Inclusion: Evaluation of the Practical Implementation of the EU Occupational Safety and Health (OSH) Directives in EU Member States, 2015. Available at http://ec.europa.eu/social/BlobServlet?docId=16895&langId=en

[2] Communication from the Commission to the European Parliament, the Council, the European Economic and Social Committee and the Committee of the Regions: Safer and Healthier Work for All – Modernisation of the EU Occupational Safety and Health Legislation and Policy, 2017. Available at https://eur-lex.europa.eu/legal-content/EN/TXT/?uri=CELEX%3A52017DC0012

[3] EU-OSHA – European Agency for Safety and Health at Work: Occupational safety and health in the textiles sector. E-Facts 30, 2008. Available at https://osha.europa.eu/en/publications/e-fact-30-occupational-safety-and-health-textiles-sector

[4] EU-OSHA – European Agency for Safety and Health at Work: Literature Review – Workplace exposure to nanoparticles, 2009. Available at https://osha.europa.eu/en/publications/workplace-exposure-nanoparticles

[5] EU-OSHA – European Agency for Safety and Health at Work: Case studies: Implementation of OSH management system in the Teofilów textile works, 2010. Available at https://osha.europa.eu/en/publications/implementation-osh-management-system-teofilow-textile-works

[6] EU-OSHA – European Agency for Safety and Health at Work: Case studies: Use of nanomaterials in textile finishing, 2012. Available at https://osha.europa.eu/en/publications/use-nanomaterials-textile-finishing
[7] Rauscher, H.; Roebben, G.; Amenta, V.; Boix Sanfeliu, A.; Calzolai, L.; Emons, H.; Gaillard, C.; Gibson, N.; Linsinger, T.; Mech, A.; Quiros Pesudo, L.; Rasmussen, K.; Riego Sintes, J.; Sokull-Klüttgen, B.; Stamm, H.: Towards a review of the EC Recommendation for a definition of the term "nanomaterial"; Part 1: Compilation of information concerning the experience with the definition. JRC Science for Policy Report, 2014. DOI: 10.2788/36237. Available at https://ec.europa.eu/jrc/en/publication/eur-scientific-and-technical-research-reports/towards-review-ec-recommendation-definition-term-nanomaterial-part-1-compilation-information?search
[8] Roebben, G.; Rauscher, H.; Amenta, V.; Aschberger, K.; Boix Sanfeliu, A.; Calzolai, L.; Emons, H.; Gaillard, C.; Gibson, N.; Holzwarth, U.; Koeber, R.; Linsinger, T.; Rasmussen, K.; Sokull-Klüttgen, B.; Stamm, H.: Towards a review of the EC Recommendation for a definition of the term "nanomaterial". Part 2: Assessment of collected information concerning the experience with the definition. JRC Science for Policy Report, 2014. DOI: 10.2787/97286. Available at https://ec.europa.eu/jrc/en/publication/eur-scientific-and-technical-research-reports/towards-review-ec-recommendation-definition-term-nanomaterial-part-2-assessment-collected
[9] Rauscher, H.; Roebben, G.; Boix Sanfeliu, A.; Emons, H.; Gibson, N.; Koeber, R.; Linsinger, T.; Rasmussen, K.; Riego-Sintes, J.; Sokull-Klüttgen, B.; Stamm, H.: Towards a review of the EC Recommendation for a definition of the term "nanomaterial": Part 3: Scientific-technical evaluation of options to clarify the definition and to facilitate its implementation. JRC Science for Policy Report, 2015. DOI: 10.2788/678452. Available at https://ec.europa.eu/jrc/en/publication/eur-scientific-and-technical-research-reports/towards-review-ec-recommendation-definition-term-nanomaterial-part-3-scientific-technical
[10] Rauscher, H.; Roebben, G.; Agniezka, M.; Gibson, N.; Kestens, V.; Linsinger, T.; Riego-Sintes, J.: An overview of concepts and terms used in the European Commission's definition of nanomaterial. JRC Science for Policy Report, 2019a. DOI: 10.2760/053982. Available at https://ec.europa.eu/jrc/en/publication/overview-concepts-and-terms-used-european-commissions-definition-nanomaterial
[11] Rauscher, H.; Mech, A.; Gibson, N.; Gilliland, Held, D.; Kestens, V.; Koeber, R.; Linsinger, T.; Stefaniak, E.: Identification of nanomaterials through measurements. JRC Science for Policy Report, 2019b. DOI: 10.2760/459136. Available at https://ec.europa.eu/jrc/en/publication/eur-scientific-and-technical-research-reports/identification-nanomaterials-through-measurements
[12] European Commission, Employment, Social Affairs & Inclusion: Guidance on the protection of the health and safety of workers from the potential risks related to nanomaterials at work. Guidance for employers and health and safety practitioners, 2014a. Available at http://ec.europa.eu/social/BlobServlet?docId=13087&langId=en.
[13] European Commission, Employment, Social Affairs & Inclusion: Working Safely with Manufactured Nanomaterials. Guidance for Workers, 2014b. Available at: http://ec.europa.eu/social/BlobServlet?docId=13088&langId=en
[14] Almeida, L.; Ramos, D.: Health and safety concerns of textiles with nanomaterials. IOP Conf. Series: Materials Science and Engineering 254, 2017. 10s2002. DOI: 10.1088/1757-899X/254/10/102002. EID: 2-s2.0–85034981671. Available at: https://iopscience.iop.org/article/10.1088/1757-899X/254/10/102002
[15] Ramos, D.; Almeida, L.; Gomes, M.: Application of Control Banding to Workplace Exposure to Nanomaterials in the Textile Industry. In: Arezes, P., et al., editors: Occupational and Environmental Safety and Health. Studies in Systems, Decision and Control. Cham: Springer, 2019; vol. 202, pp. 105–113. ISBN 978-3-030-14729-7, Online ISBN 978-3-030-14730-3 Book ID: 478838-1-En. https://doi.org/10.1007/978-3-030-14730-3_12

Ning Cui

23 China's governance, legislation and legal framework

Keywords: China, textile, safety, legislation, standards, hazards

23.1 China's legislation and standard about textile safety

China is the largest nation in textile industry globally, with the most completed industrial chain and the most integrated industrial category, including synthetic fibers, yarns, fabrics, costumes and so on. The total output of China's chemical fiber reached 50.11 million tons in the year 2018, which accounted for 50% of the total output and processing volume in the world. In the same year, China's textile and apparel exports totaled 276.73 billion US dollars, accounting for 35% of the world's total [1].

During the manufacturing of textile product, various additives are utilized for fiber production, different types of dyes and finishes are applied for textile finishing, which can bring harmful substances. Because many textiles and costumes directly contact with people skin in a long term, when harmful substance residue on the textile and accumulate to a certain amount, negative effects will be brought to the skin, or even damage people's health.

China has launched research on testing standard for hazardous substance content in textile relatively early. In 1982 and 1987, China formulated the testing method standard for textile formaldehyde and pH, which has already become two of the routine testing items of textile products.

German Government promulgated the German Food and Consumer Article Law in 1992. Certain azo dyes, which are harmful to human health, are not permitted for use in textiles by this law [2]. Under the influence of this German law, China government began to carry out research on evaluation standard for textile safety and health, testing standard for hazardous substances in textile, ecological textile performance and standards, since 1990s, to guarantee the basic safety of China's domestic consumer goods market and to adapt to international trade rules during international trade with developed countries from Europe and America. China's National Textile Standardization Technology Committee has formulated a series of testing method standards for hazardous substances in textiles, such as "Textiles –

Ning Cui, China Textile Academy, No.3 Yanjingli Middle Street, Chaoyang District, Beijing, China, e-mail: cuining@cta.com.cn

https://doi.org/10.1515/9783110670776-023

Determination of the banned azo colorants》 (GB/T17592)" and "Textiles – Test method of heavy metal ions(GB/T17593)."

After years of hard work, strict audition procedures and notification to the World Trade Organization, "National general safety technical code for textile products" (GB/T 18401–2003, now updated to GB/T 18401–2010), a national mandatory standard with the attribute of technical regulation, have been enacted in November 2003 and came into effective use on 1 January 2005. The introduction of this standard incorporates ecological and environmental protection requirements related to textile safety into national mandatory standards for the first time. It ended the historical situation that the quality standards of China's textile and apparel industry remained in only visual appearance and general physical index for many years and also indicated that China has taken a substantial step in the legalization and standardization in the area of ecological textiles. In addition, the standards have given special consideration on consumer groups of infants and children, and special limits on multiple hazardous substances have been set for these consumer groups to restrict the utilization.

Up to now, China's product standards and testing standards for textile products safety have formed a system, covering basically all organics and metal compounds applied in fiber manufacturing, textile production and post-dyeing processing. The series of standards includes

- National general safety technical code for textile products (GB18401-2010)
- Technical specifications of ecological textiles (GB/T 18885–2009)
- Limit and determination of parts of harmful substances in textile dyeing and finishing auxiliaries (GB/T20708-2019)
- Limitation of toxic and hazardous substances in textile fibers (GB/T22282-2008)
- Technical requirement for environmental labeling products – ecotypic textile (HJ/T 307–2006)

and so on.

Among the series of standards and specifications, those related to the management of hazardous substances in textile products include not only the mandatory national standards such as

- National general safety technical code for textile products (GB18401-2010)
- Restriction of hazardous materials in polyvinyl chloride artificial leather (GB 21550–2008)
- Safety technical code for infants and children textile products (GB 31701–2015)

but also the recommended technical standards such as

- Requirements of physical and chemical performance of garments (GB/T21295-2014)
- Limitation of toxic and hazardous substances in textile fibers (GB/T22282-2008)
- Technical specifications of ecological textiles (GB/T 18885–2009)

Marking, packaging, transportation and storage for garments (FZ/T80002-2016) and some recommended product standard include "Infant's wear (FZ/T 81014–2008)."

China's standards could roughly be divided into two categories: government standards and market standards. The former category consists of national standard, industry standard and local standard, while the latter consists of company standard and association standard. Although company standard and association standard have made great supplement to government standards, which have been approved to be beneficial, they still lack the influence and constraint on hazardous substance restriction like government standard.

China's restrictions on hazardous substances in textiles mainly focus on decomposable carcinogenic aromatic amine dyes, formaldehyde, phthalates, total lead, total cadmium, vinyl chloride monomer, other volatiles and extractable heavy metals (see Table 23.1), and most of which are introduced into textiles during printing, dyeing and finishing processes. Although printing and dyeing is an important way to introduce hazardous substances into textile products, hazardous substances in fiber materials should not be neglected. Taking polyester fiber as an example, during the synthesis process of polyterephthalic acid and ethylene glycol, catalyst containing antimony, currently the most efficient and economical catalyst, such as antimony acetate and ethylene glycol antimony, is needed. During this process, part of antimony will be reduced to elemental state [3], and will be distributed uniformly and remain in the fiber. When polyester fiber is transported to printing plant or weaving mill for the process of desizing, alkali treatment and so on, the elemental state antimony might be released from the fiber and discharged with the wastewater, thus causing pollution. In order to cut or minimize the negative result, restrictions on the hazardous substances in five commonly used synthetic fibers (including polyacrylonitrile fiber, man-made cellulose fiber, polyurethane elastic fiber, polyester fiber and polypropylene fiber) have been formulated in the national standard "Limitation of toxic and hazardous substances in textile fibers (GB/T22282-2008)" (see continued Table 23.1).

Table 23.1: Requirements of restricted substances in some of the China official textile standards.

Standard no.	Hazardous substances		Requirements
GB18401-2010	Decomposable carcinogenic aromatic amine dye/(mg/kg)		≤20
	Formaldehyde/(mg/kg)		Grade A ≤ 20 Grade B ≤ 75 Grade C ≤ 300
GB21550-2008	Soluble heavy metals/(mg/kg)	Lead	≤90
		Cadmium	≤75

Table 23.1 (continued)

Standard no.	Hazardous substances		Requirements	
	Vinyl chloride monomer/(mg/kg)		≤5	
	Other volatiles/(g/m^2)		≤20	
GB31701-2015	Phthalates/%	DEHP + DBP + BBP DINP + DIDP + DNOP	≤1 ≤1	
	Heavy metals/(mg/kg)	Total lead Total cadmium	≤100 ≤90	
GB/T21295-2014	Phthalates/%	DEHP + DBP + BBP DINP + DIDP + DNOP	≤1 ≤1	
	Total lead (products for children under 14)/(mg/kg)		≤90	
	Alkylphenol (AP) and alkylphenol polyoxyethylene ether (APEO)	NP + OP NPEO + APEO	≤500 1,000	
	Forbidden flame retardant (PBB, TRIS, TEPA, penta-BDE, octa-BDE)/(mg/kg)		≤10	
	Extractable heavy metals/(mg/kg)	Arsenic (As)	Baby products Other products	≤0.2 ≤1.0
		Lead (Pb)	Baby products Other products	≤0.2 ≤1.0
		Chromium (Cr)	Baby products Other products	≤1.0 ≤2.0
		Cobalt (Co)	Baby products Other products	≤1.0 ≤4.0
		Copper (Cu)	Baby products Other products	≤25.0 ≤50.0
		Nickel (Ni)	Baby products Other products	≤1.0 ≤4.0
		Antimony (Sb) Cadmium (Cd) Hexavalent chromium (CrVI) Mercury (Hg)	≤30.0 ≤0.1 ≤0.5 ≤0.02	
FZ/T81014-2008	Extractable heavy metals/(mg/kg)	Mercury (Hg) Arsenic (As) Lead (Pb) Copper (Cu) Chromium (Cr)	≤0.02 ≤0.2 ≤0.2 ≤25.0 ≤1.0	

Table 23.1 (continued)

Standard no.	Hazardous substances		Requirements
FZ/T73025-2013	Extractable heavy metals/(mg/kg)	Antimony (Sb)	≤30
		Arsenic (As)	≤0.2
		Lead (Pb)	≤0.2
		Cobalt (Co)	≤1.0
		Copper (Cu)	≤25
		Nickel (Ni)	≤1.0
		Mercury (Hg)	≤0.02
		Cadmium (Cd)	≤0.1
		Chromium (Cr)	≤1.0
		Hexavalent chromium (CrVI)	Below detection limit
GB/T22282-2008	Acrylonitrile fiber/(mg/kg)	Acrylonitrile	≤1.5
	Polyester fiber/(mg/kg)	Antimony	≤260
	Polypropylene fiber/(mg/kg)	Lead	≤1.0
	Polyurethane elastic fiber/(mg/kg)	Organotin	≤1.0
	Synthetic cellulose fiber/(mg/kg)	Absorbable organic halide	≤250
	Cotton and other cellulose seed fiber/(mg/kg)	Total pesticide	≤0.05
	Greasy wool and other protein fiber/(mg/kg)	Total pesticide	Organochlorine ≤0.5 Organophosphorus ≤2.0 Pyrethroid ≤0.5 Chitin synthesis inhibitor ≤2.0
FZ/T80002-2016	Heavy metal lead, cadmium, mercury and hexavalent chromium in package and package material/(mg/kg)	Total content	≤100

For the determination of hazardous substance limits in textiles, we can learn from the comparative analysis of Table 23.2 [4] that China has formulated its prohibition and restriction on hazardous substances in textile basically on the adoption and with reference to official regulations and standards of EU and the USA. At the same time, it has also taken into consideration of requirements of well-known civil standards from Europe, such as Oeko-Tex Standard 100. This shows that requirements of China's domestic consumer product market for the restriction of hazardous substances are consistent with the international market.

Table 23.2: Some China standards and corresponding laws and regulations abroad.

Name of China standard	Number of China standard	Corresponding laws and regulations abroad
National general safety technical code for textile products	GB18401-2010	Oeko-tex 100 standard (2003) (adopted after modification)
Safety technical code for infants and children textile products	GB31701-2015	US Consumer Product Safety Improvement Act (CPSIA) and REACH (adopted after modification)
Requirements of physical and chemical performance of garments	GB/T21295-2014	CPSIA and REACH and road map of the Zero Discharge of Hazardous Chemical (ZDHC) Group
Limitation of toxic and hazardous substances in textile fibers	GB/T22282-2008	Europe Directive 2002/37/EC Ecological textile certification
Technical specifications of ecological textiles	GB/T18885-2009	Oeko-tex 100 standard (2008)
Knitted garment and adornment for infant	FZ/T73025-2013	Oeko-tex 100 standard
Infant's wear	FZ/T81014-2008	Oeko-tex 100 standard (adopted after modification)
Restriction of hazardous materials in polyvinyl chloride artificial leather	GB21550-2008	CPSIA and REACH (partly adopted)
Marking, packaging, transportation and storage for garments	FZ/T80002-2016	Europe Directive 94/62/EC Package and package waste

23.2 China's testing standards related to functional textiles

With the progress of fiber material manufacturing technology, fiber processing technology and fabric finishing technology, the modern textile industry has long been able to produce textiles with special functions beyond those traditional functions like warming, covering and beautifying, which are called functional textiles. There are various types of functional textiles. Although there is still no unified classification criterion, the functional textiles could basically be classified into four types: protection type, comfort type, healthcare type and easy maintenance type, as shown in Table 23.3.

Table 23.3: Basic classification of functional textiles.

	Function	Examples
Protection type	Protecting human from dangerous environment, reducing or avoiding injury	Flame retardant, antistatic, antiradiation, water proof
Comfort type	Improving the dressing comfort experience	Wicking fabric, breathable fabric
Healthcare type	Improving the wearing performance of the fabric by regulating human body microcirculation	Antibacterial, far-infrared health care
Easy maintenance type	Reducing the workload of taking care of clothes, such as cleaning or ironing	Antiwrinkle, easy-cleaning

During the recent 10 years, functional textiles have maintained an upward trend in China's textile industry and market. Let us take one of the abovementioned four types, the healthcare type, as example, the total sales of it has exceeded RMB ¥13 billion (about US $1.92 billion according to current exchange rate) in 2012 [5].

Although there is a large market demand in front, when facing with different performances of most of the functional textiles, normal consumer cannot always make their judgment only by their visual and touching experience. Therefore, the determination of the performance of a textile product could only rely on a certain testing method. For functional textile products, the testing mainly focused on different functions, such as waterproof, antifouling and far-infrared performance. Let us take the testing of far-infrared performance as example. Far-infrared textile is either produced with far-infrared functional fiber prepared by coextruding of far-infrared ceramic powder and polymer, or produced by textile finishing process with ceramic powder. This kind of textile could improve the blood circulation and regulate the body's metabolism. At present, the standards related to the testing of far-infrared performance include national standard, "Textiles – Testing and evaluation for far infrared radiation properties (GB/T 30127 – 2013)," "Textiles – Testing method for thermal retention with accumulated by infrared ray (GB/T 18319 – 2001)" and textile industry standard, "Far infrared textile(FZ/T 64010 – 2000)." The scope of application, wavelength range and evaluation properties of these standards are all related to the functional component added to the fiber or the finishing method.

Due to the wide variety of functional textiles, also because of the fact that the production method, application field of the functional textile are quite different, China does not have a national standard or an industry standard on functional textile product. The formulation of relevant performance standard or testing standard has always been driven forward by the power of market demand. An impressive case is the debate about UV-resistant textiles that happened in the summer of 1999 [6]. Due to the continuous high temperature in some area of southern China at that time, the intensity of ultraviolet radiation increased sharply, and the demand for ultraviolet protection textiles increased rapidly. At that time, there was no national standard for this kind of product. Therefore, anti-ultraviolet umbrellas, hats, shawls and other products flooded into the local markets but were difficult to distinguish. With strong appeal from all walks of life, the national standard, "Textiles – Evaluation for solar ultraviolet radiation protective properties (GB/T18830)" was issued on 5 September 2002 and became effective on 1 February 2003.

In recent years, China's textile industry took an actively open attitude toward relevant international standards and advanced foreign standards. China's quality standard for textile and apparel products has shifted from single product standards to a system with commercial standards, testing standards and quality certification standard which keeps pace with the international textile market. The standard system for functional textile is being constructed and improved steadily. According to statistics [7], China already has set up more than 80 functional textile standards, covering antistatic, flame-retardant, anti-mosquito, UV protection, thermal underwear, antibacterial and deodorant textile products.

23.3 China's legislation on cleaner production

What has been discussed earlier is about China's endeavor on its standard and regulation construction for textile industry. It is committed to establishing limits of toxic and hazardous substances contained in fiber and textile products in the form of national standards and industry standards. In order to control the raw material and energy consumption in the production process, to minimize the negative impact brought to the environment and human health, China is also constructing its laws and regulation system for fiber and textile manufacturing to standardize the operation and guarantee the healthy development of this industry

In order to promote cleaner production in the industrial sector, improve resource utilization efficiency, reduce and avoid pollution, protect the environment and human health, and to promote sustainable economic and social development, China has implemented the Cleaner Production Promotion Law since January 2003. It is the first state-level law officially incorporated into the principles of pollution prevention and source reduction. The law defines the nomination of "cleaner production" as continuously adopting measures such as improving industry design, using clean energy and raw materials, adopting advanced process technology and equipment, improving management, and comprehensive utilization to achieve source pollution reduction and increase resource utilization efficiency, reduce or avoid the generation and discharge of pollutants during production, service and product consumption, in order to reduce or eliminate the harm brought to human health and the environment. This law also requires that the manufacturer takes into consideration the impact brought by the products and package design to the environment and human health throughout the whole production cycle. It provides a legal basis to reduce and eliminate the toxic and hazardous substances for the producers and users of toxic and hazardous substances.

Similar to the laws and regulations aiming at pollution source control implemented in other countries, China's "Cleaner Production Promotion Law" also focuses on the selection of green raw materials and the design and adoption of efficient production technologies. But as a "promotion law," it does not make clear provisions on what is cleaner production technology, neither does it set a mandatory reduction target for toxic and hazardous substances. It does not even require the relevant companies to report and disclose the use and discharge of toxic and hazardous substances regularly. But the law does provide a starting point for Chinese enterprises to eliminate toxic and hazardous substances, and push China toward the complete elimination of toxic and hazardous substances.

As one of the major energy consumption and pollution discharge industries in China, the textile industry takes great social responsibility in reducing and ultimately eliminating the application and discharge of toxic and hazardous substances. The policy support from the government is indispensable in realizing the ultimate goal of energy conservation and emission reduction. In 2019, "Cleaner Production Evaluation Index System for Synthetic Fiber Manufacturing" was issued by China's National Development and Reform Commission, Ministry of Ecology and Environment, and Ministry of Industry and Information Technology. The system includes the cleaner production index for six kinds of synthetic fibers: polyester, nylon 6, recycled polyester, spandex, vinylon and regenerated cellulose fiber. The system sets specific requirements for the production of synthetic fibers from six aspects: production equipment and technical indicators, resource and energy consumption indicators, resource comprehensive utilization indicators, pollutant emission control indicators, product feature indicators, and cleaner production management indicators. The compile of this system has been organized by China Chemical Fiber Industry Association. It could be used for the cleaner production audit, the judgment for potential and opportunities of cleaner production and cleaner production performance evaluation and cleaner production performance announcement system. It could also be used for environmental impact assessment and discharge permit management system.

Before the cleaner production system, China had already formulated and issued some industry standard or government management regulations in this area, such as "Technical requirement for environmental labelling products – Ecotypic textiles (HJ/T185-2006)," "Cleaner production standard – Textile industry (Dyeing and finishing of cotton) (HJ/T185-2006)," "Cleaner production evaluation index system for printing and dyeing industry (Trial)."

23.4 Occupational safety issue in textile industry in China

Textile industry occupies an important position in China's national economy. According to the fourth time national economic census, by the end of 2018, the number of corporate entities and total assets of China's textile and apparel industry accounted for 9.5% and 4.8% of the whole national industry, respectively. The employee in this industry totaled 11.03 million, which accounted for 9.6% of the national industry. From this perspective, the occupational health and safety also occupies an important position in the occupational safety issue of the national industry employee.

In China, there once have been many different names for occupational safety and health, such as labor protection, labor safety, labor safety and health, labor safety and health protection, safety and health protection, occupational safety and production safety. The central government authority which is in charge of this issue has also changed several times, from the Ministry of Labor to the Ministry of Health, from the State Administration of Work Safety to National Health Commission. Therefore China's understanding about occupational safety and health, and how to implement protection in practice, has gone through a gradual process.

From 1980s onward, China has promulgated laws and regulations such as "Labor Law," "Occupational Disease Prevention Law," "Safety Production Law" and "Work Injury Insurance Regulations." Occupational health and safety issues have been normalized to a certain extent ever since. The safety and health of labors and related legal rights have been protected to some extent.

According to statistics [8], China has up to 770 laws and regulations on occupational health and safety, involving special equipment, occupational safety, safety accidents, fire safety, chemical industry, power industry, transportation, medical industry and other various aspects of industry. But among all these laws and regulations, apart from several general laws such as "Labor law," "Labor contract law," "Occupational disease prevention law" and "Safety production law," most of them are normative acts like regulation standards formulated by occupational safety and health-related authorities. Moreover, China still lacks a state-leveled, unified "Occupational Safety and Health Law" that cannot be replaced by any other departmental law.

Textile industry in China is a labor-intensive industry. A study [9] of occupational hazards in China's cotton spinning, wool, hemp, chemical fiber, silk, printing and dyeing enterprises showed that the main occupational hazards in the textile printing and dyeing industry are dust, noise, high temperature, high humidity, chemical poisons and biological factors, together with some hazardous factors in production environment and during laboring, such as special forced body position, individual organ tension.

Dust hazards mainly come from raw material pretreatment (sieving, dehydrating and conveying of the resin chip), fiber detritus generated by friction between chemical fibers and friction disk components during texturing process, cotton and ramie (China grass) detritus or dust generated from roving and spun yarn process.

These dusts are quite easily dispersed in air and enter human body through workers' respiratory system, which could cause respiratory infection, dry nasal, itchy throat, hoarse voice and elevated body temperature to operators.

Industrial noise is the major occupational hazard of textile industry. The main pollution source is high-speed winder of chemical fiber plant, yarn spinning workshop and loom workshop. These noises are high-frequency noises which could reach 90–105 dB(A). The major noise-generating positions of textile industry are roving, spun yarn, weaving, knitting, cone winding, warping and warp knitting. With shuttle loom replaced by shuttle-less loom, the noise level has been reduced to some extent. But it is still difficult to meet the requirement of the national health standard of 85 dB(A). Overexposure to high-level noise could cause damage to various systems and functions of the human body. Research about it mainly focuses on the impact on textile worker's auditory system, nonauditory system (including nervous system, cardiovascular system, endocrine system, immune system, digestive system and reproductive system) and the working efficiency.

The printing and dyeing workshop has a typical high temperature and high humidity environment of textile industry. Because of product quality demand, the temperature in the workshop in summer always exceeds 35 °C, with the relative humidity over 60%. Especially in warp sizing workshop, the relative humidity can even exceed 80%. If the textile is damped, fiber deterioration and discoloration will occur, and harmful toadstool, such as *Aspergillus niger* and *Penicillium citrinum*, might also be generated, which not only cause deterioration of fiber and dyeing deviation but also reduce textile productivity, and bring harm to human health. The mildew on textile could possibly penetrate into the blood system through pores on the skin which could produce toxic effect. If the dust-carrying toxic mildew or *Aspergillus* is inhaled into the human body, coughing symptom disease could be caused.

Apart from the abovementioned occupational hazards (as summarized in Table 23.4) from different parts of the plant, the study also found out that laborers' consciousness to protect themselves still has some room for improvement. For example, one of the companies surveyed has prepared ear muffs for each operator, but still there are some who would rather not wear the muff only for comfort and convenience.

In order to improve the working condition of textile and chemical fiber enterprise, and to protect the physical and mental health of the worker, an occupational disease prevention standard, "Guideline for occupational hazards prevention and control in textile and dyeing industry (GBZ/T212-2008)," was drafted by Capital Medical University, China National Textile and Apparel Council, Chinese Center for Disease Control and Prevention and was implemented in January 2009. The Guideline

Table 23.4: Major occupational hazards in textile industry.

Hazardous factor	Main source of the hazard	Potential harm to health
Dust	Fiber detritus generated by friction	Respiratory infection, dry nasal, itchy throat, hoarse voice and elevated body temperature
Industry noise	High-speed winder, high-speed loom	Damage to auditory system, nervous system and cardiovascular system under longtime exposure
High temperature and humidity	Product quality demand by printing and dyeing process	Coughing symptom disease caused by the toxic mildew generated from high temperature and humidity

defines the major occupational hazards in textile workplace as dust, noise, high temperature and humidity and chemical toxicity; the major occupational hazards in printing and dyeing workplace as high temperature and humidity and chemical toxicity. In addition, the Guideline has also formulated detailed regulations on daily routine monitoring over harmful factors, health monitoring and management, and protection against major hazardous factors.

Except for those traditional issues about occupational safety and health, the discussion about the safety issue triggered by the application of micromaterial and nanomaterial has also been incorporated into occupational safety and health issue of textile industry because more and more micromaterials and nanomaterials are being used in fiber modification and fabric finishing.

China government has realized the potential harm caused to the environment and health by the production, application and disposal of nanomaterial. A research project, "Study about the biological effect mechanism and safety evaluation of important nano-material" was enlisted by China's Project 973 (National Key Basic Research and Development Plan). The overall goal of the project is to focus on nanobiological effects and carry out research study about the key science issues, such as "release of relevant nano-materials in the workplace and from consumer products, occupational exposure, and interaction with respiratory, cardiovascular, and gastrointestinal systems," "Molecular mechanism of biological effect and safety of important nanomaterials (such as nano-TiO_2, Ag and carbon nanomaterials,)," "Nanomaterial safety evaluation methods and evaluation procedures, and high-throughput screening methods for safety evaluation, etc."

References

[1] Blue book of China Chemical Fibers, analysis and forecast of China Chemical Fiber's Economy of 2019.
[2] Zheng, Y.: Safety and Standard of Textile [J]. China Standardization, 2007; (03), 14–16.
[3] Yanli, L.; Rongrui, W.: Influence of polycondensation catalyst on thermal stability of PET resin [J]. China Synthetic Fiber Industry, 1987; (04), 25–28.
[4] Ming, G.: Requirements of restricted substances in China textile official standards and their impacts on related dyeing and printing processes [J]. Dyeing and Finishing, 2017; 43(02), 50–55.
[5] Mengjuan, W.; Shanshan, H.; Yanghua, S.; Yufeng, C.; Wenliang, X.: Survey and analysis on functional textiles in Shanghai market [J]. Technical Textiles, 2014; 32(02), 35–39.
[6] Yuan, Y.: Personal view of enhancing the standard construction for functional textile [J]. China Fiber Inspection, 2010; (24), 44–45.
[7] Yani, W.: Analysis of the perfection of functional textile standard [J]. Progress in Textile Science and Technology, 2014; (03), 7–9.
[8] Yingyan, Q.: Research on regulation of occupational safety and health in China's textile industry [D]. East China University of Political Science and Law, 2018.
[9] Bianlan, L.; Dongshan, L.; Li, C.; Feng, Z.; Wei, W.; Tao, L.; Min, Z.; Xiewei, D.: Occupational hazardous factor and protection in China's textile and printing dyeing industry [J]. Industry Health and Occupational Disease, 2008; (04), 236–238.

VII Testing on micro/nanotechnology for textiles

Edith Classen

24 Test methods and labels for testing on micro/nanotechnology for textiles

Keywords: nanotextiles, standards, label, test methods, biocompatibility

24.1 Introduction

Textile material has a wide range of functions, characteristic properties and material parameters depending on the processing. Textile materials are used in various products such as clothes, home textiles up to technical products and used since thousands of years. For the last 20 years, particle technologies and, in particular, nanotechnology are being used to improve the functionality of products and textiles or to provide new properties into the products and textiles. Various textile applications are currently feasible, for example, dirt- and/or water-repellent textiles, antimicrobial textiles, textiles with UV radiation protection, "cosmetotextiles" with nanocapsules containing special body care substances, bulletproof textiles with carbon nanotubes (CNTs) and "smart textiles" in which the textile structures themselves perform electronic or electric functions [Greßler 2010, 34]. "Nanotextiles" can be produced by a variety of methods: nanofinishing, nanocoating, nanofibers and nanocomposites. The kind of used nanoparticles influences the functionality [1]. Functional textiles are engineered for a wide variety of uses and have specific performance properties. The functional textiles need to be tested to their functionality and at the same time to their general quality parameters because the new technologies can influence the quality of the raw textile materials.

Approximately 7,000 chemicals are used for the production of textiles, for example, for washing and scouring, bleaching, mercerizing, dyeing, finishing, functionalizing and printing of fibers, yarns or fabrics [2].

Textile testing is important for the assessment of product quality and product performance. In Germany, the development of the first national standard for textiles and textile machines started in 1926 to establish the state of art of textile machinery and textile products [3]. Textiles on the market should meet national and international safety standards and show high performance level depending on the specific application.

Since most of the textiles are worn directly on the skin over a long time, there is a growing concern related to the effects on health and safety. In the 1980s, potential health risks due to the presence of chemicals in the environment and consumer

Edith Classen, Life Science and Care, Hohenstein Institut für Textilinnovation gGmbH Schlosssteige 1, 74357 Boennigheim, Germany, e-mail: e.classen@hohenstein.com

https://doi.org/10.1515/9783110670776-024

products are in concern of consumers [4]. In the 1970s, 2,3-dibromopropyl sulfate was found in children sleepwear, which is found to be a potential carcinogen. Formaldehyde is discussed further. The discussion about the risks of formaldehyde in textiles started due to its emission of improperly installed foam insulation in houses. Consumers were critical to the potential risks of formaldehyde in clothing. In the 1980s, the knowledge about the emission of chemicals from textile and the transfer of chemicals from fabric to the skin was little known [4]. In 1993, the Federal Healthy Agency started the first meeting of the working "textiles" to work out a health assessment of textile products [5]. In the past years, it has been observed that certain chemical substances which are used in the production process and for the finishing of textiles can lead to health problems. However, there was often a lack of data and risk assessments. Even today, it is sometimes difficult to estimate the potential impact of certain chemicals and is assessed differently [6]. Textiles made with nanotechnologies should likewise meet these general quality standards and additional standards concerning new functionalities and the biocompatibility. The following sections show an overview of existing standards, labels and certification of textiles and important tests for comfort, functionality and biocompatibility, especially for nanotextiles.

24.2 Standards, labels and certificates for textiles

Products should not harm the consumer and the environment. Since the problems were caused in the past by emission from clothing [5], several mandatory requirements exist to bring textiles to the European market. These mandatory requirements include legal requirements concerning product safety, the use of chemicals (REACH regulations), quality and labeling. In addition, nonlegal but still mandatory requirements exit. Both categories of requirements have become stricter in recent years.

The European General Product safety directive establishes essential requirements for consumer products including textiles to protect consumer health and safety and to ensure the proper functioning of the internal European market [7]. Certain textile and apparel products have specific safety requirements, for example, baby products, medical devices, toys and personal protective equipment.

REACH (Registration, Evaluation, Authorisation and Restriction of Chemicals) is a regulation of the European Union since 2007, which was adopted to improve the protection of human health and the environment from the risks that can be posed by chemicals [8]. REACH restricts the use of a large selection of chemicals in apparel (fabrics and trims), for example, certain azo-dyes, flame retardants, waterproofing and stain-repelling chemicals and metals such as chrome or nickel (in metal trims and accessories) [9]. Additional national regulation on specific chemicals (e.g., formaldehyde, PCP and disperse dyes) must be complied in different European countries.

Switzerland has its own regulation on chemicals, the ORRChem (Chemical Risk Reduction ordinance) [10].

Many fashion brands and retailers have formulated their own restricted substance lists (RSLs), which are stricter than REACH. These RSLs are often inspired by the Zero Discharge of Hazardous Chemicals guideline on safe chemical use [11].

Nonlegal but mandatory requirements can be categorized by requirements that concern the supply chain and product requirements. In Europe, the demands regarding corporate social responsibility (CSR) are more and more important. The minimum requirement is opening up the factory for inspection and signing a code of conduct to state that local labor and environmental laws are respected and corruption avoided.

Nonlegal requirements for textile processing and fabrics set requirements that ensure that textiles and fabrics have been manufactured with respect to the environment.

Examples of different standards which are used for textile products are now given. The list of examples does not claim to be complete.

The label Standard 100 by Oekotex® ensures consumers that all materials used in a garment are tested for harmful substances. MADE IN GREEN by OEKO-TEX® label includes additional social and environmental aspects such as the socially responsible working conditions and the environmentally friendly production. The further development is the STeP by OEKO-TEX® certification which ensures sustainable textile production and offers the label "MADE IN GREEN by OEKO-TEX®" to textiles [12]. Such labeled textiles are produced from materials that have been tested for harmful substances, produced in environmentally friendly facilities, and produced in safe and socially acceptable workplaces. Another label, the EU Ecolabel–Label, ensures consumers that products and also textiles are made using less harmful substances, energy and water [13]. The label Global Organic Textile Standard "GOTS" covers everything from the production to the distribution of textiles made from at least 70% organic natural fibers [14]. bluesign® (also written as Bluesign or BLUESIGN) is based on input stream management and reduces impact on people and the environment in the entire textile supply chain [15]. In 2019, the Green Button, a government-run certification label for sustainable textiles, went on the market in Germany. The Green Button requires companies to protect human rights and the environment and shows that sustainable fashion is possible. There is currently no other label like the Green Button available. A total of 46 stringent social and environmental criteria must be met, covering a wide spectrum from wastewater to forced labor. Independent auditors check for compliance with the required standards, and they also audit the production sites in production countries, if necessary. The auditing bodies are selected based on their specialist's experience [16].

Besides these standards, a lot of tests for the quality of products are available from different institutions, for example, the "Hohenstein Quality Label" is an internationally registered trademark and has a high recognition value for the end user. The textile product has to fulfill defined requirements regarding the tested qualities.

After fulfilling the requirements, the product can be with the Hohenstein Quality Label for 1 year. Three months before expiry, renewal of the declaration of conformity and extension of validity for another year are possible. Three months before expiry of the second year, retesting of the test sample is necessary if the product is to be marked with the label for the next period [17].

24.3 Testing of textiles

Textile products have been tested to evaluate their properties and functionalities since the beginning of the textile industry. National and international standards, for example, DIN, EN and ISO standards, are developed in cooperation with experts from industry, public sector, occupational safety, users, and science and industry. Various standards are available, for example, basic, terminology, test, products, process, service, data, planning and construction standards as well as test standards for special areas have been developed and established (see table 24.1).

Table 24.1: Organization of the standardization and various kinds of standards.

Standard organization	Kind of standards
National standards – Developed in national standard committee, for example, DIN and DKE, Germany	– Basic – Terminology – Test – Product
European standards – Developed in European groups CEN CENELEC and ETSI	– Process – Service – Data – Declaration
International standards – Developed in international groups ISO, IEC and ITU	– Planning – Construction – And so on

Test methods exist to characterize the textile materials and requirements for properties, for example, mechanical behavior, textile elasticity, tensile and multiaxial tensile strength for fiber, yarns and fabrics, thickness measurement, burst strength, shear properties, flexion properties, tear and cut resistance and abrasion resistance. Textile physical measurements define the basic properties of textile materials. New standards will be developed for innovative materials, for example, 3D spacer fabrics because the new 3D construction cannot be tested with the common standard methods [18]. The implementation of nanotechnology in textile products requires also new test methods to determine the new functionality and the quality of products.

Nanotextiles can present functionalities such as antibacterial, ultraviolet radiation protection, water and dirt repellency, self-cleaning or flame retardancy. National and international test methods are available for testing most of these functions – but not for all functions. The development of test methods for new functions is important to prove and determine the quality. Up to now, no standard especially for nanotextiles are available.

24.4 Testing for nanotextiles

One of the main products based on nanotechnology are textiles with water and oil repellency [19]. Another product group is self-cleaning textiles, in which the textiles possess an ability to be cleaned with no laundry treatment [20]. Table 24.2 shows examples of various nanotextiles and test methods.

Superhydrophobic properties can be achieved via nanostructured surfaces of textiles (Greßler 2010b). With optical methods (e.g., light microscopy, scanning electron microscopy and atomic force microscopy), the surface and the cross section of textiles and fibers can be investigated, and these methods give information about (nano)-structured textiles. Surface characterization on nanoscale level can be done by atomic force microscopy. The superhydrophobic surface can be tested by measuring the contact angle of fluids like water of the surface. The performance level of repellent textiles depends on their intended use and is set by brands or retailers. There is no specific standard available wholly by focusing on stain resistance. ISO 23232, water/alcohol resistance test, can give information but do not measure the penetration and stain resistance. ISO 6520, test methods for resistance of materials to penetration of liquids, gives information about penetration, desorption and repellence. ISO 14418, test methods for oil repellency, provides a rough index of oil stain repellency [21]. All tests together provide information about the quality of textiles and knowledge about the intended use of textiles to judge the results and the performance.

Another important application is the use of nanoparticles, for example, nanosilver particles to achieve an antimicrobial performance. The antimicrobial activity of textiles with nano-silver particles can be tested with the usually used standard test for textile. Various test methods can be used to investigate the antimicrobial effectiveness of test specimen against bacteria. To record the quantitative reduction of bacteria by antimicrobial textiles, test systems have become established which specifically record this process. The two major test systems are the agar diffusion test (e.g., SN 195920-1992: Textile fabrics: determination of the antibacterial activity; SN 195921-1992: Textile fabrics: determination of the antimycotic activity; EN 14119:2003-12: Textile evaluation of the action of microfungi) and the suspension test (challenge test, e.g., ASTM E 2149-01: Standard test method for determining the antimicrobial activity of immobilized antimicrobial agents under dynamic contact conditions; JIS Z 2801: Antimicrobial products – test for antimicrobial activity and efficacy; JIS L 1902-2002: Testing for antibacterial activity and efficacy on textile products) [22].

Table 24.2: Examples of test methods for various properties of nanotextiles.

Property of nanotextiles	Test methods
– Surface investigation	– Atomic force microscopy (AFM) – Scanning electron microscopy (SEM) – Light microscopy
– Water and oil repellency – Self-cleaning – Superhydrophobic surfaces	– Contact angle measurement with liquids (water, oil, liquid foods as milk) – Water–alcohol resistance (ISO 23232) – Resistance of materials to penetrate liquids (ISO 6520) – Oil repellency (ISO 14418) – And so on
– Antimicrobial	– Determination of the antimicrobial activity (SN 195920) – Determination of the antimycotic activity (SN 19592) – Determination of the action of microfungi (EN 14119) – Suspension test (challenge test) (ASTM E 2149-01) – Test for antimicrobial activity and efficacy (JIS Z 2801) – Testing for antimicrobial activity and efficacy (JIS L 1902) – And so on

24.5 Comfort testing

Comfort of clothing is a complex, high subjective quality often defined as the absence of discomfort. Wear comfort depends on the raw material, the fiber type, the yarn construction, the fabric construction and structure, fabric thickness and the presence of additional materials like membranes. In addition, dyeing, finishing, and coating process can influence material's properties. New technologies and new functionalization for processing textiles can influence the comfort behavior of materials and this should be avoided. Therefore, the comfort aspects should be one important quality aspect of so-called nanotextiles. Wear comfort is a complex phenomenon and a quantifiable consequence of the body–climate–clothing interaction. The four important aspects of comfort in clothing are thermophysiological comfort, skin sensorial comfort, ergonomic comfort and psychological comfort [23], and these aspects can be measured because they are not entirely undefined, purely subjective individual sensations.

The comfort of nanotextiles can be determined with the existing methods of clothing physiology (see table 24.3). For materials, the heat and sweat transport through the textile materials can be determined with the sweating guarded hot plate according to ISO 11092, the measurement of thermal and water vapor resistance under steady-state condition and to CEN/TR 16422, the classification of thermoregulatory properties. The sweating guarded hotplate simulates the behavior of the human skin regarding the sweat and heat management, and various sweating rates can be simulated. Other methods to measure the behavior against water vapor are different cup methods; water absorbency can be tested with cup methods according to different standards (e.g., ASTM E 96, JIS 2 1009 and ISO 15496). For readymade garments, thermal and thermal sweating manikins can be used to determine the heat and sweat management of the whole garment after ISO 15831 (the clothing – physical effects – measurements of thermal insulation by means of thermal manikin). The use of moving manikins which simulate the movement of humans by moving arms and legs allows the consideration of the ventilation between clothes and human skin caused by movement.

Table 24.3: Comfort testing of nanotextiles.

Test material	Test methods and equipment
Material test	– Heat and moisture management with the sweating guarded hotplate (ISO 11092, CEN/TR 16422) – Moisture management with various cup tests (e.g., ASTM E 96, JIS 2 1009 and ISO 15496)
Garment test	– Heat and moisture management with thermal or thermal and sweating manikins (ISO 15831, ASTM F1930) – Wearer trials with test subjects

The results of the sweating guarded hotplate and the manikins are correlated with the human perception during wearer trials. Therefore, the judgment of the comfort is possible with material and garment measurements [24]. For nanotextiles, no further methods must be developed, and the already existing methods can apply to judge their comfort.

24.6 Biocompatibility testing

Biocompatibility is the ability of a material to perform its desired function and, at the same time, the quality of not having toxic or deleterious effects in contact with the human body and/or on biological systems. Biocompatibility is a dynamic process because there is a change in properties of material and host response over the period. Biocompatibility testing is an essential requirement for regulatory approval of medical devices such as a medical textile product. Test methods are cytotoxicity testing, sensation testing, irritation testing, systematic toxicity testing, genotoxicity testing, implantation testing and hemocompatibility testing [25].

Fabrics should not show any negative effects on the user. Fabrics should be able to work comfortably alongside a biological system – such as the human body – to aid in improving the quality of life without negative side effects or permanent damage to the life form.

In 2000, the Textile and Research Institute DITF (Denkendorf, Germany) and the University Heidelberg developed a seal of quality to test how well-tolerated textiles really are on the skin [26]. This test of the body biocompatibility of textiles uses aqueous extraction phases of the fabric and examines the cytotoxicity of the extract after transfer with fibroblasts and the skin irritation of the extract after transfer with keratinocytes. After passing the test criteria, textiles can receive the FKT label "MEDICALLY TESTED – TESTED FOR TOXINS," awarded by the Fördergemeinschaft Körperverträglicher Textilien e.V. (FKT). (see figure 24.1) The label represents proof that the skin will not be irritated or damaged by substances that may be released. The test does not look for certain substances in textiles. The focus of this test is to see whether a garment would trigger any reaction on the skin or not. The skin cells which are used in this process are extremely sensitive and if only the slightest traces of harmful substances are present, the skin cells suffer. From the vitality parameters, it can be decided that substances have been dissolved from the textile which can damage the human cells.

Figure 24.1: FKT Label "MEDICALLY TESTED – TESTED FOR TOXINS".

The label has been for 20 years on the market, and over 10,000 textiles have been tested [26].

To investigate the biocompatibility of textiles and medical textiles, the basic test of the Hohenstein Institute (Germany) is the test on cytotoxicity according to DIN EN ISO 10993-5 [22]. For this test, an extract of the test material is prepared, which is cultivated with L 929 skin cells for several days. The cell viability, respectively, potential cell-toxic effect is quantitatively determined for the treated cell culture in comparison with untreated control cultures. In this cell culture, test skin cells are used to detect cell damaging substances that may leach out of the sample material. The test allows the evaluation of the potential for cell damage as a summation parameter. The test is not an analysis for single-cell damaging substances. A growth inhibition of more than 30% in comparison with the extracting agent control is assessed as a clear cell-toxic effect [22]. The test on cytotoxicity is the base of the label "medical tested." (see figure 24.2). The label can be used for all textiles. The test methods are adopted from the tests of medical textiles.

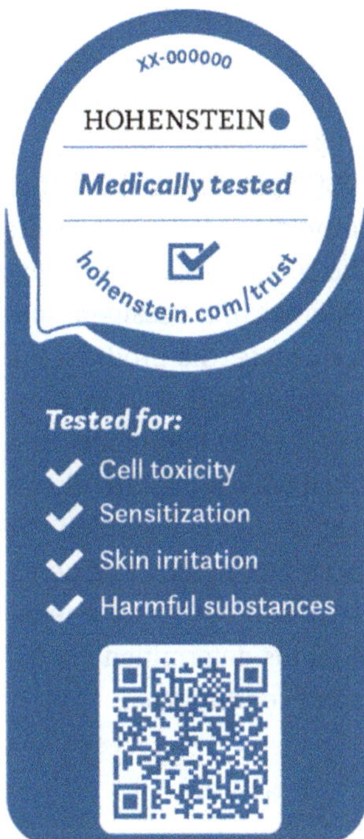

Figure 24.2: Label "medical tested" for textiles and medical textiles (Hohenstein).

The other optional test methods for the label "medical tested" are the sensitization test, the skin irritation test and the harmful substances test. The sensitization test permits the assessment of the potential of textile materials to trigger allergies and is mainly suitable for medical devices made from all types of materials and textiles in healthcare system. In this cell culture test, the allergy potential (sensitization potential) of substances which can leach out of the sample material is determined with immune cells (so-called sentry cells). The sensitization potential is recorded as a summation parameter, and no analysis for single allergens is done. The sensitization test is a screening test for the exclusion of a sensitizing effect and a useful pretest to the standard DIN EN ISO 10993-10. After the extraction of the textile material in water, the extract is incubated for 24 h with immune cells. Sensitizing substances can stimulate the cells to form surface markers that bind to the specific fluorescently labeled antibodies. The percentage of positive cells is determined from the fluorescence intensity in the flow cytometer [22].

The HET-CAM (Hen's Egg Test on the Chorioallantoic Membrane) can detect the chemical irritation of the skin. The HET-CAM (according to DB-ALM Method Summary no. 96) is a recognized alternative to the animal test on the rabbit eye (Draize Test) according to DIN EN ISO 10993-10. A prerequisite for performing the HET-CAM is passing the cytotoxicity test. For the test, an extract of the sample material is prepared and placed on the chorioallantoic membrane (CAM) of incubated hen's eggs for a few minutes. Certain textiles can be applied directly to the CAM. The strength of three different reaction types of CAM (coagulation, hemorrhage and lysis of blood vessels) is determined by the end-point method under stereomagnifier. As a result, three degrees of irritation can be recorded (no or slight, moderate and strong irritation).

The chemical characterization according to DIN EN ISO 10993-18 is the fourth optional test, which gives basic information to the toxicological risks of the textile material and is part of the tests for biocompatibility of the DIN EN ISO 10993 standard series for medical devices. It provides data for the biological assessment and the evaluation of the toxicological risk of a medical device or a textile product. For this test, the sample is extracted at 37 °C for 24 h in methanol as well as in hexane. The chemical substances are determined by gas chromatography combined with mass spectrometry. For the qualitative analysis, the detected substances are identified using a combination of calibrated data and a database (NIST) in which the mass spectra of over 180,000 different organic substances are stored. The semiquantitative determination of the chemical substances is carried out using a reference substance (toluene). The analytical evaluation threshold is determined via the estimated exposure to the chemical substances under clinical conditions.

24.7 Health aspects

Nanoparticles are very small particles and can be released from the textile materials due to different effects (abrasion and other mechanical stresses, sweat, irradiation, washing, temperature changes, etc.). Nanotextiles may release individual nanoparticles, agglomerates of nanoparticles or small particles of textile with or without nanoparticles, depending on the type of integration of nanoparticles in textiles. The most important exposure route of the human body to nanoparticles in case of textiles is skin contact. After abrasion, the inhalation is one way of nanoparticles to enter the human body. Investigation of textiles with nano-silver incorporated in the fiber and nano-silver at the surface after finish and coating processes of two research projects show that the release of nano-silver from the textile into air is small and depends on the incorporation of nanoparticles in and on the fiber [27, 28]. In 2017, a test method was developed to evaluate the skin exposure to nanoparticles, and to evaluate the transfer of nanoparticles from the textile to the skin by the effect of abrasion and sweat [29]. The dangerous effects of nanotextiles against human health have not been broadly demonstrated, and the database is not sufficient [1]. More research is necessary to give an overview in the textile production and the various use scenarios of textile products.

24.8 Labels for nanotextiles

In the beginning of the century, many textile products were advertised with nanotechnology. "Nano" has been used as a catchword. However, there was no possibility for customer and consumers to check this advertising message and find information or to assess the quality of the product. This was the reason for initiatives in different countries for voluntary labeling with special labels and so-called quality seals.

For example, the Hohenstein Institute (Germany) developed the "Hohenstein Quality for nanotechnology." It established the examination of textiles containing nanomaterials. The functionality of nanotextiles should be based on nanotechnology (e.g., antimicrobial and soil repellent), and these textiles are generally safe for use and there are no possible biological risks [30]. (see table 24.4).

Another quality seal label for self-cleaning textiles was developed at the German Institut for Textil- and Fiber Research Denkendorf (Germany) ("Selfcleaning inspired by nature") [30]. The test procedure includes whether superhydrophobia occurs, whether the material is durable and whether the surface structures are present.

The certificate, the "CENARIOS," was developed from the "Die Innovationsgesellschaft" together with TÜV-SÜD. "CENARIOS" is a risk-management and monitoring system and can be compared with the ISO 9000 certification, thus a complicated and tedious process. Following a successful evaluation, the corporation receives a certificate, which however must be refreshed at regular periods. The certificate should help guaranteeing workplace safety and product safety; it documents responsible conduct in the sense of CSR; and in the instance of legal actions, the corporation can prove that its production process is in coherence with the state of technical knowledge (CENARIOS 2013).

Table 24.4: Various nanolabels on the German market.

Nanotechnology label	
Self-cleaning (Hohenstein)	TESTED QUALITY HOHENSTEIN INSTITUTES SAMPLE TESTED FOR: NANOTECHNOLOGY ✓ SOIL REPELLENCY ✓ SKIN COMPATIBILITY ✓ ABRASION RESISTANCE ✓ HOT COLOUR FASTNESS TEST-NO.: FI XX.X.X.XXXX Nanotechnology covers all applications resulting from nanoscience. Nanotechnology is dealing with functional systems based on the use of sub-units with specific size-dependent properties of the individual sub-units or of a system of those. **This product was tested and evaluated following the guidelines of the independent Hohenstein Institutes.** Fluorine-based **nanotechnology** is used. In comparison to untreated materials, a **soil-repellent effect** is confirmed. Tests on tissue compatability prove **skin compatability**. The resistance of the finish to **abrasion** according to the specific product requirements is proven. The **hot colour fastness** is warranted.
Selfcleaning Textiles (DITF) Denkendorf	
Nano Innovationsgesellschaft	**CENARIOS®**

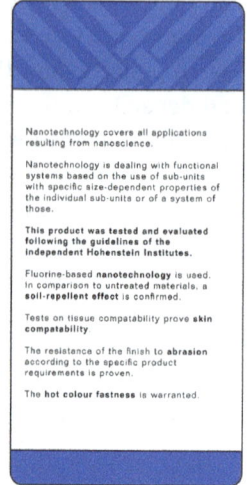

The development of nanotextiles and the proof of quality of such textiles request a lot of effort to guarantee the product quality. Nowadays, there are not so much labeled products on the market. There is also no binding, labeling or reporting requirement for textiles containing nanomaterials.

24.9 Conclusion and outlook

For the last 20 years, nanotechnology has been one of the used technologies to improve the functionality of products and textiles or to provide new properties into the products and textiles. Only products with a high quality and usability as well as production costs lower or similar to conventional products can be successful on the market. The quality of such nanotextiles can often be tested with common test methods; however, sometimes new test methods have to be developed. Quality labels exist to certify the quality of nanotextiles with certain properties. Only nanotextiles with very good properties can survive on the market, for example, properties with self-cleaning properties showed often after-use damages on the surface and the self-cleaning properties decrease already after some use periods. Textiles with nano-silver as antimicrobial substances show often not better effects than conventional with silver substances finished textiles. The production of raw materials, the manufacturing of nanotextiles, and the use and disposal of nanotextiles have to be safe and to bring no risks to human health and environment. Today, CNTs and graphene are often used to improve the material properties. Treatment of textiles with CNTs leads to a wide variety of conductive textiles with different electrical properties and can be used for multifunctional wearable electronic textiles [31]. Graphene is a material that has lightweight and thermal and electrical conducting properties. A type of graphene ink is developed, which can be incorporated into clothing and can create smart athletic clothing to monitor the performance and health and the optimum of movement [32]. The discussion about the release of microparticles from textiles during use and washing is ongoing, and in future, the behavior of nanotextiles will be an important topic for which specific tests will be requested.

References

[1] Saleem, H.; Zaidi, S.: Sustainable use of nanomaterials in textiles and their environmental impact. Materials, 2020; 13: 5134.
[2] Nimkar, U.: Sustainable chemistry: A solution to the textile industry in a developing world. Current Opinion in Green and Sustainable Chemistry, 2018; 8, 13–17.
[3] DIN Jahresbericht Textilnorm, 2016, Beuth Verlag.
[4] Hatch, K. L.: Chemicals and textiles: Part II. Textile Research Journal, 1984; 11: 721–732.
[5] BGA, 1993: Arbeitsgruppe "Textilien" beim BGA Bericht über die 1. Sitzung des Arbeitskreises "Gesundheitliche Bewertung" am 22.6.1993
[6] Leist, H. J.: Wie problematisch sind Chemikalien in Bekleidungstextilien?. Pädiatrische Allergologie, 2016; 03: 42–25.
[7] European parliament: https://eur-lex.europa.eu/legal-content/EN/ALL/?uri=celex%3A32001L0095 (Status 19.05.2021).

[8] REACH: Full title: Regulation (EC) No 1907/2006 of the European Parliament and of the Council of 18 December 2006 concerning the Registration, Evaluation, Authorisation and Restriction of Chemicals (REACH), establishing a European Chemicals Agency, 2006.
[9] Echa: https://echa.europa.eu/regulations/reach/understanding-reach (19.05.2021).
[10] Fedlex: https://www.fedlex.admin.ch/eli/cc/2005/478/en (Status as of 1. March 2021).
[11] ZDHC Foundation: https://www.roadmaptozero.com/input (Status 19.05.2021).
[12] OEKO-TEX: www.oeko-tex.com (Status 19. 05.2021).
[13] EU-Ecolabel: https://eu-ecolabel.de/en/ (Status 19.05.2021).
[14] Global standard: https://global-standard.org/ (Status 19.05.2021).
[15] Bluesign: https://www.bluesign.com/en (Status 19.05.2021).
[16] Green Button: Bundesministerium für wirtschaftliche Zusammenarbeit und Entwicklung, https://www.gruener-knopf.de/en (Status 19.05.2021).
[17] Hohenstein: https://www.hohenstein.com/en/certification/hohenstein-quality-label (19.05.2021)
[18] 3D spacer fabric: https://www.din.de/de/mitwirken/normenausschuesse/textilnorm/aktuelles/einrichtung-des-arbeitsausschusses-pruefgeraete-und-pruefmethoden-fuer-abstandstextilien-im-din-normenausschuss-textil-und-textilmaschinen-textilnorm–309284 (Status 19.05.2021)
[19] Asif, A. K. M. A. H.; Hasan, M. Z.: Application of nanotechnology in modern textiles: A review. International Journal of Current Engineering and Technology, 2018; 8: 227–231.
[20] Katiyar, P.; Mishra, S.; Srivastava, A.; Prasad, N. E.: Preparation of TiO2–SiO2 hybrid nanosols coated flame-retardant polyester fabric possessing dual contradictory characteristics of superhydrophobicity and self-cleaning ability. Journal of Nanoscience and Nanotechnology, 2020; 20: 1780–1789.
[21] Parlidou, S.; Paul, R.: Soil Repellency and Stain Resistance through Hydrophobic and Oleophobic Treatments. In: Williams, J., editor: Waterproof and Water Repellent Textiles and Clothing. Woodhead Publishing, 2018.
[22] Hoefer, D.: Antimicrobial Textiles – Evaluation of Their Effectiveness and Safety. In: Hipler, U.-C.; Elsner, P., editors: Biofunctional Textiles and the Skin. Curr. Probl. Dermatol. Basel: Karger, 2006; vol. 33, pp. 43–50.
[23] Meechels, J.: Körper-Klima-Kleidung: Wie Funktioniert Unsere Kleidung?. Berlin: Schiele & Schiele, 1998.
[24] Bartels, V. T.: Improving Comfort Vote in Sport and Leisure Wear. In: Song, G., editor: Improving Comfort in Clothing. Woodhead Publishing Series in Textiles: Number 106. Cambridge: Woodhead Publishing, 2011; pp. 385–412.
[25] Qin, Y.: Biocompatibility Testing for Medical Textile Products. In: Qin, Y., editor: Medical Textile Materials. December 2016. DOI: 10.1016/B978-0-08-100618-4.00014-5.
[26] Doser, M.: Fördergemeinschaft Körperverträgliche Textilien e.V. (FKT), Presseinfo 9.11.2020
[27] Bremer Umweltinsitut: Abschlussbericht des BMBF Projektes Umsicht, 2013.
[28] Stegmaier, T.; Hammer, T.: Kurzbericht des Projektes Technotox, 2013.
[29] Ameida, L.; Ramos, D.: Health and safety concerns of textiles with nanomaterials. 17th World Textile Conference 2017, Textiles – Shaping Future, 2017.
[30] Fact sheet nano products use of nanomaterials in Textiles, UBA, 4. 2013.
[31] Lima, R. M. A. P.; Espinoza, J. J. A.; da Silva, F. A. G., Jr.; de Oliveira, H. P.: Multifunctional wearable electronic textiles using cotton fibers with polypyrrole and carbon nanotubes. ACS Applied Materials & Interfaces, 2018; 10(16), 13783–13795.

[32] ACS: The Role of Graphene in the Textile Industry, 2019. 11, ACS MATERIAL LLC.
[33] CENARIOS® - Weltweit erstes Nano-Gütesiegel bringt mehr Sicherheit für Mensch, Umwelt und Unternehmen, Innovationsgesellschaft, Medienmitteilung St. Gallen 2007 http://innovationsgesellschaft.ch/wp-content/uploads/2013/08/Medienmitteilung_CENARIOS.pdf
[34] Greßler, S.; Simko, M.; Gazso, A.; Fiedeler, U.; Nentwich, M.: Nanotextiles, Nanotrust dossieres No. 015en, December 2010a.
[35] Greßler, S.; Simko, M.; Gazso, A.; Fiedeler, U.; Nentwich, M.: Self-cleaning, dirt and water repellent coating on the basis of nanotechnology, Nanotrust dossieres No. 020en, December 2010b.

VIII Sustainable practices

Dominic Berndt, André Matthes*, Holger Cebulla

25 Circular economy and recycling

Keywords: sustainability, textile waste, textile recycling processes, circularity, microplastics

25.1 Introduction

The textile industry is currently undergoing a fundamental change for several reasons. Firstly, the countless suppliers of fast fashion are facing huge challenges from a growing group of environmentally conscious consumers who are increasingly focusing on slow fashion and sufficiency, as well as the changes in consumption in the wake of the corona pandemic (source of changes in consumption). On the other hand, the topic of sustainability has not only reached the trade fairs of the entire textile sector, including technical textiles [1, 2], but is now playing a decisive role in the highest political circles for shaping our future on this planet. A positive aspect is that the comprehensive change to a more sustainable way of doing business is now seen as an opportunity for economic development [3].

This will require new business models that will allow for a longer use of textiles, for example, through higher quality products with a longer life cycle or through intelligent design that increases the reparability and recyclability (e.g., Worn Wear by Patagonia® or Design for Circularity by Budde). In addition, the reuse of usable textiles or of resources that can be used as secondary raw materials for textile products will become even more important. At this point, new labeling and traceability methods are the decisive success factors as it is discussed in "Alte Fasern in Neuen Textilien Verwandeln" – Bioökonomie.de [4] and by Weber [5]. Existing companies and start-ups will prove to be highly innovative in making this change.

The recycling of textiles is one of more frequently discussed topics regarding sustainability and achieving a circular economy. As of now, the textile sector is known to be one of the most polluting industries worldwide. The textile industry

***Corresponding author: André Matthes,** Professur Textile Technologien, Fakultät für Maschinenbau, Technische Universität Chemnitz, Reichenhainer Straße 70 | 09126 Chemnitz, Germany, e-mail: andre.matthes@mb.tu-chemnitz.de
Dominic Berndt, Professur Textile Technologien, Fakultät für Maschinenbau, Technische Universität Chemnitz, Reichenhainer Straße 70 | 09126 Chemnitz, Germany, e-mail: dominic.berndt@mb.tu-chemnitz.de
Holger Cebulla, Professur Textile Technologien, Fakultät für Maschinenbau, Technische Universität Chemnitz, Reichenhainer Straße 70 | 09126 Chemnitz, Germany, e-mail: holger.cebulla@mb.tu-chemnitz.de

https://doi.org/10.1515/9783110670776-025

accounts for around 10% of the global carbon emission and almost 20% of the global wastewater [6]. Considering that over the last years the textile market steadily increased, those numbers are not expected to decrease. They will increase due to a growing world population and a rise of the gross domestic product in emerging countries. Most of the textile products end up as waste within the first year of use, and the resources for their production are lost. At the same time, the quality of textiles decreases which makes it more difficult for them to be re-used or brought back through any kind of recycling process. Furthermore, the aforementioned waste leads to social, economic and ecological challenges. Military conflicts take place for resources, waste is being exported to the Global South and burdens its people and their local environment. Due to nonexisting or insufficient local waste management systems, waste is being brought directly into the environment and will endanger different organisms. Since most of the textile products end up being landfilled or incinerated, the harmful effect on the environment is being strengthened. Moreover, since textiles are subject to wear and tear (especially during washing) which causes the emergence of microparticles, the environment in every part of the world is affected by textiles. Also, a lot of virgin feedstock is irrecoverably lost. In a world with limited feedstock and a rising awareness for the necessity of a circular economy, the existing material flow for textiles is inadequate. However, the recovery process of textiles is presently not in a state, where the economical, ecological and social parts always benefit.

Particle-enhanced textiles represent a new challenge in regard to recycling and circular economy but also an opportunity to learn from past mistakes as well as improving the recycling industry and the circular economy. As for now, there are no clear solutions and guidelines on how to handle particle-enhanced textiles after their usage phase for different reasons. One of them is that particle-enhanced textiles are fairly new in comparison to classical apparel or home textiles. Therefore, the level of knowledge for the recycling of particle-enhanced textiles is lower. Another reason is limited quantity within the material flow of particle-enhanced textiles which makes the recycling process currently uneconomic. The aim of this chapter is to point out various already existing recycling processes which could be suitable to a certain degree for particle-enhanced textiles and deduce practical requirements for particle-enhanced textiles in order for them to exist in a circular economy.

25.2 Recycling of textiles

In 2015, the worldwide fiber production was 96.80 million tons and it is expected for 2020 that the fiber production exceeds 115 million tons [7]. The amount of fibers for the apparel industry in 2015 was around 53 million tons [8].

This great material flow may seem like a perfect foundation for a well-established recycling system. But the reality looks different. As of 2015, the apparel industry uses around 97% of virgin feedstock for its products. Figure 25.1 shall give an overview over the worldwide material flow for textiles in the apparel industry.

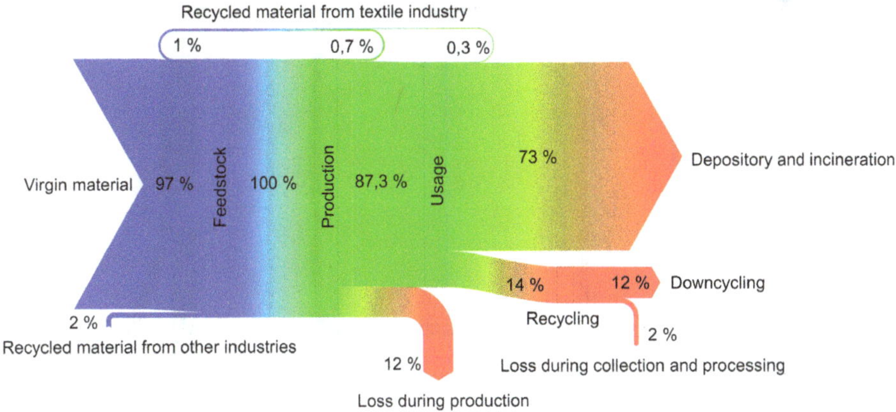

Figure 25.1: Material flow within the apparel industry [8].

As can be seen, the apparel industry has a considerable amount of annual material turnover. But out of the 53 million tons of fibers every year, 46.6 million tons end up either as pre- or post-consumer waste. Even though in some statistics it is said that around 64% of the collected textiles are rewearable it does not mean that they will be worn again [9]. More than 87% of the feedstock is turning to waste at the end of the product life phase or is lost during production, collection or processing. However, this does not include any necessary resources for the production and distribution process, for example, water, electrical and thermal energy, chemicals, land area or fossil fuels, nor does it include the negligence of quality from the fast fashion industry. Especially the low quality hinders the possibility of reusing or further processing into high-quality recyclates which will lead to even more textile waste within the next few years. However, the apparel industry only uses about 55% of the annual fiber production, and the rest goes to other industries like automotive, aerospace, construction, medical and hygiene. Some of them are already designed to be single-use products with a short product life phase (e.g., sanitary and medical products) or are being processed to composites and therefore difficult to recycle. Taking the other 43.8 million tons of annually produced fibers into account, the material flow for textiles becomes even greater and it can be assumed that the amount of incoming waste will increase as well.

To fully understand the part of the product life cycle after the usage phase and to put statistics regarding this topic into perspective, it is important to clarify its terminology. Figure 25.2 shall give an overview over reusing and different recovery options, whereas recycling is only one of them.

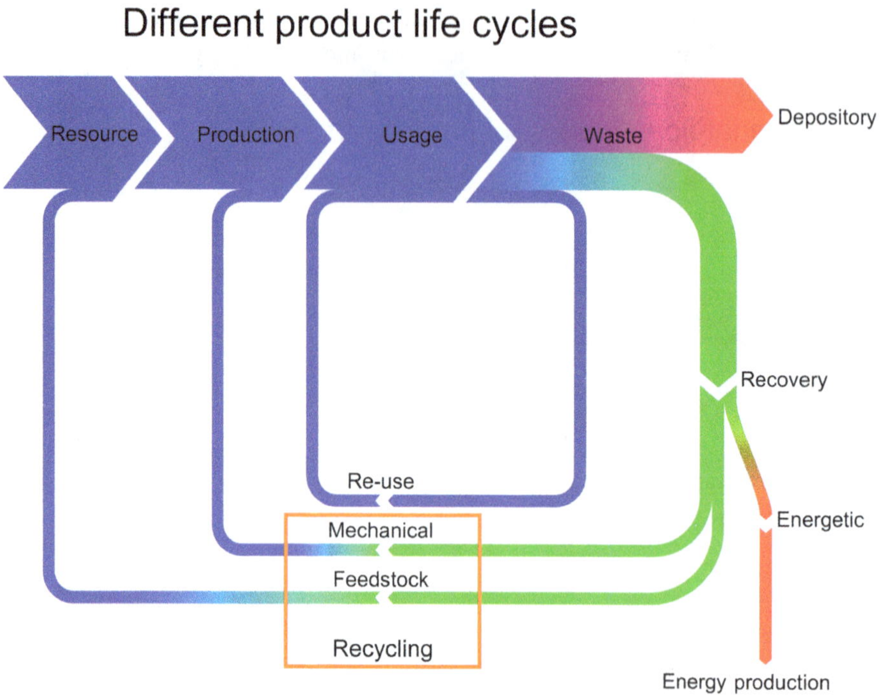

Figure 25.2: Different ways of product recovery.

In order to talk about the process of recovery, the once used feedstock must be declared as waste. Waste on the other side is defined as something that a holder throws away or needs to throw away [10]. If a product is being thrown away but brought directly back into the usage phase with the same purpose as originally designed, it is called reuse and not labeled as waste. A typical example for reusing is secondhand products. Waste is separated into waste for recovery and waste for deposition, which is the least desirable option and not according to the idea of circular economy. Waste for recovery is divided into three branches for further processing. However, if the waste runs through one of the recovery processes, it is not labeled as waste anymore. The most common procedures for textile waste are:

- landfilling,
- incineration and
- recovery for energetic purpose.

Other options are mechanical and feedstock recycling. Mechanical recycling brings the material back into the cycle without changing its basic structure mostly in a macroscopic scale, for example, shredded fibers for insulation or cleaning rags. Feedstock recycling is the process in which the product is dissolved into its polymer or monomer components. The most common example of feedstock recycling is melting old PET bottles and spinning fibers out of the polymer melt or dissolving used cellulosic fibers to spin new regenerated fibers.

Especially the last mentioned recycling process is within the plastic industry an established method. One reason why the plastic recycling is in an advanced state compared to the textile recycling is the larger waste stream which makes the process more profitable and therefore helps to realize a commercial scale. In 2015, the worldwide plastic production was around 407 million tons with estimates to be above 600 million tons by 2025 [11]. The majority is being used in the packaging industry and labeled as waste within the first few months of use. In 2017, the German waste management systems collected 5.2 million tons of post-consumer plastic waste. But only 15.6% of it was brought back as a resource for new plastic products [11]. This is still 5 times more than the 3% of recycled feedstock used in the global textile industry. Another reason why a successful recycling process in the textile industry presents such a great challenge is the great variety of textile materials and their finishing. There are not only synthetic fibers but also natural fibers and blends of different fiber types with the combination of dyes and finishes which have to be taken into account when considering recycling. To fully understand the complexity of the recycling process for textiles, it is important to know about all the necessary individual steps:

1. collecting,
2. transport to recycling facilities,
3. sorting and classification and
4. deciding whether recovery or landfilling/incineration is more suitable

The first step after the usage phase is collecting the used garment, mostly in recycling banks. But this assumes that textiles are brought to those recycling banks and are not just thrown away into residual waste. From then on, the textile waste is transported to recycling facilities where the waste is sorted out in a multistage process. In most cases, this is still carried out as manual labor due to the complexity of textile products. In recent years, developments have been made to automatize the sorting process of reusable and recyclable products (e.g., FIBERSORT project funded by Interreg NEW of the European Commission: https://www.nweurope.eu/projects/project-search/bringing-the-fibersort-technology-to-the-market/). However, as of now they are still limited to sort disposed textiles by fiber material and broad shades of colors. Finishes, impurities and damages remain a challenge. The following classification of the sorted textiles includes textiles for reuse, fibers to be recycled for new apparel or insulation, downcycling for cleaning rags and thermal recovery for energy purpose [12]. The most commonly used recycling technology for textile to textile recycling is shredding or defibering.

Gloy et al. [13] give a comprehensive and detailed insight into fiber separation technologies and processes. They describe the known efficient systems for collecting, sorting and recycling used textiles. The article also provides detailed information on

- Box and belt sorting systems,
- the Fibersort system with near-infrared spectroscopy
- the loading of systems,
- cutting machines and
- the process of tearing.

They show how using such techniques textile waste can be economically processed into tearing fibers. The authors see the current areas of application for the tear fibers obtained in this way almost exclusively in nonwovens for insulation, upholstery and automotive textiles. High-quality applications can so far only be found in niche solutions.

However, during the process of mechanical recycling, fibers get damaged, fiber dust emerges, fiber blends are still mixed together and only large yarn counts are achievable. These limitations lead to a textile product of lower quality. Therefore, other options that are more suitable for the recycling of textiles will be now reviewed.

In the following paragraph, selected individual projects are presented, which relate to special requirements in the recycling process and whose intelligent solutions contribute to increasing the reuse of fiber materials.

One project which was carried out by Felber [14] researched a process for the recycling of textile machine elements made of high-strength synthetic fibers. Their objective was to utilize preconsumer waste from the production of narrow textiles other than for the energetic recovery. They focused mainly on the recycling of fabrics with aramid and UHMW-PE fibers and their life cycle assessment. Those fiber types are not only expensive but also in the case of recovery other than the energetic option they cannot be processed with an extruder that easily. Consequently, the raw material and most of the manufacturing energy are lost. Furthermore, the mainly used shredding process causes a fibrillation of the fibers which leads to clogging of the card clothing [14]. However, with a combination of guillotine cutters, a modified and patented hammer mill from the Thuringian Institute of Textile and Plastics Research (TITK) Rudolstadt, it is possible to achieve a complete fiber separation of staple fibers from woven or braided fabrics without damaging the fiber structure. In the next step, the staple fibers are processed with a card to a nonwoven fabric. It is even possible to process postconsumer waste if the textiles are not too much contaminated or cleaned in a previous step. But as of now, this process is limited to textiles without any seams. The results of the life cycle assessment show that the aforementioned recycling process has a lower environmental impact in comparison to the energetic recovery and manufacturing process for virgin stable fibers [14]. However, it must be taken into consideration that this process does not solve the problem of fiber blends or any sort of finishing and requires an extensive sorting process in order to be successful. Fiber blends, dyes or any other kind of finished textiles are still a great challenge for the recycling system. At first, they must be sorted into fractions which is mainly manual labor and even they are not 100% correct due to finishes which cannot be detected.

However, even if it is possible to sort textiles by their basic color, there will always be a slight difference due to different color shades. This makes it impossible to manufacture a yarn from recycled fibers and meet the customer demands regarding a specific color. Recover® is a company which takes on the challenge to manufacture a reproducible yarn from recycled materials with a desired color. Therefore, mechanically recycled cotton is blended with recycled fibers, for example, from PET bottles. In a process called "ColorBlend," the color of the recycled cotton fibers is analyzed and afterward blended with a certain colored carrier fiber in order to manufacture the required yarn color [15]. Even though this process is a promising achievement toward a circular fashion model, a few unsolved challenges remain. One of the challenges is that the recycled yarn is a blend of recycled cotton and synthetic fibers. Those blends are impossible to be separated through a mechanical recycling process and leaves feedstock recycling as the only option as the next recycling process. Another challenge is the limited number of recycling processes a cotton fiber can be put through. Since every mechanical recycling causes a reduction of the fiber length, the amount of added virgin fiber material must be increased in order to be able to manufacture a processable staple fiber yarn.

At this point, another innovative process comes into place. Natural Fiber Welding (NFW) technology offers the opportunity to increase the number of possible mechanical recycling processes without adding virgin fiber material or a qualitative deterioration of the recycled yarn. A detailed insight into the NFW technology and process is given in Haverhals [16]. In short, the technology enables the fusion of cotton fibers within a spun yarn. The advantages are a much more durable cotton yarn, and even more importantly, the possibility to manufacture a processable yarn from mechanically recycled fibers without changing the natural microstructure of the fiber [16]. NFW is limited by the fiber material since the NFW technology is unable to process petrol-based synthetic fibers. This is especially important because most textile fabrics consist of fiber blends which cannot be separated through mechanical recycling technologies.

This is where feedstock recycling presents an opportunity which allows to separate certain fiber materials from others. But due to the low price for virgin feedstock, they are mainly carried out in pilot plants. Since PET and cotton fibers represent the most commonly used textile fibers in the apparel industry, a few feedstock recycling processes have been established for the implementation of a circular economy within the textile industry. The process consists of two steps. First, the textile waste needs to be separated into its basic components. Therefore, PET is being depolymerized into its monomers and cotton is being purified into a dissolving pulp. These raw materials can be processed into new fibers. Some of the existing feedstock recycling processes for the most commonly used fibers and fiber blends are Mistra, Worn Again, Refibra™, Infinite fiber, Clarus™, solvoPET, only to name a few. But as long as the price for fibers made of virgin feedstock remains lower than fibers from secondary raw material, a serious move toward circular economy remains hard to accomplish. Another major challenge for those aforementioned processes are impurities (e.g., metal and plastic components such as zippers and buttons) and finishing chemicals which affect the process and its yield in a negative way. This means that in order to achieve good recyclates and a high yield rate, an extensive sorting and cleaning process is necessary. A current large-scale research project regarding the topic of circular economy for textiles is RESYNTEX. It is an EU-funded project with partners from 10 different countries. One of their aims is to provide secondary raw material from textile waste with low-value material and establish a closed-loop textile recycling with the focus on feedstock recycling [17]. The following table 25.1 shall provide an overview of some of the aforementioned recycling processes, their processed material, results and limitations.

Table 25.1: Overview of different recycling processes.

Developer	Type of recycling	Processed material	Recycled product	Limitations
ZIM Project: Development of a recycling process for machine elements made out of high-strength synthetic fibers	Mechanical recycling	Textile fabrics made out of filament yarns	Staple fibers processible to nonwoven fabrics	– No separation of blends – Only seamless fabrics – Limited numbers of recycling repetitions
Natural Fiber Welding®	Feedstock recycling	Cellulosic short staple fibers within a yarn or silver	Yarn or silver with long staple fibers	– Processes only for cellulosic fibers – Requires at least parallel align fiber silver
Refibra™, Lenzing	Feedstock recycling	Cellulosic preconsumer waste	Man-made cellulosic fiber	– Sensitive to impurities – Natural fibers turn to synthetic fibers
Worn Again Technologies	Feedstock recycling	Cotton PET blends	PET resin and cellulosic pulp	– Batch procedure – Varying yield
BMBF Recycling project: solvoPET	Feedstock recycling	PET and blended PET waste	TPA and MEG (raw material for PET production)	– Uneconomic procedure – Additional expense due to disposal of waste chemicals

Another example for the necessity of a working recycling process is the increasing usage of carbon fibers and its composites. Over the past 6 years, the global carbon fiber market increased by 40% [18]. They possess great properties for applications in lightweight construction such as tensile strength, density in comparison to metals and the possibility of load path-oriented design, especially the CO_2 footprint of the transportation industry benefits from lightweight construction with carbon fibers. However, the expense for production and inadequate recycling processes outweigh the saving of carbon emission [19]. Several recycling processes are being developed such as mechanical and different variants of pyrolysis and solvolysis. The latter two focus on the separation of carbon fibers and inserts from the resin where the resin is mostly considered as waste and ends up being burnt for energetic recovery purpose [18]. The production of carbon fibers is a very energy-intensive process. The recovery options from composites on the other hand require 10–20 times less energy depending on the chosen process. However, the challenges that the different recycling options face are insufficient knowledge about the properties of the recyclates, only a small size market for recovered carbon fibers and limited availabilty of waste; for further processing of the recovered fibers, they need additional preparation in order for them to be handled like virgin material [18]. Since products made of carbon fibers are designed for a long usage phase, it is difficult to predict a constant retrievable amount of waste which is required for a continuously working recycling process.

25.3 Circular economy and cradle to cradle

> *A circular textiles economy describes an industrial system which produces neither waste nor pollution by redesigning fibres to circulate at a high quality within the production and consumption system for as long as possible and/or feeding them back into the bio- or technosphere to restore natural capital or providing secondary resources at the end of use.* (Hemkhaus et al. [20])

The current predominant linear economy of “take–make–dispose” combined with the mass of textile and plastic waste poses a challenge for the environment and mankind. The main incitements during the design phase of consumer products are economic reasons. The feedstock is supposed to be as cheap as possible, easy to process and not necessarily long-lasting, for a more sustainable future the circular economy seems to be one possible solution. Figure 25.3 illustrates the idea of a circular economy.

The main goal is to keep products as long as possible in the life cycle and at the same time reducing the resource used, limiting the greenhouse gas emission and solving the global waste problem, especially the last aim is a major challenge for our currently predominant “throwaway society.” Products designed for a linear economy use mainly cheap and easily accessible resources and are not primarily meant to be reused or recycled, which makes it difficult to keep them in the life cycle. Other challenges that hinder the realization of a circular economy within the textile industry are

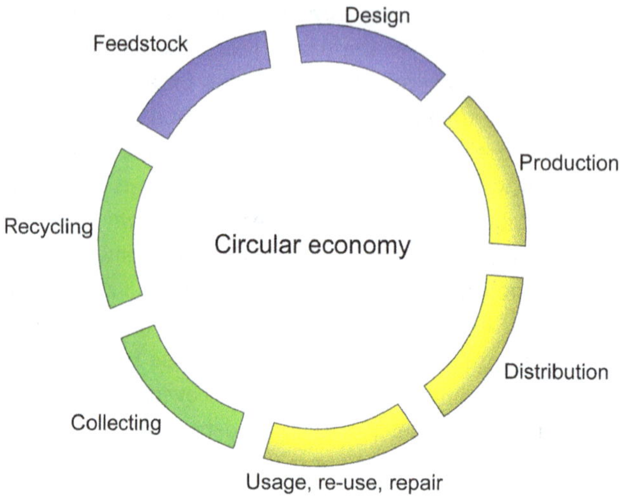

Figure 25.3: Model of the circular economy.

missing standards for the collecting and processing of textiles, lack of consumer awareness and lack of knowledge among designers about recycling processes and sustainability. Moreover, additional machines to handle recycled materials are required in textile factories and also the willingness of consumers to act more sustainable is not synonymous with actual doing so [20].

As one of the first steps for achieving a circular economy, it is important to recognize waste as a resource which requires a change of mind. Because as of now, disposal is still the most performed step. Since 2008, there is a five-step waste hierarchy (Figure 25.4) in place consisting of prevention, reuse, recycling, other recovery and disposal from top to bottom with prevention being the most desirable in order to limit the waste problem [10].

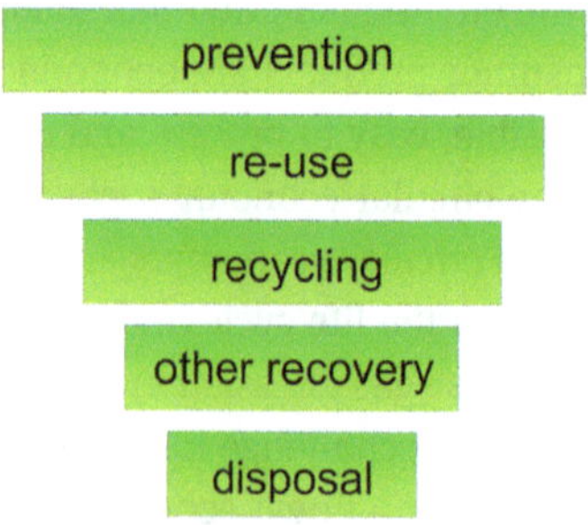

Figure 25.4: Five-step waste hierarchy [10].

This however is easier said than done since there are challenges along the life cycle of a textile. The fast fashion industry brings the exact opposite of waste prevention and floods the textile market with cheap apparels. The low quality of the textiles prevents an extended usage phase and hinders the possibility for reuse and repair. Furthermore, the low quality causes a fee-based disposal of the textile waste for the recycling companies [21]. An additional challenge which occurred in the last couple of years for the recycling companies in the EU is the prohibition of plastic bags. Since most textiles are collected through containers, the contamination of textiles among each other is much greater and leads to a higher percentage of nonrecyclable textiles [21].

The textile industry is trending in two different directions. On the one hand, the already mentioned fast fashion sector is on the rise and leads to an increasing amount of textile waste [21]. On the other hand, the awareness among manufacturers, retailers and consumers for more sustainability also increases which is indicated through the establishment of closed-loop manufacturing process for a textile product. Haeggblom and Budde [22], co-founder of circular.fashion, summed the specification for a circular textile economy up as follows:

1) Design for circularity. Designers need to have the right tools, resources and knowledge to be able to make right decisions
2) Extended product life and customer knowledge about channels for reuse and recycling
3) Sorting systems for closed-loop recycling

All three challenges can only be mastered with a sound knowledge of the components of textile products and a functioning communication and data exchange. This is the only way to successfully realize unlimited cycle times for clothing in the future. All actors in the supply chain must be integrated into the system and the essential information, such as recycling requirements for designers, take-back options for consumers and material specifications for sorting companies, must be exchanged transparently [22].

The system postulated by Haeggblom and Budde is fundamentally based on the cradle-to-cradle (C2C) principle developed by Braungart and Mc Donough [23]. The idea of C2C is to focus not only on products that are harmless to the environment and people, but at the same time to consider all other resources and auxiliary materials such as chemicals, water and energy used in the manufacturing process. With the result that everything that is used and produced can be returned to the same or a different life cycle according to the principle "feedstock to feedstock" [24], a distinction is made between the natural and technical cycles; therefore, the concept requires a completely new approach to product design, as all materials used in the overall process must be viewed and analyzed holistically.

The implementation of the concept for the production of a C2C jeans [25] highlighted these challenges and identified six key factors for successful C2C product development in the textile sector:

1. selection of suitable suppliers,
2. organizational complexity,
3. complexity of the assessment,
4. limited availability of the evaluated components,
5. limited availability of evaluated chemicals and
6. compliance with commercial performance indicators.

This illustrates the difficulties that still exist at present in spreading the concept widely. With each additional C2C product, however, the effort required also decreases, as existing experience and certified preliminary products can be drawn upon.

Other critics point out that there are some areas where there is a discrepancy between theory and reality. One of these is insufficient weighting of criteria within the assessment, such as energy flow, especially as a large amount of energy is needed for the process of turning waste into a new product to keep materials in continuous cycles [26]. It is proposed to combine the C2C assessment, which focuses on the material, with a life cycle assessment.

25.4 Circular economy and recycling for particle-enhanced textiles

As pointed out in the aforementioned content, the realization of a circular economy and implementation of extensive recycling process is a challenge for the apparel industry with a material stream greater than the stream of specialized textiles like the particle-enhanced ones. Another problem that occurs is additional contamination of other textiles through the released particles from the enhanced textiles. Those contaminated textiles might be useless for further process, unable to be processed by the current recycling plants or affect the yield of feedstock recycling processes. This requires separated collection, recycling processes less sensitive to impurities or recycling plants that are adjusted to the additional load of enhanced textiles caused through the different properties of the particles. Furthermore, the work safety needs to be re-evaluated when processing particle-enhanced textiles due to possible release of particles into the air. For a successful realization of a closed-loop textile production, unanswered questions need to be clarified. A few currently unknowns regarding particle-enhanced textiles are:

- possible material streams,
- identify buyers and applications for recyclates,
- adapt regulations and standards on how to handle reused products and recyclates and
- identify if new safety and health regulations are necessary.

Another important aspect which needs to be considered is the respective recycling branch since particle-enhanced textiles cannot be viewed as classic textiles. To solve this issue, an interdisciplinary recycling process (e.g., recycling of textile and electronic goods) needs to be installed.

On the other side, particle-enhanced textiles and their production processes are fairly new in comparison to apparel manufacturing. They present an opportunity to implement the concept of a circular economy from the beginning. One of the most important steps in the circular economy is the design phase, where the most economical, ecological and social impacts are defined. During the design phase, it already needs to be determined on how to produce, use and recycle a specific textile. Therefore, manufacturers, potential users and recyclers need to be involved in the designing process [27]. Another critical step is the distribution phase, where innovative business models like textile leasing and renting are worth mentioning. This would make it easier for a possible reuse or recycling process since the material composition and intention of use are known and a possible contamination and sorting process can be omitted. For the usage phase, it is important to provide a long period of use, including the possibility of repair and reuse. After the usage phase, the textiles are collected and separated for an ensuing recycling process. A thorough collecting and separation are important for high-quality recyclates [27].

This is where another possible advantage of particle-enhanced textiles comes into play. The particles cannot only bolster the properties of the textile but can also be used to carry information [5]. There are already some options available regarding the transparency and traceability of apparel products, for example, TextileGenisis, Haelixa and bioRe. The information the particles are carrying can be material composition, used chemicals, manufacturer, care instruction but also responsible recycler where the collected textile needs to be processed. All those information are required for an extensive sorting and choosing a customized recycling process for the highest yield rate. With those types of intelligent labels, leasing and renting of textiles also become a more feasible option [4]. Other advantages that can be achieved through the use of particle-enhanced textiles are a separation of composite materials or a more efficient performance of textile filters with a positive influence on wastewater treatment or controlling the air pollution [28]. In the first case, particles can be used to separate an adhesive bonding through the introduction of a magnetic field which the particles convert into heat [28]. Particle-enhanced textiles can be used in gaseous and fluid media to remove pollutants from wastewater and exhaust air all the way down to nanoparticles [28]. However, advantages as well as disadvantages need to be subject to further research projects to gather more knowledge regarding their quality, behavior and properties of particle-enhanced textiles, especially a release of nano- or microparticles and a potential risk for humans and the environment needs to be evaluated as well as a waste strategy for the accruing particles.

25.5 Microplastics

As already mentioned, little is known of particles in context with production, usage and recycling of particle-enhanced textiles. However, particles and especially microplastics emerging from classic textiles have been subject to a greater number of studies. Microplastics are solid, insoluble, not biodegradable and synthetic polymers. The size of the particles ranges below the diameter of 5 mm, including nanoparticles smaller than 0.1 μm [29]. There are three different types of microplastics:

- Type 1: produced on purpose
- Type 2: released during usage phase
- Type 3: emerged through degradation and fragmentation

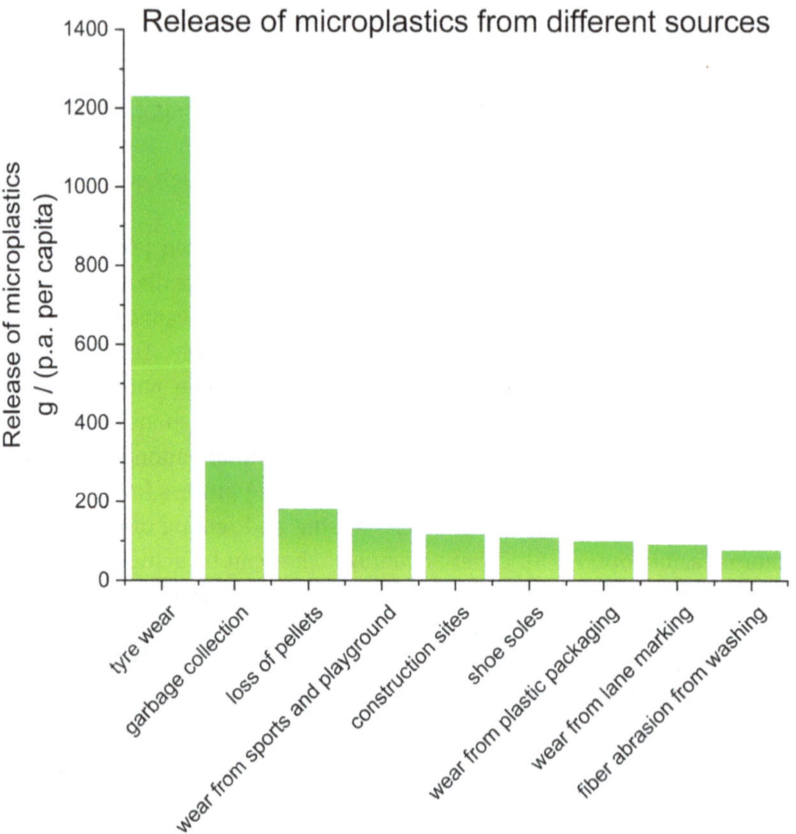

Figure 25.5: Sources of microplastics [30].

Type 1 microplastics are cosmetic/personal care products (e.g., skin care and tooth paste) and washing liquids, whereas type 2 particles are often released through wear and tear of synthetic products like textiles or tires. Type 3 microplastics are often emitted after natural exposure like UV radiation and abrasion through erosion. Figure 25.5 gives an overview of different sources and the annual resulting amount of microplastic in Germany per capita [30].

As it is shown, most of the microplastics in Germany are released into the terrestrial environment where they are exposed to wind and rainwater. Those two cause the transportation of microplastics into the maritime environment. Either it is washed directly into the groundwater and later into rivers, seas and oceans or it is brought into the wastewater system. Modern sewage treatment plants are able to filter microplastic from the wastewater. But the microplastic-contaminated sewage sludge is used as fertilizer and thus microplastics are brought back into the terrestrial environment where they are able to contaminate the groundwater [30]. From a global perspective, the greatest maritime pollution through plastics occurs in South East Asia where waste is being landfilled. During the monsoon season, the plastics are washed away into the rivers and brought directly into the ocean. However, most of the landfilled waste is imported from industrial countries, namely the USA, Japan and Germany.

As mentioned earlier, microplastic pollution occurs in maritime as well as in terrestrial environment. But there is a difference in the number of researches which have been carried out in the past regarding the topic of microplastic pollution. Plastics in the maritime environment are fairly well documented in comparison to the microplastic in their terrestrial counterpart but both have a negative impact nonetheless [29]. Many reports show how microplastics are taken up by small maritime organisms and make their way up the food chain until they end up on our table and in our body. The effect of microplastics in our body is still considered a work in progress. But it is known that microplastics tend to bind harmful substances which also make their way into the food chain [31].

An often-discussed topic is the release of microfibers from textiles through washing. In every washing cycle, fibers smaller than 1 mm are released into the wastewater. Some of them are of synthetic origin and others are natural fibers contaminated with synthetic finishing or dyes. As the graphic indicates textile washing and usage only accounts for a small fraction of the microplastic pollution. However, it is important to understand that those numbers only show the microplastic release during the usage phase of the textile product. The total amount of microplastic release relating to textiles which includes production, distribution and recycling is expected to be much greater.

25.6 Conclusion

An interesting concept is design for recycling. This is an approach which already started in the plastic industry called RecyClass where different parties along the value chain came together to elaborate guidelines on how to design plastic products that are processable with current recycling technologies. A measure is also relevant for the textile industry but not easily transferable due to greater variety of used raw material sources. One way is to focus on designing textiles made out of one type of fiber material. Other ways are upscaling the already existing feedstock recycling processes to a commercial level and to work on technologies which are able to process shortened textile fibers due to mechanical recycling. This, however, remains a challenge as long as the economic necessity is missing due to lower virgin feedstock prices compared to recycled feedstock prices.

Solving the waste problem or achieving a circular economy is not something that one country, company, consumers or government can implement alone. It needs all the aforementioned parties to work together on that matter especially in a global acting textile industry. If in one country the collection and sorting processes work well but those textiles are exported into other countries where they might be worn again and then landfilled/incinerated or incinerated right away, it only shifts the responsibility from one place of the Earth to another place without solving the problem. Most of this does not show up in the statistics in the country of origin. Exporting waste with the purpose to praise a locally good working waste management system might be labeled as green-washing.

However, only focusing on the recycling process can lead to a rebound effect, where virgin and recycled materials are combined and used as feedstock for new textile products. This ultimately leads to an increasing amount of textile waste which needs to be recycled. In order to be successful, a lot of things need to come together such as

- Regulations from legislators which make single used goods of low quality less attractive and the use of long-lasting recyclable products worthwhile
- Educating and raising the awareness of the consumer toward a more sustainable consumption
- A willingness of companies to install a manufacturing process focusing on circularity of its products
- Traceability of all used materials and energy sources

Regarding the need for circularity, there is no difference between classic textiles and particle-enhanced textiles. Since the latter are fairly new in comparison to the classic textile and its market is not as big, the implementation of the necessary criteria for a circular economy might be easier to realize.

References

[1] ITMA 2019: SUSTAINABILITY@ITMA, 2019. Available at: https://www.itma.com/highlights/sustainability@itma (Accessed: 01 September 2020).

[2] Techtextil 2019: Sustainability at Techtextil, 2019. Available at: https://techtextil.messefrankfurt.com/frankfurt/de/programm-events.html#nachhaltigkeit (Accessed: 01 September 2020).

[3] European Commission: The European Green Deal, 2019. COM(2019) 640 final. Available at: https://eur-lex.europa.eu/legal-content/EN/TXT/?uri=CELEX:52019DC0640 (Accessed: 24 September 2020).

[4] Alte Fasern in Neuen Textilien Verwandeln – Bioökonomie.de, 2019. Available at: https://biooekonomie.de/nachrichten/alte-fasern-neue-textilien-verwandeln (Accessed: 30 June 2020).

[5] Weber, B.: Neues System Zur Rückverfolgbarkeit von Textilien. EU-Recycling Und Umwelttechnik, 2019; 36(8), 6.

[6] UNECE: UN Partnership on Sustainable Fashion and SDGs. In: High-Level Political Forum on Sustainable Development, New York, USA, 9–18 July. New York: UNECE, 2018.

[7] Opperskalski, S.; Siew, S.; Tan, E.; Truscott, L.: Preferred Fiber & Materials – Market Report 2019, 2019. Available at: https://store.textileexchange.org/product/2019-preferred-fiber-materials-report/ (Downloaded: 10 June).

[8] Ellen MacArthur Foundation: A New Textiles Economy: Redesigning Fashion's Future, 2017. Available at: https://www.ellenmacarthurfoundation.org/assets/downloads/publications/A-New-Textiles-Economy_Full-Report_Updated_1-12-17.pdf (Downloaded: 23 June 2020).

[9] Circle Economy: Recycled Post-Consumer Textiles, an Industry Perspective, 2020. Available at: https://www.nweurope.eu/media/9453/wp-lt-32-fibersort-end-markets-report.pdf?utm_source=0.+Master+list&utm_campaign=dabe7daddf-EMAIL_CAMPAIGN_2019_09_10_10_16_COPY_01&utm_medium=email&utm_term=0_d023026741-dabe7daddf-128051745&mc_cid=dabe7daddf&mc_eid=f80d25aa47 (Downloaded: 10 June 2020).

[10] European Parliament, Council of the European Union: Directive 2008/98/EC of the European Parliament and of the Council of 19 November 2008 on Waste and Repealing Certain Directives, 2008. Available at: https://eur-lex.europa.eu/legal-content/EN/TXT/PDF/?uri=CELEX:32008L0098&from=EN (Accessed: 24 June 2020).

[11] Heinrich-Böll-Stiftung: Plastikatlas: Daten und Fakten über eine Welt voller Kunststoff, 2019. Available at: https://www.boell.de/de/2019/05/14/plastikatlas (Downloaded: 17 June 2020).

[12] Effiziente Sortierung, 2020. Available at: https://www.texaid.de/de-DE/produkte-leistungen/sortierung.html (Accessed: 23 June 2020).

[13] Gloy, Y.-S.; Gulich, B.; Hofmann, M.: Textile Waste Management and Processing. In: Matthes, et al., editors: Sustainable Textile and Fashion Value Chains. Cham: Springer International, 2020. DOI: 10.1007/978-3-030-22018-1.

[14] Felber, A.; Zusammenfassung – Recycling Hochfester Synthetischer Faserseile. Chemnitz: Technische Universität Chemnitz, 2020.

[15] Smits, H.: Recycled Cotton – Building a Circular Fashion System Today. In: Sustainable Textile School. Chemnitz, Germany, 18 September. Chemnitz: STS, 2017.

[16] Haverhals, L. M.: Natural Recycled Super-fibers: An Overview of a New Innovation to Recycle Cotton. In: Matthes, et al., editors: Sustainable Textile and Fashion Value Chains. Cham: Springer International, 2020. DOI: 10.1007/978-3-030-22018-1.

[17] The Project – RESYNTEX, 2020. Available at: http://www.resyntex.eu/the-project (Accessed: 10 June 2020).

[18] Job, S.; Leeke, G.; Mativenga, P. T.; Oliveux, G.; Pickering, S.; Shuaib, A.: Composites Recycling – Where Are We Now?, 2016. Available at: https://compositesuk.co.uk/system/files/documents/Recycling%20Report%202016.pdf (Accessed: 30 June 2020).

[19] Brünglinghaus, C.: Die Schattenseiten von CFK. ATZ – Automobiltechnische Zeitschrift, 2015; 117(7–8), 8–13. DOI: 10.1007/s35148-015-0093-7.

[20] Hemkhaus, M.; Hannak, J.; Malodobry, P.; Janßen, T.; Griefahn, N. S.; Linke, C.: Circular Economy in the Textile Sector – Study for the German Federal Ministry for Economic Cooperation and Development (BMZ). Eschborn: Deutsche Gesellschaft für Internationale Zusammenarbeit (GIZ) GmbH, 2019.

[21] Henkel, R.: Recycling Am Limit: Die Altkleiderbranche Erstickt Im Müll, 2019. Available at: https://fashionunited.de/nachrichten/business/recycling-am-limit-die-altkleiderbranche-erstickt-im-textilmuell/2019052231946 (Accessed: 16 June 2020).

[22] Haeggblom, J.; Budde, I.: Circular Design as a Key Driver for Sustainability in Fashion and Textiles. In: Matthes, et al., editor: Sustainable Textile and Fashion Value Chains. Cham: Springer International, 2020. DOI: 10.1007/978-3-030-22018-1.

[23] Braungart, M.; Mc Donough, W.: Cradle to Cradle: Remaking the Way We Make Things. New York: North Point Press, 2002.

[24] Stakeholder Reporting GmbH: Eine runde Sache – die weltweit erste Jeans, die im Kreislauf bleibt. Bonn: Bündnis für nachhaltige Textilien, 2018.

[25] FashionForGood: Developing C2C CertifiedTM Jeans, 2018. Available at: https://fashionforgood.com/wp-content/uploads/2018/08/FashionForGood_Denim-Case-Study-FINAL-1.pdf (Accessed: 24 September 2020)

[26] Toxopeus, M. E.; de Koeijer, B. L. A.; Meij, A. G. G. H.: Cradle to Cradle: Effective Vision vs. Efficient Practice?. The 22nd CIRP conference on Life Cycle Engineering, Sidney, Australia, 7–9 April. Red Hook: Curran, 2015; pp. 384–389, DOI:10.1016/j.procir.2015.02.068.

[27] Brüggemann, A.: Circular Economy Als Schlüssel Für Nachhaltiges Wirtschaften Und Ressourcensicherheit. KfW Research – Fokus Volkswirtschaft, 258, 2019. Available at: https://www.kfw.de/PDF/Download-Center/Konzernthemen/Research/PDF-Dokumente-Fokus-Volkswirtschaft/Fokus-2019/Fokus-Nr.-258-Juli-2019-Kreislaufwirtschaft.pdf (Accessed: 6 July 2020).

[28] Becker, H.; Dubbert, W.; Schwirn, K.; Völker, D.: Nanotechnology for Humans and the Environment – Promote Opportunities and Reduce Risks. Dessau-Roßlau: Umweltbundesamt, 2009.

[29] Machado, S.; Abel de, A.; Kloas, W.; Zarfl, C.; Hempel, S.; Rillig, M. C.: Microplastics as an Emerging Threat to Terrestrial Ecosystems. Global Change Biology, 2018; 24(4), 1405–1416. DOI: 10.1111/gcb.14020.

[30] Bertling, J.; Hamann, L.; Bertling, R.: Kunststoffe in der Umwelt: Mikro- und Makroplastik. Ursachen, Mengen, Umweltschicksale, Wirkungen, Lösungsansätze, Empfehlungen. Kurzfassung der Konsortalstudie. Oberhausen: Fraunhofer-Institut für Umwelt- Sicherheits- und Energietechnik UMSICHT, 2018. DOI: 10.24406/UMSICHT-N-497117.

[31] Galloway, T. S.: Micro- and Nano-Plastics and Human Health. In: Bergmann, M.; Gutow, L.; Klages, M., editors: Marine Anthropogenic Litter. Cham: Springer International Publishing, 2015; pp. 343–366. DOI: 10.1007/978-3-319-16510-3_13.

Amrei Becker*, Thomas Gries

26 Sustainability in the textile industry

Keywords: sustainability, circular economy, recycling, biopolymers, bioeconomy

26.1 Sustainability in the textile industry: challenges and opportunities for the use of particles in fibers

Since the Industrial Revolution, the world has undergone profound economic and technological developments that have impacted many aspects of life worldwide. The rapid and radical changes have led to a steadily growing, and in some cases more affluent, global population and thus increasing demand for food, medical care and other consumer goods such as textiles. This global economic growth and the accompanying increase in resource consumption come at a price. Each year humans consume the equivalent of 1.7 earths to provide the resources necessary to produce goods and absorb waste. This means that it takes the earth 1.7 years to regenerate what was consumed in one year [1]. Exceeding planetary or ecological limits threatens the stability of the ecosystem and thus the livelihood of humanity and wildlife [2, 3].

In addition to ecological factors, sustainability also includes financial and social aspects. The United Nations (UN) presented the 2030 Agenda with 17 Sustainable Development Goals, which came into force in January 2016. As shown in Figure 26.1, the goals serve to promote social equality, health and ecological wellbeing worldwide [4].

In 2019, the global fiber production was at an amount of 110 million tons. Of these, 72% are man-made fibers based on natural or synthetic polymers [5]. Fibers are used in clothing, home and household textiles, as well as in technical textiles. Due to diversity in the use of raw materials, wide range of product applications, fast-fashion trend, and its size, the textile industry is one of the worldwide biggest contributors to environmental and social challenges. It is expected that the textile and fashion industry's percentage of the carbon budget could increase from 2% in 2015 to 26% in 2025 based on the 2-degree scenario [6].

*Corresponding author: **Amrei Becker,** Institut für Textiltechnik of RWTH Aachen University, Otto-Blumenthal-Straße 1, 52074 Aachen, Germany, e-mail: amrei.becker@ita.rwth-aachen.de
Thomas Gries, Institut für Textiltechnik of RWTH Aachen University, Otto-Blumenthal-Straße 1, 52074 Aachen, Germany

https://doi.org/10.1515/9783110670776-026

Figure 26.1: 17 Sustainable Development Goals presented by the United Nations [4].

26.2 Systematic approaches to sustainability

There are two possible systematic approaches making economic systems more sustainable, called circular economy and bioeconomy; see Figure 26.2. Both concepts define the transformation from a linear economy with the use of (fossil) resources to an economy with closed material cycles and are further described below.

26.3 Circular economy and recycling

The circular economy already starts with the design or construction (so-called design for recycling) of products. The design plays a decisive role in whether the product or raw material can be recycled or not. The following design features make it difficult to recycle textiles:

- Use of fiber blends
- Multilayer constructions
- Haberdashery (e.g., zippers, buttons, ribbons)
- Dyes and prints
- (Functional) coatings
- Additives and particles for functionalization

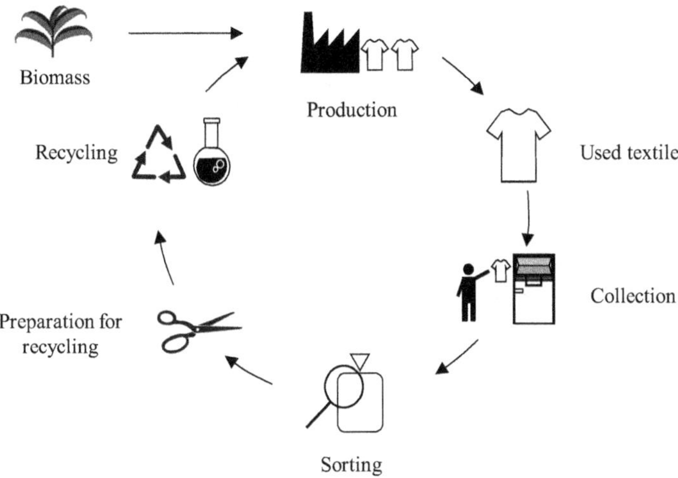

Figure 26.2: Circular (bio-)economy.

In Germany, approximately 1.5 million tons of used textiles from private households were collected separately in 2019 [7]. Besides the collected textiles, many textiles from private households were disposed via residual waste and there are also textiles from commercial and technical applications. According to the "Waste Framework Directive" of the European Union, which will come into force in 2025, textiles will have to be collected separately and compulsorily throughout the EU in the future. It is expected that this will increase the separately collected quantity by several million tons within a few years. About half of the collected home and clothing textiles are considered no longer usable or wearable and therefore cannot be resold on the second-hand market. Currently, non-reusable textiles are mostly processed into so-called downstream products such as cleaning rags or insulation mats.

For high-quality mechanical or chemical recycling, pure and clean materials are required. Since no suitable treatment processes are currently available for industrial use, most mixed textiles are incinerated or landfilled [6]. Various recycling processes already exist for (mixed) textile waste on a laboratory and pilot plant scale (see Table 26.1).

26.4 Recycling and particles

High-quality mechanical or chemical recycling require input materials that are as pure as possible. Impurities of the polymer or textile can be macroscopic as well as microscopic. Macroscopic impurities include sewn or bonded haberdashery, multilayer textiles, fiber blends, coatings, prints and so on. Particles and other additives,

Table 26.1: Overview of (polymer-independent) recycling processes for synthetic textiles.

Recycling process		Operating principle
Mechanical	Tearing	Synthetic or natural textiles are torn down to the single fiber. These fibers can be processed into yarns or nonwovens.
Thermomechanical	Regranulation	Material-homogeneous, thermoplastic fibers or textiles are shredded, melted, filtered and regranulated. New polymer granules are recovered.
Physicochemical	Solvent-based separation	From mixed fiber material/ textiles, the desired polymer is separated selectively by solvents. Pure polymer, precipitated from solvent, is recovered.
Chemical	Depolymerization (back-to-oligomer)	Fibers or textiles, with high content of the to-be-recycled material can be processed. The polymer is broken down into short polymer chains (oligomers) and then reconstituted into a new polymer. The polymer quality depends on the input material and process.
	Depolymerization (back-to-monomer)	Polymers are broken down into their basic building blocks (monomers) and are then repolymerized. Fibers or textiles, with high content of the to-be-recycled material can be processed. The polymer is broken down into the basic building blocks (monomers) and then re-polymerized. The polymer quality depends on the input material and process.
Thermochemical	Pyrolysis	Mixed organic materials are decomposed in the absence of oxygen into pyrolysis oils, cokes and gases.

drafts or dyes are belonging to the microscopic impurities. In principle, particles are considered as contaminants in recycling processes, and the resulting challenges differ greatly depending on the recycling process chosen.

Since fibers are retained during *mechanical recycling (tearing)*, particles and other microscopic impurities are no challenge for the process. However, the quality of the manufactured tearing fibers is strongly dependent on the impurities, so that, above all, efforts are made not to use fibers with harmful additives as input material for the tearing process in the first place.

During *thermomechanical recycling processes (regranulation)*, impurities can greatly reduce the quality of the recycled granules. Volatile substances can be extracted by means of degassing in the extruder. It is also possible to filter solid impurities, such as particles or gels, from the melt. Metal mesh or nonwoven filters with a pore size of up to 10–20 µm can be used here. This makes it possible to filter even up to half of the colorants from the melt. However, due to the high-pressure buildup and the short filter life, much coarser filters of at least 40–60 µm pore size are usually used in industrial regranulation. One of the most used particles in melt-spun yarns is titanium dioxide (TiO_2), which is added at up to 1% to yarns for the apparel industry for matte finishes. However, TiO_2 easily forms agglomerates due to its polarity and is difficult to filter from the melt due to its electrostatics. Carbon blacks or other dyes can also be added mainly to industrial yarns as a masterbatch in the melt spinning process. Complete decolorization and removal of these particles from spinneret-dyed yarns is not possible at present [8].

A filter test has been developed to evaluate the quantity, dispersion and dispersibility of pigments and fillers in the melt. In EN 13900-1 to EN 13900-4, various methods for evaluating the dispersibility properties are specified. The methods enable to determine the filler content and the possible challenges in recycling. The standardized methods allow comparison between similar pigments (e.g., between a test sample and a reference pigment). The results can give an indication on the relative dispersibility under practical application conditions.

In chemical and physicochemical recycling processes, particles and other insoluble impurities can be filtered from the monomer, oligomer or polymer solution in a manner analogous to their removal from the melt. The basic polymer building blocks should be as pure as possible prior to re-polymerization. Today, only a few industrial plants exist worldwide for the (physico) chemical recycling of (textile) plastics.

Organic impurities and particles do not interfere with the thermo-chemical recycling process, as they are pyrolyzed in the same way as the polymer. Inorganic particles, on the other hand, cannot be pyrolyzed in the process and remain in the slag.

26.5 Biopolymers

According to the European Bioplastics e.V., biopolymers are polymers that are either bio-based and/or biodegradable [9]. Therefore, biopolymers are divided into three different groups; see Figure 26.3. The first group includes biodegradable polymers based on fossil raw materials (petro-based). These include for example polycaprolactone and polybutylene adipate terephthalate. Biodegradable polymers made from renewable raw materials belong to the second group. Representatives of the second group include polylactic acid, polyhydroxyalkanoates and naturally occurring polymers such as polysaccarides and proteins. Polymers in the third group are made from renewable raw materials and are not biodegradable. The third group includes, for example, naturally occurring rubber or amber. In addition, the so-called drop-in polymers, which are marked with the prefix "bio-," belong to the third group. The drop-in polymers can be bio-based (e.g., Bio-PE) or partially bio-based (e.g., Bio-PET, Bio-PTT, Bio-PBT, Bio-PA and Bio-PUR). The molecular structure of these bio-based polymers is not different from that of petro-based polymers [9, 10].

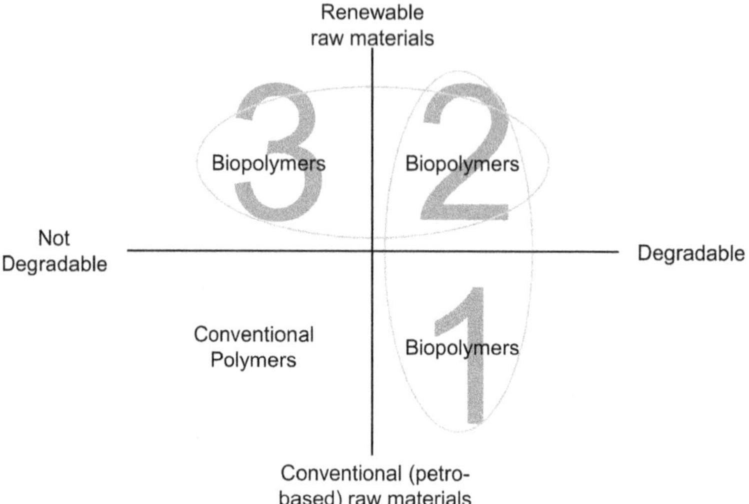

Figure 26.3: Three groups of biopolymers [9].

Until the 1930s, plastics were produced from renewable raw materials. Biopolymers were thus the first industrially processed polymers. In the period after the Second World War, the extraction of coal, crude oil and natural gas began to increase. This enabled the production of synthetic polymers based on petrochemicals. Between 1975 and 2020, the production of man-made fibers grew from 24 to 108 million tons per year [11]. Today, Biopolymers are used in various fields. In the packaging industry, for example, carrier bags, film packaging, bottles, cups, containers for plants or fruit and vegetable trays are made from biopolymers. Biopolymers are also used in medicine and hygiene, mainly because of their degradability and body compatibility.

Despite the wide range of potential applications, today the market volume of biopolymers is limited due to their limited economic competitiveness. Compared to conventional synthetic polymers, biopolymers are currently still usually more expensive [9, 12].

The possibility of composting products made from biopolymers at the end of their life opens markets and applications that cannot be served by synthetic polymers. According to DIN EN 13432, complete biodegradability means the "decomposition of a chemical compound or organic material by microorganisms in the presence of oxygen into carbon dioxide, water and salts of other elements present [. . .]" and new biomass [13]. In the absence of oxygen, the chemical compounds or organic material can be decomposed "into carbon dioxide, methane, mineral salts, and new biomass." This is illustrated in Figure 26.4.

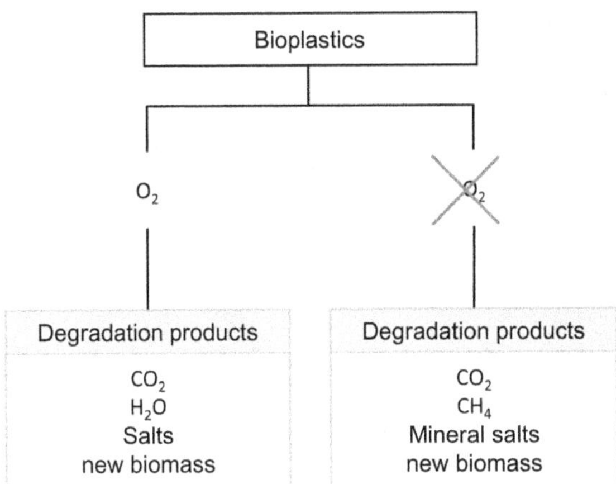

Figure 26.4: Complete biodegradability of packaging made of biopolymers in the presence (left) and in the absence (right) of oxygen according to DIN EN 13432 [13].

26.6 Biopolymers and particles

For better processability in the melt spinning process or functionalization, biopolymers are often modified and/or processed as blends. Therefore, organic and inorganic fillers, adhesion or flow promoters and plasticizers are used. Typical textile finishes are dyes and matting agents for visually appealing textiles and flame retardants for technical textiles [14].

Drop-in biopolymers (group 3, see Figure 26.3) do not differ chemically from their fossil counterparts. Therefore, the same particles and additives can be used here for modification. *(Partially) bio-based polymers* (group 1 and 2), on the other hand, often have different properties. Nevertheless, it is possible to use similar particles as in conventional polymers. However, by adding further substances, the *biodegradability* or the *(industrial) compostability* of the polymers can be influenced. Biodegradation is a process in which a material is converted to naturally occurring metabolic end products.

In addition, particles can influence the *ecotoxicity* of a material. Thus, particles can be studied to have (negative) effects on living organisms in the environment even after the polymer has degraded. Ecotoxicological tests are performed on organisms that can be easily grown in the laboratory, such as algae or water fleas. The organisms are exposed to the substances under investigation, and it is observed whether, for example, growth is inhibited, or the mobility of the organisms is restricted. This can be influenced, for example, by the mechanical shape of particles in the organism's system [15].

26.7 Summary and outlook

Particles and additives play an important role in functionalizing fibers and filaments. In textile recycling, they are usually seen as contamination in the process to recover preferably pure recycled material. The addition of particles into fibers made of bio-based and/ or biodegradable polymers has potentials as well as certain challenges. Many approached and solutions are currently being developed to enable the transformation towards a textile circular and bio-based economy.

References

[1] Global Footprint Network, 2020. URL: https://www.footprintnetwork.org/

[2] Rockström, J., et al: A safe operating space for humanity. Nature, 2009; 461, 472–475.

[3] Steffen, W., et al.: Planetary boundaries: Guiding human development on a changing planet. Science, 2015; 347(6223).

[4] UN Communications materials. URL: https://www.un.org/sustainabledevelopment/news/communications-material/

[5] The Fiber Year Consulting: The Fiber Year 2020: World Survey on Textiles and Nonwovens. Freidorf, Switzerland: The Fiber Year GmbH, 2019; pp. 19.

[6] Ellen Mac Arthur Foundation: A new textiles economy: Redesigning fashion's future. Available from: URL: https://www.ellenmacarthurfoundation.org/assets/downloads/publications/A-New-Textiles-Economy_Full-Report.pdf

[7] bvse e.V: Textilstudie 2020 – Bedarf, Konsum, Wiederverwendung und Verwertung von Bekleidung und Textilien in Deutschland, 2020. URL: https://www.bvse.de/dateien2020/1-Bilder/03-Themen_Ereignisse/06-Textil/2020/studie2020/bvse%20Alttextilstudie%202020.pdf

[8] Stefan Vandendijk: Expert Talk. Company: NV Bekaert SA, Belgium, 31.05.2021.

[9] Endres, H.-J.; Siebert-Raths, A.: Technische Biopolymere – Rahmenbedingungen. In: Marktsitutation, Herstellung, Aufbau Und Eigenschaften. Germany, München: Hanser, 2009; pp. 5–8.

[10] Dolmans, R.: Bewertung kommerziell erhältlicher biobasierter Polymere in der textilen Filamentextrusion, Aachen, RWTH Aachen University, Dissertation, Shaker, 2014.

[11] Fernández, L.: Production volume of textile fibers worldwide 1975–2020. Available from: URL: https://www.statista.com/statistics/263154/worldwide-production-volume-of-textile-fibers-since-1975/. last accessed 6.12.2022.

[12] Beier, W.: Biologisch Abbaubare Kunststoffe. Germany, Dessau-Roßlau: Umweltbundesamt, 2009.

[13] DIN EN 13432: Verpackung – Anforderungen an die Verwertung von Verpackungen durch Kompostierung und biologischen Abbau – Prüfschema und Be-wertungskriterien für die Einstufung von Verpackungen.

[14] Mourgas, G.: Schweizer, M.: Expert Talk. Company TECNARO GmbH, Germany, Ilsfeld, 26.07.2021.

[15] Ilvonen, O.: Ökotoxikologische Bewertung von Bauprodukten. Germany, Dessau-Roßlau: Umweltbundesamt. Available from: URL https://www.umweltbundesamt.de/themen/wirtschaft-konsum/produkte/bauprodukte/studien-zur-messung-vermeidung-bewertung-von/oekotoxikologische-bewertung-von-bauprodukten. publication date: 23.02.2016, last accessed: 6.12.2022.

IX Industrial innovation and market opportunities for nano- and micro-enabled textiles

Ning Cui

27 Industrial innovation and market opportunities for nano- and micro-enabled textiles

Keywords: nanotechnology, textile, environmental problem, ethical problem, market opportunity, costs, commercialization, innovation, technology acceptance, nanosafety

- Global & future trends & challenges in science, R&D and production (fiber demand, applications and markets, China)
- Industrial commercialisation issues: large scale production and costs, impact on the environment, philosophy & ethics on the wide spread use – Market/customer acceptance

27.1 The industry innovation brought by the application of micro- and nanotechnology in textile field

Micro- and nanotechnology has been applied in textile industry in three ways. The first is the spinning technology of nanofiber with nanodimensional scale. The second is to combine micro- and nanopowders with textile materials. The third is the application of nanotechnology in textile printing and dyeing area.

Nanofiber generally refers to fiber materials that have nanoscale dimensions, especially in the diameter direction. There are roughly two types of preparation methods for it. One is the preparation of in-organic nanofiber by molecular technology, such as carbon nanotube fibers [1] with diameter of 1–50 nm and length from a few microns to a few hundred microns, and some other metal, semiconductor and alloy nanofibers developed with this technology. The other is the preparation of organic nanofibers with diameter from 10 to 1,000 nm by technologies such as electrospinning [2], island and sea composite spinning [3], catalytic extrusion polymerization [4] and molecular spinneret spinning [5]. In addition, some other efforts and researches have been carried out for new preparation methods and for enlarging the polymer family of nanofiber raw material.

The latter two ways of application for micro- and nanotechnology in textile industry have been commercialized successfully with functional textile products such

Ning Cui, China Textile Academy, No.3 Yanjingli Middle Street, Chaoyang District, Beijing, China, e-mail: cuining@cta.com.cn

https://doi.org/10.1515/9783110670776-027

as antibacteria, fire-proof, anti-ultraviolet and so on. Compared with textile technology which utilizes micro additives, the amount of nanograde functional components could be reduced by 50%. Moreover, during fiber spinning process, the cycle for spinning pack could be extended for one to four more weeks, which greatly increase the made-up rate, reduce the raw material loss, production cost and the energy consumption [6].

Various types of functional textiles have been introduced to market, bringing upgraded consumer experience. The textiles could be roughly classified into the following categories [7] according to the additional functions brought to fiber materials or textile materials by the nanopowders:

(1) Finished type nanotextile. The natural fiber normal fabrics such as cotton, wool, silk and bast with waterproof, antiwrinkle, antifouling or anticorrosion performances, which have been finished by corresponding technologies.
(2) Protective nanotextile. Fabrics with protective functions such as anti-ultraviolet, antiradiation, antistatic, flame retardant, high-temperature resistant, thermal insulation and sound insulation.
(3) Comfort-type nanotextile. Fabrics with functions like super comfort, quick dry, breathable and so on.
(4) Healthcare-type nanotextile. Fabrics with antibacterial, far infrared emission, negative ion release and other medical and health care functions.
(5) Intelligent nano-textile. Breathable fabrics capable of adjusting body temperature, phase-change temperature regulating fabrics, signal-response color-changing fabrics, bionic sports fabrics and so on.

27.2 Development trend of microtextile and nanotextile technology

After nanotechnology was applied in chemical fiber material and textile material, it has brought brand new experience in the aspects of comfort, safety and function. Nanotechnology will definitely create huge economic value in textile industry and show its special social value in the future. At the same time, questions such as how to evaluate and ensure the safety of nanomaterial during application, how to reduce the production cost of nanofiber and nanotextile, how to develop more varieties of nanomaterials and new type of nanodyes and additives, together with how to increase the added value of textile products, will be the future development trend of nanofiber textiles. To achieve the abovementioned goals, the following issues about technology and safety should be seriously considered.

(1) Nanosafety issue

The introduction of nanotechnology has input additional function to traditional textiles. On the other hand, due to the potential danger and harm like biological toxicity brought by the dimension effect of the nanomaterial, whether it will cause negative influence to human health or bring disadvantage to the environment during its application or exposure, need to be faced and answered. Up to now, many researches have been carried out to testify the safety of various nanomaterials. During a study of inhalation experiment [8] of nano-TiO_2 particles through mulberry silkworm skin, it was found that they could penetrate into silkworm body through the skin and exhibit toxic side effects which cause growing stop and death. However, unified conclusion about the nanosafety has not yet been reached. In addition, in the manufacturing process of nanomaterial, various type of organic solvents, surfactant, metal and heavy metal will be applied, which leaves more or less residue or impurity on the material. While discussing the safety of nanomaterials, the hazards of all substances involved in the synthesis of nanomaterials should be fully considered.

(2) Dispersion and agglomeration issue of nanomaterial

When nanomaterial is applied on traditional textiles, the dispersion and the agglomeration issue has been one of the largest technical barriers, whether the nanomaterial is added by blending during fiber production or added as a micro-emulsion functional finishing agent during fabric finishing. Because of the large specific surface area and high surface activity, nanomaterial inevitably has an inclination of self-agglomeration. For example [9], particle size uniformity and dispersion uniformity of the functional copolymerization component, mostly inorganic nanopowder, for in-situ polymerization will influence the quality and performance stability of polymer functional fibers.

Therefore, the morphology control and modification of functional copolymerization components is important for the diffusion of nanomaterial and the interface matching between nanomaterial and fiber polymer. Donghua University [10] effectively solves the agglomeration problem of the delustrant TiO_2 in nylon 6 fiber by superfine-processing and surface modification with compound coupling agent. The tensile strength and elongation of nylon 6 dull fiber has thus been increased.

(3) Surface functional construction for general-type synthetic fiber

By blending the functional components into spinning system, functional fiber could be prepared by traditional melt spinning or solution spinning. And this has been proven to be the efficient production method for functional fiber up to now. However the additional performance of functional fiber prepared by this method sometimes could not totally meet the demand of the end user. Taking conductive fiber as an example, which is the essential fiber material for various smart and wearable textiles, this kind of functional fiber prepared by melting spinning, including with raw material produced by in-situ polymerization, or solution spinning, mostly could only reach the technical level of antistatic fiber. This is not enough for the application requirement of smart textiles. Combining nanograde conductive material such as carbon black, carbon nanotube or graphene with fiber material, by different coating technique like solution coating, inkjet printing or chemical vapor deposition, new type of conductive fiber could be prepared by coating a thin layer on the fiber surface. This kind of layer not only introduces high electric conductivity, but also maintains the flexibility and elasticity of the fiber.

Surface functionalization of fiber could be realized by improving the binding force between the coating and fiber material through technical methods like plasma surface modification [11] or introducing covalent bond [12].

(4) Diversification of micro- and nanotechnology applications

The application of nanomaterial in textile industry shows some diversified tendency, such as the functionalization of various type of fiber and multifunctionalization of synthetic fiber.

① Functionalization of various types of fiber
After the successful functionalization of synthetic fibers by technical method such as in-situ polymerization, multiple efforts have been tried to introduce additional functions like anti-ultraviolet, antibacteria to natural fibers, such as cotton [13], silk [14] and bast fiber.

② Multifunctionalization of synthetic fiber
In some cases, two or more functional performances like anti-ultraviolet, antibacteria and cool feeling will be combined into one textile to design and develop high performance sportswear and casual wear [15]. At the same time, multifunctional finishing technique will be applied on pure cotton fabric or cotton/chemical fiber blend fabric with nanomaterial to produce multifunctionalized textile which have large potential market demand.

(5) Cost performance analysis of nanomaterial

Up to now, the nanomaterials have only been utilized by their inherent characteristics. Regardless of their nanograde dimension, the materials could also be utilized if they are only micro grade. So the real properties of the nanograde material have not been applied or displayed. Of course, the physical and chemical properties of nanograde material display differently from those of the micro grade material, and nanomaterials have undoubtedly the reinforcing effect on the various functions. This leads to the cost-effectiveness of nanomaterials and about how to balance between the function improving and the cost increasing during industrialized production. The price of nanograde material is much higher than that of normal powder after all.

27.3 Technology development and market opportunity of micro- and nanotechnology in textile industry in China

As a traditional textile production and consumption country, China has invested great efforts on the research and development of micro- and nanofunctionalized fiber material and textile, promoting China's textile innovation and leading a new generation consumption tendency.

27.3.1 Technology progress in China

The four key factors of "safety, environmental protection, energy and new generation material" have become the direction of China's textile new material industry during its current critical period of transformation and upgrading. Therefore, realizing the industrialization of high-performance fibers and their composites, improving the differentiation and functionalization of normal type fiber, developing new type biomass fiber material as substitution for petroleum resources, promoting the green manufacturing technology of modified functional fiber, are the key tasks of accelerating the industrial structure optimization and the sustainable development of this industry.

Nanomodified fiber and textile material is one of the most dynamic industrial sectors in the material area, in which China has been involved comprehensively and has obtained extensive achievements in the following four aspects:

(1) Organic–inorganic hybrid modified material, and corresponding in-situ polymerization technology, realizing flexible manufacturing with products of diversity and high quality.
(2) Scaled-up, continuous and low-cost production of high performance nanocomposite fiber, realizing low carbon emission, high value for the fiber and improving the cost-effective of fiber and its products.
(3) Research and technology integration of nano-intelligent fiber and textiles, greatly extending the function and application for fiber material.
(4) Organic–inorganic hybrid-modified material and nanofiber for biomedical use, broad applications in the fields of oral restoration, tissue engineering and new medical devices could be expected.

27.3.2 Market opportunity of functional textile in China

Fiber products in China are mainly applied in the area of garment textile, decorative textile and technical textile. According to statistics, China's total output of chemical fiber in 2018 amounted to 50.11 million tons, 31% of which has been consumed for technical textile. In other words, 70% of the total chemical fiber went to garment and decorative area, which is also the major application area for functional fibers.

A survey [16] was carried out about the functional textiles in Shanghai market in 2013 by Donghua University and Shanghai Entry-Exit Inspection and Quarantine Bureau. The functional textile specimen surveyed were classified as business shirt for gentleman, sportswear for men and women, underwear for men and women, accessories including socks, umbrella and hats, and maternity wear. The most densely populated and economically dynamic city in China is Shanghai. It is located in the Yangtze River Delta region where the textile industry is most concentrated. Thus, the result of this survey is to some extent representative of the direction of China's textile and garment industry.

The result of the survey showed that the request for men's business shirt focused on the performance of comfort represented by moisture wicking and perspiration, and performance of easy care represented by ironing-free and antiwrinkle. The function that is requested most by the market is ironing free and antiwrinkle, which accounted for 94% of all the specimen, followed by performance of comfort, including moisture wicking and perspiration, smoothness and antistatic, which accounted or 34.4% of all the specimen.

The result of the survey illustrated the tendency of garment fabric, which is also the future direction of business shirt. The request for hygienic, protective and environmental protection lowered sequentially, accounted for 15.9%, 6.8% and 2.3%. In addition, there are very few samples with other special additional functions. For example, only one piece of aromatic smell sample was collected by this survey.

The survey result for sportswear showed that the function requirement for this type of garment is mainly reflected on comfort. Twenty-nine pieces of specimen have the performance of moisture wicking and perspiration, accounted for 93.55%。 The specimen with quick dry performance accounted for 54.8% of the all. Other functions such as cool feeling, waterproof and wind proof, temperature adjustment, averagely accounted for 12.9% of all specimen, while only 6.5% is about super light texture.

Comfort, including moisture wicking and perspiration, quick dry, is the basic requirement for underwear. The performance of hygiene, including antibacteria and deodorizing which is quite popular in this market, is another basic requirement. The survey also found that the price for underwear and sock with additional functions is two to four times higher than normal products, which showed the economic value brought by functions.

In China, pregnant office ladies' requirements on garments are not limited to looseness and comfort. They cared more about the security of their fetus and themselves and therefore paid more attention to the health and protective function of the garments. 11 pieces of pregnancy wear were investigated during the survey. Almost all the functions focused on antiradiation, while some of them had antibacteria performance.

In addition, 12 types of umbrellas and sun hats were investigated. They are employed to protect human body from ultraviolet. Because they had no direct contact with the skin and would not be worn for long time, their additional functions are relatively simple. Only a few of the sun hats had some comfort performance, because of their contact with head.

Analyzing the result of this survey, it was found that functional textiles have entered many walks of people life, solving people's problem. Shanghai, the most globalized metropolis in China, represents the current situation and developing trend of China's garment market. So, the result of this survey also revealed that functional textiles have entered deeply into the Chinese market.

27.4 Issues brought about by the industrial commercialization of micro/nanotechnologies

27.4.1 Production costs and market acceptance of large-scale production of micro- and nanotechnology

The application of micro- and nanotechnology in textiles has produced functional textile products with more types and varieties than traditional textiles. On the one hand, it gives consumers more choices, but on the other hand, it also makes the functional textiles enterprises face a more complex market competition situation.

(1) The competition situation faced by the enterprises

Functional textiles are emerging products in textile market. Different types of new textiles with properties such as flame retardant, sunscreen, antiwrinkle, ironing free, water and oil repellent have been quickly recognized and accepted by the public through all kinds of marketing channels. However, from the perspective of enterprises, functional textiles have quite different characteristics on development, production and sales from those of traditional textiles. So, the market experience and competitive advantage accumulated by enterprises in traditional textiles area might not be directly borrowed to serve the development and sale of new products.

Firstly, very few enterprises have the ability of original innovation. So, for most enterprises, the information about functional textiles comes from market demand. When faced with the demand for a new functional product, it is difficult for the enterprise to make an accurate judgment on the volume of the demand and persistence of the market in a short time. Whereas once the enterprise decides to participate in the market competition of this product, it will face the pressure from many aspects such as the market competitors, the necessity of product upgrade and competitiveness improving.

Secondly, the additional functions brought by nanotechnologies are the main feature of the functional textiles to be distinct from traditional products and other functional products, and is the only way for enterprises to win consumers in this market. The above-mentioned pressures upon the enterprises also originated mainly from the additional function of textiles. If the enterprise cannot attract and satisfy the consumers by the functions of the products, the market share will be taken away by competitors who can provide products with superior performance. Enterprises will not only lose their original market share, but also be faced with overstock problem or fail to earn back the investment cost.

Thirdly, from the perspective of product life cycle, the life cycle of functional textiles is different from traditional textiles. In addition to the life cycle of fabrics themselves, it depends more on the life cycle of the function. For example, the life cycle of an antimicrobial sock depends largely on whether its antimicrobial property can be maintained after being worn and washed for many times. So the degradation or expiration of the function means that the product could be withdrawn from the market and be replaced by the products with superior performance. Therefore, it is crucial for functional textile enterprises to strengthen their research and technological innovation ability and to accelerate the product update so as to maintain the market share and enhance the market competitiveness.

(2) Cost problems brought by the application of nanotechnology

In order to support the enterprise's innovation and technical development and to maintain its market competitiveness, a large amount of capital and intellectual cost will be needed.

During the production practice, the capital cost may become the constraint of technological innovation. The technology innovation has the characteristics of high risk, high profit and advance payment. This determines that the volume of the funds needed for innovation will be large and the occupancy time will be long.

As for the intellectual cost, the enterprises specialized in development and production of new type fiber and textile material will ask for more research staff and higher technical level of the researcher. This will also increase the cost of the enterprises.

In addition, it should be noted that due to the individualization of functional textiles, it is not easy for all products to be accepted by the market in a short time. A relatively long adaptation period is needed. Therefore, the cost consumption in brand construction and marketing is also significantly different from that of traditional textiles. At the same time, during the marketing period, enterprises will also face the situation of low productivity and relatively high production cost.

27.4.2 Environmental and ethical problems brought by the promotion of nanotechnology

(1) Environmental and health issues faced by the promotion of nanotechnology

Traditional fiber and textile materials have been endowed with additional functions by the application of nanotechnology. Higher performance and consumption experience beyond the traditional textiles could be enjoyed by the customer. However, safety issue brought by the potential risks of nanomaterial and nanotechnology has attracted more and more concern from the academe. Professionals, including Chinese scholars, have studied the possible impacts of the production and consumption of new textile materials containing nanomaterials on the environment and human health, so as to analyze and evaluate the level of impact to the environment and human being.

In the production process of functional textiles, nanomaterials are combined with textile fibers either by blending or by means of adhesives, resins or crosslinking agent. In the process of production, wearing, washing and recycling, the binding fastness between nanomaterials and textile material will be degraded due to friction, UV irradiation, water washing, perspiration or temperature change. And the nanoparticles attached to fiber material may be peeled off and released into the surrounding environment, resulting in potential danger.

DING [17] studied the release of nanoparticles from nanosilver functionalized textiles exposed to different environments, including tap water, pond water, rainwater, artificial sweat, detergent solution, with deionized water as contrast group. It turned out that the largest releasing volume was found in deionized water, while the releasing rate in rain water was the fastest. After the textiles have undergone aging treatment, the total release of silver nanoparticles in deionized water, artificial sweat and detergent increased by 75.7–386%. Both dissolved silver and granular silver were observed in all media except artificial sweat, whereas only dissolved silver was observed in artificial sweat. With the natural aging of textile materials, the release rate is significantly accelerated. And the release volume in deionized water increased the most after aging, especially for the granular silver part.

At the same time, because the functional textile is one of the most widely used skin contact products, whether the releasing of nanoparticles from it will affect human health has also become the research focus.

Churg and others [18] implanted 120 and 21 nm TiO_2 on rat trachea and studied the influence. The results show that inhalation of small-sized nanoparticles can significantly increase the degree of inflammation, and that different sizes of particles of the same component affect their toxic kinetics in organism. Other research [19] has been carried out to prove that the inhalation toxicity of TiO_2 particles increased sharply with the decrease of particle size.

These results show from different angles that nanomaterials will inevitably enter environment and human body through various ways during the production, use and treatment of nanoproducts, including nanomodified functional textiles. What is more, the degree of biological toxicity in the environment and the influence on human health and ecology is still difficult to predict.

(2) Social and ethical topic about the promotion of nanotechnology

Research on potential health hazards, and concerns about the ecological environment, are not the only problems posed by nanotechnology applications. With the attention on the safety of nanomaterials, the research on the issues of health and environmental ethics brought by nanotechnology, and the impact on social ethics has gradually come into people's vision.

Stepping into twenty-first century, more and more social and nongovernmental organizations, research institutions and many scholars, including philosophers, sociologists, jurists and scientists, began to get involved in the research of the ethical issues of nanotechnology and its social impact. In December 2003, a report [20] was posted by the National Science Foundation of the United States and the Subcommittee on Nanoscience, Engineering and Technology, entitled "Nanotechnology: Societal Implications I – Maximizing Benefits for Humanity." In July 2004, another report [21] about the possible health, safety, environmental, ethical and social issues arising from the application of nanotechnology was published by the Royal Society and the Royal Academy of Engineering, entitled "Nanoscience and technologies: Opportunities and Uncertainties."

But ethical problems brought by nanotechnology are different from the ethical problem posed by other technologies. Taking bioethics for example [22], many of the issues discussed in the field are emerging from long-term development of modern biotechnology. Whereas for the ethics of nanotechnology, nanotechnology itself is in its early stage, many of the topics of nanoethics can only be based on speculation and possibilities. In this sense, like nanotechnology, nanoethics is also facing a world of possibility and unknown. The so-called nanoethics issue is a possible ethic of possible future world ethical risks.

For the ethical problems brought by nanotechnology, different scholars have different classification methods or emphasis. Some scholars have attributed the social and ethical issues of nanotechnology to six aspects [23]: health and safety, impact on legal and regulatory systems, social dimensions, impact on science and technology policies, and broader philosophical issues. With regard to the focus areas of nanoethics, Chinese scholar WANG Guoyu [24] believes that it should include the issues of safety of nanomaterials itself, nanotechnology and personal privacy, the problems caused by medical applications, the military applications of nanotechnology, and fair distribution of the benefit originated from nanotechnology.

Here we take the problems caused by medical applications for example. Nanotechnology can be applied in the medical area such as disease diagnosis, targeted drug delivery and gene therapy. But the most controversial issue is the application of nanotechnology in human enhancement. The so-called human body enhancement refers to the improvement of human physical and mental abilities through technical means. Being a kind of life existence under the natural influence of heredity and environment, can we human being use technology to make up for our born difference from others, or surpass the ability of others? This will be the nanoethics issue that can trigger deep discussion. Indeed it is hard to draw a clear dividing line between healing a disease and physical enhancement beyond health.

Nanotechnology also bears the pressure of commercialization like other new technologies. The risk of nanotechnology, especially when combined with other technologies, such as nanomodified fiber or functional textiles, always tends to be long-term, cumulative and potential. Therefore, public understanding of nanotechnology, and the risks posed by nanotechnology is not only a subject that should be studied in academia, but also a component part of nanoethics.

A survey [25] was conducted by Chinese scholars in Dalian, China, in 2011 about public awareness of nanotechnology, its risks and ethical issues.

The survey found that China's recognition of nanotechnology is relatively high, 79.4% of the respondents said "they have heard of nanotechnology," 42.8% said they have had contact with nanotechnology products, but only 38.7% have thought about the safety of nanotechnology. About 57.4% of the respondents considered nanotechnology to be "both risky and beneficial." The survey also revealed that the risk awareness and safety awareness of the Chinese public is relatively weak. People who claim to have contact with nanotechnology products acquired their nanoknowledge mostly from various advertising and product promotion, but they have no authoritative judgment on whether or not they have ever contacted real nanotechnology products. This will not only disturb the scientific sense of the public but also mislead people's daily consumption. From a long-term perspective, it will impede the development and application of real nanotechnology products.

In fact, it is not only that the public understanding about potential risks after the marketization of nanotechnology is unsatisfactory, also the academic concern about nanosafety and nanoethics is far less than the concern about the development and the application of the technology itself.

Chinese scholar [26] has carried out a contrast study about the high influence researcher's concern about the status of nanoscience development and about emergence and development of nanoethics and social research with reference to the data from the Web of Science database of Thomson Reuters, from 1961 to 2010. The result turns out that, in terms of the number of papers, the research on nanoethics and society by international community cannot match that of nanoscience. Between 1991 and 2010, the total number of papers on nanoscience added up to 599,207. Correspondingly, there are only 476 articles on nanoethics and social studies, accounting for 0.08% of the entire nanoscience. And the studies related to real reflections on ethical issues is even less.

Looking back upon the impact and reflection on human activities brought by biotechnology and artificial intelligence technology, we know that research on nanoethics must be carried out in sync with the development of nanoscience and technology. It is gratifying that the uncertainty and hazards brought by nanotechnology have attracted the attention of academia and the public. Nanotechnology is likely to be the first new technology that has been carefully studied before it may have any negative effects, and eventually can benefit our mankind.

With the evolution of nanotechnology, human body and environment are being more frequently exposed to nanomaterials and nanoproducts. Moreover, evaluation about its safety, research about the ethical and social attributes of the technology is still far behind the development of nanotechnology itself. From the perspective of sustainable development, standardization and legislation about the development, production, consumption and post-treatment of the technology and production should be seriously concerned. And this asks for deep participation of the governments to construct a robust protective barrier to ensure the healthy growth of this promising technology.

References

[1] Iijima, S.: Helical microtubules of graphitic carbon [J]. Nature, 1991; 354, 56–58.

[2] Tsai, P. P.; Yurong, Y.: Present development of nanofiber technology. Technical Textiles, 2008(03) 1–5 + 14.

[3] Yamanaka, H.; Masuda, M.; Aranishi, Y.: Advanced nanofibers arising from pursuit of advanced melt spinning technology [J]. Melliand China, 2019; 47(02), 12–14.

[4] Watanabe 1, K.; Kageyama, T.: Self-assembled formation of GaAsP nano-apertures above InAs/GaAs quantum dots by the thermal diffusion of phosphorus. Physica Status Solidi B, 2016; 253(4), 659–663.

[5] Jacob, K. I.; Polk, M.: Molecular spinnerets for polymeric fibers. National Textile Center Annual Report, M98-G8, November 2010, USA.

[6] Yinbg, J.; Haiqi, W.: Application of nanomaterials and nanotechnology in Textile Printing and Dyeing Industry [J]. Textile Science Research, 2004; (04), 5–8 + 53.

[7] Bingyao, D.: Current status and development trend of functional textiles [C]. Changzhou, China. Proceedings of National Symposium on Weaving New Product Development, China Textile Engineering Society, 2008; pp. 14–17.

[8] Shifan, W.; Jiayou, G.; Peiyun, L.: Synthesis of coated nano TiO2 and study of its inhalation through mulberry silkworm skin [J]. Chemistry World, 2004; (09), 500–501.

[9] Meifang, Z.; Yuyuan, S.; Bin, S., et al. In situ polymerization and application of PET/Ag composite resin materials: 201010513405. 7[P]. 2011-0 0.

[10] Li, G.: Effect of Surface Modification of Titania on the Properties of Full Dull PA6 Fibers [D]. Shanghai: Donghua University, 2014; pp. 39–53.

[11] Garg, S.; Hurren, C.; Kaynak, A.: Improvement of adhesion of conductive polypyrrole coating on wool and polyester fabrics using atmospheric plasma treatment[J]. Synthetic Metals, 2006; 157(1), 41–47.

[12] Mičušík, M.; Nedelčev, T.; Omastová, M.; Krupa, I.; Olejníková, K.; Fedorko, P.; Chehimi, M. M.: Conductive polymer-coated textiles: The role of fabric treatment by pyrrolefunctionalized triethoxysilane [J]. Synthetic Metals, 2007; 157(22), 914–923.

[13] Li, D.; Lirong, Y.; Sijun, X., et al.: Preparation and property of bionic super-hydrophobic cotton fabric based on TiO2 nanotubes [J]. China Textile Leader, 2019; (04), 30–32 + 34.

[14] Bingshuo, W.; Hong, L.; Yuyue, C.: Research on anti-ultraviolet finishing of silk fabrics treated with modified zinc oxide nanoparticle [J]. Silk, 2017; 54(07), 18–23.

[15] Kuiyong, D.: Development of multi-functional composite finishing of textiles [J]. Printing and Dyeing, 2004; 23, 46–49.

[16] Mengjuan, W.; Shanshan, H.; Yanghua, S.; Yufeng, C.; Wenliang, X.: Survey and analysis on functional textiles in Shanghai market [J]. Technical Textiles, 2014; 32(02), 35–39.

[17] Churg, A.; Stevens, B.; Wright, J. L.: Comparison of the uptake of fine and ultrafine TiO2 in a tracheal explant system [J]. The American Journal of Physiology, 1998; 274, 1 Pt 1.

[18] Bing, W.; Weiyue, F.; Yuliang, Z., et al.: Status of study on biological and toxicological effectof nanoscale material. Science in China Series B-Chemistry, 2005; 48, 385–394.

[19] Dahu, D.; Lulu, C.; Shaowei, D., et al.: Natural ageing process accelerates the release of Ag from functional textile in various exposure scenarios. Scientific Reports, 2016; 6, 37314.

[20] Roco, M. C.; Bainbridge, W. S.: Nanotechnology:Societal Implications Ⅰ – maximizing Benefits for Humanity. Berlin: Springer, 2006.

[21] Royal Society: Nanoscience and nanotechnologies:opportunities and uncertainties[M]. London: Royal Society United Kingdom&Royal Academy of Engineering, 2004; pp. 51–54.

[22] Wang, G. Y.; Gong, C.; Zhang, C.: Nanoethics: Research progress, problems and challenges (in Chinese). Chinese Science Bulletin (Chinese Ver), 2011; 56, 96–107.

[23] Sparrow, R.: The social impacts of nanotechnology: An ethical and political analysis [J]. Journal of Bioethical Inquiry, 2009; 6(1), 13–23.

[24] Guoyu, W.; Yuliang, Z.: Study on Safety and Ethics of Nanotechnology [M]. Beijing: Science Press, 2015; pp. 102–106.

[25] Li, L.; Guoyu, W., et al.: Analysis of public perception of nanotechnology in China – based on the empirical investigation in Dalian area [J]. Journal of Dalian University of Technology (Social Sciences Edition), 2012; 33(04), 59–64.

[26] Haiyan, H.; Guoyu, W., Xianwen, W.; Chunjuan, L.; Chao, G.: The Emergence and Development of the International Ethic and Social Studies of Nano Science and Technololgy [J].]. Journal of Engineering Studies, 2011; 3(04), 352–364.

X Future prospects for particle technology in textiles

Franz Pursche*, Thomas Gries

28 Future prospects for particle technology in textiles

Keywords: particle technology, nanoparticles, functionalization, fibers, textiles

Functionalization of material systems is one of the key developments nowadays in order to overcome technical deficits and open up new fields of applications. Particularly in the textile sector, nanoparticles are frequently used for functionalization. Here the effect is used that the properties of materials change significantly when their size is changed to the nanoscale range (1–100 nm). When such nanoparticles are introduced into existing material systems, new functionalities can be created or existing properties can be improved. Man-made fibers are particularly suitable for this purpose since particles can be introduced easily into the polymer fiber manufacturing process.

Industrially, the use of nanoparticles in man-made fibers is state of the art and well established on the market. This is particularly evident in the example of polyester (PET) based fibers. Approximately 60% of all existing fibers on the market are man-made fibers, of which polyester fibers have a share of about 50%. Polyester fibers are used to a large extent in the clothing sector [1]. Titanium dioxide (TiO_2) is added to a large proportion of the available PET fibers to reduce the gloss of the fibers and make them appear darker.

Away from mass applications, particles can be used to open up further fields of application. Figure 28.1 shows existing and future applications of nanomodified fibers. In the agricultural sector, deterrents can be incorporated into fibers to keep pests away. This reduces the use of harmful pesticides and ensures food security for the world's population. Through the use of conductive particles such as carbon blacks or carbon nanotubes, man-made fibers can achieve electrical conductivity for use as fully integrated sensor fibers. The potential applications are extensive and range from smart textiles to condition monitoring in hydrogen pressure tanks. In the field of filtration, the use of antimicrobial and antiviral nanoparticles is currently being investigated in order to increase filter performance as well as filter service life. This topic has received increased attention, especially since the beginning of the global corona pandemic in 2020. Other potential applications of antimicrobial nanoparticles are in the field of sports textiles, for example, to prevent odors or to influence the water absorption capacity of textiles.

***Corresponding author: Franz Pursche,** Institut für Textiltechnik of RWTH Aachen University, Otto-Blumenthal-Straße 1, 52074 Aachen, Germany, e-mail: franz.pursche@ita.rwth-aachen.de
Thomas Gries, Institut für Textiltechnik of RWTH Aachen University, Otto-Blumenthal-Straße 1, 52074 Aachen, Germany

https://doi.org/10.1515/9783110670776-028

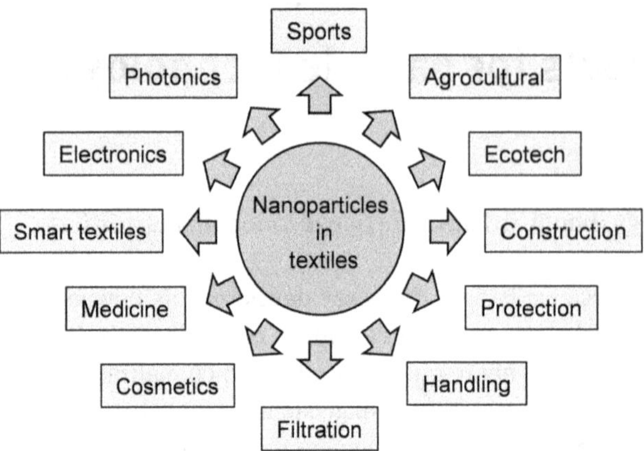

Figure 28.1: Current and future application areas for nanoparticles in textiles.

The medical sector also benefits from nanoparticles. Drug release concepts can be used to release targeted active ingredients, for example to promote wound healing, or in the field of regenerative medicine to promote regenerative processes in the body by means of a constant and targeted release of active ingredients. The area of personal protection equipment offers, great potential for use of nanoparticles in a wide range of application areas. Flame retardancy, flame protection, insulation, protection against radiation, dissipation of electrical charge, fluorescence, neutralization and insulation of hazardous substances as well as increased mechanical resistance are solutions that can be achieved in this field by means of nanoparticles.

However, there are also drawbacks to the use of particle technology in fibers that should not be ignored. One negative aspect is the toxicity of the particles, especially due to their small diameter. Due to the respirable nature, protective measures must be used during processing. In addition, there is the possibility of dust formation and corresponding dust explosions that can occur during processing. Especially in low-wage countries, restrictions in operator safety are conceivable due to cost pressure.

Another weak point of particle technology is currently the sustainability aspect. In principle, particles initially hinder the recycling process and make separation more difficult, which is especially true for mechanical recycling. In general, recycling processes depend on a high purity of the input material. Particles increase the complexity accordingly and make recycling more difficult. However, chemical or pyrolytic recycling processes are conceivable and possible, which will have to be researched more intensively in the future.

Nevertheless, the more difficult recycling process is also countered by clear sustainability advantages of particle technology in the field of spin dyeing. In spin dyeing, fiber dyeing is achieved by the addition of nanoparticles, without the use of further costly chemical and water-based dyeing processes. Spin dyeing saves process time and avoids dye losses due to wastewater, energy and chemicals. There is also the advantage of continuous color uniformity and high color fastness. Energy, chemicals and water use can thus be reduced by 50–80% compared to conventional dyeing processes, making the technology highly relevant, especially for mass applications in the apparel and home and living sectors [2–4].

Nevertheless, the key issue of sustainability continues to be countered by price pressure, particularly in the textile technology. The price pressure due to competition from low-wage countries with low environmental standards and low energy costs is very high. Functionalization usually comes at a price. Of course, this also applies to sustainability aspects. The use of materials from renewable resources or improved design for recycling of existing products is already conceivable today for many technical products. Unfortunately, in the past, consumer willingness to accept a higher price for a sustainable product with lower performance has been low. However, there are recent signs that this consumer attitude is changing, especially among young people. For instance, there is a high willingness to pay a higher price for sustainable food [5]. This has not yet been the case for technical non-food products. One possible reason for this could be that the sustainability aspect in technical products is very complex and is often discussed emotionally. A key task of product development and research therefore also lies in the creation of understanding, the early involvement of stakeholders and users, to clearly demonstrate the advantage of sustainable products and increase the willingness for higher prices. Legislators also have a key role to play in the areas of education and information, rules, regulations and standards but also in influencing prices. It is clear that, due to its complex structure, the issue cannot be solved overnight and requires the involvement of all parties: research – industry – legislator – end user.

In summary, nanoparticles offer the possibility of functional integration into textiles for a wide range of everyday applications. Many application areas are discussed in this book. Future applications will be developed. Functionalization and functional integration is one of the mass topics of today's society and is visible in nearly every aspect of our daily life. Due to the very large number of conceivable material combinations and the associated bundling of properties of materials, more new functionalities are conceivable. These can be used as input parameters in the development process and aligned accordingly with the application and the need. This technology trend must be aligned with the highly relevant mass topic of sustainability. In the near future, robust recycling technologies for nanoparticle functionalized fibers need to be developed. By means of spin dyeing, particles can contribute a great sustainability aspect in terms of CO_2 emissions.

References

[1] Preferred Fiber & Materials: Market Report, TextileExchange, 2021.
[2] Ikea Fact Textile Dyeing: Inter IKEA Systems B.V., 2015. https://preview.thenewsmarket.com/Previews/IKEA/DocumentAssets/436953.pdf
[3] Sardana A., Dayal Am.: Dope-Dyed fibres: A solution to environmental pollution, Fibre2Fashion, 2017. https://www.fibre2fashion.com/industry-article/7896/dope-dyed-fibres-a-solution-to-environmental-pollution
[4] Arthinun B.: DOPE DYED – The Future of Textile Industry, FaserhionHometex, 2022. Access von April 2022. https://www.fashionhometex.com/articles/2018/08/dope-dyed-future-of-textile-industry#:~:text=Here%20are%20the%20Benefits%20of%20DOPE%20DYED&text=By%20adopting%20the%20dope%20dyeing,footprint%20of%20the%20final%20products
[5] Monitor Deloitte: Sustainability as a value driver – How sustainability elevated product innovation and price differentiation.

Index

3D printing 329
– 3D printing of garments 333
– 3D printing onto textiles 333

Acoustic properties 322
Additives 74
Additive manufacturing 329
Adhesion 50, 334
Agglomerate 4, 485
Aggregate 4
Agricultural textiles 123
Antimicrobial Properties 101, 191, 276, 441
Antistatic. See electrical properties
Application diversification 486

Bi-component fiber. See multicomponent filament
Bi-component spinning 202, 310
Biocompatibility 282, 444
Biodegradable material 381
Biopolymers 477

Carbon black 296
carrier for biomass 156
Centrifugal extrusion
China 421, 487
Chinese market 488
Circular economy 463, 474
CNT 296, 350
Coalescence separator. See Filter, aerosol
Coating 220
Color 97–100
Comfort 442
Commercial production 95, 489
Concrete 136
Construction 131, 162
Cradle-to-cradle 465
Cross Section 70
Complex coacervation 249
Conductivity 60, 189, 289, 296
Cosmetotextiles 235
Cost performance analysis 486
Cyclodextrin 244

Digital light processing 332
Dispersion 22, 28, 485

Degradation 345
Drug delivery 279
Dyeing 97, 243
– Dyeing, vat 98
– Dyeing, pigment 99

Electrical properties 82, 110
Electrospinning 31, 70, 203, 306
– Electrospinning, AC 316
– Electrospinning, centrifugal 312
– Electrospinning, gap 312
– Electrospinning, gas assisted 313
– Electrospinning, magnetic 312
– Electrospinning, neddleless 313
– Electrospinning, nozzle-based 308
Electrospraying 313
Encapsulation 25, 106, 245
End-of-life 345
Ethics 492
EU 387, 396, 407
Exposure 360

Facade 132
Fiber discharge 380
Fiber properties 35
Fiber-reinforced composites 300
Fiber-reinforced plastics 134
Fiber-reinforcement bars 135
Fiber separation 461
Fillers 16, 335
– Fillers, inorganic 19
– Fillers, organic 22
Filtration 198, 323, 380
– Filter, aerosol 148
– Filter, fine dust 152, 205
– Filter, water 155, 207
Finishing 45
Five-category scheme 221
Flame-retardant properties 75, 189
Flue gas desulfurization 151
Fluorescent particles 188
Fog catcher 160
Functionality of Nanoparticles 73, 184
Functionalized synthetic fibers 68
Fused filament fabrication 330

https://doi.org/10.1515/9783110670776-029

Geotextiles 140, 301

Heat collection 163
Heat exchange 169
Heat storage 164
High-performance fiber 217
HM-HT fiber ropes 213
Hot pressing 206
Hydroentangling 205
Hydrophobic properties 72, 109, 441

Insect-repellent 126
Insulation 139, 166
Interfacial polymerization 248
Ion-release 349, 354

Labeling 447
Life cycle assessment 457
Lightweight construction 134
Legal framework 387
– Legal framework, Chinese 421
– Legal framework, cosmetic products 402
– Legal framework, European 396, 407
– Legal framework, testing 438
Legislation. See legal framework
LLCP 217
Luminescent particles 188

Market acceptance 489
Mechanical properties 74, 101
Mechanical resistance 192
Medical application 79, 275, 321, 444
Melt-blown process 203
Melt dispersion method 251
Meltspinning 28, 70
Membrane 131, 133
Microcapsule binding 252
Microencapsulation 25, 106, 246, 279
Microfiber 68
Microplastics 373, 467
Multicomponent filament 70

Nanocomposite 72
Nanofiber 69
Nanofillers. See Fillers
Nanomaterial 5, 414, 447
Nanoparticle 4, 15, 46
Nanoplastics 374
Nanoregisters 395
Nanotechnology 14
Natural fiber welding 461
Needling 205

Occupational safety and health 407, 430
Optical effect 188

Particle 4, 46
– Particle Surface 52
– Polymeric particle 56
Particulate matter 197
Percolation 296
Phase change materials 107, 189
Photopolymerization 332
Physicochemical properties 348, 353
Physicochemical recycling 477
Piezoelectricity 37, 84
Pigment 50
Polyjet technique 333
Powder fusion. See selective laser sintering
PPE 177
Prefiltration 158
Printed electronics 291
Printing 100
Product life-cycle 457
Protection 181

Radiopaque 278
REACH 387, 438
Recycling 456
– Recycling methods 458, 476
– Recycling process 462, 476
Regenerative medicine 281
Regulation. See legal framework
Release 257, 345, 447
Reuse 457
Risk 177, 343, 407
– Risk categories 180
– Risk, environmental 352, 491
– Risk, human health 347, 484
– Risk mitigation 179
Rope 226
– Rope, braided fiber 215

Selective laser sintering 332
Selfcleaning 60, 447
Sensor 295

Silver 104
single-pot strategy 69
Slip-flow effect 202
Smart textiles 287, 334
Sol-gel technology 53
Solution spinning 29
Solvent evaporation method 249
Solvent spinning,
See solution spinning
Spacer textile 152, 158
Spray drying 251
Standards 416, 426, 439
Steel wire ropes 213
Stereolithography 332
Structural health monitoring 299
Surface activity 59
Surface functionalization 485
Sustainability 455, 473

Taylor Cone 306
Testing 438
– Testing standards 440
– Testing methods 442
Textile Coating 49
Thermal properties 75, 189
Thermomechanical recycling 476
Tissue engineering. See regenerative medicine
TLCP 217
Toxicity 347, 374
Tri-component fiber. See multicomponent filament

UV protection 82, 104, 111

Washing stability 50
Wastewater treatment 155, 379
Water harvesting 161
Wear mechanism 223
Wet spinning 30
Workplace safety 351, 408, 415, 430

X-rays 277

www.ingramcontent.com/pod-product-compliance
Lightning Source LLC
Chambersburg PA
CBHW060956210326
41598CB00031B/4847